Lecture Notes in Electrical Engineering

Volume 694

The book series *Lecture Notes in Electrical Engineering* (LNEE) publishes the latest developments in Electrical Engineering - quickly, informally and in high quality. While original research reported in proceedings and monographs has traditionally formed the core of LNEE, we also encourage authors to submit books devoted to supporting student education and professional training in the various fields and applications areas of electrical engineering. The series cover classical and emerging topics concerning:

- Communication Engineering, Information Theory and Networks
- Electronics Engineering and Microelectronics
- Signal, Image and Speech Processing
- Wireless and Mobile Communication
- Circuits and Systems
- Energy Systems, Power Electronics and Electrical Machines
- Electro-optical Engineering
- Instrumentation Engineering
- Avionics Engineering
- Control Systems
- Internet-of-Things and Cybersecurity
- Biomedical Devices, MEMS and NEMS

For general information about this book series, comments or suggestions, please contact leontina.dicecco@springer.com.

To submit a proposal or request further information, please contact the Publishing Editor in your country:

China

Jasmine Dou, Associate Editor (jasmine.dou@springer.com)

India, Japan, Rest of Asia

Swati Meherishi, Executive Editor (Swati.Meherishi@springer.com)

Southeast Asia, Australia, New Zealand

Ramesh Nath Premnath, Editor (ramesh.premnath@springernature.com)

USA, Canada:

Michael Luby, Senior Editor (michael.luby@springer.com)

All other Countries:

Leontina Di Cecco, Senior Editor (leontina.dicecco@springer.com)

**** Indexing: Indexed by Scopus. ****

More information about this series at http://www.springer.com/series/7818

Pradeep Kumar Singh · Arti Noor ·
Maheshkumar H. Kolekar ·
Sudeep Tanwar · Raj K. Bhatnagar ·
Shaweta Khanna
Editors

Evolving Technologies for Computing, Communication and Smart World

Proceedings of ETCCS 2020

Springer

Editors
Pradeep Kumar Singh
Department of Computer Science
and Engineering
Jaypee University of Information
Technology
Solan, Himachal Pradesh, India

Maheshkumar H. Kolekar
Department of Electrical Engineering
Indian Institute of Technology
Patna, Bihar, India

Raj K. Bhatnagar
Department of Electrical Engineering
and Computer Science
University of Cincinnati
Cincinnati, OH, USA

Arti Noor
CDAC Noida
Noida, Uttar Pradesh, India

Sudeep Tanwar
Department of Computer Engineering
Institute of Technology, Nirma University
Ahmedabad, Gujarat, India

Shaweta Khanna
JSSATEN Noida
Noida, Uttar Pradesh, India

ISSN 1876-1100 ISSN 1876-1119 (electronic)
Lecture Notes in Electrical Engineering
ISBN 978-981-15-7803-8 ISBN 978-981-15-7804-5 (eBook)
https://doi.org/10.1007/978-981-15-7804-5

This Springer imprint is published by the registered company Springer Nature Singapore Pte Ltd.
The registered company address is: 152 Beach Road, #21-01/04 Gateway East, Singapore 189721, Singapore

Preface

The International Conference on Evolving Technologies in Computing, Communications and Smart World (**ETCCS-2020**) targeted the researchers from different domains of recent technologies, computing, communication and smart world innovations at a single platform to show their research ideas. Conference covers the novel ideas based on algorithms, surveys, policies, architectures, communication challenges and future research aspects. There are five technical tracks to include the topics of interest, but are not limited to the following: (i) emerging computing technologies, (ii) network and computing technologies, (iii) wireless networks and Internet of Everything (IoE), (iv) communication technologies, security and privacy and (v) next-generation computing technologies. The conference provides a total of five technical tracks to all authors to identify the most suitable tracks for their manuscript. So, each author may identify the most suitable theme matching to their paper for submission.

As the theme of the conference is recent and tracks are as per the evolving technologies, it is expected that the submissions will attract the good citation in future, and proceedings will emerge as one of the good collections for download. The organizing team is confident that it will evolve as an intellectual asset in the long term year after year. The International Conference on Evolving Technologies in Computing, Communications and Smart World (ETCCS-2020) was held at CDAC Noida on 31 January to 1 February 2020 in association with Southern Federal University, Russia, and Jan Wyzykowski University, Polkowice, Poland, as academic partners. We are highly thankful to our valuable authors for their contribution and our Technical Programme Committee for their immense support and motivation towards making the ETCC-2020 a grand success. We are also grateful to our keynote speakers: Prof. J. K. Bhatnagar from University of Cincinnati, USA; Prof. Narayan C. Debnath, Eastern International University, Vietnam; Dr. Arpan Kumar Kar, DMS, IIT Delhi; Mr. Aninda Bose, Senior Editor, Springer; Dr. Zdzislaw Polkowski, Jan Wyzykowski University, Polkowice, Poland; and Industry Expert Sh. Alok Varshney, for sharing their technical talks and enlightening the delegates of the conference. We are thankful to various session chairs for chairing the session and giving invited talks whose names include: Dr. Vivek

Sehgal, Dr. Yugal Kumar, Dr. D. Vaithiyanatha, Dr. R. K. Saini, Dr. Rohit Tanwar, Dr. Akshi Kumar, Dr. Praveen Kumar Malik, Prof. Pravin Chandra, Prof. Arvinder Kaur and Dr. Anupan Singh.

The volume editors are thankful to Sh. Vivek Khaneja, Executive Director, CDAC Noida, India, for providing the CDAC Noida as a venue. The CDAC Noida is thankful to MeitY for providing a grant to CDAC to support the conference. The CDAC Noida is also thankful to Dr. B. K. Murthy, Scientist G and Group Coordinator CDAC.

We express our sincere gratitude to our publication partner, LNEE Series, Springer, for believing in us.

Solan, India	Pradeep Kumar Singh
Noida, India	Arti Noor
Patna, India	Maheshkumar H. Kolekar
Ahmedabad, India	Sudeep Tanwar
Cincinnati, USA	Raj K. Bhatnagar
Noida, India	Shaweta Khanna

Contents

About the Editors

Dr. Pradeep Kumar Singh is currently working as an Associate Professor in the Department of CSE at Jaypee University of Information Technology (JUIT), Waknaghat, Himachal Pradesh. Dr. Singh is a Senior Member of CSI and ACM. He is an Associate Editor of the IJAEC, IGI Global USA, SPY, Wiley & IJISC Journals. He has published 90 research papers. He has received three sponsored research projects grant worth Rs. 25 Lakhs. He has edited a total of 10 books from Springer and Elsevier and also edited several special issues for SCI and SCIE Journals from Elsevier and IGI Global.

Dr. Arti Noor is presently working as Senior Director at CDAC, Noida. She has done her Ph.D. from IIT, BHU, in 1990. He has 20 years of teaching VLSI design related courses to M.E. students of BITS, Pilani, and CDAC Noida. She has guided six Ph.D. and guided 200 student's projects of B.Tech./M.Tech./M.E. and examined 100 M.Tech. theses. She has published 81 research papers in journals and conferences including monographs.

Dr. Maheshkumar H. Kolekar is working as an Associate Professor in the Department of Electrical Engineering at Indian Institute of Technology Patna, India. From 2008 to 2009, he was a Postdoctoral Research Fellow with the Department of Computer Science, University of Missouri, Columbia, USA.

Dr. Sudeep Tanwar is an Associate Professor in the Computer Science and Engineering Department at the Institute of Technology of Nirma University, Ahmedabad, India. He has specialization in WSN, IOT and 5G Technology. He has authored or co-authored more than 90 technical research papers and five books. He has edited several special issues from IGI and Elsevier Journals.

Prof. (Dr.) Raj K. Bhatnagar is currently working as a Professor of Computer Science, Department of Electrical Engineering and Computing Systems, University of Cincinnati, USA. He completed his B.Tech. at IIT Delhi, India (1979), and Ph.D.

at the University of Maryland, USA (1989). His research interests include data mining, AI and pattern recognition problems, knowledge discovery and big data and cloud computing environments.

Dr. Shaweta Khanna is presently working as an Assistant Professor at JSS Academy of Technical Education, Noida, India. She has her expertise in the area of VLSI design, semiconductors, electronics design and circuits and IoT. She has done her Ph.D. from Guru Gobind Singh Indraprastha University, Delhi, India. She has published many papers in SCI and SCIE Journals from Taylor and Francis. Dr. Khanna has worked as Publicity Chair of many conferences of IEEE and Springer and organized several special sessions in conferences.

Toward Response-Type Identification for the Real Time

Chaman Verma, Veronika Stoffová, Zoltán Illés, and Mandeep Singh

Abstract The present study focused on applying machine learning techniques to recognize the student's responses provided in the educational technology survey in higher educational institutions. The present survey was piloted to investigate the differential analysis in two universities residing in India and Hungary. The survey's dataset was comprised of 37 features and 331 instances spotting technological questions awareness among students. The responses of students were a hybrid type such as binary, ordinal and nominal. The authors trained and tested the four datasets separately with four machine learning classifiers using two testing techniques. Both the binary and multiclassification problems are solved to identify the response type. To identify the student's response type, the strength of the classifiers is also compared with the statistical T-test at the 0.05 confidence level. To support the real-time implementation of the predictive models, the CPU user training time is also estimated and compared with the T-test. The cross-validation method significant enhanced the prediction accuracy of each classifier. To identify the response type, it is proved that the multinomial regression (MLR) outperformed others with the uppermost prediction accuracy of 98.3% and prediction time (lowest 0.03 and the highest 0.07 s) is computed. The present study may help to social science students to predict the response type of the respondents in the real time with different features.

Keywords Supervised machine learning · Real time · Response-type prediction · Educational technology

---

C. Verma (✉) · Z. Illés
Eötvös Loránd University, Budapest, Hungary
e-mail: chaman@inf.elte.hu

Z. Illés
e-mail: illes@inf.elte.hu

V. Stoffová
Trnava University, Trnava, Slovakia
e-mail: NikaStoffova@seznam.cz

M. Singh
Chandigarh University, Mohali, India
e-mail: mandeeptinna@gmail.com

P. K. Singh et al. (eds.), *Evolving Technologies for Computing, Communication and Smart World*, Lecture Notes in Electrical Engineering 694,
https://doi.org/10.1007/978-981-15-7804-5_1

1 Introduction

Nowadays, the application of supervised machine learning algorithms is increasing in every domain. These are so many lazy and early machine learning classifiers available to develop several predictive models using educational datasets. Traditionally, the differential analysis of the educators' demographic features was conducted with some major statistical techniques [1–3]. In spite of differential analysis, a new concept to predict the of stakeholder's demographic features toward educational technology is provided with various machine learning algorithms. To propose the real-time, prediction predictive model, CPU user time also played an important play to identify the data patterns. The latest, few important features such as educator's gender [4–7] are predicted with machine learning techniques. Other major features are also framed in the form of predictive models such as student's national identity [8], locality scope [9], national-level status [10] and locality such as rural and urban [11]. The survey's responses are also identified based on student's age [12] and a residency [13, 14], study level [15], institutions [16] with feature filtering. Further, classification algorithms were used appropriate to classify the student's attitude [17], awareness level toward technology [18]. The development and the availability of the latest technology for real time are also predicted [19]. Also, the residential status (rural or urban) has been identified [20]. The schematic theme of present paper also belongs to the scientific research literatures [21–24]. Enthusiasm among students is also predicted with the help of various learner algorithms [25].

The present chapter is framed into five sections. Section 1 deliberates the introductory theory with tiny literature. Section 2 directs the way to device the idea. Section 3 frames the experimental analysis and Sect. 4 about the performance measures to the justification of tasks. Section 5 briefs the inference of experiments.

2 Research Procedures

2.1 Preprocessing Dataset

The authors have collected the primary data samples using stratified random sampling with Google Form and personal visits. Table 1 depicts that two universities entitled Chandigarh University (CU) and Eotvos Lorand University (ELTE) participated with a total of 331 students and distribution is also shown.

Table 1 Participated in universities

ELTE	CU
169	162

Table 2 Dataset features

DAICTMT	UICTMT	EBICTMT	AICTMT
16	06	09	06

Table 2 demonstrates the four major technological awareness parameters spotting Information Communication and Mobile technology (ICTMT) such as Development and Availability (DAICTMT) with 16 features, Usability (UICTMT) with 06 features, Educational Benefit (EBICTMT) with 09 features and Attitude (AICTM) with 06 features. A total of 09 features belongs to the demographic are eliminated using self-reduction.

Table 3 displays the response type which was asked for filled up during the survey. The DAICTM parameter has three types of response types; UICTMT has five response types; EBICTMT and AICTMT have also five types of response. Only six missing values are found and substituted with the mode and mean values of the training dataset. Figure 1 shows that the frequency of responses is also defined with response type which is an actual number of records in each dataset. The four dissimilar datasets are normalized on the scale of 0–1 to put all responses on the same scale. In each dataset, the response type is considered as a response or class variable and features responses are considered as dependable variables or predictors.

Table 3 Response type and instances

DAICTMT	UICTMT	EBICTM and AICTMT
Yes (163)	Never (29)	Strongly disagree (01 and 05)
No (165)	Rarely (02)	Disagree (10 and 15)
Don't know (3)	Sometimes (157)	Undecided (86 and 75)
	Often (119)	Agree (155 and 155)
	All the time (24)	Strongly agree (79 and 81)

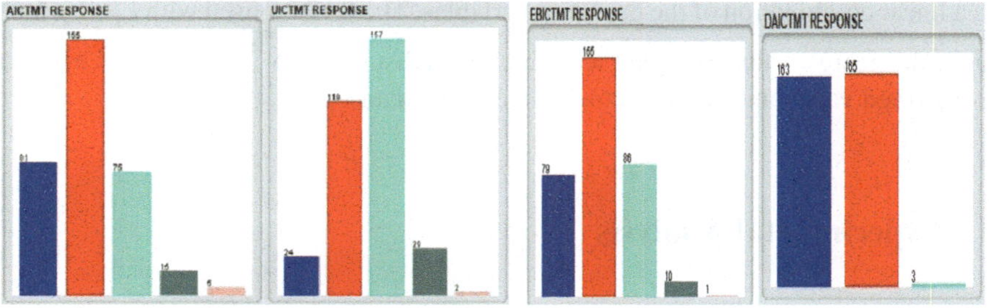

Fig. 1 Response-type distribution

2.2 Testing Dataset

After the preprocessing, to identify the response type, firstly, k-fold cross-validation is used with the fixed value of k with 10. In the cross-validation, each dataset is classified into k number of test subsets and $k - 1$ is considered as train subsets that are used to be tested against the k test set. This process goes on until all the train subsets are not tested. In the second method, the holdout technique is applied with the 66:44 ratios, whereas 66% is training dataset and 44% testing dataset. This splitting is performed randomly as compared to the preserved order splitting. Further, four datasets are trained and tested with holdout and k-fold paralleled with a statistical T-test at 0.05 significant level. Besides the prediction time, major performance measures such as kappa statistic (K), F-score (F), root mean square error (E), Matthews correlation (c) 333 CPU user training time (t) are also tested to check the significant difference among classifiers.

2.3 Tool with Classifiers

Two early learning classifiers such as support vector machine (SVM) and random forest (RF) are applied, and one regression method called multinomial logistic regression (MLR) is used. Also, one lazy learner classifier named k-nearest neighbour (KNN) is modeled. The SVM plots each response as a point in n-dimensional space where n is equal to 37 with the value of each feature being the value of a particular coordinate. Afterward, classification is performed by discovering the hyperplane that adequately differentiates target classes. In the RF model, several classification trees are made up to classify the responses. To classify the 37 features based on multiclass, each tree gives a classification and it is called the tree "votes" for that specific class. The forest picks the group of trees with the highest votes. The MLR is a type of classification used to classify the data patterns when we have more than two nominal response variables in which the log odds of the consequences are modeled as a linear combination of the predictor variables. The KNN is used with $k = 1$ which picks the nearest value provided with Euclidean distance $\sqrt{\sum_{i=1}^{k} (x_i - y_i)^2}$ where x is observed value and y is the actual value in the dataset.

3 Experimental Analysis

This section is related to the experimental analysis with two major experiments on the four types of datasets. Also, the accuracy of each classifier is tested with the statistical T-test at 0.05 level of confidence. The four classifiers are modeled and compared in the popular machine learning tool Weka 3.9.3. Figure 2 shows the experimental environment used for it. The total number of iterations is 10 to compare the four

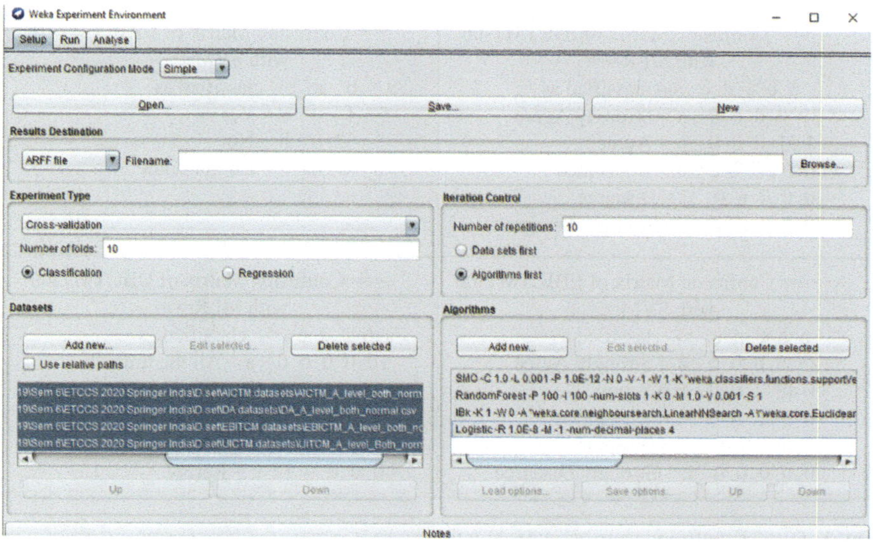

Fig. 2 Experimental environment

algorithms on four different datasets. Two testing approaches are used: k-fold with $k = 10$ and random holdout with 66:44 splits. The results of each experiment are shown with confusion matrices and comparative graphs.

3.1 *Using Hold-Out*

Figure 3 presents the classification accuracy provided using the holdout technique with 66:44 splitting ratios. The outcomes of this method revealed that the SVM

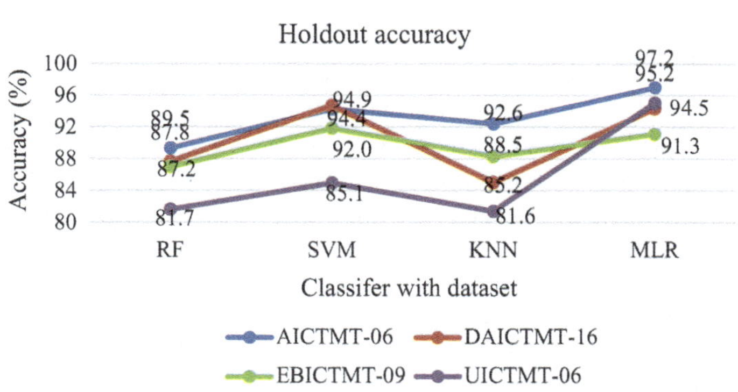

Fig. 3 Response-type prediction with varying training ratio

=== Confusion Matrix of AICTMT-06 with MLR=== a b c d e <-- classified as 30 0 0 0 0 | a = Strongly Agree 1 51 0 0 0 | b = Agree 0 0 20 4 0 | c = Undecided 0 0 0 6 0 | d = Disagree 0 0 0 0 1 | e = Strongly Disagree	=== Confusion Matrix of DAICTMT-16 with SVM=== a b c <-- classified as 57 1 0 | a = Yes 3 52 0 | b = No 0 0 0 | c = Do not know
=== Confusion Matrix of EBICTMT-09 with SVM=== a b c d e <-- classified as 25 5 0 0 0 | a = Strongly Agree 0 48 2 0 0 | b = Agree 0 5 25 0 0 | c = Undecided 0 0 1 2 0 | d = Disagree 0 0 0 0 0 | e = Strongly Disagree	=== Confusion Matrix of UIICTMT-06 with MLR=== a b c d e <-- classified as 10 0 0 0 0 | a = All the time 0 37 1 0 0 | b = Often 0 2 50 0 0 | c = Sometime 0 0 0 10 2 | d = Never 0 0 0 0 1 | e = Rarely

Fig. 4 Dataset confusion matrices at 66:44 ratios

classifier outperformed others with the highest accuracy of 94.9% in DAICTMT-16 and 92% in the EBICTMT-09 datasets. The MLR classifier proved outstanding with the accuracy of 97.2% in the AICTMT-06 and 95.2% accuracy in the UICTMT-06 dataset.

Figure 4 displays the confusion matrix of predicted versus actual values of each dataset provided by holdout experiment execution. It is proved that out of a total of 113 responses, 108 are correctly identified in AICTMT-06. The SVM classifier predicted 109 and 100 from the datasets DAICTMT-16 and EBICTMT-09, respectively. From the dataset UICTMT-06, the MLR identified accurately 108 responses.

3.2 Using k*-Fold*

Data from Fig. 5 shows the classification accuracy provided with *k*-fold cross-validation having $k = 10$. It is observed that in the AICTM dataset testing, MLR classifier outperformed others with a maximum accuracy of 98.3%. For the DAICTMT dataset, the SVM outperformed others with the highest accuracy of 96.5%. Also, the MLR classifier attained the peak accuracy of 93.8% for the EBICTM and 97.1% accuracy for the UICTMT.

Figure 6 depicts the joint confusion matrix created with the help of Weka explorer utility. On one hand, in the dataset AICTMT-06, a total of 326 responses are accurately predicted out of total 331, and on another hand, SVM predicted 318 responses from the dataset DAICTMT-16. Also, it is found that MLR predicted 311 from the EBICTMT-09 and 321 from the UICTMT-06 datasets. Hence, the misclassification

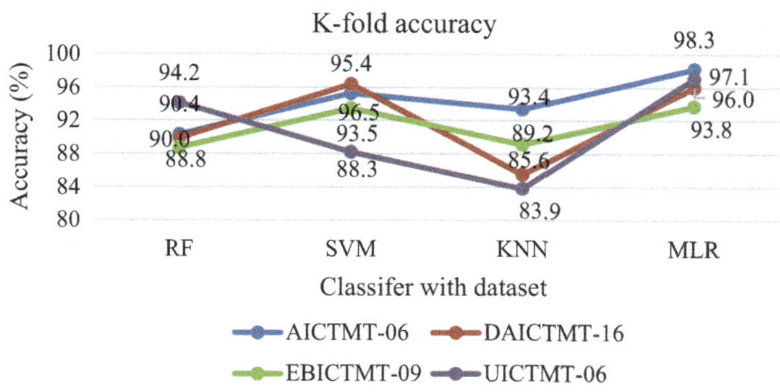

Fig. 5 Response-type prediction with k-folds with $k = 10$

=== Confusion Matrix of AICTMT-06 with MLR=== a b c d e <-- classified as 81 0 0 0 0 | a = Strongly Agree 1 154 0 0 0 | b = Agree 0 0 73 2 0 | c = Undecided 0 0 0 13 2 | d = Disagree 0 0 0 0 5 | e = Strongly Disagree	=== Confusion Matrix of DAICTMT-16 with SVM=== a b c <-- classified as 159 4 0 | a = Yes 7 158 0 | b = No 0 2 1 | c = Do not know
=== Confusion Matrix of EBICTMT-09 with MLR=== a b c d e <-- classified as 74 5 0 0 0 | a = Strongly Agree 1 153 1 0 0 | b = Agree 0 0 81 5 0 | c = Undecided 0 0 4 3 3 | d = Disagree 0 0 0 1 0 | e = Strongly Disagree	=== Confusion Matrix of UIICTMT-06 with MLR=== a b c d e <-- classified as 24 0 0 0 0 | a = All the time 0 115 4 0 0 | b = Often 0 5 151 1 0 | c = Sometime 0 0 0 29 0 | d = Never 0 0 0 0 2 | e = Rarely

Fig. 6 Dataset confusion matrix at tenfolds

is found very less in each matrix, and also, it is proved that the k-fold method has significantly upgraded the accuracy of each classifier.

3.3 Feature Relationship with k-*Fold*

To significant prove the influence of binary classification in machine learning, the Mathews correlation coefficient (MCC) plays a vigorous activity. Likewise, the correlation coefficient, it computes the correlation coefficient between the actual values and predicted values. The following formula of MCC is mentioned below:

Table 4 MCC with tenfolds

Dataset	RF_c	SVM_c	KNN_c	MLR_c
AICTMT-06	0.91	0.94	0.94	0.99v
DAICTMT-16	0.82	0.94	0.74	0.94v
EBICTMT-09	0.89	0.94	0.91	0.95v
UICTMT-06	0.84	0.79	0.79	1.00v

$$\text{MCC} = \frac{\text{TP} \times \text{TN} - \text{FP} \times \text{FN}}{\sqrt{(\text{TP} + \text{FP})(\text{TP} + \text{FN})(\text{TN} + \text{FP})(\text{TN} + \text{FN})}}$$

where TP is true positive rate, TN is true negative rate, FP is false positive rate and FN is false negative rate. Based on the zero value of both the FP and the FN, the MCC value will be 1 which reveals a perfect positive correlation. It values always lies between −1 and 1.

Table 4 reflects the MCC comparisons of each classifier with a *T*-test at 10-folds with 0.05 significant level. On one hand, the MLR classifier found a victor (V) to prove the positive correlation among features of each dataset. Another hand, the RF and the KNN classifier's MCC values are found smaller than the MLR and the SVM classifiers. The peak MCC value signifies the best positive correlations between predicted and actual values.

3.4 User CPU Training Time with k-Fold

This section expresses the concept of the prediction time of each predictive model. To present a real-time predictive model, the time measurement unit t (second) is compared with each classifier on the datasets. A total of 10 iterations are performed to find a significant difference.

Table 5 illustrates the statistical difference among classifier's training time with a *T*-test at 0.05 significant level at tenfolds. It is discovered a significant variation (*) between the training time of MLR and KNN in the case of the AICTMT-06 and UICTMT-06 datasets. Except for the RF classifier, each classifier found a significant difference in the DAICTMT-16 dataset. In the dataset EBICTMT-09, the SVM and KNN time is found significantly different.

Table 5 Response prediction training time with *k*-fold

Dataset	RF_t	SVM_t	KNN_t	MLR_t
AICTMT-06	0.04	0.05	0.00*	0.03*
DAICTMT-16	0.07	0.02*	0.00*	0.04*
EBICTMT-09	0.06	0.05*	0.00*	0.07
UICTMT-06	0.05	0.05	0.00*	0.04*

4 Performance Measures

This section discusses various important measures needed to justify the comparative and predictive strength of predictive models. All of these measures such as kappa, accuracy, error and F-score are computed using tenfolds with the T-test at 0.05 confidence level applicable.

Table 6 signifies the strength among dataset features to predict the response type as strong predictors using kappa statistic (k) value. Also, differentiate the observed prediction accuracy versus expected prediction accuracy.

$$k = \frac{p_o - p_e}{1 - p_e} = 1 - \frac{1 - p_o}{1 - p_e},$$

where p_o is observed accuracy and p_e is expected accuracy and $k \leq 1$. We found the highest bonding (v) among instances of each dataset calculated by MLR except the SVM in the dataset DAICTMT-16-K. Also, the KNN classifier is found significant in the same dataset DAICTMT-16-K.

Table 7 compares the F-score (F) of provided with the classifiers, and the MLR classifier found victor (v) among all datasets, and the SVM classifier has v status in the DAICTMT-16-F. Below is the formula to calculate the F.

$$F1 = 2 \times \frac{\text{Presision} * \text{Recall}}{\text{Precision} + \text{Recall}}$$

where precision is the proportion of positive prediction of response type and recall is the proportion of actual positive prediction of response type. No significant difference is found between the classifiers considering F value.

Table 6 Attribute bonding with k-fold

Classifier	RF_k	SVM_k	KNN_k	MLR_k
AICTMT -06-K	0.86	0.93	0.90	0.97v
DAICTMT-16-K	0.80	0.93v	0.72*	0.92v
EBICTMT-09-K	0.83	0.90	0.83	0.91v
UICTMT-06-K	0.74	0.81	0.74	0.95v

Table 7 Testing of F-score with k-fold

Classifier	RF_f	SVM_f	KNN_f	MLR_f
AICTMT -06-F	0.93	0.95	0.95	0.99v
DAICTMT-16-F	0.91	0.97v	0.87	0.97v
EBICTMT-09-F	0.91	0.95	0.93	0.96v
UICTMT-06-F	0.84	0.76	0.79	1.00v

Table 8 Residual error with k-fold

Classifier	RF_e	SVM_e	KNN_e	MLR_e
AICTMT -06-E	0.16	0.32v	0.13*	0.05*
DAICTMT-16-E	0.23	0.29	0.28v	0.14*
EBICTMT-09-E	0.17	0.32v	0.19	0.15
UICTMT-06-E	0.21	0.32v	0.22	0.11*

Table 8 differentiates the root mean square error (E) of each classifiers having the below formula:

$$\text{RMS Error} = \sqrt{1 - r^2}\text{SD}_y$$

where r is the residual variations between the predicted values and the actual values and SD_y is the standard deviation of the observed value. It can realize that the MLR classifier has the lowest E in all datasets. Also, the T-test proved that the MLR is significantly different in the three datasets except for the EBICTMT-09-E. Also, the KNN classifier is found significantly different in the AICTMT06-E.

5 Conclusion

This chapter enlightened on the identification of response type having a different scale in the technological survey held in Indian and Hungarian universities. Two different dataset testing techniques were applied to individual datasets having numerous features. The cross-validation testing method significantly boosts the prediction accuracy. The findings of the first experiments proved that the MLR classifier defeated all with a maximum accuracy of 98.3% in the AICTMT-06. The SVM classifier provided the top accuracy of 96.6% in the DIACTMT-16. Also, the MLR proved victor classifier with an accuracy of 93.8% and 97.1% in EBICTMT-09 and UICTMT-06 datasets, respectively. The outcomes of the second experiment concluded that the MLR classifier outperformed others in the AICTMT-06 with 97.2% accuracy and in the UICTMT-06 with 95.2% accuracy. Subsequently, the SVM also proved as a winner with an accuracy of 94.9% in the DAICTMT-16 and with an accuracy of 92% in the EBICTMT-09. A statistical T-test found all the classifiers significant except the RF classifier in CPU training time. Further, to differentiating agreement of instances, only the KNN finds significant as compared to others. The root means square error value E is significantly different from the KNN and the MLR classifier. The ending remark of this chapter concludes that the MLR classifier is best suited to identify the response type of students toward educational technology. Also, future work is suggested to apply fully cross-validation, feature filtering, feature mapping and novel robust classifiers.

Acknowledgements The present chapter is written with the support of the Hungarian Government and European Social Funds under the project "Talent Management in Autonomous Vehicle Control Technologies (EFOP-3.6.3-VEKOP-16-2017-00001)."

References

1. Verma C, Dahiya S, Mehta D (2016) An analytical approach to investigate state diversity towards ict: a study of six universities of Punjab and Haryana. Indian J Sci Technol 9:1–5. https://doi.org/10.17485/ijst/2016/v9i31/87426
2. Verma C, Dahiya S (2016) Gender difference towards information and communication technology awareness in Indian universities. SpringerPlus 5(370):1–7. https://doi.org/10.1186/s40064-016-2003-1
3. Verma C, Stoffová V, Illés Z (2018) Analysis of situation of integrating information and communication technology in Indian higher education. Int J Inf Commun Technol Educ 7(1):24–29. https://doi.org/10.2478/ijicte-2018-0003
4. Bathla Y, Verma C, Kumar N (2020) Smart approach for real time gender prediction of European school's principal using machine learning. In: Proceedings of ICRIC 2019. Lecture notes in electrical engineering (LNEE). Springer, Berlin, pp 159–175. https://doi.org/10.1007/978-3-030-29407-6_14
5. Verma C, Tarawneh AS, Stoffová V, Illés Z, Dahiya S (2018) Gender prediction of the European school's teachers using machine learning: preliminary results. In: Proceeding of 8th IEEE international advance computing conference, pp 213–220. https://doi.org/10.1109/IADCC.2018.8692100
6. Verma C, Stoffová V, Illés Z (2019) An ensemble approach to identifying the student gender towards information and communication technology awareness in European schools using machine learning. Int J Eng Technol 7(4):3392–3396
7. Verma C, Illés Z, Stoffová V (2019) Gender prediction of Indian and Hungarian students towards ICT and mobile technology for the real-time. Int J Innovative Technol Exploring Eng 8(9S3):1260–1264
8. Verma C, Tarawneh AS, Illés Z, Stoffová V, Singh M (2019) National identity predictive models for the real time prediction of European school's students: preliminary results. In: IEEE international conference on automation, computational and technology management, pp 418–423. https://doi.org/10.1109/ICACTM.2019.8776842
9. Verma C, Stoffová V, Illés Z (2020) Ensemble methods to predict the locality scope of Indian and Hungarian students for the real time. In: 4th international conference on advanced computing and intelligent engineering. Advances in intelligent systems and computing. Springer, Berlin, pp 1–13
10. Verma C, Illés Z, Stoffová V (2020) Real-time classification of national and international students for ICT and mobile technology: an experimental study on Indian and Hungarian university. J Phys Conf Ser 432:1–8. IOP Science. https://doi.org/10.1088/1742-6596/1432/1/012091
11. Verma C, Stoffová V, Illés Z (2019) Real-time prediction of student's locality towards information communication and mobile technology: preliminary results. Int J Recent Technol Eng 8(1):580–585
12. Verma C, Stoffová V, Illés Z (2019) Age group predictive models for the real time prediction of the university students using machine learning: preliminary results. In: 2019 IEEE third international conference on electrical, computer and communication, pp 1–7. https://doi.org/10.1109/ICECCT.2019.8869136
13. Verma C, Stoffová V, Illés Z (2020) Prediction of residence country of student towards information, communication and mobile technology for real-time: preliminary results. In: International

conference on computational intelligence and data science, vol 167. Procedia computer science. Elsevier, Amsterdam, pp 224–234. https://doi.org/10.1016/j.procs.2020.03.213
14. Verma C, Stoffová V, Illés Z (2020) Feature selection to identify the residence state of teachers for the real-time. In: IEEE international conference on intelligent engineering and management, London, pp 1–6 (Accepted)
15. Verma C, Illés Z, Stoffová V (2020) Study level prediction of Indian and Hungarian students towards ICT and mobile technology for the real-time. In: IEEE international conference on computation, automation and knowledge management, UAE, pp 215–219. https://doi.org/10.1109/ICCAKM46823.2020.9051551
16. Verma C, Illés Z, Stoffová V, Singh M (2020) ICT and mobile technology features predicting the university of Indian and Hungarian student for the real-time. In: IEEE system modeling & advancement in research trends, pp 85–90
17. Verma C, Illés Z (2019) Attitude prediction towards ICT and mobile technology for the real-time: an experimental study using machine learning. In: The 15th international scientific conference e-learning and software for education, Romania, pp 247–254. https://doi.org/10.12753/2066-026X-19-171
18. Verma C, Stoffová V, Illés Z (2019) Prediction of students' awareness level towards ICT and mobile technology in Indian and Hungarian University for the real-time: preliminary results. Heliyon 5(6):1–7. https://doi.org/10.1016/j.heliyon.2019.e01806
19. Verma C, Illés Z, Stoffová V (2020) Real-time prediction of development and availability of ICT and mobile technology in Indian and Hungarian university. In: Proceedings of ICRIC 2019. Lecture notes in electrical engineering (LNEE). Springer, Berlin, pp 605–615. https://doi.org/10.1007/978-3-030-29407-6_43
20. Verma C, Illés Z, Stoffová V (2020) Predictive modeling to predict the residency of teachers using machine learning for the real-time. In: Proceedings of FTNCT 2019. Communications in computer and information science (CCIS). Springer, Berlin, pp 1–11. https://doi.org/10.1007/978-981-15-4451-4_47
21. Singh P, Paprzycki M, Bhargava B, Chhabra J, Kaushal N, Kumar Y (2018) Futuristic trends in network and communication technologies. In: FTNCT 2018. Communications in computer and information science, vol 958, pp 3–509. https://doi.org/10.1007/978-981-13-3804-5
22. Singh PK, Bhargava BK, Paprzycki M, Kaushal NC, Hong WC (2020) Handbook of wireless sensor networks: issues and challenges in current scenario's. In: Advances in intelligent systems and computing, vol 1132. Springer, Cham, pp 155–437. https://doi.org/10.1007/978-3-030-403 05-8
23. Singh PK, Kar AK, Singh Y, Kolekar MH, Tanwar S (2020) Proceedings of ICRIC 2019. Recent innovations in computing. Lecture notes in electrical engineering, vol 597. Springer, Cham, pp 3–920. https://doi.org/10.1007/978-3-030-29407-6
24. Sharma A., Singh PK, Kumar R (2019) An efficient architecture for the accurate detection and monitoring of an event through the sky. Comput Commun 148:115–128. ISSN 0140-3664. https://doi.org/10.1016/j.comcom.2019.09.009
25. Singh M, Verma C, Kumar R, Juneja P (2020) Towards enthusiasm prediction of Portuguese school's students towards higher education in real-time. In: IEEE international conference on computation, automation and knowledge management, UAE, pp. 215–219. https://doi.org/10.1109/ICCAKM46823.2020.9051459

Automatic Leaf Species Recognition Using Deep Neural Network

Deepak Kumar and Chaman Verma

Abstract Automatic plant species identification system that designed and implemented by the researcher of computer vision experts assists botanist in speedy identification of unknown plant species. Nowadays, deep neural networks are being used in various fields such as speech recognition, human motion identification, and others. It is obvious for researchers to explore other areas to automate that are working on traditional features. In this article, a deep learning transfer learning method is used for recognizing the species of a plant leaf, in which foremost layers of pre-trained AlexNet deep neural network model extract the useful leaf characteristics directly from the input image. Then, the last three layers for classification are customized according to requirements. The projected methods are assessed on 15 leaf species that have 240 images in total from 100 leaves species dataset from the UCI machine library Web site and have succeeded in achieving an accuracy of 95.56% with fine-tuning of hyperparameters. The accuracy is also checked with other parameters like by changing the volume of images and hyperparameters like L2Regularization, minibatch sizes that exhibits high performance despite large changes.

Keywords Leaves species · AlexNet · L2Regularization · Minibatch size · Execution time

D. Kumar (✉)
Guru Kashi University, Talwandi Sabo, Punjab, India
e-mail: dr.d.k.mehta81@gmail.com

C. Verma
Eötvös Loránd University, Budapest, Hungary
e-mail: chaman@inf.elte.hu

P. K. Singh et al. (eds.), *Evolving Technologies for Computing, Communication and Smart World*, Lecture Notes in Electrical Engineering 694,
https://doi.org/10.1007/978-981-15-7804-5_2

1 Introduction and Problem Statement

We often see many kinds of tress when departing from cities or going on any journey in the suburbs of cities. We rarely able to distinguish those species grown on the trees and plants, and most of the species are unfamiliar to us. It is believed that on the earth, there are around 10,000 species of plants and identifying the species among them is a very tedious process. Lots of the trees are in humid regions, as limited botanical research has been conducted out in these areas. Plant leaves species classification can be performed by various trending machine learning algorithms. Machine learning algorithms [1–5] are enhancing day by day; therefore, the accuracy of problem solution sharply and we know deep learning is a type of machine learning which is very progressive technique today for the tedious classification assignments [4, 6–12]. Much more technique exists for image classification like a bag of features, histogram techniques [13, 14]. AlexNet is one of the pre-trained classification models among GoogleNet, ResNet, VGGNet and many more. AlexNet and deep learning architectures developed by Krizhevsky et al. [15] classify ImageNet dataset images with 1000 classes of 1.2 million images. It contains the stack of convolutional layers with a changing receptive field. In this, a total of five convolution layers, a rectified linear unit (ReLU) with batch normalization and max pooling in some layers. The stacked architecture performance is significantly better than the traditional shallow machine learning algorithm with proper regularization techniques [16]. Mohanty et al. [17] also took benefit of the transfer learning approach in which pre-trained AlexNet process for predicting the new classes of images. In this article, for plant leaves classification, we also used a transfer learning approach with AlexNet. With this help of the model, we able to identify 15 different species in 240 leaves species images.

Plant leaf species identification interest is not only for botanist, but also for a large part of society people can benefit from it. Moreover, society is concerned about changing climate and in-line changes in the geographic distribution along with a wealth of species. The novel crop breed development often relies on the integration of genes from wild relatives of the existing crops, and therefore, it is our keen interest to retain path of the scattering of all plant taxonomy. With the passage of deteriorating environments, most plants which are of rare species already dead; still, many more of the rare species are at the margin of loss, so plant species study can contribute to the protection of the environment. Conventional plant species identification for a novice is more time consuming and full of a hurdle; therefore, automatic identification with the help of image processing with deep learning ideas can make the task very easy.

The present study aims to build the classifier that can determine the type of species in an image. As an initial work, there are 100 types of species are available: Acer campestre, Acer capillipes, Acer palmatum, Alnus cordata, Cornus chinensis and many more exist, and some are depicted in Fig. 1.

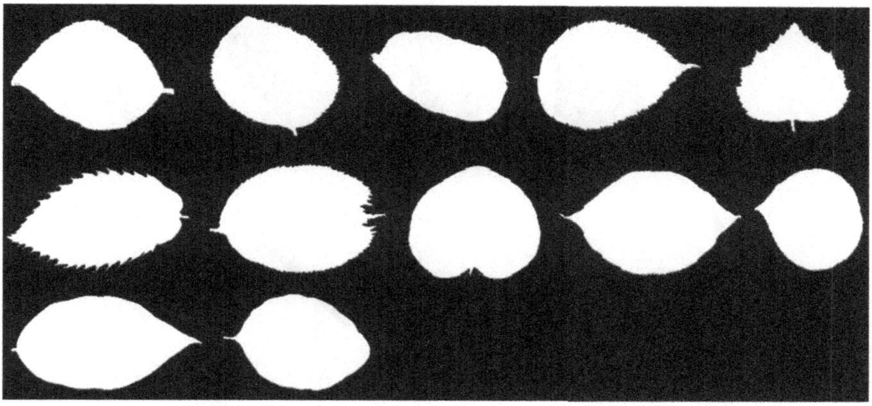

Fig. 1 Plant leave spices

2 Material and Methods

An ephemeral description on the architecture of the deep neural network model with h/w and s/w configurations is explored in this segment.

2.1 System Configuration

The model of deep neural network depends most significantly on the GPU processor with the support of CUDA core. This experimental article study is conducted on 4 GB of GPU 1050 with CUDA support, 8 GB of random-access memory (RAM). The pre-trained AlexNet model in MATLAB 2019a is used for this research article.

2.2 AlexNet

As depicted in Fig. 2, AlexNet pretrained model mainly consists of eight total layers from which five are convolutional layers and rests are fully connected. In the first layer, i.e., convolutional layer, a filter of size $11 \times 11 \times 3$ that characterizes height, width and depth correspondingly is passed over the given image of size $227 \times 227 \times 3$. When the applied filter is passed to a corresponding pixel, the filter matrix dot product with the corresponding value of pixel in the image receptive field is taken out. As a result, 96 filters are passed in the foremost layer. Therefore, 96 activation maps are created from the Rectified Linear Unit (ReLU) layer of the basic foremost convolution layer. Analogously, convolution layers with 2 with number of 256 filters of size $5 \times 5 \times 48$, layer 3 with 384 filters of size $3 \times 3 \times 192$ are required for

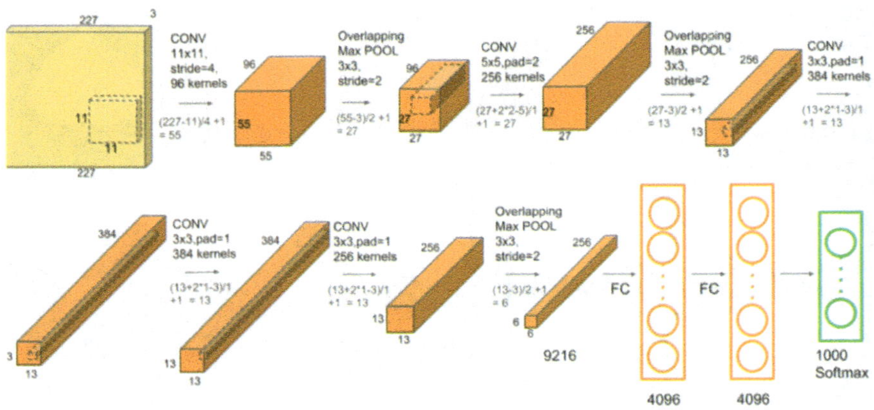

Fig. 2 AlexNet architecture

executing operation of convolution and activation maps are produced with various neurons stimulated in respectively map [18, 19].

In the classification model, various convolutional layers are pursued by the rectified linear unit, max pooling, and normalization layers. The ReLU is considered as a nonlinear, non-saturating map function which activates all the convolution layers plus also the last two fully connected layers. In the max pooling layer, the output size from the former convolution layer is decreased by exploring and holding the max value in the corresponding field. Fully connected layers 6 and 7 have in total of 4096 neurons that are connected.

In turn, a dropout layer has been adopted to randomly prevent the number of connections in a network for training which have displayed to increase the network performance during the test phase. The ending fully connected layer has been changed with the whole amount of classes, i.e., 15 species classes with bias and weight learning rate [20].

2.3 Dataset

The images for 15 different species of plant leaves are taken from the UCI machine library Web site [21]. A total of 240 images are available for selecting 15 different species from the dataset. The original input image of size 370 × 429 is resized to a size 227 × 227 for the given input to the AlexNet model [22].

3 Results and Discussion

In this study, we used deep learning pre-trained transfer learning methods for categorizing the class of objects with the MATLAB development environment tool. The augmented images with resizing parameters according to required input (227 × 227) are provided to the pre-trained AlexNet model. AlexNet model has been trained from ImageNet dataset for identifying thousand (1000) classes of objects; therefore, the last layer that is fully connected layer has been replaced with an equal number of fifteen identifying leaf species. First, 15 species dataset is fragmented into two parts: The primary part is training and the secondary part is for testing with 80–20 ratio with hold out method. The initial learning rate of the model is set to 0.0001, and further, the fine-tuning of the training model of the last three classifications layers' parameters—the weight learning rate—is set to 10 and the bias learning rate is tagged with 20. Table 1 shows another parameter.

The training progress is monitored with the plot parameter as shown in Fig. 3. An adaptive learning rate method (adam) specially is used in training the model in deep neural networks. It can be considered as a feature mixture of stochastic gradient descent with momentum (SGDM) and RMSDrop. It exhausts the feature of RMSDrop in which squared gradient scales the rate of learning, and it also takes gain of momentum by using a moving average of gradient-like SGD with momentum [23]. The overfiting problem is handled using (L2Regularization) for the loss function $E(\theta)$ with a weight of 0.0005. The weight decay is another name of regularization [24]. In Fig. 1, the accuracy obtained after training and testing the AlexNet model was 95.56 form 15 classes of species with 240 leaves images in total Fig. 4 also displaying the corresponding confusion matrix.

Here, the authors discussed the results achieved on the confusion matrix to analyze the property and behavior of the measures [3, 10, 25]. The diagonal confusion matrix exhibits the accuracy in results as described in Fig. 4 of perfect classification. It explores 95.6% actual classification and 4.4% misclassification ratio.

The model enactment is assessed by changing the number of images and by changing the training options (hyperparameters). These hyperparameters used as updating minibatch size, and L2Regularization. We also noted the execution time on

Table 1 Parameter and value

Parameter	Value
Max epoch	10
Minibatch size	32
L2Regularization	0.0005
Gradient threshold	1
Learning rate	Piecewise
Solver name	Adam

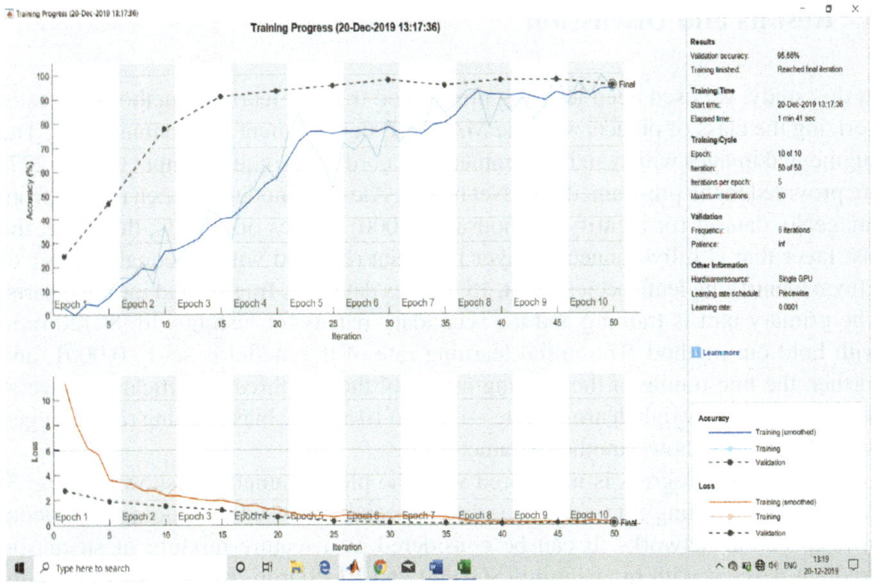

Fig. 3 Training progress with validation frequency

various minibatch sizes to assess the accuracy with global initial learning rate 0.0001 and max epoch as 10 epochs.

From Figs. 5 and 6, when minibatch sizes are increased from 2 to 128, the accuracy is noteworthy. Only on 22 and 32 batch size gives optimal accuracy, i.e., 95.56% and on 32 batch size takes less execution time, i.e., 28 s. As a total of 240 images in total in 15 classes therefore in one epoch, 32 batch size is optimal.

The classification accuracy is assessed with equal number of sample images for each class label. In the dataset, the leaf species class has 16 images in each and a total consists of 15 classes—240 images. Therefore, the highest limit is set as 16 images for each class which is given as input to training the AlexNet model. We trained the network in each phase with passing images with an interval of 4, and the accuracy shown in Fig. 7.

The accuracy gets low on the reduction of samples in a single class as compared to others. The most accuracy of 95.56% observed using 16 images, and the least accuracy of 88.64% noted using 8 images. We did not see the difference between accuracies on the use of 12 and 4 images. Based on the image reduction, the prediction accurateness drops. The accredited reason for the less accuracy is the complexity in making differentiation with the pictures of extra classes.

In the next scenario, the hyperparameter, namely L2Regularization (L2Reg), is varied to analyze its implication on the classification performance. The accuracy of this model is analyzed with L2Regularization.

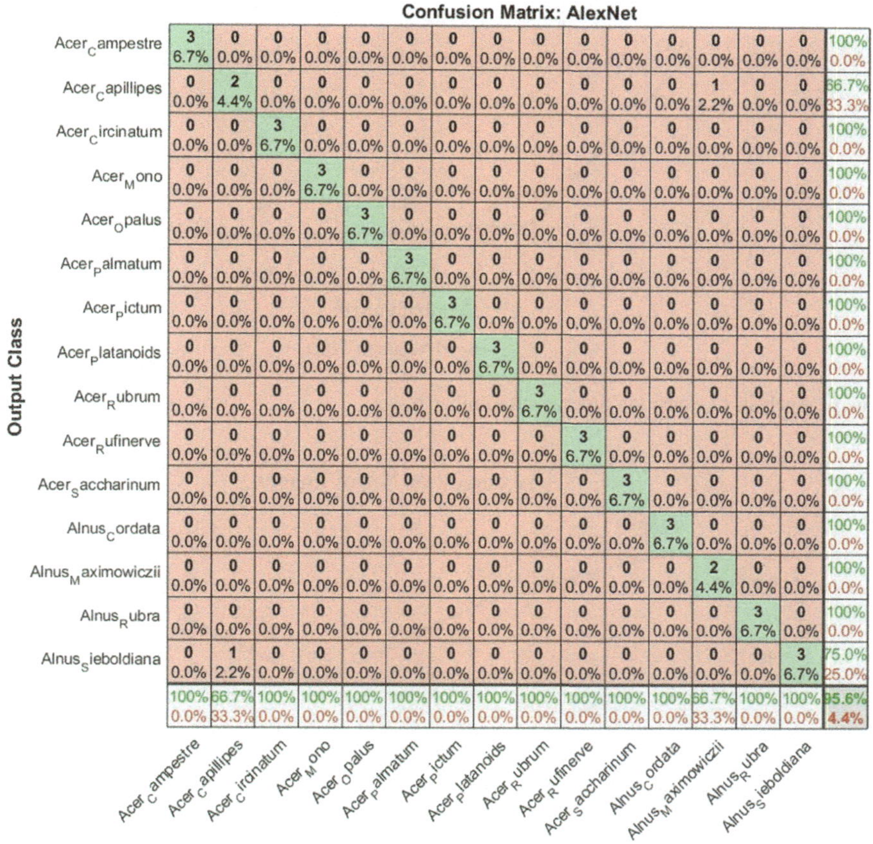

Fig. 4 Analysis of confusion matrix

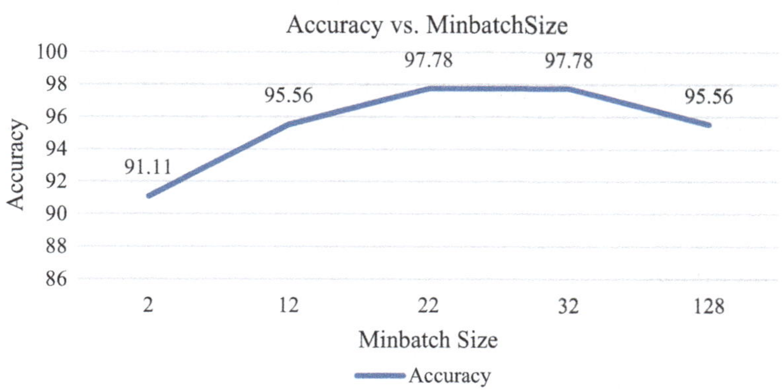

Fig. 5 Analysis of accuracy versus minibatch size

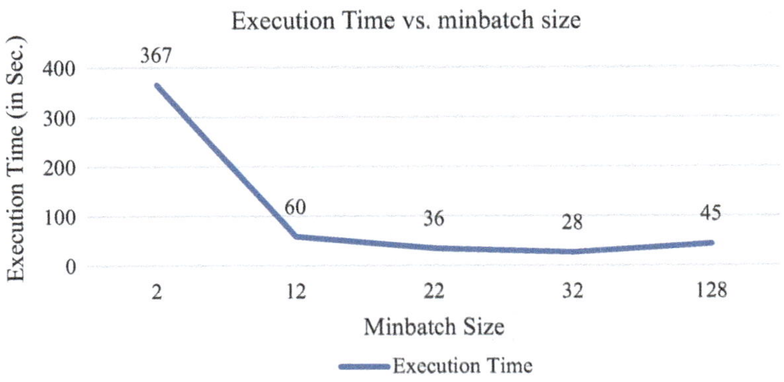

Fig. 6 Analysis of execution time versus minibatch size

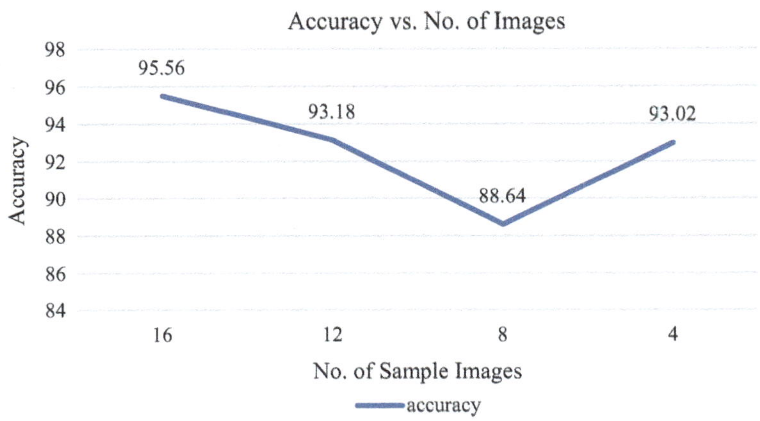

Fig. 7 Analysis of accuracy versus number of sample images

When the L2Reg rate is 0.0001, the highest accuracy results, i.e., 97.78%, but when the rate is decreased at an odd gap, accuracy got decreased, and on 0.001, the same accuracy again resulted as shown in Fig. 8.

4 Conclusion

Plant leaves classification has been performed with images from the UCI machine library dataset with deep learning. The categorization correctness resulted using 240 images is 95.56% for the given model. The model performance has been assessed by changing the number of sample images, fine-tuning of various minibatch sizes, and varying L2Regularization rate results are obtained. As from this article, several

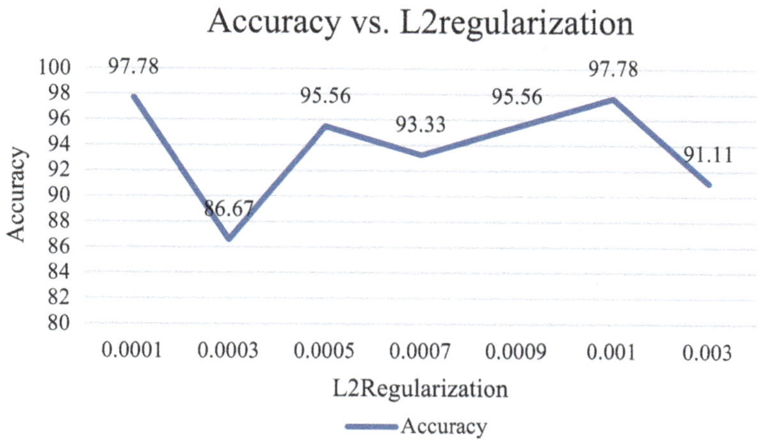

Fig. 8 Analysis of accuracy versus L2Regularization rate

sample images considerably affect model performance. The highest prediction accuracy is achieved when the sample size is 16. With setup, the minibatch size in tuning the performance of this model exhibits full accuracy correlation to the classification. Similarly, the L2Regularization rate initially exhibits the highest accuracy and fluctuates to lower accuracy slowly. In terms of computation load, the AlexNet model resulted in the best accuracy when minibatch is set to 32, L2Regularization rate to 0.0001 with minimum execution time 28 s when compared to another hyperparameter of this model. Future work can be expanded and compared with other models like ResNet50 and GoogleNet and with supervised classification algorithms like SVM and XGBOOST with rest species.

References

1. Cybenko G (1989) Approximation by superposition of a sigmoidal function. Math. Controls Signals Syst 2(4):303–314
2. Verma C, Illés Z, Stoffová V (2020) Real-time classification of national and international students for ICT and mobile technology: an experimental study on Indian and Hungarian university. J Phys Conf Ser 432:1–8. IOP Science. https://doi.org/10.1088/1742-6596/1432/1/012091
3. Verma C, Tarawneh AS, Stoffová V, Illés Z, Dahiya S (2018) Gender prediction of the European school's teachers using machine learning: preliminary results. In: Proceeding of 8th IEEE international advance computing conference, pp 213–220. https://doi.org/10.1109/IADCC.2018.8692100
4. Bathla Y, Verma C, Kumar N (2020) Smart approach for real time gender prediction of European school's principal using machine learning. In: Proceedings of ICRIC 2019, Lecture notes in electrical engineering (LNEE). Springer, Berlin, pp 159–175. https://doi.org/10.1007/978-3-030-29407-6_14
5. Mehdipour M, Yanikoglu B, Aptoula E (2017) Plant identification using deep neural network via optimization of transfer learning parameters. Neurocomputing 235(C)

6. Verma C, Illés Z, Stoffová V (2020) Predictive modeling to predict the residency of teachers using machine learning for the real-time. In: Proceedings of FTNCT 2019, Communications in computer and information science (CCIS). Springer, Berlin, pp 1–11. https://doi.org/10.1007/978-981-15-4451-4_47
7. Singh P, Paprzycki M, Bhargava B, Chhabra J, Kaushal N, Kumar Y (2018) Futuristic trends in network and communication technologies. In: FTNCT 2018. Communications in computer and information science, vol 958, pp 3–509. https://doi.org/10.1007/978-981-13-3804-5
8. Singh PK, Bhargava BK, Paprzycki M, Kaushal NC, Hong WC (2020) Handbook of wireless sensor networks: issues and challenges in current scenario's. In: Advances in intelligent systems and computing, vol 1132. Springer, Cham, pp 155–437. https://doi.org/10.1007/978-3-030-40305-8
9. Singh PK, Kar AK, Singh Y, Kolekar MH, Tanwar S (2020) Proceedings of ICRIC 2019. Recent innovations in computing. Lecture notes in electrical engineering, vol 597. Springer, Cham, pp 3–920. https://doi.org/10.1007/978-3-030-29407-6
10. Sharma A, Singh PK, Kumar R (2019) An efficient architecture for the accurate detection and monitoring of an event through the sky. Computer Commun 148:115–128. ISSN 0140-3664. https://doi.org/10.1016/j.comcom.2019.09.009
11. Verma C, Tarawneh AS, Illés Z, Stoffová V, Singh M (2019) National identity predictive models for the real time prediction of European school's students: preliminary results. In: IEEE international conference on automation, computational and technology management, pp 418–423. https://doi.org/10.1109/ICACTM.2019.8776842
12. Singh M, Verma C, Kumar R, Juneja P (2020) Towards enthusiasm prediction of Portuguese school's students towards higher education in real-time. In: IEEE international conference on computation, automation and knowledge management, UAE, pp 215–219. https://doi.org/10.1109/ICCAKM46823.2020.9051459
13. O'Hara S, Draper BA (2011) Introduction to the bag of features paradigm for image classification and retrieval
14. Blachnik M, Laaksonen J (2008) Image classification by histogram features created with learning vector quantization
15. Krizhevsky A, Sutskever I, Hinton GE (2012) ImageNet classification with deep convolutional neural network. In: Advances in neural information processing system, pp 1097–1105
16. Pasupa K, Sunhem W (2016) A comparison between shallow and deep architecture classifier on small dataset. In: IEEE international conference information technology and electrical engineering
17. Mohantly SP, Hughes DP, Salathe M (2016) Using deep learning for image based plant disease detection. Front Plant Sci 7, article id 1419
18. Yamashita R, Nishio M, Do RK, Togashi K (2018) Convolutional neural networks: an overview and application in radiology
19. Shinde PP, Shah S (2018) A review of machine learning and deep learning applications. In: 2018 fourth international conference on computing communication control and automation (ICCUBEA), Pune, India, pp 1–6
20. Almisreb AA, Jamil N, Din NM (2018) Utilizing AlexNet deep transfer learning for ear recognition. In: 2018 fourth international conference on information retrieval and knowledge management (CAMP), Kota Kinabalu, pp 1–5
21. UCI: https://archive.ics.uci.edu/ml/datasets/One-hundred+plant+species+leaves+data+set. Accessed on 25 Nov 2019
22. Shaha M, Pawar M (2018) Transfer learning for image classification. In: 2018 second international conference on electronics, communication and aerospace technology (ICECA), Coimbatore, pp 656–660
23. Tijmen T, Geoffrey H (2012) Divide the gradient by a running average of its recent magnitude. In: COURSERA: neural networks for machine learning, vol 4, no 2, pp 26–31
24. Bishop CM (2006) Pattern recognition and machine learning. Springer, New York, NY
25. Verma C, Stoffová V, Illés Z (2019) Real-time prediction of student's locality towards information communication and mobile technology: preliminary results. Int J Recent Technol Eng 8(1):580–585

Injecting Power Attacks with Voltage Glitching and Generation of Clock Attacks for Testing Fault Injection Attacks

Shaminder Kaur, Balwinder Singh, Harsimranjit Kaur, and Lipika Gupta

Abstract Fault injection attacks pose serious threat in security of embedded devices since they require less expertise to conduct them. Suitable countermeasures can only be built if there effects are studied and analyzed in detail. This paper presents circuits which produces power fault injection attack (positive, negative, positive/negative) through voltage glitching on cadence 180 nm technology node. Additionally, it presents a VHDL code for generating clock fault (overclocking and underclocking attacks) injection attack. Effect of power and clock fault injection attacks is analyzed on combinational circuits. Techniques presented in this paper can be used as testing platforms for doing analysis of fault injection attacks on different combinational and sequential circuits so that countermeasures can be built in future in order to have attack-resistant devices. The paper emphasizes the perspective from the attacker, rather than the perspective of countermeasure development.

Keywords Fault injection attack · Clock glitch attack · Overclocking · Underclocking · Positive fault attack · Negative fault attack · Cadence virtuoso

1 Introduction

While fault injection attacks are known to be serious security threat, among them clock glitch attack and supply attacks are the one which requires in depth explanation. There are various ways of injecting faults, viz. tampering the power supply, variation

S. Kaur (✉) · L. Gupta
Chitkara University School of Engineering and Technology, Chitkara University, Baddi, Himachal Pradesh, India
e-mail: er.shaminderkaur@gmail.com

B. Singh
Centre for Development of Advance Computing, Mohali, India

H. Kaur
Chitkara University Institute of Engineering and Technology, Chitkara University, Rajpura, Punjab, India

P. K. Singh et al. (eds.), *Evolving Technologies for Computing, Communication and Smart World*, Lecture Notes in Electrical Engineering 694,
https://doi.org/10.1007/978-981-15-7804-5_3

in clock signal and EM generation. Tampering voltage supply or clock signal is well-known method of injecting faults, which cause a device to operate incorrectly or produce erroneous output. Variation in clock signal is one of the simplest measures of injecting faults and is commonly used by adversaries to inject faults. Another category of attack is power glitch attack. Power glitch injection setup is, however, less accurate than clock glitch, but it is easier to inject fault on power supply line [1–3]. Tampering with external signal affects the timings properties of gates, which produces faults. These variations further produce setup and hold time violation leading to faulty or erroneous outputs. Most of the work done so far focuses on analyzing and study effects of these attacks on embedded devices. Handfuls of authors have touched on implementation part of these attacks. Our work is centric toward how to generate these attacks rather than to study their effects, after attack is conducted on various devices. We worked from attacker point of view rather than providing countermeasures against side-channel attacks. This paper demonstrates two attacks, viz. clock attack and supply attack.

This paper presents circuits implemented practically on cadence, and results show generation of positive and negative power attacks. Further, VHDL codes for generation of overclocking and underclocking attacks are presented. Very few literatures reveal how to generate less complicated overclocking and underclocking attack. Various methods make use of external instruments [4, 5] such as pulse generator or variable power supply, phase-locked loop (PLL) which are quite complex.

Our method of generating clock glitch attack is implemented easily, and it is less costly. This analysis attack can be used as means of low cost and less complicated fault injection attack to study effects of injection attacks on various modules such as smart cards, microcontrollers, etc. in order to build suitable countermeasures against side-channel attacks. This paper is organized into following sections as described. Section 2 presents related work. Section 3 gives contribution of this paper. Algorithms for generation of overclock and underclock glitch attacks are presented in Sects. 4.1 and 4.2. Section 5 presents glitch-producing circuits using cadence 180 nm along with their waveforms. This section also presents effect of glitches on combinational circuits.

2 Related Work

Various mechanisms to produce fault injection attacks such as variation in power supply voltage, clock signal, temperature, etc., are proposed. Authors have used various platforms for creating and analyzing effects of clock/power attacks, viz. field programmable gate array (FPGA), cadence, mentor graphics, etc.

Zussa et al. [1] have studied both negative power glitch attack and overclock glitch attacks and compared their results, further found them to be identical. Implementation of these attacks is missing. Djellid-Ouar [6] have deliberated induced short supply voltage to study and analyze the behavior of CMOS circuits; however, less information on inducing short supply variation is given. Gomina et al. [7] presented mechanisms to detect glitches, but no information is given on generating them. Mostly, the work focuses on analyzing effects of glitch attack on embedded devices. However, the concept of generating glitches is not properly explained in these papers.

This paper bridges the gaps given above.

Agoyan et al. [8] used of delay-locked loop (DLL) of FPGA to generate clock glitch attack using FPGA Xilinx Virtex-5 is shown. Two clocks (CLK_DELAYED i) with programmable skews are generated from CLK. Faulty clock is obtained by switching between the CLK_DELAYED I using trigger signal. Method proposed is quite complex for generating clock glitch attack using FPGA. Balasch et al. [9] observed the effects of clock glitches on low-cost processor by performing large number of experiments on five devices. They have generated clock glitch attack using two mechanisms. One is by using normal CLK and high-frequency CLK. Other is using nominal CLK and shifted CLKs. These methods require lot of precision and accuracy.

The methods presented in this paper for generating clock glitch using VHDL code is quite simple and less complex.

3 Contribution

In this work, we focused on points, which are missing in the literature so far. Less literature specifies the concept of generation of clock/power attacks. In addition, the techniques given so far are quite complex and time-consuming.

Specifically this paper provides following contributions:

1. Authors focuses on either clock or power attack. This study focuses on both attacks (power attack and clock attack) which researchers can use and test it on various embedded devices.
2. VHDL-based algorithm for generating clock glitch attack for "***under***" clock glitch attack.
3. VHDL-based algorithm for generating clock glitch attack for "***over***"-clock glitch attack.
4. Overclock and underclock glitch attacks do not require any external equipment to generate them.
5. VHDL code presented in this paper adds further functionality to control shape, delay and width of glitchy clock signal.

Future Work: The future prospective would be that the algorithms given for over- and underclock glitch attacks could tested practically on some module. The VHDL codes can be interfaced with PC and FPGA kit, and clock glitch attack can be studied practically on some modules such as microcontrollers, smart cards, etc. FPGAs are flexible, and their time-to-market is low; hence, they become desirable to implement digital systems [10].

4 Concept of Generation of Clock Glitch Attack

Figure 1 shows generation of clock glitch attack. As observed, it creates a glitchy clock signal [11–13]. Glitch parameters are defined as width T_w and glitch period T_p, where T_w is the width of glitch and T_p is time of glitch. Algorithms are given for generating such clock glitch attacks as shown below. T_w and T_p parameters can be varied and controlled by making modifications in the VHDL codes given.

4.1 Generation of Underclock Glitch Attack

Algorithm 1: VHDL code generation for underclock glitch attack

```
module sample_clock ( );
reg clock=1'b0;
integer a=0;
always
begin
    a=a+1;
    if (a==8)
    clock = #5 ~ clock;
    else
    clock = #10 ~ clock;
end
end module
```

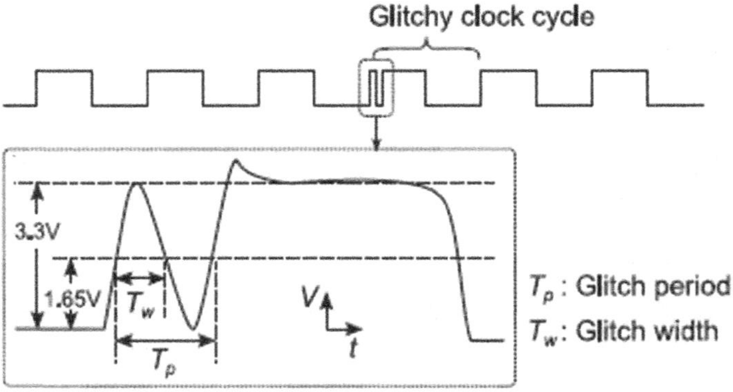

Fig. 1 Concept of glitchy clock cycle [12]

Algorithm 1 shows the code for generation of clock glitch attack. This attack is ***underclock glitch attack***. Figure 2 shows the output of Algorithm 1. It produces a 1-bit fault injection attack. The clock glitch attack can be generated anywhere along the clock signal. Width and position of clock glitch can be varied by modifying the code. As shown in Fig. 2 underclock glitch attack is generated at 81.00 ns. By increasing or decreasing the period of clock signal, these attacks can be generated.

Algorithm 2 shows the generation of overclock glitch attack. The outputs of this code are shown in Fig. 3. Figure 3a, b shows overclocking attack for different glitch positions. As observed from Fig. 3a glitch attack occurs between 65 and 80 ns at 15th clock pulse, i.e., when $a = 15$. In Fig. 3b, glitch attack occurs around 40 ns at 8th clock pulse, i.e., when $a = 8$.

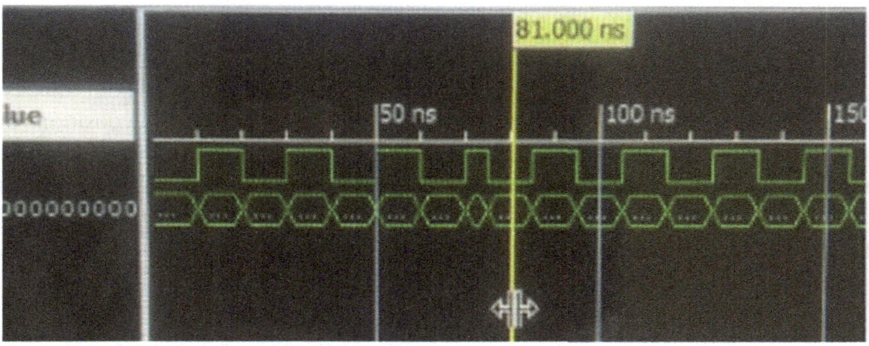

Fig. 2 Underclock glitch attack at 81.000 ns

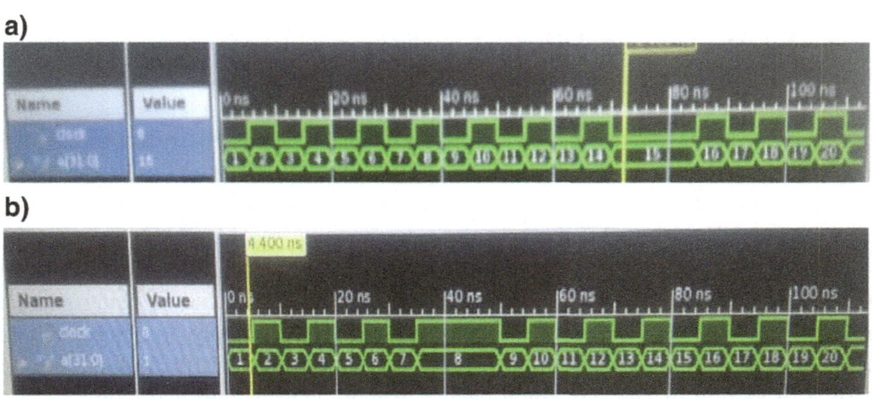

Fig. 3 **a** Overclock glitch attack at glitch position 1 (65–80 ns). **b** Overclock glitch attack at glitch position 2 (35–50 ns)

4.2 Generation of Overclock Glitch Attack

Algorithm 2: VHDL code generation for overclock glitch attack

```
module sample_clock ( );
reg clock=1'b0;
integer a=0;
always
begin
    a=a + 1;
    if (a==15)
    clock = #15 ~ clock;
    else
    clock = #5 ~ clock;
end
end module
```

The basic characteristics of clock signal: *width* and *positions* of both attacks, viz. "overclock attack" and "underclock attack" can be controlled. The positions of overclock glitch attack are varied as shown in Fig. 3. Further, these attacks can be implemented practically on some testing modules such as smart cards, and analysis can be done easily. These methods are less complex to do analysis and find appropriate measures against fault injection attacks.

5 Power Glitch Attack-Producing Circuits on Cadence Virtuoso

Section 5.2 presents power glitch attack-producing circuit. Various circuits are implemented on cadence 180 nm, and their corresponding outputs clearly explain the tampering phenomena.

Figures 4 and 5 generate negative power glitches, whereas Fig. 6 generates positive/negative glitches. Figure 7 produces only positive power glitch. Overall, these circuits provide generation of positive power glitches, negative power glitches and combination of both attacks.

As clearly observed form the waveforms, there is spike in continuous output waveform, which produces supply fault injection attack. Instead of using normal power supply, the attacker may use this tampered power supply and can generate faults into the circuit. Circuit will further produce erroneous or faulty output. During normal execution of code, if glitch occurs, it may cause some instruction to skip allowing adversary to steal secret information.

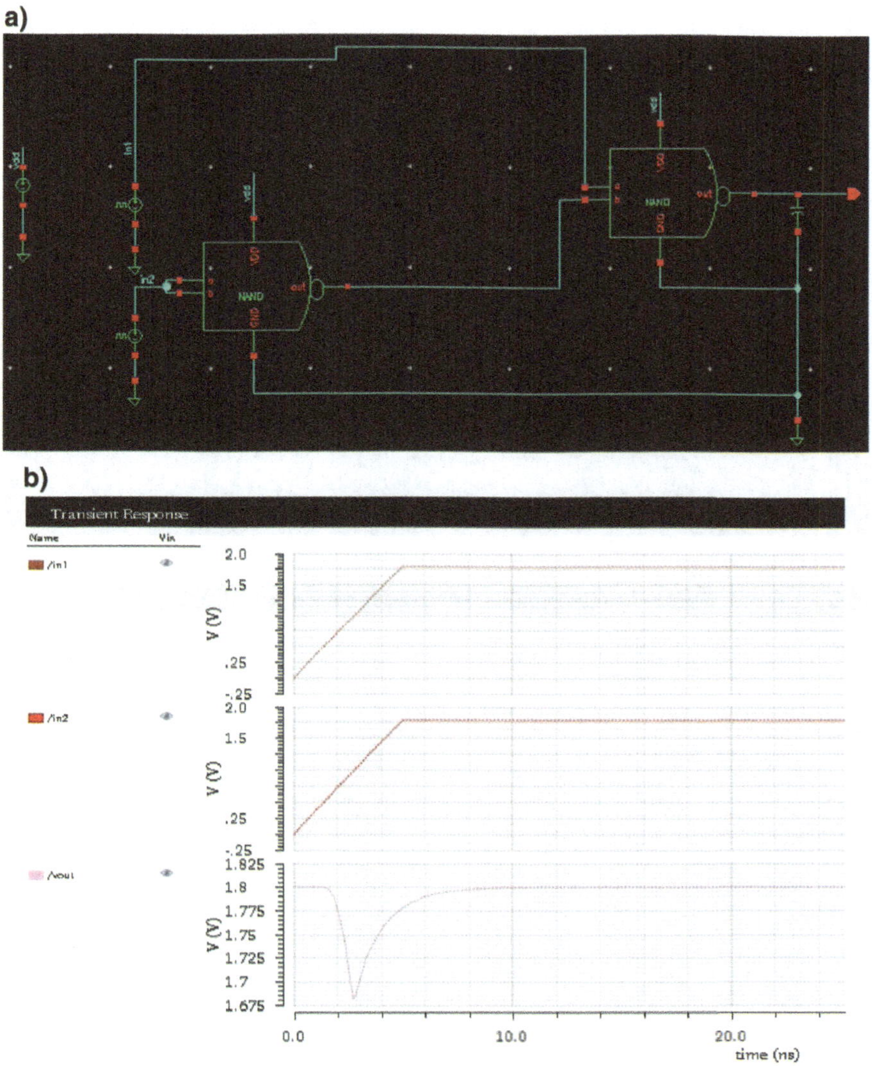

Fig. 4 **a** Power glitch attack-producing circuit 1. **b** Waveform of glitch-producing circuit 1

5.1 Power Supply Attack Versus Propagation Delay

Propagation delay largely depends on RC time constant of device and on power supply. If the power supply of a device is tampered, then it may produce fault, and due to this fault, propagation delay may be introduced into the circuit.

$$\text{power supply} \propto \frac{1}{\text{propagation delay}}$$

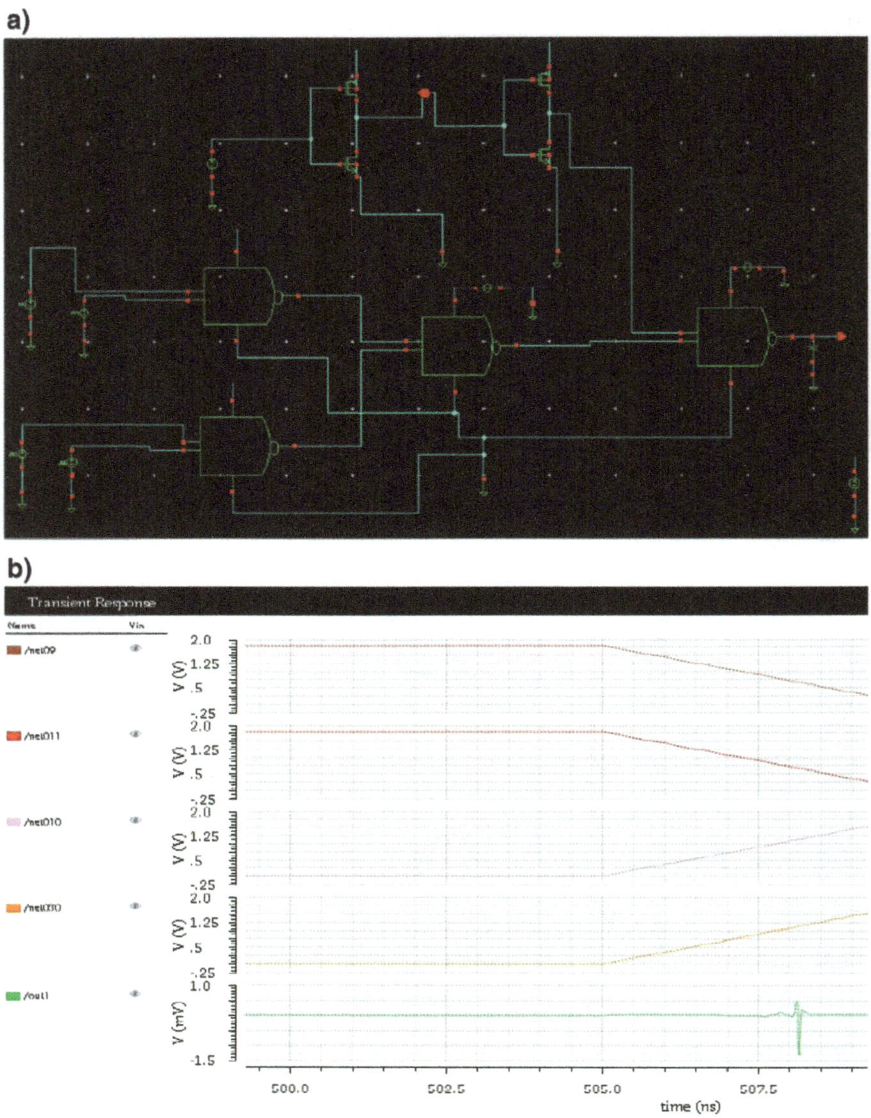

Fig. 5 **a** Power glitch attack-producing circuit 2. **b** Waveform of glitch producing circuit 2

Further propagation delay causes glitches due to change in timing properties of various gates. With increase or decrease in power supply, propagation delays decrease or increase, respectively. Therefore, by varying the external signal or by tampering the supply line, adversary can easily steal the information not intended to him.

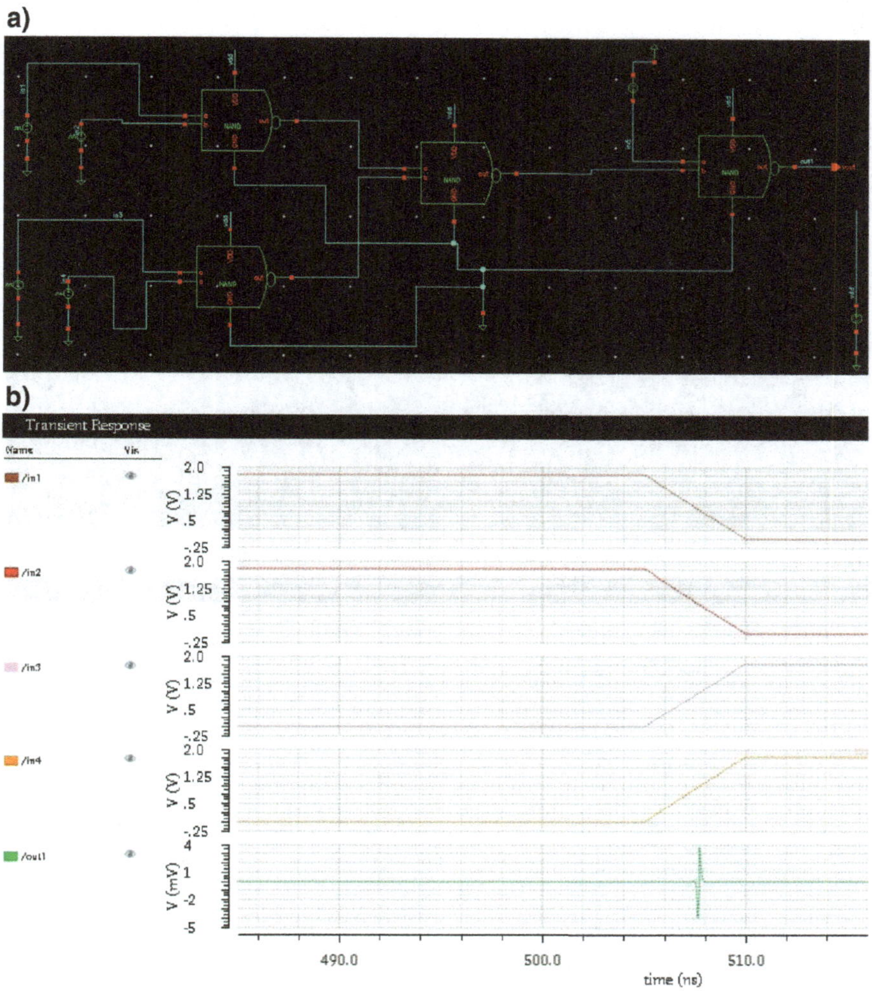

Fig. 6 **a** Power glitch attack-producing circuit 3. **b** Waveform of glitch producing circuit 3

5.2 Types of Power Glitch Attack

There are two types of power glitch attack:

1. Underpowering of device
2. Overpowering of device

Underpowering of device: The chip while doing any cryptographic operation is operated with depleted power supply. Ranging from single-bit error to multiple errors, transient faults are then introduced, and they become more invasive if the supply is further reduced. These attacking techniques such as clock attack and power attack do

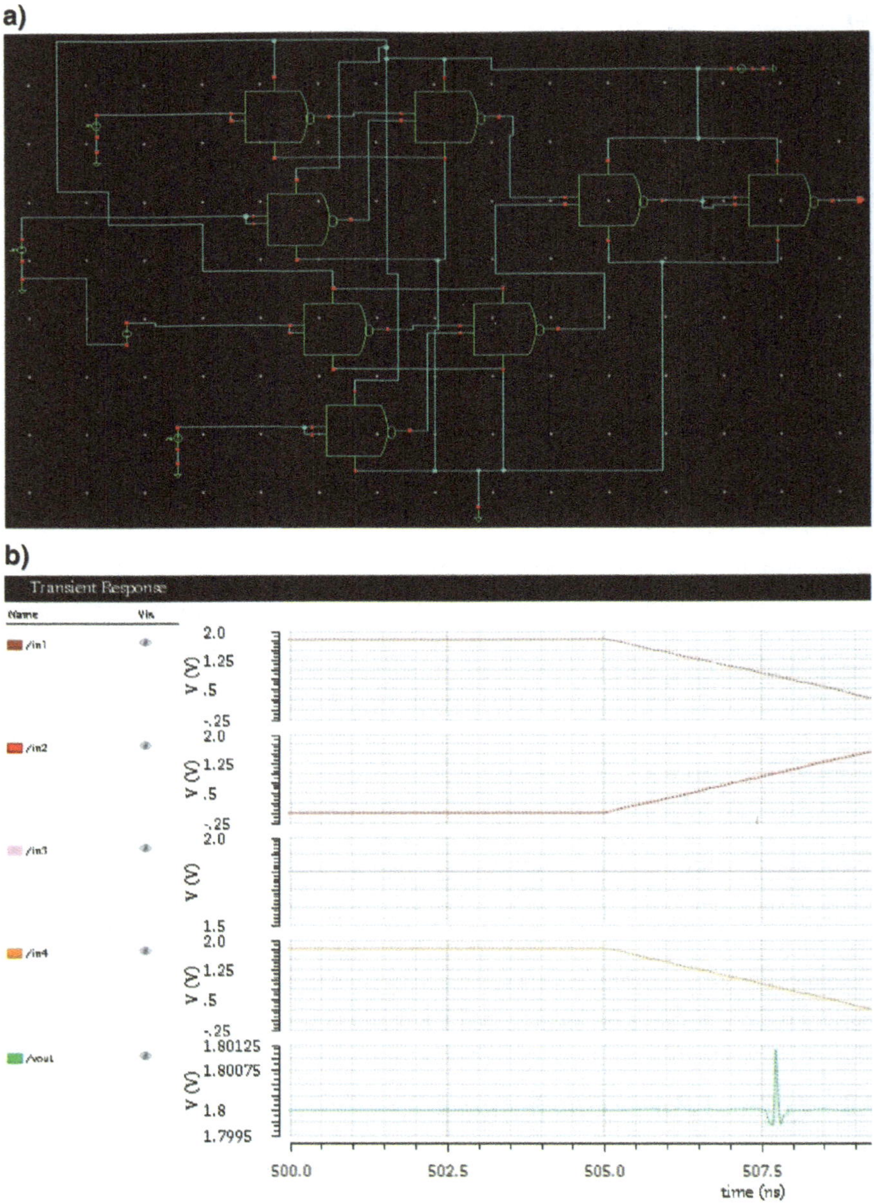

Fig. 7 a Power glitch attack-producing circuit 4. b Waveform of glitch producing circuit 4

not require any expertise to conduct them. These attacks are mostly performed. The voltage underfeeding, achieved by employing a precise power supply unit, requires the attacker to be able to tap into the power supply line of the device and connect his power supply unit [14].

This requires only basic skills and can be easily achieved in practice without leaving evidence of tampering. Moreover, no knowledge of the implementation details of the device is needed. These attacks are refined now by generating spikes at specific location and skip particular instruction of algorithm. Attacker needs a custom circuit in order to generate spike in power supply. This supply is fed into device under attack instead of original power supply line [15, 16].

This shows the simple circuit for generation of power attack. As observed from the figure, it contains two NAND gates: NAND gate 1 and NAND gate 2. Inputs of NAND gate 1 are short. First input of NAND gate 2 is given directly, whereas second input is given through NAND gate 1. This causes propagation delay in the arrival time of both the inputs. Due to this, glitch is produced as can be observed from the figure below: Fig. 4b.

Output waveform shows the generation of glitch at around 2.8 ms. The parameters of the glitch generate are:

$$\textbf{Glitch period} = 2.8\,\text{ms (approx.)}$$
$$\textbf{Glitch width} = 1\,\text{V (approx.)}$$

These parameters can be varied by making modifications in the circuits. The success rate of attack increases if the adversary can precisely choose the glitch attack parameters.

The circuits in Figs. 4 and 5 show the generation of negative power attacks. As observed from the waveforms of above circuits, there is a negative spike in the output waveform. The adversary uses this tampered power supply instead of original power supply and generates faults in the embedded device. This kind of attack is known as "***under-power-glitch-attack***".

The circuit shown in Fig. 6 is used for generation of positive/negative power attack. This circuit can test the effects of both attacks. Power attacks can be categorized as: positive attacks/overpower attack and negative attacks/underpower attack. Various authors have studied the effects of either positive or negative attacks individually. Very few have studied the combined effect of these attacks. It is important to study the combined effects, since embedded devices must be secure against all types of side-channel attacks [17].

5.3 Effects of Power Glitch Attack on Combinational Circuits

Figures 8 and 9 show the effect of power glitch attack on combinational circuits. The combinational circuit we used is simple NAND gate and inverter. As observed from

a)

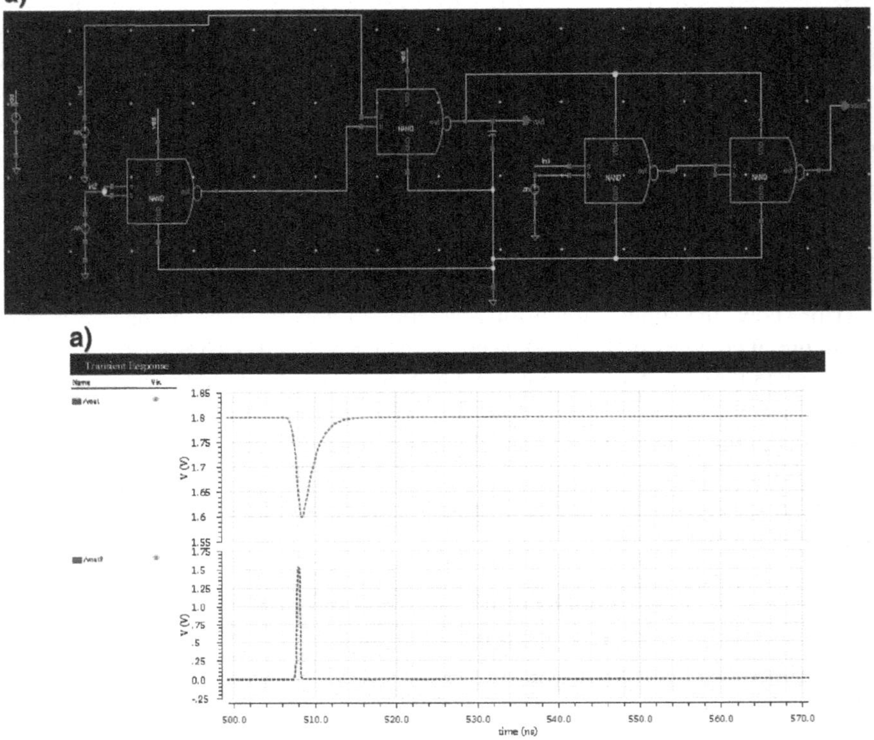

Fig. 8 **a** Circuit producing power glitch attack on NAND gate. **b** Waveform of effect of power glitch attack on NAND gate

Figs. 8b and 9b, there is a spike in the output of NAND gate and inverter output. As can be observed from waveforms at the point of glitch/spike, there is a fault at the output voltage and after sometimes the output becomes stable. In Fig. 8b, the output goes from 0 to 1 at the point of glitch, and then, it is restored to its original value 0, i.e., its pre-glitch value. Combinational circuit gets affected by glitches and may produce erroneous output which may lead to serious threat (Table 1).

6 Conclusion

We have introduced various circuits that produce power fault injection attacks and presented VHDL code for generating clock fault injection attacks. We have tried to work from attacker's point of view rather than providing countermeasures against side-channel attacks. Through this study, we have modeled various circuits, which produce combined attacks: positive/negative so that their combined effect is studied. Further, this research work concludes that if glitches whose amplitude is less than

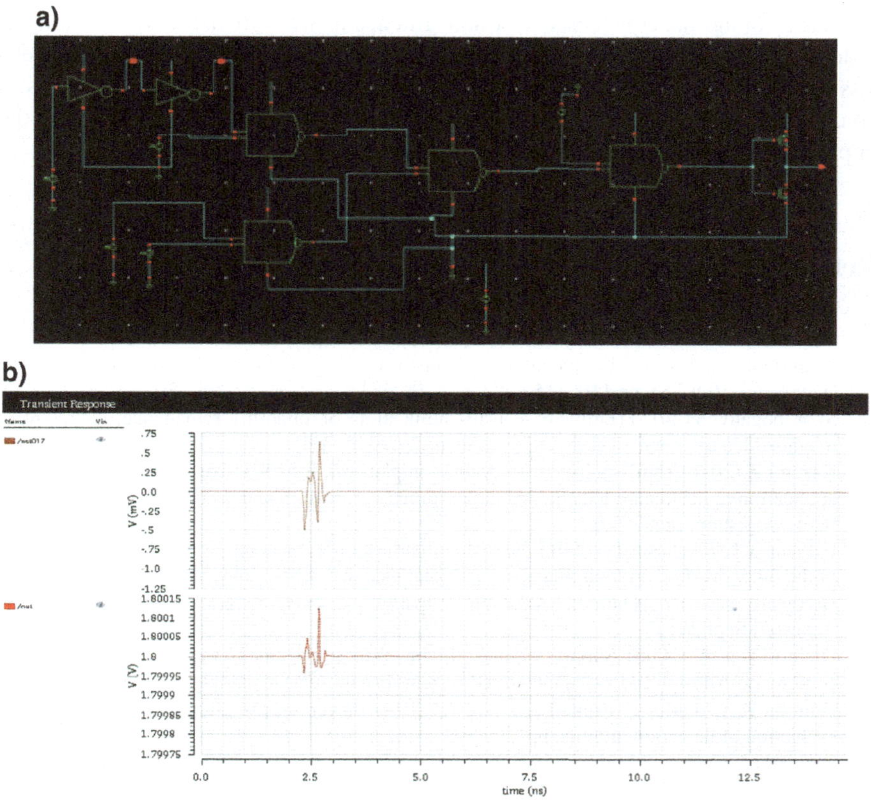

Fig. 9 **a** Circuit producing power glitch attack on inverter. **b** Waveform of effect of power glitch attack on inverter

Table 1 Related work in power attacks through voltage glitches

Paper	Year	Level	Applicability	Results
TSW [18]	2016	Architectural	ARM	Faults modeled by corrupted instructions
ZDCT [1]	2013	Digital logic	Clock glitching	Fault caused by setup/hold violation
DCB [6]	2006	Gates/elements	Voltage glitching	D flip-flops not susceptible to voltage glitches
Lu [19]	2018	Transistor	Voltage glitching	Effects are data dependent
This paper	2019	Gates	Voltage glitching	1. Glitches whose amplitude is less than reference voltage are considered to be safe glitches, and their effect does not last long 2. Combinational circuits are easily affected by these attacks 3. Faults caused by propagation delay

reference voltage are considered to be safe glitch, and their effect does not last long. It is demonstrated through various waveforms as discussed above. This paper provides less complex and low-cost platforms for testing temporal fault injection attacks in practice, so that suitable countermeasures can be built. As such, it underlines the importance of further research in countermeasure design.

References

1. Zussa L, Dutertre JM, Clediere J, Tria A (2013) Power supply glitch induced faults on FPGA: an in-depth analysis of the injection mechanism. In: IEEE 19th international on-line testing symposium (IOLTS), pp 110–115
2. Skorobogatov S (2018) Hardware security implications of reliability, remanence, and recovery in embedded memory. J Hardware Syst Secur 2(4):314–321
3. Kazemi Z (2018) Hardware security evaluation platform for MCU-based connected devices: application to healthcare IoT. In: 3rd international verification and security workshop, IVSW, IEEE workshop, pp 87–92
4. Fukunaga T, Takahashi J (2009) Practical fault attack on a cryptographic LSI. In: Proceedings of the 6th workshop on fault diagnosis and tolerance in cryptography, pp 84–92
5. Hardware security [Online]. Available https://www.coursera.org/learn/hardware-security. Accessed Sept 2019
6. Djellid-Ouar A (2006) Supply voltage glitches effects on cmos circuits. In: International conference in design and test of integrated systems in nanoscale technology (DTIS), pp 257–261
7. Gomina K, Gendrier P, Riguad JB, Tria A (2014) Power supply glitch attacks: design and evaluation of detection circuits, pp 136–141
8. Agoyan M, Dutertre JM, Naccache D, Robisson B, Tria A (2010) When clocks fail: on critical paths and clock faults. In: International conference on smart card research and advanced applications, pp 182–193
9. Balasch J, Gierlichs B, Verbauwhede I (2011) An in depth and black box characterization of effects of clock glitches on 8 bit MCU. In: Workshop on fault diagnosis and tolerance in cryptography, pp 105–114
10. Shum W, Anderson JH (2011) FPGA glitch power analysis and reduction. In: IEEE/ACM international symposium on low power electronics and design, Fukuoka, pp 27–32
11. Schmidt J, Herbst C (2008) A practical fault attack on square and multiply. In: 5th workshop on fault diagnosis and tolerance in cryptography, Washington, pp 53–58
12. Endo S, Sugawara T, Homma N, Aoki T, Satoh A (2011) On chip glitchy clock generator for testing fault injection attacks. J Crytopgraphic Eng 1(4)
13. Bozzato C, Focardi R, Palmarini F (2019) Shaping the glitch: optimizing voltage fault injection attacks. IACR Trans Cryptographic Hardware Embed Syst (2):199–224
14. Singh P, Paprzycki M, Bhargava B, Chhabra J, Kaushal N, Kumar Y (2018) Futuristic trends in network and communication technologies. In: FTNCT, Communications in computer and information science, vol 958, pp 3–509
15. Singh PK, Bhargava BK, Paprzycki M, Kaushal NC, Hong WC (2020) Handbook of wireless sensor networks: issues and challenges in current scenario's. In: Advances in intelligent systems and computing, vol 1132. Springer, Cham, Switzerland, pp 155–437
16. Singh PK, Kar AK, Singh Y, Kolekar MH, Tanwar S (2020) Proceedings of ICRIC. Recent innovations in computing. Lecture notes in electrical engineering, vol 597. Springer, Cham, pp 3–920

17. Ge Q, Yarom Y, Cock D, Heiser G (2018) A survey of micro-architectural timing attacks and countermeasures on contemporary hardware. J Cryptographic Eng 8(1):1–27
18. Timmers N, Spruyt A, Witteman M (2016) Controlling PC on ARM using fault injection. In: Workshop on fault diagnosis and tolerance in cryptography (FDTC), pp 25–35
19. Lu Y (2019) Injecting software vulnerabilities with voltage glicthing. arXiv: 1903.08102

Traffic Jam Minimization and Accident Avoidance System Using IoT

Mohsin Khan and Bhavna Arora

Abstract Traffic jam is one of the biggest challenges of twenty-first century, and hence, there is an urgent need to combat the issue of vehicular congestion. Traffic congestion can be classified by the Department of Transportation as 'recurring' which occurs on regular basis due to bottlenecks, narrow roads, etc. and 'non-recurring' which consists of temporary disruptions like accidents, bad weather, etc. With the emergence of 4G/5G technology, the Internet of Things (IoT) has grown exponentially, thus facilitating real-time transfer of data bits. The indulgence of the IoT for the vehicular movement can be used to minimize the traffic congestions. A two-way communication is established between the vehicles in its local vicinity or zone and the critical information flows between them, in real time. In this paper, a concept has been proposed that will attempt to minimize the traffic congestion by using IoT. The proposed concept will handle the safety of the vehicles, thus reducing the probability of accidents and will minimize the traffic jam especially 'phantom jams or jamitons.' To elucidate the proposed concept, algorithms and flowcharts for maintaining a minimum or safe distance between vehicles, overtaking and lane changing are presented in this paper.

Keywords Internet of Things (IoT) · Traffic jam · Vehicular congestion

1 Introduction

According to the statistics, 78.6 million automobiles have been sold in 2018 worldwide [1]. A mathematical explanation for the occurrence of traffic congestion was proposed by Boris Kerner. The proposed traffic theory comprises three phases, namely free flow, synchronized flow and wide-moving jams [2]. During the transition from the free flow to the synchronized flow as the density of the vehicles increases, the speed of the vehicles drops considerably, but there is no noticeable change in

M. Khan (✉) · B. Arora
Department of Computer Science and IT, Central University of Jammu, Rahya Suchani, Samba District, Bagla, Jammu and Kashmir, India
e-mail: khann.mohsin@icloud.com

P. K. Singh et al. (eds.), *Evolving Technologies for Computing, Communication and Smart World*, Lecture Notes in Electrical Engineering 694,
https://doi.org/10.1007/978-981-15-7804-5_4

flow rate. Hence, there is a synchronized flow of automobiles present in different lanes. While the traffic moves through the bottleneck, transition to wide-moving jam emerges through synchronized flow when both flow rate and velocity drop significantly as the density of the vehicles per unit area increases. The widening of the roads has been established in order to minimize traffic jams, but recent studies have shown that after the widening of the roads, the density of the vehicles or automobiles increases, thus grasping a threshold point where the jams are again meant to happen [3]. Sometimes there are no causes such as closed lanes, accidents or bad weather; still there are congestions on the roads; this phenomenon is called 'phantom jams' [4] which was coined by the researchers at Massachusetts Institute of Technology (MIT). It is caused due to hard hitting of brakes when two or more vehicles progress too much close to each other maintaining a small space in between. According to the Department of Transportation, traffic jams are classified as 'recurring' and 'non-recurring.' Recurring traffic congestion is caused due to high density of vehicles than the roads can accommodate. Non-recurring traffic congestion is triggered due to environment, mechanical issues and humans [5]. The environmental issues incorporate rain, fog, snowstorm, etc. The mechanical issues involve breaking down of car or accidents, and the humans are the main reason as there are many cases of drunk driving or emotional driving according to the National Highway Traffic Safety Administration [6]. With the emergence of Internet of Things (IoT), it is possible to minimize the congestion as the vehicles will be connected to each other in its local zone sharing critical information and calculating some predefined variables [7]. Those variables include the limit of speed up to which the driver must drive the automobile at specific situations so that a safe distance is maintained between the vehicles, the current speed of the vehicle, distance covered from a reference point [8]. The transfer of that critical information or data bits needs to be real time, i.e., the transference of data prerequisites to be fast enough with minimal transmission delay which can be attained by 4G/5G network communication [9].

1.1 Background and Research Motivation

The automated vehicles use mechatronics, artificial intelligence and multiagent system, and its autonomy levels are from 0 to 5, where level 0 means no automation, level 1 means driver assisted, level 2 means partial automation, level 3 means conditional automation, level 4 means high automation and level 5 means full autonomy [10]. Levels 0, 1 and 2 are more prone to accidents and jams as the vehicle is under the direct control of the driver, and most of the decisions are taken by the driver, while as the levels 3, 4 and 5 are less prone to accidents and jams because the vehicles are programmed and not under the direct influence of the driver. This means that the decisions taken in those levels are executed by the system. Levels 3, 4 and 5 vehicle automation cannot be completely applied in the urban and rural areas of developing countries like Brazil, India, Pakistan, South Africa, [11] etc. The main cause of accidents is risky overtaking manoeuvre. According to Brake and Direct line's survey [12],

1. 80% drivers have felt threatened by an overtaking manoeuvre.
2. 94% drivers have observed a risky overtaking manoeuvre, and more than half, i.e., 53% have spotted those manoeuvre on the monthly basis.
3. 18% drivers have acknowledged that they have overtaken without making sure if there are other vehicles that could have smashed.

A dynamic system needs to be fabricated in order to minimize those fatal situations regarding overtaking manoeuvre. Thus, an innovative touch needs to be integrated in the level 2 automation where the vehicle is under the direct control of the driver. The proposed model adds an innovative touch to the level 2 automation which is vastly used in the developing countries, consisting of algorithms and flowcharts. Those algorithms and flowcharts are explained in the later section of this paper, which when implemented will reduce the probability of traffic jams especially phantom jams or jamitons and will also minimize the probability of accidents caused due to overtaking and lane shifting on a one-way road structure with many lanes, as the driver has to take only action based on the decisions which are taken by the collaboration of the vehicles using IoT, sharing data and calculating the results.

2 Related Work

If every vehicle is replaced with the automated vehicle, the traffic jam will be completely eradicated, but this scenario is not possible. The automated vehicle is programmed according to the rules and regulations of the Department of Transportation. Rules are transformed into equations which are later imposed via any programming language. On the basis of levels 3, 4 and 5, self-driving car or robotic cars are capable of sensing its environment with little or no human input [13]. In self-driving cars, common positioning system is based on the satellite technologies, such as GPS, Galileo, GLONASS [14]. The technology proposed as connected vehicle technology is advantageous for the movement of traffic at the intersections [15], which provides an efficient signal control strategy. The authors have also proposed a fast branch and bound algorithm for the improvement of the computational efficiency leading to reduction in delays and stops [16].

Fuzzy logic can be used to control the trajectory of automated vehicles with a centralized controller. Alternative track prediction, keeping a safe distance between vehicles and integrating night vision using infrared cameras can be achieved using fuzzy logic [17]. The author has developed ethical crashing algorithms using guided track systems for avoiding collisions. The rational approach is integrated with artificial intelligence approach and a natural language requirement. The algorithms related to automated automobile predict various alternative crash paths, and a path is selected on the basis of lowest damage or likelihood of collision [18].

Waymo [19] was set in motion as a project of Google self-driving technology development company. The technology is in testing phase; it has passed the tests and has been employed in certain environments, yet it has to pass the tests in harsh environments.

3 Proposed Model

The proposed model can be implemented via MATLAB and its components related to the driving scenario simulation. The MATLAB code created during the implementation of the proposed algorithms presented in this paper needs to be tested. The first testing phase can be conducted using MATLAB performance testing framework which includes performance measurement-oriented structures [20], and the second testing phase can be conducted by integrating Google Map API. The infrastructure of roads based on the Google Maps will be used to simulate the scenario, and the road structures are selected in the series of distinct stages. First those structures are selected which will be easy to simulate the scenario, and after passing the previous structure, more difficult environment will be selected on the Google Maps. The efficiency of the algorithm is developed and enriched after deep deliberation and analysis of scientific research studies regarding to this field. Thus, a model is proposed that is going to work on the following algorithms and flowcharts.

3.1 Proposed Algorithm and Flowchart for Maintaining a Threshold Distance Between Two Vehicles

Algorithm 1. Maintaining a minimum and a safe distance between the vehicles not only maintains a free flow but also reduces the probability of phantom jams or jamitons. The algorithm regarding to the maintenance of a minimum distance is given as follows:

```
Vehicle at front = v1
Vehicle at back = v2

Speed of v1= Spv1
Speed of v2= Spv2

Velocity of v1= vel1
Velocity of v2= vel2

Distance between v2 and v1 = dist

A point p is selected by v1 as reference point
While, Time tn where n ≥ 0
Distance, dv1 from point p = dv1 + ∫_{tn}^{tn+5} vel1 dt
Distance, dv2 from point p = dv2 + ∫_{tn}^{tn+5} vel2 dt
dist = dv2 − dv1

If dist ≤ length of v2
Then "v2 has to decrease the speed"
Else
"Depends upon the driver of v2"

Time is incremented, tn = tn+5
End While
```

Flowchart 1. The flowchart regarding to the algorithm for maintaining a threshold distance between vehicles is given as (Fig. 1):

3.2 Proposed Algorithm and Flowchart for Overtaking

Algorithm 2. Controlling overtaking will reduce the probability of fatal accidents, thus an algorithm has been fabricated which is as follows:

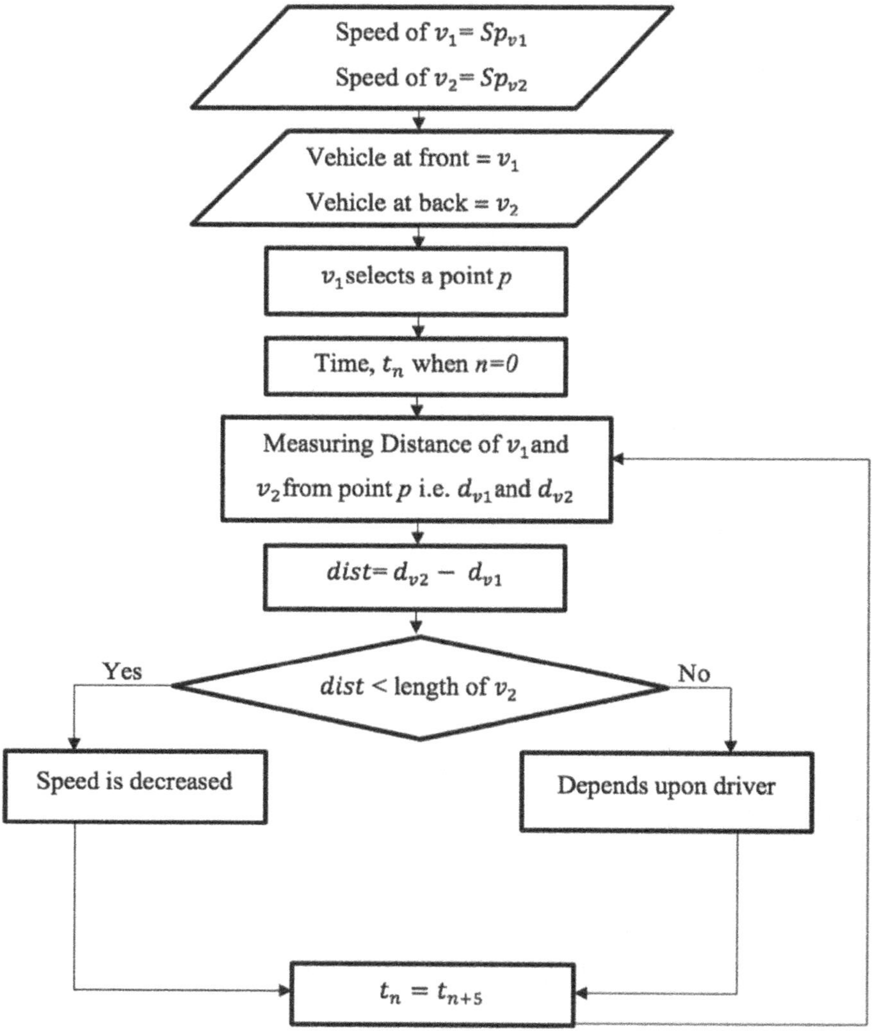

Fig. 1 Maintaining a threshold distance between two vehicles

Vehicle that wants to overtake= v_0
Vehicles in other lane= v_l
Number of vehicles, v_0 has to overtake= n
Vehicle in the front-end of v_0= v_n
Distance of v_0= d_{vo}
Distance travelled by vehicles present at the front-end of v_0 = d_{vn}

Initial value, $n = 1$
While $n \leq 3$
Algorithm 1 between v_n and v_{n+1}
If
$dist > (length\ of\ v_0 + \frac{length\ of\ v_0}{2}) = \delta$
Overtaking is initiated at time T
Else $n++$
End While

If $n > 3$
"overtaking not possible"
Else
"overtaking is possible"

In other lane at time T,
Applying algorithm 1 to v_0 , v_1 and v_l
Distance covered by any vehicle in other lane= d_{vl}

For
$d_{vn} + length\ of\ v_n < d_{vl}$
Or
$d_{v0} - length\ of\ v_0 > d_{vl}$
"Permission for overtaking is granted to v_0"
Else
"Permission for overtaking isn't granted to v_0"
End For

If $d_{v0} > d_{vn}$
"overtake successful"

Flowchart 2. The flowchart considering the algorithm for overtaking is given as follows (Fig. 2):

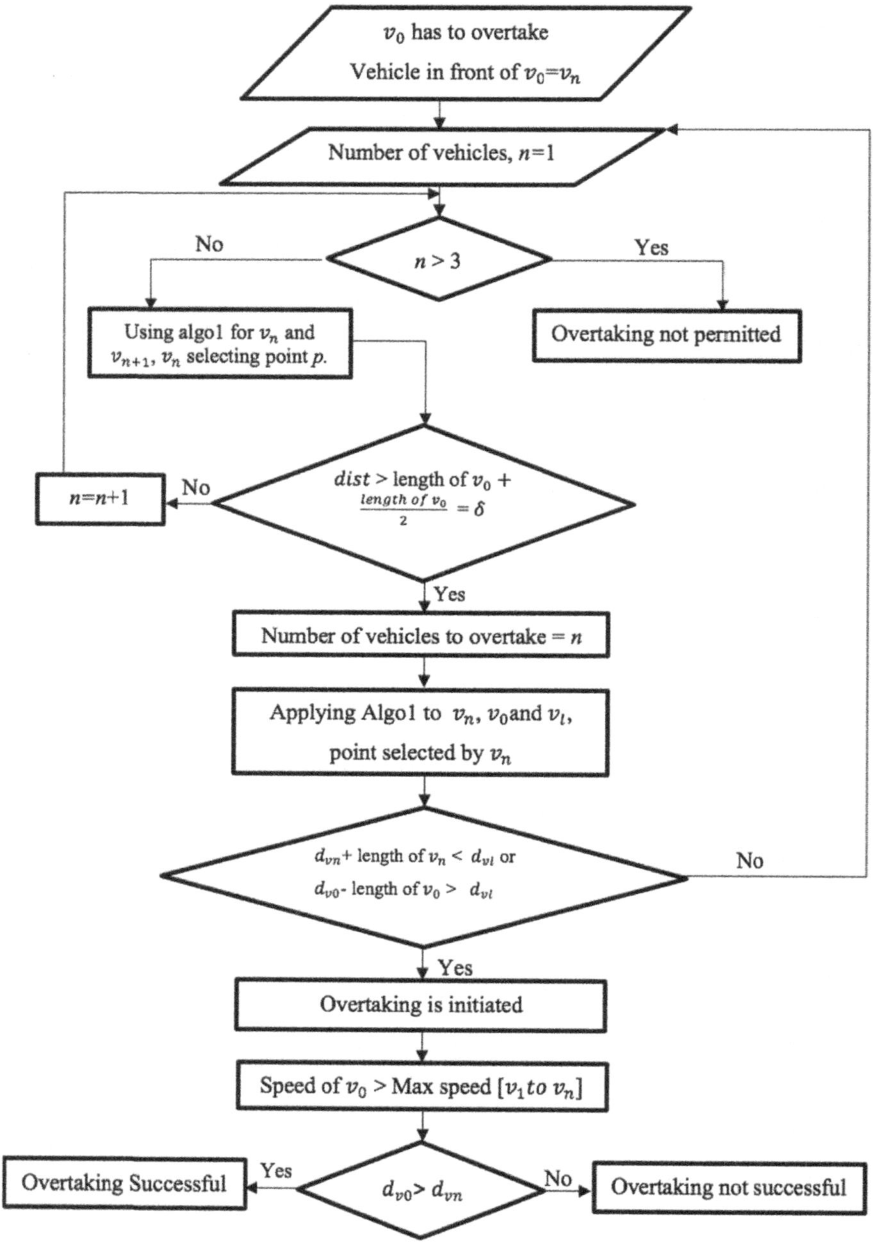

Fig. 2 Overtaking on a one-way road with more than one lane

3.3 *Proposed Algorithm and Flowchart for Lane Shifting*

Algorithm 3. Controlling the changing of the lanes can further help in minimizing traffic jams and accidents. The algorithm for lane shifting is as follows:

```
Vehicle that wants to change lane= v0
Vehicles in other lane in the zone of v0= vl
In other lane
Applying algorithm 1 to v0 and vl;reference point selected
by v0
Distance travelled by v0 = dvo
Distance travelled by vl = dvl

While
dvo+ length of  v0 < dvl or
dvo- length of  v0 > dvl
"Lane changing is permitted to v0"
End While
```

Flowchart 3. The process flow for shifting the lane is given in following flowchart (Fig. 3):

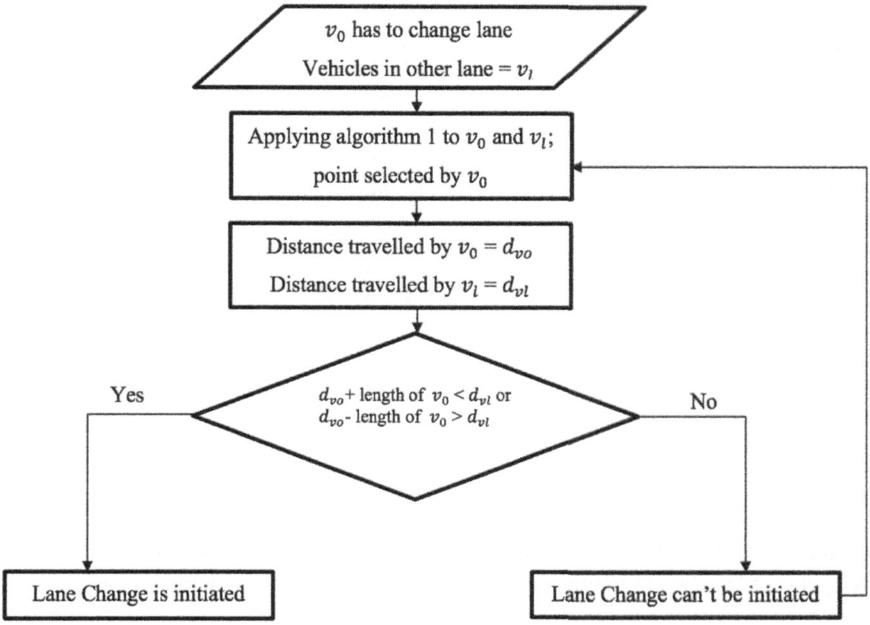

Fig. 3 Lane shifting on a one-way road

4 Discussion

The main reason for traffic jams is overtaking and lane changing which can be avoided by the communication of vehicles in their vicinity using IoT, sharing the real-time value of variables for individual calculation in each component or vehicle [21]. Hard hitting of brakes arises jamitons which can be avoided by keeping a safe distance between the vehicles. The further explanation on the basis of previous algorithms for the maintenance of threshold distance between vehicles, overtaking and lane changing in this paper is as follows:

4.1 *Maintaining a Threshold Distance Between Two Vehicles*

The vehicle at the front is denoted by v_1, and the vehicle at the back is denoted by v_2. The speed of both vehicles will be given as Sp_{v1} and Sp_{v2}, respectively. Similarly, vel_1 and vel_2 are velocities of vehicles v_1 and v_2, respectively. v_1 select a reference point p as a starting point, and this point will be the latitude and longitude in real world which is selected on the GPS connected to the system at a specific time t_n, where initial value of n is equal to zero. The loop starts at this point applying the increment after obtaining the distance first time. The distance d_{v1} of the vehicle v_1 is measured from that point using equation.

$$d_{v1} = d_{v1} + \int_{t_n}^{t_n+5} \mathrm{vel}_1 \mathrm{d}t \tag{1}$$

Other vehicles v_2 travel through same reference point, i.e., same latitude and longitude, and its distance d_{v2} is measured by the equation

$$d_{v2} = d_{v2} + \int_{t_n}^{t_n+5} \mathrm{vel}_2 \mathrm{d}t \tag{2}$$

Distance dist between v_1 and v_2 is measured by subtracting d_{v1} from d_{v2}.

$$\mathrm{dist} = d_{v2} - d_{v1} \tag{3}$$

The dist should always be greater than length of v_2 between v_1 and v_2. If the distance is less than the threshold value, then the driver has to apply brakes decreasing the speed and maintaining the dist of more than threshold value which is length of v_2. The time t is incremented by the value of five seconds, i.e., t_{n+5} after which the loop starts again.

4.2 *Overtaking*

The vehicle that wants to overtake is denoted by v_0, and the vehicles present in front are denoted by $v_{1\,\text{to}\,n}$ where max value of n depends upon the vehicles, v_0 has to overtake. Let T be the time of the initiation of the process. Using algorithm 1, distance is measured between v_n and v_{n+1} where initial value of $n = 1$.

$$\delta = \left(\text{length of } v_0 + \frac{\text{length of } v_0}{2}\right) \quad (4)$$

If the dist is greater than δ, then the number of vehicles to overtake will be equal to the recent value of n and if the dist is less than δ, n is incremented by 1 which means the next vehicle is checked for δ. The loop continues until δ is found or the condition of $n \leq 3$ is met which means max number of vehicles to overtake is 3. When $n > 3$, v_0 doesn't get any acknowledgement. If the conditions are satisfied, an acknowledgement is received by v_0 and an acknowledgement is sent by the vehicle that finds the δ, i.e., v_n. The last value of n denotes the number of vehicles that will be overtaken by v_0. The vehicles in the other lane in the zone of v_0 is checked for distance, and algorithm 1 is applied, where the reference point is selected by v_n. The distance covered by v_0 is denoted by d_{v0}, and the distance covered by v_n is denoted by d_{vn}. In the other lane, if the distance covered by every vehicle in that local zone is greater than (d_{vn}+length of v_0) and less than (d_{v0}−length of v_0),, then the overtaking is possible and the driver will get the permission of overtaking else the overtaking is not possible. So, if the criteria are met, then in the other lane during the process of overtaking, the vehicles which have covered distance more than (d_{vn} + length of v_n) and less than (d_{v0} − length of v_0) have to maintain the speed that was recorded at time T, i.e., the initialization of the overtaking process. The maximum of the speed of vehicles from v_1 to v_n is taken in consideration and v_0 has to maintain the speed greater than that during overtaking. For checking the successful completion of the overtaking process, d_{v0} must be greater than d_{vn}.

4.3 *Lane Changing*

The vehicle that wants to change lane is denoted by v_0, and its local zone or local vicinity is selected. The distance covered by v_0 is denoted by d_{v0}, and the distance covered by vehicles in other lane, v_l is denoted by d_{vl}. In the other lane, if the distance covered by every vehicle in that zone is greater than (d_{v0} + length of v_0) and less than (d_{v0} − length of v_0), then the lane shifting is possible and the driver will get the permission for changing the lane else it is not possible. The condition is checked in a loop until the lane changing procedure is completed.

5 Conclusion and Future Scope

The changing of lanes, overtaking and upholding a safe distance between vehicles can help the transportation department to minimize the traffic congestion or 'jamitons.' If any driver of the vehicle does not follow the result of the algorithms, then an alert message will be received by Department of transportation with information about the vehicle and its owner. The system can be implemented in real life by programming Raspberry Pi in an IoT environment [22] integrating Google Map's API. In future, the algorithm can be developed for the intersection and for the turning points. The speed of the transfer of critical information which has been discussed in this paper must be in real time or with a minimal delay. This can be attained by using 5G/4G where delay is much less and transfer happens in real time [23].

References

1. Automotive industry—statistics & facts. https://www.statista.com/topics/1487/automotive-industry/. Last accessed 9/12/19
2. Kerner BS (2009) Introduction to modern traffic flow theory and control: the long road to three-phase traffic theory. Springer Science & Business Media, Berlin
3. Traffic congestion reconsidered. https://theconversation.com/traffic-congestion-reconsidered-111921/. Last accessed 20/12/19
4. Flynn MR, Kasimov AR, Nave JC, Rosales RR, Seibold B (2008) On "jamitons," self-sustained nonlinear traffic waves. arXiv preprint: 0809.2828
5. McGroarty J (2010) Recurring and non-recurring congestion: causes, impacts, and solutions. Neihoff Urban Studio–W10, University of Cincinnati
6. Risky driving. https://www.nhtsa.gov/risky-driving. Last accessed 30/12/19
7. Mahmood Z (2020) Connected vehicles in the IoV: concepts, technologies and architectures. In: Connected vehicles in the internet of things. Springer, Cham, pp 3–18
8. Liu X et al (2020) Optimizing the safety-efficiency balancing of automated vehicle car-following. Accid Anal Prev 136:105435
9. Singh P, Paprzycki M, Bhargava B, Chhabra J, Kaushal N, Kumar Y (2018) Futuristic trends in network and communication technologies. In: FTNCT 2018. Communications in computer and information science, 958, pp 3–509
10. Meyer G, Beiker S (eds) (2014) Road vehicle automation. Springer International Publishing, Cham
11. Nantulya VM, Reich MR (2002) The neglected epidemic: road traffic injuries in developing countries. BMJ 324(7346):1139–1141
12. Reports and statistics regarding overtaking manoeuvre. https://www.brake.org.uk/assets/docs/dl_reports/DLreport-ariskybusiness-sec1-howsafeisyourdriving-apr15.pdf
13. Montemerlo M, Becker J, Bhat S, Dahlkamp H, Dolgov D, Ettinger S, Haehnel D et al (2008) Junior: the Stanford entry in the urban challenge. J Field Rob 25(9):569–597
14. Bhatta B (2010) Global navigation satellite systems: insights into GPS, GLONASS, Galileo, Compass, and others. BS Publications
15. Guler SI, Menendez M, Meier L (2014) Using connected vehicle technology to improve the efficiency of intersections. Transp Res Part C Emerg Technol 46:121–131
16. Singh PK, Kar AK, Singh Y, Kolekar MH, Tanwar S (2020) Proceedings of ICRIC 2019. Recent innovations in computing. Lecture notes in electrical engineering, vol 597. Springer, Cham, pp 3–920

17. Naranjo JE, Sotelo MA, Gonzalez C, Garcia R, De Pedro T (2007) Using fuzzy logic in automated vehicle control. IEEE Intell Syst 22(1):36–45
18. Draganjac I et al (2020) Highly-scalable traffic management of autonomous industrial transportation systems. Robot Comput Integr Manuf 63:101915
19. Gibbs S (2017) Google sibling waymo launches fully autonomous ride-hailing service. The Guardian 7
20. Performance testing framework. https://in.mathworks.com/help/matlab/performance-testing-framework.html. Last accessed 8/1/20
21. Minovski D, Åhlund C, Mitra K (2020) Modeling quality of IoT experience in autonomous vehicles. IEEE Internet Things J 7(5):3833–3849
22. Zhao CW, Jegatheesan J, Loon SC (2015) Exploring IoT application using raspberry pi. Int J Comput Netw Appl 2(1):27–34
23. Klymash M, Beshley H, Panchenko O, Beshley M (2017) Method for optimal use of 4G/5G heterogeneous network resources under M2M/IoT traffic growth conditions. In: 2017 international conference on information and telecommunication technologies and radio electronics (UkrMiCo), pp 1–5. IEEE

Software Reusability Estimation Using Machine Learning Techniques—A Systematic Literature Review

Deepika and Om Prakash Sangwan

Abstract Software engineering is an application of engineering which is more focused on original development, but reusability plays a very significant role in order to produce good quality, error free, and less maintainable software. Software reusability is an attribute of quality which helps in selecting beforehand acquired notions in new statuses. Software reusability not only advances productivity, but it also provides a good quality software and has also optimistic effect on maintainability. Software reusability is advantageous in the manner that it provides high reliability, low cost of maintenance, and reduction in development time. In this paper, we have discussed and analyzed various machine learning techniques used for estimation of software reusability. It is found that machine learning techniques are competitive in nature with other reusability estimation techniques and can be used for estimation of reusability. This study will help software developers and information industry to elucidate that how software reusability can assist them in selecting advanced quality of software.

Keywords Software reusability · CBSD · OOSD · AOSD · Machine learning techniques

1 Introduction

The key systematic plan of action to overcome the problem of software crisis is to adopt reusability. Software reusability is an attribute of quality in which new software system is implemented from pre-existing software systems. It can be applied at any scale and also at any level of abstractions. Software development process contains various phases, and developing a software requires more time and expensive

Deepika (✉) · O. P. Sangwan
Department of Computer Science and Engineering, Guru Jambheshwar University of Science and Technology, Hisar, Haryana, India
e-mail: deepikagodara92@gmail.com

O. P. Sangwan
e-mail: sangwan0863@gmail.com

P. K. Singh et al. (eds.), *Evolving Technologies for Computing, Communication and Smart World*, Lecture Notes in Electrical Engineering 694,
https://doi.org/10.1007/978-981-15-7804-5_5

also. It is widely believed that software reusability improves productivity, flexibility, and yields quality product. Software reusability is one methodology which is given prominence due to the advantages that it brings to the industry. An appropriate software reuse design principle gives increased productivity, requires lesser development time, and reduces cost. Maintenance of the software takes 80% cost of the software development Adekola et al. [1]. Less maintainability helps in adding new features, easier removal of bugs, and increasing strength of software. Software reusability and maintainability are related to each other. Reduced maintainability gives added software reusability which brings down the rate of software crisis. Multiple reuse in different version of products maintains those products having higher quality. With multiple reuses, productivity increases because it requires reduced work for bringing the product into the market and requires to create fewer work product from scratch which automatically brings down the time span of products to market. It avoids redundancy in development efforts and reduces the amount of maintenance required. Reusability advances productivity and also reduces cost of maintenance. In order to keep product operational, the performance of all activities is required to be maintained which require correcting faults and adapting to changed environment. There are various similarities in reuse process and maintenance work required for a product. Maintenance is typically reuse-oriented. Effective reuse consists of certain steps that need to follow as shown in Fig. 1. It represents the various phases of software reuse process.

First phase describes the plan of reuse, second phase prepares for solution, third phase analyzes the solution structure for reuse, fourth phase acquires the predefined

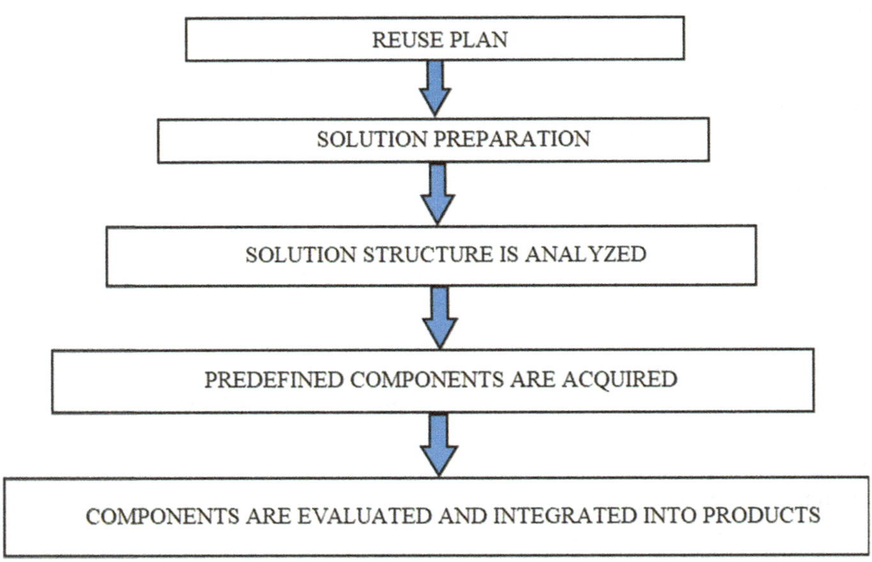

Fig. 1 Software reuse plan

components, and in last phase, the components are evaluated and integrated into products.

This paper is structured as follows: The subsequent section in brief explains the various approaches to software reuse. Section 3 briefs machine learning techniques. Section 4 describes the literature survey and estimation techniques of software reusability. Section 5 represents discussion, and conclusion is presented in Sect. 6.

2 Software Reuse Approaches

In order to improve software productivity and lessen software crisis, reuse plays a very important role. There are various approaches used for reusing a software. Here we provided a brief explanation of the commonly used software reuse approaches. These approaches can be broadly divided into three categories as shown in Fig. 2.

2.1 Component-Based Software Development (CBSD)

Component-Based Software Development is a reuse-based approach which composes loosely coupled independent components into a system. CBSD is part of software engineering which emphasizes on the concept of components. Component can be a module, a software package, or a web service. Components communicate

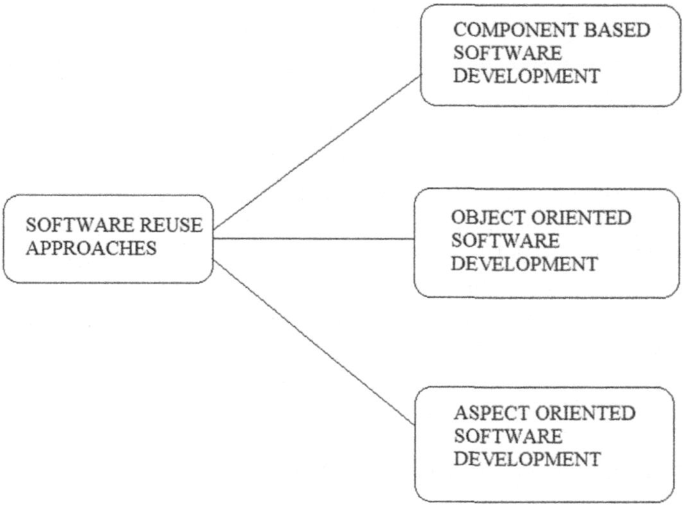

Fig. 2 Approaches to software reuse

with each other via interfaces. Components are cohesive in nature, because each system process is placed into a separate component so that data and functions inside a component are semantically related. They communicate via interfaces and are substitutable, i.e., one component can be replaced by another component of updated version or an alternative. To make a component reusable, significant effort is required that component should be fully documented, thoroughly tested, robust, and can back-propagate error messages. Components can be tightly or loosely coupled. In loosely coupled components, it makes use of little knowledge of other components because loosely coupled components have distributed memory. On the other hand, tightly coupled components are strongly interrelated as they are having shared memory.

Wangoo and Singh [2] presented an architectural model for prediction of component reusability level, that is based on classification by decision tree induction which enhances cost, quality, time, and level of reuse of components. Code and non-code components are considered for determining size and level of reuse.

Fuzzy multi-criterion approach was used by Singh and Tomar [3] for measuring reusability of component metrics in context of web services. Reusability is a non-functional quality attribute, and proposed fuzzy multi-criterion approach easily quantifies the reusability. Component reusability was divided into two perspectives, i.e., developers' and users' perspective.

2.2 *Object-Oriented Software Development (OOSD)*

Object-oriented programming focuses on objects of problems through development phase. OOSD is an important technical approach for analysis and design for any application, or business by using object-oriented programming. It enhances reusability as components designed in object-oriented programming can be easily reused as compared to those designed in conventional programming. It is an iterative and incremental development approach.

Padhy et al. [4] proposed model for analyzing the reusability of object-oriented software metrics. The proposed model for reusability estimation uses fuzzy logic, neural network, and genetic algorithm techniques of soft computing. The research work was divided into two phases. In first phase, evolutionary computing-enriched artificial intelligence-based regression model is used for estimating reusability. In second phase, reusability estimation model which is aging and survivability aware is proposed. It was concluded that proposed AGAANN reusability estimation model is better than other approaches.

2.3 *Aspect-Oriented Software Development (AOSD)*

AOSD is evolved from OOSD in order to provide good modularization in the process of software development. It is an optimistic software development approach. It

Table 1 Comparison of software reuse approaches

CBSD	OOSD	AOSD
Focuses on the concept of components	Focuses on data rather than procedure	Focuses on the concept of advanced separation of concerns
Program is divided into separate components which are related semantically	Program is divided into entities called objects	Aspects are cross-cutting concerns that affect many parts of the program
Components communicate with each other via interfaces	Objects communicate with each other via functions, no cross-cutting concerns are there	Cross-cutting concerns cause system interdependencies
Reuse-based approach	Iterative and incremental approach	Provides parallel development environment

provides benefits inferring from advanced separation of concerns. A concern represents primary purpose of program, describing functional concerns, and on other hand secondary concerns represent QoS and non-functional requirements. Separation of concerns should be known to and supported by program abstractions. These enable software development at a higher semantic level. AOSD can also be used with other conventional approaches like OOSD. Key focuses of AOSD are system concerns, factors affecting the system and maintainability. AOSD provides parallel development environment.

Aspect-Oriented Software reusability estimation was done by Singh et al. [5] using fuzzy logic approach. They explored various metrics that affect the reusability of software and concluded that proposed model will help software developer in selecting software with better quality in terms of reusability for aspect-oriented software. Here is the table representing common differences between reuse approaches (Table 1).

3 Machine Learning

Nowadays, machine learning techniques are widely used in software reuse. Machine learning algorithms help in dealing with various problems of software engineering and can also be used in development and maintenance task. It makes software product adaptive and self-configuring as discussed by Zhang [6]. Lounis and Ait-Mehedine [7] explored machine learning techniques for producing predictive models for three different quality characteristics and concluded that the accuracies obtained by machine learning models are comparable to other techniques. Artificial intelligence in SE application levels taxonomy, i.e., how AI can be applied in the field of software engineering is explained by Feldt et al. [8]. They categorize application into three facets. The results showed the risks associated with distinct AI applications. It

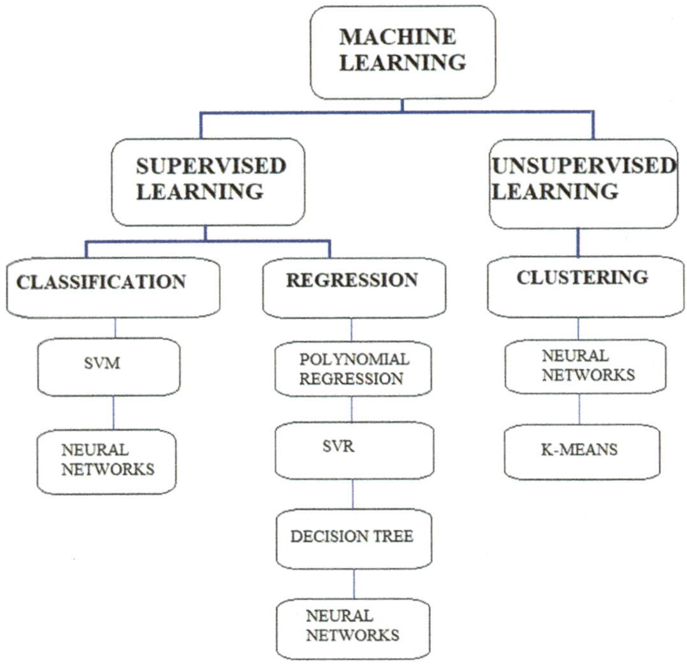

Fig. 3 Classification of machine learning techniques for software reusability

also helps companies in selecting particular AI technique for their software applications and also in creating strategies. Figure 3 represents the classification of various machine learning techniques used for estimation of software reusability.

3.1 Supervised Learning

Supervised learning trains the model on the basis of both input and output data and makes future predictions. In order to develop predictive models, it uses classification and regression techniques.

Classification Techniques. Classification techniques classify the available input into categories. If the available data can be categorized or can be separated into different class, use classification algorithms.

For example, whether an email is spam or genuine, it also finds application in speech recognition, face recognition and medical imaging.

Regression Techniques. Regression techniques predict the output of input data in continuous form or response. If the output is in the form of real numbers or we are working with ranges, then regression techniques are used. Applications of these techniques include electricity load forecasting, handwriting recognition, forecasting stock prices, and for showing fluctuations in power demand.

Srinivasan and Fisher [9] used regression and neural networks machine learning approaches for the estimation of efforts required for software development and compared CartX and backpropagation learning methods to traditional methods for estimating efforts of software. Proposed techniques are found to be competitive with SLIM, COCOMO, and function points.

3.2 Unsupervised Learning

Unsupervised learning trains the model on the basis of input data only and helps in finding hidden patterns and intrinsic structures. It explores the data available and gives good internal representation of data and reduces dimensions of data.

Clustering. Clustering is an unsupervised learning technique. It is used for exploratory analysis of data and for finding hidden patterns in data. It finds application in market research and object recognition. On the basis of similarity characteristics, data is portioned into groups. Clusters are formed so that objects in one cluster are highly similar to each other and objects in different clusters differ.

4 Machine Learning Techniques Used for Software Reusability Estimation

A methodical literature review was conducted on various databases. The databases were ACM Digital Library (DL), IEEE Explore Digital Library, Science Direct, Scopus, and Springer online journal collection. The terms used were: Software Reusability, Machine Learning, Component Based Software, Object Oriented and Aspect Oriented Software Development. Here, we commenced a review of various machine learning techniques used for reusability estimation.

Sangwan et al. [10] put forward a model grounded on four parameters, i.e., changeability, understandability, documentation quality, and interface complexity of software for assessing software reusability level using soft computing techniques. In order to predict the level of reusability, they used fuzzy model, neural network model, and neuro-fuzzy model. The study shows that soft computing techniques can efficiently predict the reusability level and neuro-fuzzy model can be further extended as cost estimation model and for quality prediction.

Sanz-Rodriquez et al. [11] proposed a model for evaluating learning object reusability which is based upon common metadata and structure that describes learning objects. The reusability of learning objects depends on the difficulty category of elements, portability, cohesion and coupling. Aggregation methods used for reusability are weighted mean, Choquet's integral and multiple linear regression.

Estimated reusability is then compared by using eLera's evaluators and analyzing MERLOT repositories. The results show that estimated reusability provides useful information for selecting reusable learning objects and improves productivity and quality of eLearning systems.

Sanz-Rodriquez et al. [12] evaluated reusability of aprioristic learning objects. They have identified metrics which are used for evaluating the reusability of learning objects and formulated aprioristic model for evaluating reusability on the basis of aggregation of metrics according to their significance level. The aggregated metrics are then compared with learning object review instrument (LORI) reusability evaluations as done by experts. It was found that aprioristic reusability evaluations are approximate to LORI evaluations.

Nesbit and Li [13] reviewed learning object evaluation approaches and also introduced eLera. It is a website designed for supporting researchers, students, teachers and media developers. It maintains searchable learning object database and provides LORI to evaluate resources. They also reviewed cooperative learning object exchange (CLOE), MERLOT and DLNET approaches which can be used for evaluation of learning objects. Neven and Duval [14] surveyed learning object repositories and compared their features and architectures. Learning object repositories contain learning objects, LOM, and learning object references. They also checked whether LOR's follow client–server-based or peer-to-peer approach.

Zimmermann et al. [15] presented an approach for improving retrieval of learning resources by taking into account the difference between desired and original usage scenario. They ranked learning resources on the basis of adaptation efforts. Papamichail et al. [16] relates popularity of software with reusability using GitHub Stars and Forks repos. GitHub Stars and Forks formulated a reusability score. They identified the static analysis metrics at both class and package levels and used these metrics to train reusability estimation models using state-of-the-art machine learning algorithms as perceived by developers.

Mao and Lounis [17] proposed an experiment using machine learning techniques to verify three hypotheses about the influence of the internal metrics. Machine learning approaches used are C4.5 algorithm, windowing, and cross-validation technique. The result shows that selected metrics can predict the reusability of classes with high level of accuracy.

Prakash et al. [18] described the software reuse process. In order to mine the useful data from software repositories using various software metrics, we can efficiently and effectively apply data mining techniques such as classification, clustering, and visualization for evaluating reusable components of software and concluded that these techniques yield better understanding and evaluation of software reuse in software development process. Lounis et al. [19] presented assessment models which are based on efficient machine learning techniques to assess the maintainability and reusability of software. Support vector machine and artificial neural network were used as an efficient alternative to classical machine learning algorithms to build the assessment models, and results showed that proposed approaches performed efficiently. Di Stefano and Menzies [20] performed a case study on reused dataset and used three different styles of learners. The results showed that using learners of different variety on a dataset produces more definitive and useful results. They also presented various factors that affect the success or failure of a software reuse program.

de Almeida et al. [21] invested machine learning algorithms to accurately assess the correctability of faulty software components and presented the result of empirical study. Three different algorithm families were analyzed and concluded that the machine learning algorithms can easily generate adequate prediction models. Their work addresses two different domains, namely software engineering and machine learning. It is concluded that model built by them helps software components to reduce maintenance efforts, which will give positive effect on reusability. Zahara et al. [22] considered the machine learning regression algorithms to evaluate the reusability of object-oriented software components and performed comparison using different parameter values. Object-oriented-based metrics and four different regression algorithms as multi-linear regression, model tree M5P, IBK (instance-based learner) with no distance weighting, and additive regression with M5P were used.

Object-oriented reusability metrics and CK metrics are analyzed by Padhy et al. [23]. Why research is needed in the area of reusability was explained and stated the necessity of reusability. They also studied the relationship between reusability and CK metrics. Deepika and Sangwan [24] presented a review of different software reusability estimation techniques and also found out the various factors that affect the estimation of software reusability. They concluded that neuro-fuzzy technique can effectively predict the level of reusability of software components.

Padhy et al. [25] proposed a fault proneness software reusability prediction model of object-oriented software classes to ensure optimal web service software design. For the estimation of reusability, object-oriented metrics was used. For reducing computational overheads, rough-set-analysis-based feature extractions were used and reuse proneness prediction was done by using different classification algorithms.

Vinobha et al. [26] proposed an aspect-oriented reusability evaluation model. Reusability of aspect-oriented software is evaluated using inheritance metrics. An automated tool to calculate value of proposed metrics called as aspect-oriented reusability evaluation measurement (AOSRM) was proposed, and it was found that reusability of aspect elements was higher than elements of classes.

The current practices in Malaysia related to software reusability approaches were studied by Ahmaro et al. [27]. Quantitative analysis method was used by collecting data from 183 industries in Malaysia through online questionnaire and concluded that the major reusability approaches used in IT industries of Malaysia are design patterns, component-based development, COTS integration, service orientation system, and application framework.

Fazal-e-Amin et al. [28] presented a conceptual model for reusability estimation and studied reusability during evaluation of software. Attributes of reusability related to different versions are also analyzed and compared. Evolutionary study of reusability was performed.

Fuzzy multi-criterion approach was used by Singh and Tomar [29] for measuring reusability of component reusability metrics in context of web services. Reusability is a non-functional quality attribute, and proposed fuzzy multi-criterion approach easily quantifies the reusability. Component reusability was divided into two perspectives, i.e., developers' and users' perspective, and taken into account five metrics coupling, interface complexity, security, response time, and statelessness. Imoize et al. [36] gave an exposition to the importance of software reuse and metrics of software reuse in the field of software engineering processes and also for stakeholders. The study reveals software reuse benefits and how software development stakeholders can harness these benefits.

Kaur and Kaushal [37] used fuzzy logic approach to assess reusability, maintainability, and understandability for estimating quality of aspect-oriented software using internal attributes of quality at package level. The result acquired from proposed approach has been validated using AspectJ and AJHotDraw software. In Table 2, different research papers which have taken into consideration different algorithms and models in order to estimate reusability on the basis of different parameters have been presented:

Based on the findings, it is concluded that estimation of reusability can be done efficiently with the help of different machine learning techniques, and researchers have also done lot of work regarding the metrics for the evaluation of reusability. However, little work has been done on software models used for predicting software reusability using machine learning techniques.

5 Discussion

A total of 35 research papers have been recognized to find different machine learning techniques used for estimation of reusability and also provided concise explanation of those techniques. Our main emphasis is to identify efficient machine learning technique for reusability estimation. Even after machine learning techniques have outstanding scope, the researchers have largely worked on software reusability metrics only, and less work is done on reusability factors. So, in order to derive an efficient reusability estimation model, certain research questions needed to be answered:

- RQ1: Can certain factors affecting reusability will be helpful in estimating reusability?
- RQ2: Is it feasible to derive an efficient and robust reusability prediction model?
- RQ3: Can machine learning techniques enhance the accuracy of reusability prediction?

In our further research work, these questions will be explored and investigated to ensure an optimal software reusability prediction model.

Table 2 Summary of software reusability estimation techniques

S.No.	Author (Year)	Algorithms/models used	Input parameters	Validation criteria	Dataset used	Experimental results
1	Padhy et al. [30]	LM, ELM, evolutionary computing-based ANN	CK metrics	MAE, MMRE, RMSE, SEM	100 web of service software projects taken	AGA-ANN outperforms other techniques
2	Sanz-Rodriguez et al. [12]	LORI tool	Cohesion, size and complexity, portability metrics	Weighted mean, Choquet's integral, multiple regression	MERLOT and eLera repositories	Aprioristic reusability estimation approximates to expert evaluations
3	Srinivasan and Fisher [9]	CARTX's and backpropagation	Product, personnel and project attributes	MRE, R^2	COCOMO database of 63 projects and Kemmerer's 15 projects	Both learning techniques performed well
4	Papamichail et al. [16]	SVR, random forest, polynomial regression	Complexity, coupling, cohesion, documentation, inheritance, size	NRMSE	100 Star and GitHub Java projects	Proposed approach can be effective for estimating reusability at class and package levels
5	Prakash et al. [18]	Decision tree, ANN, clustering, classification	Software metrics	Precision, recall, MAE, RMSE	167 open-source projects	Applied techniques provide better understanding and evaluation of software reuse components
6	Vinobha et al. [26]	Aspect-oriented software reusability measurement tool (AOSRM)	Inheritance metrics	AdIF, PIF, AIF	Java and AspectJ versions of UAS	Reusability of aspects was found to be more than elements of classes

(continued)

Table 2 (continued)

S.No.	Author (Year)	Algorithms/models used	Input parameters	Validation criteria	Dataset used	Experimental results
7	Nagpal et al. [31]	Fuzzy AHP and fuzzy TOPSIS	Response time, ease of use, ease of navigation and informative	Closeness coefficient	Three website developers as decision-makers	Gives satisfactory result for ranking different websites
8	Mao et al. [17]	C4.5, Windowing, Cross validation	Correlation b/w reusability and inheritance, coupling, complexity metrics	Accuracy	Open multi-agent system development environment (LALO)	Selected metrics predict with a high level of accuracy the class which is reusable
9	Singh and Tomar [29]	Fuzzy multi-criteria approach	Coupling, IC, security, response time, statelessness	Weighted average	Weather forecast service and Google Maps	Provide ranking and importance weights to selected attributes of web services components
10	Zahara et al. [22]	Regression algorithms	CK metrics	MAE, RMSE, RRSE	Dataset of object-oriented metrics values and reusability values	IBk with no decision weighting has better accuracy of prediction than other regression algorithms
11	Singh et al. [32]	Fuzzy logic, neural network	Adaptability, maintainability, modularity, interface complexity, flexibility	MSE	Rule base of 243 rules	Neural network is more stable than fuzzy model

(continued)

Table 2 (continued)

S.No.	Author (Year)	Algorithms/models used	Input parameters	Validation criteria	Dataset used	Experimental results
12	Maggo and Gupta [33]	LMNN	Cyclomatic complexity, halstead software science indicator, regularity metric, reuse frequency, maintainability index	Accuracy, recall, precision, error rate	From 165 object-oriented code fragments	Efficient model for selecting reusable components from software resources
13	Shri et al. [34]	K-means and decision tree	CK metrics	Accuracy, precision, recall, RMSE	–	Developed reusability model produced high-precision results
14	Lounis and Ait-Mehedine [7]	ANN, SVM	Coupling, cohesion, complexity, inheritance, and size metrics	Accuracy	Data collected from C++ and ADA medium-size applications	Presented efficient assessment model based on machine learning techniques
15	Fazal-e-Amin et al. [28]	Exploratory (interview method)	Coupling, cohesion and size metrics	Pearson correlation	–	Determined correlation between identified factors and reusability
16	Deepika and Sangwan [35]	ANFIS	Understandability, Interface complexity, portability, customizability, and maintainability	RMSE, correlation	243 rules are designed in the system on the basis of 5 I/P variables	Proposed ANFIS model can help the software developer in selecting better quality of software in terms of reusability

6 Conclusion

In this paper, we have discussed various papers on software reusability estimation using machine learning techniques. A brief explanation of software reuse approaches and machine learning techniques has also been provided. Software reuse technology has been evolving through generations. Reusability allows us to make rapid changes and exploit new features. It is effective only when it is done in systematic way and is based on reuse design principles. Machine learning algorithms can be used in dealing with different software engineering challenges, in building software development tools, and in maintenance work, etc. Most of the work done for software reusability estimation considers metrics only, and little work has been done on software models. It is therefore concluded that more efforts need to be done to predict software reusability using machine learning techniques to enhance the accuracy of reusability prediction. Major findings have been discussed, and certain research questions have been raised which can be used for future research work.

References

1. Adekola OD, Idowu SA, Okolie SO, Joshua JV, Akinsanya AO, Eze MO, Seun E (2017) Software maintainability and reusability using cohesion metrics. IJCTT 54:63–73
2. Wangoo DP, Singh A (2018) A classification based predictive cost model for measuring reusability level of open source software. 5:19–23
3. Singh AP, Tomar P (2018) Component reusability metrics to measure reusability of web services using fuzzy-multi-criteria approach. J Softw Evol Process 1–16
4. Padhy N, Singh RP, Satapathy SC (2019) Enhanced evolutionary computing based artificial intelligence model for web-solutions software reusability estimation. Clust Comput 22:9787–9804
5. Singh P, Sangwan O, Singh A, Pratap A (2015) A framework for assessing the software reusability using fuzzy logic approach for aspect oriented software. IJITCS 02:12–20
6. Zhang D (2000) Applying machine learning algorithms in software development. In: Proceedings of the 2000 monterey workshop on modeling software system structures in a fastly moving scenario
7. Lounis H, Ait-Mehedine L (2004) Machine-learning techniques for software product quality assessment. In: Fourth international conference on quality software (QSIC 2004) proceedings, pp 102–109
8. Feldt R, Neto FG, Torkar R (2018) Ways of applying artificial intelligence in software engineering. In: RAISE '18: proceedings of the 6th international workshop on realizing artificial intelligence synergies in software engineering, pp 35–41
9. Srinivasan K, Fisher D (1995) Machine learning approaches to estimating software development effort. IEEE Trans Softw Eng 21(2):126–137
10. Singh Y, Bhatia P, Sangwan OP (2011) Software reusability assessment using soft computing techniques. ACM SIGSOFT Softw Eng Notes 36:1–7
11. Sanz-Rodriguez J, Dodero JM, Sanchez-Alonso S (2011) Metrics-based evaluation of learning object reusability. Software Qual J 19(1):121–140
12. Sanz-Rodriguez J, Dodero JM, Sanchez-Alonso S (2010) Metrics-based evaluation of learning object reusability. Springer Science + Business Media, LLC, Berlin, pp 121–140
13. Nesbit JC, Li J (2004) Web-based tools for learning object evaluation. https://pdfs.semanticscholar.org/f1a5/e157937e377c65c53c3e26089e1c691f90c9.pdf

14. Neven F, Duval E (2002) Reusable learning objects: a survey of LOM-based repositories. In: Proceedings of the tenth ACM international conference on multimedia, New York, NY, USA, pp 291–294
15. Zimmermann B, Meyer M, Rensing C, Steinmetz R (2007) Improving retrieval of reusable learning resources by estimating adaptation effort. Presented at the proceedings of the first international workshop on learning object discovery & exchange, pp 46–53
16. Papamichail M, Diamantopoulos T, Chrysovergis I, Samlidis P, Symeonidis A (2018) User-perceived reusability estimation based on analysis of software repositories. In: 2018 IEEE workshop on machine learning techniques for software quality evaluation (MaLTeSQuE), pp 49–54
17. Mao Y, Sahraoui HA, Lounis H (1998) Reusability hypothesis verification using machine learning techniques: a case study. In: Proceedings 13th IEEE international conference on automated software engineering (Cat. No.98EX239), pp 84–93
18. Prakash BVA, Ashoka DV, Aradhya VNM (2012) Application of data mining techniques for software reuse process. Procedia Technol 4:384–389
19. Lounis H, Gayed TF, Boukadoum M (2011) using efficient machine-learning models to assess two important quality factors: maintainability and reusability. In: 2011 joint conference of the 21st international workshop on software measurement and the 6th international conference on software process and product measurement, pp 170–177
20. Di Stefano JS, Menzies T (2002) Machine learning for software engineering: case studies in software reuse. In: 14th IEEE international conference on tools with artificial intelligence (ICTAI 2002), Proceedings, pp 246–251
21. de Almeida MA, Louis H, Melo WL (1998) An investigation on the use of machine learned models for estimating correction costs. In: Proceedings of the 20th international conference on software engineering, pp 473–476
22. Zahara SI, Ilyas M, Zia T (2013) A study of comparative analysis of regression algorithms for reusability evaluation of object oriented based software components. In: 2013 international conference on open source systems and technologies, pp 75–80
23. Padhy N, Panigrahi R, Baboo S (2015) A systematic literature review of an object oriented metric: reusability. In: 2015 international conference on computational intelligence and networks, pp 190–191
24. Deepika, Sangwan OP (2016) Software reusability estimation using neuro-fuzzy technique—a review. Cyber Times Int J Technol Manag 9:40–46
25. Padhy N, Panigrahi R, Satapathy SC (2019) Identifying the reusable components from component-based system: proposed metrics and model. In: Information systems design and intelligent applications, pp 89–99
26. Vinobha A, Senthil Velan S, Babu C (2014) Evaluation of reusability in aspect oriented software using inheritance metrics. In: 2014 IEEE international conference on advanced communications, control and computing technologies, pp 1715–1722
27. Ahmaro IYY, Bin Mohd Yusoff MZ, Mohd Abualkishik A (2014) The current practices of software reusability approaches in Malaysia. In: 2014 8th Malaysian software engineering conference (MySEC), pp 172–176
28. Fazal-e-Amin AKM, Oxley A (2011) A review of software component reusability assessment approaches. Res J Inf Technol 3:1–11
29. Singh A, Tomar P (2016) Web service component reusability evaluation: a fuzzy multi-criteria approach. I.J. Inf Technolo Comput Sci 40–47
30. Padhy N, Singh RP, Satapathy SC (2018) Software reusability metrics estimation: algorithms, models and optimization techniques. Comput Electr Eng 69:653–668
31. Nagpal R, Kumar Bhatia P, Sharma A (2015) Rank university websites using fuzzy AHP and fuzzy TOPSIS approach on usability. IJIEEB 7(1):29–36
32. Singh C, Pratap A, Singhal A (2014) Estimation of software reusability for component-based system using soft computing techniques. In: 2014 5th international conference—confluence the next generation information technology summit (Confluence), pp 788–794

33. Maggo S, Gupta C (2014) A machine learning based efficient software reusability prediction model for java based object oriented software. Int J Inf Technol Comput Sci 6:113
34. Shri A, Sandhu PS, Gupta V, Anand S (2010) Prediction of reusability of object-oriented software system using clustering approach. World Acad Sci Eng Technol 43:853–856
35. Deepika, Sangwan OP (2016) Neuro-fuzzy based approach to software reusability estimation. IJCTA 151–159
36. Imoize AL, Idowu D, Bolaji T (2019) A brief overview of software reuse and metrics in software engineering. World Sci News 122:56–70
37. Kaur PJ, Kaushal S (2018) A fuzzy approach for estimating quality of aspect oriented systems. Int J Parallel Program. https://doi.org/10.1007/s10766-018-0618-2

Unsupervised Learning-Based Sentiment Analysis with Reviewer's Emotion

Harsh Jigneshkumar Patel, Jai Prakash Verma, and Atul Patel

Abstract The sentiment analysis performed using the general methodologies, i.e., lexicon and neural networks based mainly on the content written by the user. They all are mainly content-centric methodologies. The aspect of the user's mindset and sentiment for writing the reviews is never considered and the emotions of the writer. In this paper, we are proposing the consideration of these aspects and their impact. They are accommodated on the basis of the sentiment score of the review written by the user. The intensity of the words used to describe the product or an issue matters significantly in the classification of the product features. Unsupervised learning methods were used to calculate more precise sentence-level sentiments with the help of contextual dependencies. They are more suitable for the aspect-based sentiment analysis as they are found to be more adaptable to different contexts and domains with the change in information rather than changing the entire model structure. The clustering algorithms are used for segregating the different types of groups related to viral sharing of ads. Various factors can be analysed to decide whether the ad is shared by a user or not.

Keywords Sentiment analysis · Text summarisation · Unsupervised learning · Machine learning · Text mining

H. Jigneshkumar Patel · J. Prakash Verma (✉)
Institute of Technology, Nirma University, Ahmedabad, Gujarat, India
e-mail: jaiprakash.verma@nirmauni.ac.in

H. Jigneshkumar Patel
e-mail: 17bit028@nirmauni.ac.in

A. Patel
CMPICA, Charusat University, Changa, Gujarat, India
e-mail: atulpatel.mca@charusat.ac.in

P. K. Singh et al. (eds.), *Evolving Technologies for Computing, Communication and Smart World*, Lecture Notes in Electrical Engineering 694,
https://doi.org/10.1007/978-981-15-7804-5_6

1 Introduction

Text data analysis is the area of processing text data for useful information, and it may be summary of product reviews or opinions on any issues. This helps in getting insights of the product features and issues of any specific cause [19]. Sentiment analysis helps in unveiling the attitude taht an individual holds toward an issue or a product. E-commerce platforms use sentiment analysis for improving the customer outreach, so that the opinions of the users can be used for the improvisation of the product or product features. The aims of sentiment analysis are extracting three major attributes from the text data: polarity, subject and the opinion holder. With the help of sentiment analysis, the unstructured data is converted to statistical information which makes it more interpretable for the analysts to interpret the views and experiences of the users. Sentiment analysis is performed mainly on three main levels of text data, namely document, sentence and sub-sentence level. The approaches used for text summarisation in sentiment analysis are lexicon-based, machine learning or deep learning methodologies [17]. It can also be a hybrid of more than of these. Aspect-based sentiment analysis is a crucial innovation in the past years which helps in analysing the text sentiment based on the whole sentence or the text data rather than the word itself. This is achievable with the LSTM classifiers used which consider the whole text data for analysing the context of the words being analysed. Other machine learning algorithms are used for particular purposes. Unsupervised machine learning algorithms are also used for sentiment classification of the data.

2 Motivation

The reviews of products and issues related to different agendas are mainly judged and scored only based on the content written. This cannot be used to judge the issue or a product because the reviews written may vary in the sentiment intended by the experience or an ideology. This is where the way of writing of the writer plays the most importance as there may be various types of writers from varying from calm minded people to people with extremities of their state of their minds. Calculating the sentiment score, keeping the writer's way of writing would lead to better and improvised results without any bias. This may help in better analysis and improvisation of the products and the reformation of the ideologies [7].

The techniques used for resolving most of the problems are parts of speech (POS) tagging, lexicon-based technique (bag of words) followed by the maximum entropy technique, SVM and the N-gram technique [2, 10]. News-based supervised sentiment analysis for prediction of futures buying behaviour—The vector space model (VSM) with binary weights—is used to convert unstructured news headlines to a structured feature vector. Capturing of the subjectivity is done using the dictionary-based sentiment metrics from the full news text using Loughran and McDonald's finance sentiment dictionary [5, 9, 20].

Clustering algorithms for grouping the reviews' features according to the respective classes. Social network analysis is implemented as the reviews will not be analysed and summarised individually but as a group sentiment. Different lexicons used for identifying the topics discussed in the review. Stemming and lemmatisation algorithms are used with the TF-IDF approach for topic classification. Kullback–Leibler divergence compares the topic distributions. Senti-strength tool is used to determine the polarity of the tweets. Impact scores are calculated for the analysis of the survey data [3]. Naive Bayes algorithm is used to classify tweets into their corresponding sentiments and emotion classes which are then fed into the calculation of user influence score. K-means clustering algorithm is also applied to cluster the sentiments into the three classes—positive, negative and neutral [18]. An evaluation metric, agreement score is evaluated to check if the text supports a statement [8].

The fuzzy-based model is made with the help of K-means clustering algorithm, principal component analysis (PCA) and fuzzy trapezoidal membership function. The Naive Bayes classifier is used to classify the sentiment of tweets. Three lexicons namely SentiWordNet, VADER and AFINN are used in isolation of each other [15]. Using any lexicon includes preprocessing of the text like removal of stopwords, removal of punctuations, small-casing the alphabets, lemmatisation, POS tagging and word sense-disambiguation. Then, the formulation of the rule-base is done by calculating the values of the nine rules. Then, the defuzzification of the values is done through the centroid fuzzification method which gives the final output of the sentiment of the text data [14].

As shown in Fig. 1, the growth of use of Internet has increased enormously in the past 10–12 years. China itself has experienced a huge increase in the use of about 600 million users over the few years. As per Fig. 1, there is a steady rise in the number of Internet users all over the world which in turn will increase the amount of data generated and handled. According to Forbes, there accounts of 2.5 quintillion bytes of data being generated every day. IBM states that the total amount of data stored in the world in the year 2000 was around 800,000 petabytes (PB). It is expected that the number will reach around 35 zettabytes by 2020. 90% of the total data was produced in the last two years with Google processing around 3.5 billion search results every day. But the companies, may it be large or small, can benefit from the text data being generated by the customers or any other publication printing data related to the same field of work. The sentiments and the opinion regarding any product or service can be known which may help the company work in the particular area.

Rest of paper is organised as per following way. Section 2 shows work done by different researchers in the area of sentiment analysis. Section 3 is showing the proposed research work. Section 4 is showing the methodology used to achieve objectives defined in proposed research work section. Section 5 shows the execution and implementation of this research work. In Sect. 6, we are discussing the result generated.

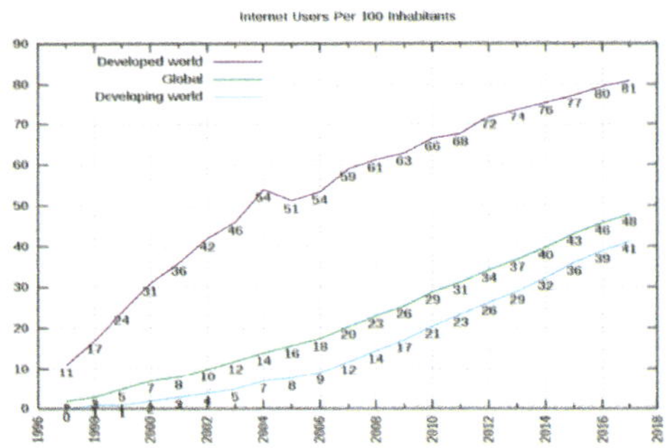

(a) Internet Users per 100 Inhabitants.

Worldwide Internet users

	2005	2010	2017*
World population	6.5 billion	6.9 billion	7.4 billion
Users worldwide	16%	30%	48%
Users in the developing world	8%	21%	41.3%
Users in the developed world	51%	67%	81%

(b) Worldwide Internet Users

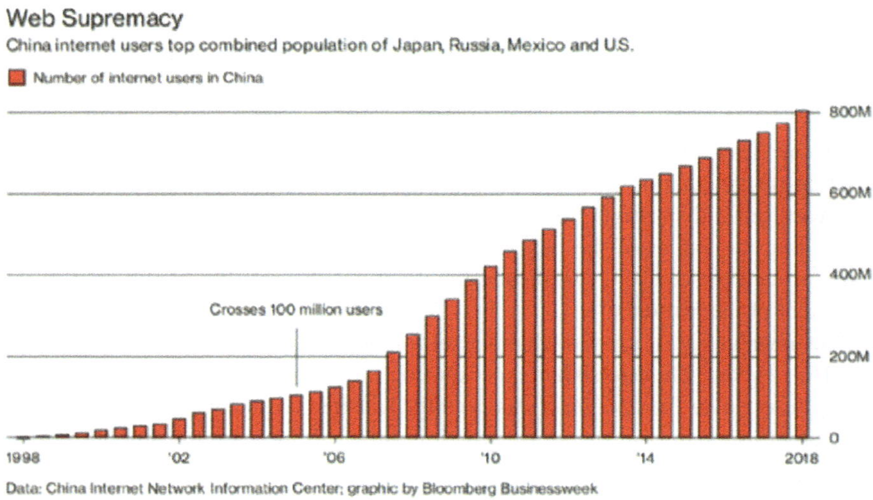

(c) Internet Users in China

Fig. 1 Statistical analysis of Internet users and effect data generation

3 Related Work

The research stated the importance of identifying the domain or context of the text corpus to increase the sentiment score by scoring the words according to the relevance to the context of the text piece. Doaa Mohey El et al. [2] emphasised the problems caused due to particular datasets or data forms and the evaluating algorithms with the constraints related to each of them. It also posed a relationship between the review structure and the sentiment challenges, namely: structured sentiments, semi-structured sentiments and unstructured sentiments. The research states in [11] the emergence of the audio and visual analysis for generation of information. It also briefed about the challenges faced during text-based analysis which consists of negation, ironies and ambiguous use of words in the different context. The audio text data analysis is done from the MFCC audio features extracted. The visual analysis is implemented with the help of tags associated with them [16].

Claudia Diamantini et al. [1] proposed the real-time analysis of large amount of heterogeneous data from different domains. The method is mainly based on lexicon-based and statistical-based techniques. The exploratory data analysis involves data visualisation techniques which help in identifying patterns in data. As per [13], the research revolves around the inconsistencies in the sentiment scores graded by the users and the ones generated by the sentiment analysis algorithm. This claims that the negative aspect in a positive review must have a different score than the sentiment score of the positive aspect in a negative review and vice versa. Kalpak K. Kulkarni et al. show that the research involves information discovery about the seeding influencers of the viral ad shares by studying the cognitive responses of the users.

Qing Sun et al. states the different types of sentiment analysis namely context-sensitive and features based. The traditional sentiment analysis fails at analysing the sentiments of the words with respect to the context of the product domain or a particular issue. Dongmin Hyun et al. [4] proposed a target-dependent convolutional neural network for the target-level sentiment analysis. TCNN is used for classifying the importance between target words and the corresponding neighbouring words. Feature extraction from the text data is executed with the help of deep learning methodologies. The study shown in [20] is about the predictability of the real-time news data on investors' buying behaviour in the futures' market using supervised sentimental analysis. The process includes various stages such as news-trend alignment, market-trend identification, news representation, classification and evaluation.

This study presented by Tamer El-Diraby et al. [3] aims to develop a methodology for analysing opinion dynamics to give a base for the other agencies in the same field, and this research is based on the case study of Translink railway services with it being compared to the others. The overall view of public opinions is considered rather than the individuals for better sentiment analysis regarding an issue. The research aim shown is [8] which is analysing the sentiment of the text data by different machine learning, lexicon-based methods. Srishti Vashishtha et al. [14] shown the aims of finding the sentiment of text data with the help of machine learning algorithms and fuzzy rules based system. The feature extraction of the process is done with the help of machine learning methods, and the numerical values of the corresponding sentiment words related to it are calculated between 0 and 1 by the fuzzy logic-based system. Table 1 is showing the comparative analysis of work done by different researchers in the area of sentiment analysis.

4 Proposed Research Work

The normal sentiment analysis system works on the words and words-intensity lexicon to classify a text document as negative, positive or neutral in sentiment. The process takes place by creating the word-intensity embeddings of the word sequences.

Table 1 Comparative analysis of work done by different researcher in the area of sentiment analysis

Approach	Year	Objective	Technique used	Highlights	Challenges
Machine learning, Lexicon-based [2]	2016	Overcome challenges in SA	SVM, N-grams, BOW	Resolving domain and structure relationship challenges	Accuracy depends on the lexicon used
Deep learning, lexicon-based, unsupervised learning [11]	2017	Analyse multimodal sentiment analysis and the problems	SVM, POS, Alexnet, BiLSTM, Sentistrength, Vader and Umigon, GI, LBP	Insights of the analysis of different modes of text data	Difficulty in neutral classification and inaccuracy in visual SA
Machine learning, deep learning [1]	2018	Reduce problems of irrelevance, negations and word disambiguation	BiLSTM, ETL, Inference rules, lemmatisation, stemming, SentiWordNet	Increases accuracy of the model and reduces noise	Uses of various methods make the model less adaptive to new topics
Machine learning, deep learning, Web scraping [13]	2018	To find the actual sentiment score of the text data	Polarity Aggregation model, CNN, RVest package	Analysis also considers aspect rather than just features	Large amount of data needed to train the deep learning models
Unsupervised learning algorithms, psychographic segmentation, deep Learning [6]	2019	Segment users according to the cognitive responses for seeding ads	Four cluster method, Semantria	Used to measure both the direction and intensity of the text data	Accuracy may not be stable for all types of data
Fuzzy-based, lexicon, deep Learning [12]	2019	Calculate feature-based and context-sensitive sentiment	Improvised LDA methods, GATE, HL-SOT, POS tagging	Implicit features are also learnt for the review	Lot of training data and computation power is required
Machine learning, deep learning, lexicon-based [4]	2019	Development of target-aware CNNs	Bi-LSTM, TCNN hard, TCNN soft, AF-LSTM, ATAE-LSTM, TD-LSTM	Distance-related aspect of target analysed better than LSTM	Ignores the explicit information features
Machine learning, deep learning, lexicon-based [20]	2019	Analysing and reducing the time lag between news and market interests	VSM, Naive Bayes, SVM, TF-IDF, info-gain, chi-square, term-frequency	Helps in acquiring maximum profits by reducing the time lag	VSM does not account for word order and semantic roles of news text
Deep learning, lexicon-based [3]	2019	Develop a methodology for analysing opinion dynamics	Clustering algorithms and Sentistrength algorithm	Helps in comparison of each feature of the service	Classification of topics is limited and does not provide a WHY to the problem
Deep learning, Lexicon-based [8]	2019	Analysis of sentiment of users from Twitter data	Naive Bayes, K-means, maximum entropy, POS, SVM, lexicon-based method	Content is reviewed thoroughly with minimal human interaction	Large amount of data and computation power used
Fuzzy rules based methods, machine learning [14]	2019	Finding sentiment using fuzzy rules system	Naive Bayes, K-means, PCA, fuzzy trapezoidal membership function	Can detect neutral sentiment and can handle large datasets efficiently	Accuracy depends on the fuzzy grades which increases the rule exponentially

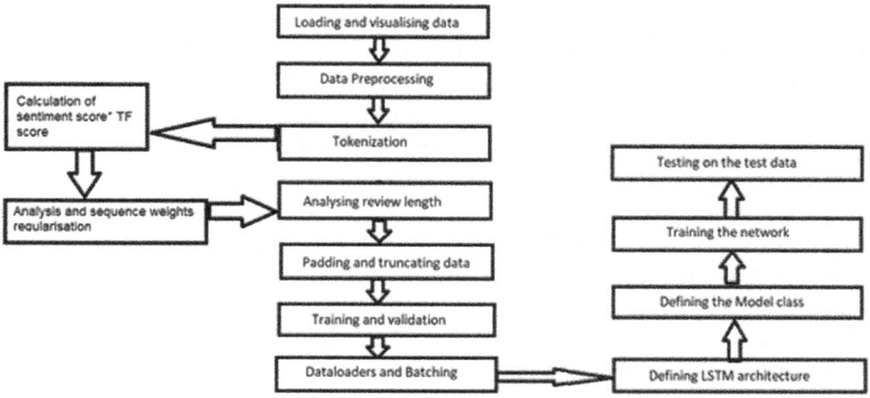

Fig. 2 Full system architecture of the model

The sentiment analysis may be on different levels of text: word, target, sentence and document level. This process takes place through the methods and evaluation processes of deep learning. The conflict in the sentiment analysis's present procedure is the user's emotions or the nature of the person to which he/she is inherent to. This adversely affects the sentiments of the opinion being generated by the user as they tend to spill off their behaviour in the reviews given by them. The evaluation metrics classify the text document on the basis of the average values. Thus, the intensity of the descriptive words used for the features matter because the method used for classifying is mainly based on the intensity of the words used (please refer Fig. 2).

Case Study If two people are considered for giving a review for a particular product, their reviews largely depend on their inherent nature. The first taken to be a calm and composed one who does not lash out with his/her words at anyone and the other one being very aggressive with the language and the intensity of the words that he/she uses. If both are affected negatively by the product to the same extent, the reviews that they given will be of different intensities. The first one will write a normal review and will not be harsh with his words unless he is affected by certain features very badly. And the review that had been given would not be of the same harsh intensity at all notes. However, the other aggressive one would react and write the review in an aggressive tone even if he is not affected to a great extent. So, we have developed a methodology to efficiently analyse with these kind of comments. If the writer is seen to be using harsh language in the whole text document, we check the past reviews of the user if he has been writing the reviews in the similar manner. This is executed by checking the average difference between the sentiment score of them with respect to the other reviewers. If the user is found to be writing reviews in the same manner, we try to mild down the intensity of the words used by reducing the weights of each word in the word sequence. This is done with the process of regularisation. By utilising this way, we tend to minimise the weight of the review given by that particular user. This is done so that the impact of the aggressive user's review does not dominate that of the other users.

We start the process by creating the word embeddings of the text input with the lexicon based on intensity. Preprocessing of the dataset is executed which segregates the users who have written reviews previously or not. We calculate the sentiment score of the feature description with the intensity values by the LSTM used. We count the average for all the user input text data and generate the average of all the users reviewing. If the difference of the sentiment score value of the average of a particular user and that of the whole user-set exceeds a threshold set keeping in mind the previous reviews' intensity level of the user, we regularise the weights of the words by decreasing the value of the weights of the words in the sequence. If the person is writing his first review, then we cannot judge him on the basis of his previous reviews. Then, we count the average sentiment score of all the other users and compare if there is a large difference in the average sentiment score. If the average tends to variate on a scale more than the threshold specified, then the regularisation of the weights of the embedding takes place.

5 Methodology

The text data is preprocessed by in-built functions of the NLTK library. Some of the functions used for the preprocessing are discussed in Algorithm 1 for process flow. The LSTM neural network uses many activation functions like ReLu and sigmoid activation functions. The optimiser used for the neural network is the RMSProp optimiser. Dropout and L2 regularisation are used too. Cross-entropy loss function is used to calculate the loss and accuracy of the model.

6 Execution and Implementation

6.1 Dataset Selection

The dataset used is the Amazon reviews text dataset with ID of the reviewers. The dataset consists of three columns, i.e., the ID column of the dataset which depicts the unique ID of the user writing the review, the label depicting the sentiment of the review text data in the binary form, i.e., 1 and 0. The third column consisted of the text data of the review which is to be used for the sentiment analysis algorithms.

6.2 Data Preprocessing

Figure 3 shows text data preprocessing steps applied. The text data is preprocessed using many methods provided by Python libraries. Lowercasing—The method is

Algorithm 1 Process Execution Steps

Input: The Amazon reviews text dataset with ID of the reviewers.
Output: The LSTM model which are the same values of the sigmoid activation layer.
while Termination condition not reached **do**
  for Each word in given set of reviews *i* **do**
    Step1: Data Preprocessing
    Removal of the stop-words.
    Lowercasing the text data.
    Lemmatization
    Stemming
    Tokenization
    Punctuation removal
    Stop-words removal
    Removal of numbers
  end for
  The execution of the sentiment analysis includes advances text processing. Thereby listed are some of the methods:
  N-grams
  Term frequency
  TF-IDF
  LSTM neural network
  Word embedding
  The LSTM neural network uses many activation functions like ReLu
end while

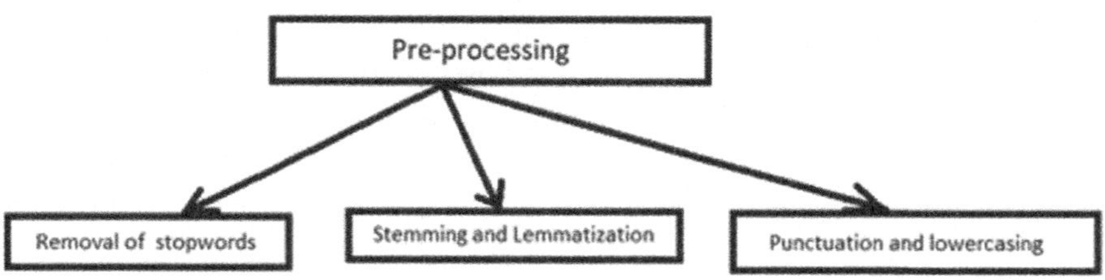

Fig. 3 Preprocessing system architecture

used to convert the text data of the CSV files to lowercase characters. This is used to avoid having multiple copies and analysing the text data. Removing punctuation—This method is used to remove the punctuation in the text data which does not add any extra information to the analysis. Removal of stopwords—This is used to reduce the processing load as the frequently used words are not significant to have sentiment. Tokenisation—This method is used to convert the text data of the text corpus into word sequences for feeding them into the model for processing. Stemming—This refers to the removal of words having the same meanings but different suffixes for different forms. Lemmatisation—This is an improvised method of stemming as it retains the meaning of the words after the stemming process and does a morphological analysis to obtain the root word.

6.3 Text Processing After Tokenisation

Figure 4 shows text tokenisation step of text data preprocessing. Following is the stepwise execution plan for tokenisation. The steps are as follows. Term frequency score: The TF score of the words in the review text data is calculated with the help of TF function. This method is used to count the inverse frequency of the words in a number of documents. This helps in analysing the importance of the words for the sentiment score calculation. N-grams: N-grams are the combination of words that are used for analysing the context of the text data being analysed. The more the longer an n-gram, more will be the context of the text be considered. Sentiment score: The sentiment score of the words is calculated by the VADER lexicon-based methodology. The score is retrieved from the files by mapping it with the corresponding word. Multiplication of the scores: The TFIDF score and the sentiment score of the words are multiplied to generate a new parameter for the LSTM model.

6.4 Execution

Figure 2 is showing execution steps used to implement the model to achieve proposed objectives. Following are the stepwise execution plan for the model building and evaluation. The steps are as follows. Loading and visualisation data: The text data is loaded as a CSV file and is then analysed by visualising the format of the data loaded. Data preprocessing: This methodology is used for improvising the process of sentiment analysis by applying cleaning processes. Tokenisation: This method is

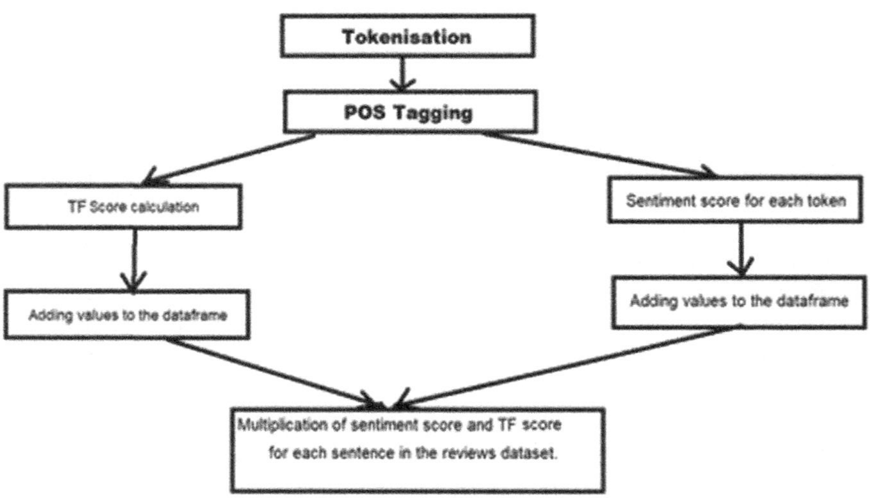

Fig. 4 Text processing for tagging after tokenisation

used for converting the text data into word sequences for processing the sentiment of the text data. Analysing the review length: This method is used for analysing and discarding the reviews having length more than the length specified by the user. Padding and truncating: This method is used for padding the word sequences before feeding it into the LSTM RNN model. Training and validation: Training and validation of the text data takes place which is used for hyperparameter tuning of the model for improving the result accuracy. Dataloaders and batching: This method involves the usage of the generator functions for batching the dataset into batches and the usage of PyTorch takes place for the same. Defining the LSTM architecture: This method is used for defining the components of the LSTM model, namely embedding layer, LSTM layer, fully connected layer, sigmoid activation layer and the output.

Deep learning allows computational models that are composed of multiple processing layers to learn representations of data with multiple levels of abstraction. The layers are as follows. Embedding layer: This layer is used for conversion of the word tokens into word embeddings. LSTM layer: Used to define the hidden units and the number of layers. Fully connected layer: This is used to map the LSTM output to the desired output size. Sigmoid activation layer: This is used to convert the output values between 0 and 1. Output layer: This layer is the final output layer for the LSTM model which are the same values of the sigmoid activation layer.

7 Results and Discussion

The training accuracy of the model is 84.735218% (please refer Fig. 5). This result states that the user sentiment of the review is a crucial factor for calculation of the sentiment score. The improvisation in the cross-validation accuracy states that not only

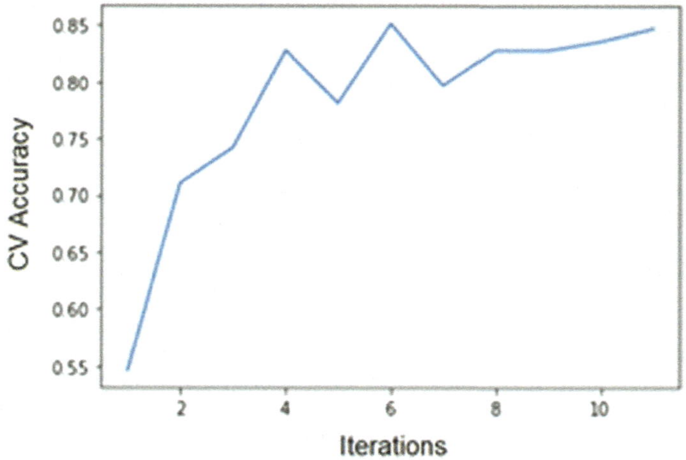

Fig. 5 The training accuracy of the model

does the model fit in well but it also calculates the sentiment score accurately. This result is on the real-time data of the Amazon reviews which proves the authenticity of the model trained.

8 Conclusion and Future Work

The model and the idea for the sentiment analysis states that the review of the user is not only based on the experience while using the product but also due to the mental state while writing the text data. The analysis of the user's writing habits and the way of writing affects the sentiment score of the review. This would help the company segregate the reviews and have a more realistic and beneficial data for the improvisation of the product and the services being provided by it. The future work includes the scope of the idea of reviewing the other reviews written by the same user to generate the perfect sentiment score the text review data. The unique ID can help in finding the other reviews by the user written. This scope of improvement in the model would bring about significant changes in the model accuracy. It also includes the use of sarcasm detection from the text data which adversely affects the sentiment depicted by the writer. Rather than implication from just one review of the user, analysis and weight regularisation would be applied on the basis of more than one review.

References

1. Diamantini C, Mircoli A, Potena D, Storti E (2019) Social information discovery enhanced by sentiment analysis techniques. Future Gener Comput Syst 95:816–828
2. El DM, Hussein DM (2018) A survey on sentiment analysis challenges. J King Saud Univ Eng Sci 30(4):330–338
3. El-Diraby T, Shalaby A, Hosseini M (2019) Linking social, semantic and sentiment analyses to support modeling transit customers' satisfaction: towards formal study of opinion dynamics. Sustain Cities Soc 49:101578
4. Hyun D, Park C, Yang MC, Song I, Lee JT, Yu H (2019) Target-aware convolutional neural network for target-level sentiment analysis. Inf Sci 491:166–178
5. Kaushal NC, Paprzycki M, Bhargava BK, Singh PK, Hong WC (2020) Ditzinger: handbook of wireless sensor networks: issues and challenges in current scenarios. Springer International Publishing
6. Kulkarni KK, Kalro AD, Sharma D, Sharma P (2019) A typology of viral ad sharers using sentiment analysis. J Retail Consum Serv
7. Ruz GA, Henríquez PA, Mascareño A (2020) Sentiment analysis of twitter data during critical events through bayesian networks classifiers. Future Gener Comput Syst 106:92–104
8. Sailunaz K, Alhajj R (2019) Emotion and sentiment analysis from twitter text. J Comput Sci 36:101003
9. Singh PK (2019) Futuristic trends in network and communication technologies: first international conference, FTNCT 2018, Solan, India, February 9–10, 2018: revised selected papers. Springer

10. Singh PK, Kar AK, Singh Y, Kolekar MH, Tanwar S (2020) Proceedings of ICRIC 2019: recent innovations in computing. Springer
11. Soleymani M, Garcia D, Jou B, Schuller B, Chang SF, Pantic M (2017) A survey of multimodal sentiment analysis. Image Vis Comput 65:3–14 (Multimodal Sentiment Analysis and Mining in the Wild Image and Vision Computing)
12. Sun Q, Niu J, Yao Z, Yan H (2019) Exploring ewom in online customer reviews: sentiment analysis at a fine-grained level. Eng Appl Artif Intell 81:68–78
13. Valdivia A, Hrabova E, Chaturvedi I, Luzón MV, Troiano L, Cambria E, Herrera F (2019) Inconsistencies on tripadvisor reviews: a unified index between users and sentiment analysis methods. Neurocomputing 353:3–16 (Recent Advancements in Hybrid Artificial Intelligence Systems)
14. Vashishtha S, Susan S (2019) Fuzzy rule based unsupervised sentiment analysis from social media posts. Expert Syst Appl 138:112834
15. Verma JP, Mankad SH, Garg S (2018) Big data analytics: performance evaluation for high availability and fault tolerance using mapreduce framework with HDFS. In: 2018 Fifth International Conference on Parallel, Distributed and Grid Computing (PDGC), pp 770–775
16. Verma JP, Mankad SH, Garg S (2019) A graph based analysis of user mobility for a smart city project. In: Next Generation Computing Technologies on Computational Intelligence. Springer Singapore, Singapore, pp 140–151
17. Verma JP, Patel A, Evaluation of unsupervised learning based extractive text summarization technique for large scale review and feedback data. Ind J Sci Technol 10(17) (2017)
18. Verma JP, Patel B, Patel A (2013) Article: web mining: opinion and feedback analysis for educational institutions. Int J Comput Appl 84(6):17–22 (full text available)
19. Verma J, Patel A (2016) An extractive text summarization approach for analyzing educational institution's review and feedback data. Int J Comput Appl 143:51–55
20. Yadav R, Kumar AV, Kumar A (2019) News-based supervised sentiment analysis for prediction of futures buying behaviour. IIMB Manage Rev 31(2):157–166

Event-Triggered Share Price Prediction

Jay Pareshkumar Patel, Nikunj Dilipkumar Gondha, Jai Prakash Verma, and Zdzislaw Polkowski

Abstract The stock market price analysis or the prediction of the stock prices has always been a classical problem because of the fluctuating prices of the stocks for a particular company based on the economy. This stock market price analysis/prediction problem has attracted researchers from various fields like statistics, machine learning (ML), deep learning (DL), etc. The analysis or prediction of the stock prices will help the individuals/customers to buy/sell shares of a particular company in order to incur profit. The aim of the proposed paper is to accurately predict the future prices of shares of a company. The prediction on prices can be done through various techniques or methods of machine learning and deep learning. In this paper, we are proposing a hybrid approach of deep learning neural network long short-term memory (LSTM) with sentiment analysis to predict the variations in share prices. First, we apply sentiment analysis on the various company news, market sentiments and get the values. Then, the sentiment results are combined with the LSTM features and observed the results. We predicted the variations in the prices of the stocks for one day and for 30 days long-time period. The observation that we got is very helpful to get the idea of a particular company's future variations in share prices.

Keywords Share price prediction · Deep learning · Sentiment analysis · Machine learning · Long short-term memory (LSTM)

J. Pareshkumar Patel · N. Dilipkumar Gondha · J. Prakash Verma (✉)
Institute of Technology, Nirma University, Ahmedabad, Gujarat, India
e-mail: jaiprakash.verma@nirmauni.ac.in

J. Pareshkumar Patel
e-mail: 17bit033@nirmauni.ac.in

N. Dilipkumar Gondha
e-mail: 17bit024@nirmauni.ac.in

Z. Polkowski
Wroclaw University of Economics and Business, Wroclaw, Poland
e-mail: zdzislaw.polkowski@ue.wroc.pl

P. K. Singh et al. (eds.), *Evolving Technologies for Computing, Communication and Smart World*, Lecture Notes in Electrical Engineering 694,
https://doi.org/10.1007/978-981-15-7804-5_7

1 Introduction

The stock market is a market designated for an activities like selling and buying of stocks of publicly-held firms takes place. In stock market, stock exchanges takes place among hundreds and thousands of participants by ensuring fair pricing and under a set of rules which are predefined. The values of the stocks for an individual firm/company depends on the selling and buying of its stock. If more number of customers buy a particular firm/company's stocks, the values of that stock increases; and if more number of participants sell stocks, the value of the stock decreases, and also the stock price variation depends on various other economic factors. The stock market prediction is a very interesting and classical problem, which has never failed to attract researchers from various multidisciplinary fields like statistics, economics, machine learning, deep learning, etc. [1]. The stock market prediction system is a system which does the prediction of future values of stocks of a firm for a given interval of time. The analysis or prediction of the stock prices will help the individuals/customers to buy/sell shares of a particular company in order to incur profit. Although a great amount of effort is given for this in previous decades, the accurate prediction of the values, and their movement is still a challenging problem. Various models and strategies have been used for the prediction of share values.

The machine learning-based methodologies include support vector machine (SVM), forest algorithm, fuzzy system, etc. Neural network-based approaches like recurrent neural network (RNN), long short-term memory (LSTM) [2] are used for the prediction of future values of stocks [3, 4]. The accuracy achieved through machine learning-based techniques is better than that achieved from statistical techniques. For example, the accuracy achieved through support vector machine in daily stock value movement is approximately 56%.

Other than these, hybrid models are also used for the prediction. Tsai and Wang et al. [5] combined decision tree and neural networks for achieving a 70% accuracy in predicting the stock values. Though machine learning-based methods have shown good accuracy, but they still have limitations. The machine learning-based approaches cannot handle the non-stationarity of the values of stocks, in which the deep learning-based methods have shown considerable amount of accuracy [6]. The methodology used in this proposed paper is based on a hybrid model. This hybrid model consists of neural network-based technique LSTM along with sentiment analysis from the data obtained from news and market. Many researchers have given a lot of efforts in previous decades [7].

As per paper [8], the proposed work is based on the prediction of stock market which predicts the future values of the stocks of various companies. In this proposed paper, we have done the prediction of the prices of stocks using DL approach of long shot-term memory (LSTM) along with the news data sentiment analysis and also with market indexes. The paper predicts the variation of the stock prices which helps for the buying/selling of stocks. Stock market help the entrepreneurs for raising the funds for their businesses. Companies can also gain funds through stock market for various operational and strategic reasons, establishing new markets and building

(a) Impact of stock market on the Indian economy

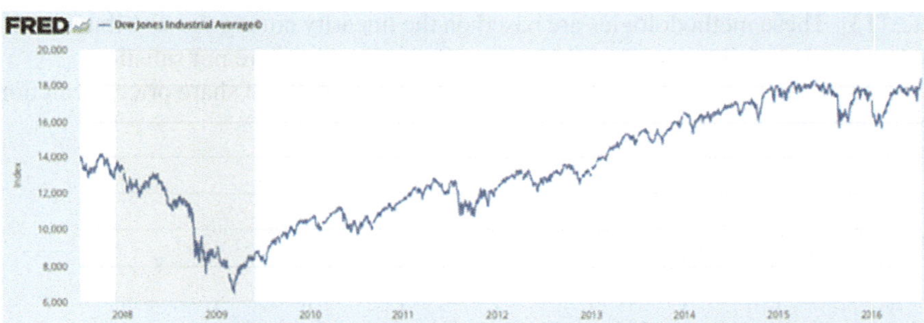

(b) The effect of the stock market crash in 2009

Fig. 1 Some facts of stock market analysis

infrastructures. Through this work, successful prediction of stocks can be done which helps various individuals and companies to yield financial profit [9, 10].

The stock market has a profound and a huge impact on a country's economy. The stock market affects the wealth prosperity of the individual customers. The share market affects the gross domestic product (GDP) of the country to a great extent. If the share market is in bull mode, people tend to have more money which increases the GDP and in bear mode, due to lack or loss of money, people tend to spend less money which has a negative effect on the GDP. The daily stock movements do not affect the economy to a great extent but a crash in the stock market can have a large effect on the economy [11, 12].

Figure 1a shows the increasing impact of the stock market on the Indian economy over the years. Figure 1b also shows the effect of the stock market crash in 2009 which led to the recession.

2 Related Work

Number of people involved in trading /investing in stocks have increased significantly, in last few decades, which includes both professionals and non-professionals. The financial market is a complex, evolutionary, and nonlinear dynamical system [13]. It has become crucial to be accurately predict the stock prices to avoid risks. Lijuan Cao et al. [14] studying and have introduced stock market forecasting methods based on machine learning techniques, where ANN and SVM models been the most widely discussed, compared, and used. Tay et al. [15] state that the SVM gives the best performance for financial time series data. It is observed that the implemented support vector machine algorithm performs very well in predicting the stock price downwards, irrespective of market trend but performance reduces significantly in predicting the stock price upwards [16].

In general, these methodologies can be broadly classified into two categories: statistical techniques and machine learning-based strategies [17]. The statistical-based techniques includes auto-regressive integrated moving average approach (ARIMA) [18], linear regression-based approach, smooth transition auto-regressive (STAR), etc. [13]. These methodologies are based on the linearity among the distributed variables, but in real stock market, these linearity assumptions are not satisfied.

Table 1 shows some legacy case studies done in the area of share price prediction and analysis.

3 Proposed Research Work

The proposed model predicts the value of the stocks for a firm/company from the data sources like news Website, market indexes and from the past values of the stock trends available from Google Finance or Yahoo Finance (Please refer Fig. 2). The prediction of the model is based on two parameters, i.e., sentiment analysis of the news information and past value analysis through long short-term memory (LSTM)-based on neural networks. In the sentiment analysis module, data is collected from news displayed on the general news Websites and also from companies Website. The data collected is preprocessed, and a binary polarity is assigned to each word. At the end of the day, overall sentiment is calculated by average of all the sentiments. The value prediction for the stock is carried out through LSTM deep learning algorithm. Along with that, the prediction of the value from sentiment analysis of the data collected from news Websites and market indexes is taken into consideration. The cumulative effect of both the modules is taken, and the final value of the stocks is predicted. Thus, the proposed hybrid model minimizes the error which occurred in the single long short-term memory (LSTM) model without considering sentiment analysis.

Table 1 Different case study analysis in the area of stock market analysis

Case study	Year	Objective	Accomplishment	Highlights	Challenges
Moving average By R.H. Hooker and published as "instantaneous averages" in Journal of the Royal Statistical Society	1901	Predict the stock movement by moving average of past stock value	The predicted closing value for each day will be the average of a set of previously observed values. For each subsequent step, the predicted values are taken into consideration while removing the oldest observed value from the set	It is the easiest method for stock movement prediction. The graph is the average of closing values of the firm/company in past years	The result of this technique is not very promising. The result can not show the effect of crest and troughs in the stock values
Linear regression from a lecture presented by Sir Francis Galton	1877	Predicting the values of the stocks by basic linear regression method on the previous years values of the stock	In this method, we create the features such as month, year, week, date, quarter start, quarter end, etc. And then, perform linear regression by splitting the data into training and validation sets	This method is the most basic and easy method of machine learning. This method can perform well where the independent features are useful. Like, in big market sales	In this method, the model is overfitting the date and month column. It will consider the value from the same date a month ago. It will not consider all the characteristics of the previous values
K-nearest neighbors By, Fix and Hodges in an unpublished US Air Force School of Aviation Medicine report	1951	Predict the stock values by KNN algorithm by finding the similarity between new data points and old data points	In this method, we find exact k no. of nearest neighbors of a particular point in the graph and then take the average value from that neighboring points	In this method, the plot for the predicted and actual values should provide a more clear understanding than above others	It shows somewhat same results as linear regression model
Auto ARIMA By, Box, George; Jenkins, Gwilym	1970	Predicting the time series predictioning of stock values with Auto (Automatic parameter tuning) ARIMA	ARIMA is auto-regressive integrated moving average. These all terms are showing the various components of the method	It uses the past data to understand the pattern in time series. This technique is far better than any above. The reason is a capacity to capture a trend	This model can capture a trend in the series, but does not focus on the seasonal part
Prophet By, Facebook	Open Source	prediction of time series on stock data by Prophet, the method provided by Facebook	Prophet is the library which requires no data preprocessing. The input is a dataframe with date and target columns	It can capture a trend of the graph and also seasonality from past data	As, stock values cannot depend upon particular trend or seasonality. Prophet is somewhat unusable in stock prediction

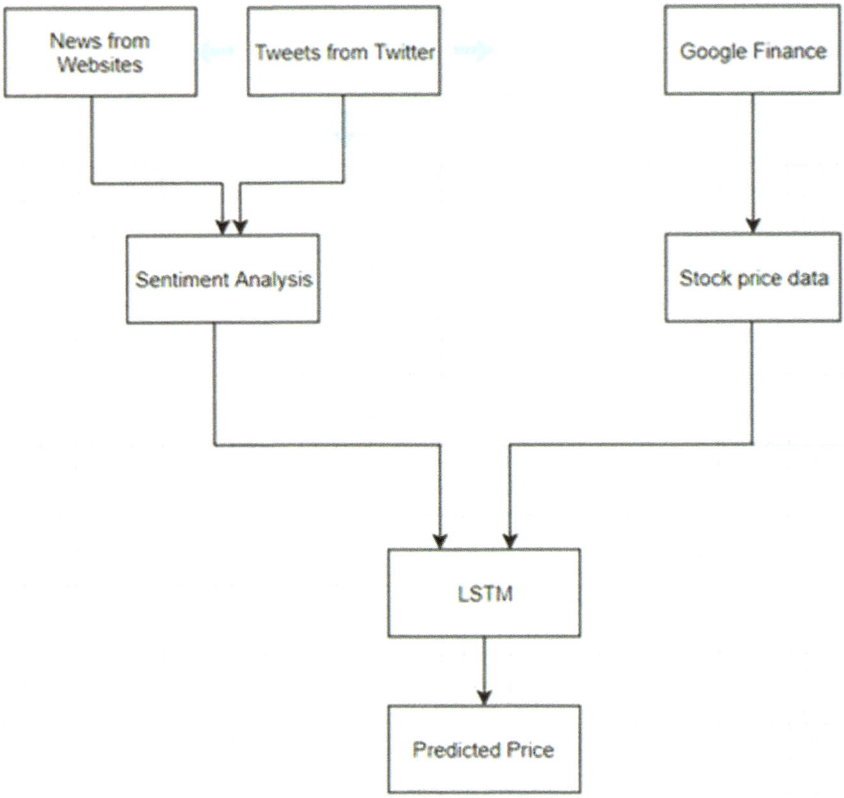

Fig. 2 System architecture

4 Methodology

4.1 Long Short-Term Memory (LSTM)

Long short-term memory (LSTM), in the field of deep learning, is an architecture which is basically based on recurrent neural networks (RNN) [19]. The recurrent neural networks are efficient, but this architecture has long-term dependencies problem. It can only remember data from small duration of time in the sequence of data. LSTM approach was introduced to overcome the problem of long-term dependencies [20]. As shown in Fig. 3, LSTM can not only process single data points, but it can process data from entre sequence; remembering information for long time is their natural behavior. Another problem with RNN was that, if a single information has to be updated or removed, it changes the entire information by applying the function to the entire sequence of data. In the case of LSTM, this problem is taken care of.

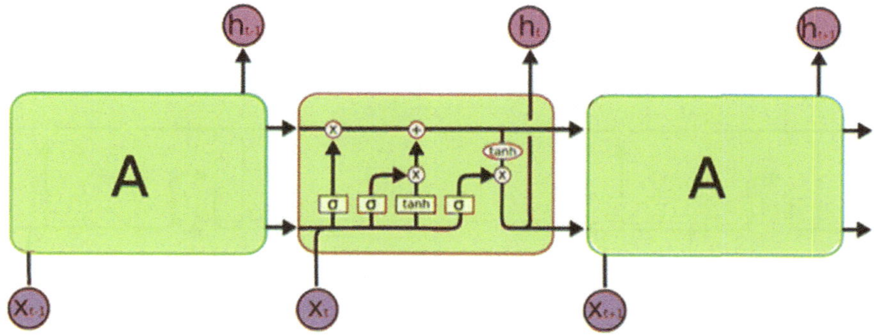

Fig. 3 Execution steps long short-term memory (LSTM)

LSTM approach uses a conveyor belt mechanism, which updates or removes only the desired information or data. The following diagram shows the basic architecture of the LSTM cell.

Figure 3 displays the different memory blocks called "cells" which are present in the LSTM network. There are two states from each cell that are transferred to the next cell, koonthe present cell state, and the hidden state. These memory blocks are solely responsible for the updation and removal of the information which is done through gates mechanism. Each LSTM cell has in particular three gates which are discussed below:

Input Gate: The input gate is responsible for the updation of the information in the cell. It specifies whether the cell is updated or not.

Forget Gate: The forget gate specifies the memory which is set to 0 or not. It is responsible for the removal of the information or data in the cell.

Output Gate: The job of the output gate is to select only the useful information or data from the cell which is to be displayed as an output. As LSTM can process the entire sequence of data, it is applicable to tasks like speech recognition or handwriting recognition, etc. In our proposed work, we use the LSTM-based deep learning approach for the prediction of stock's values.

5 Execution and Implementation

5.1 Dataset Selection

As per Fig. 4, data for the model is taken from three different places. For the LSTM model, the opening and the closing values of stocks for a particular firm/company is taken from Google Finance. Other than that, the data for the sentiment analysis of the data from news information and information obtained from market indexes is

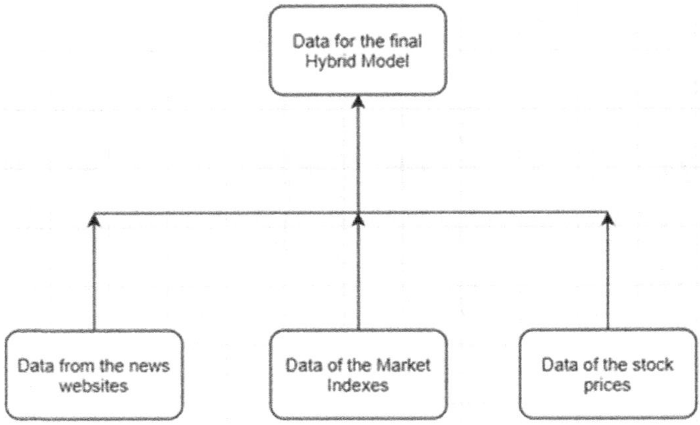

Fig. 4 Dataset selection and extraction

also taken into consideration. The news sentiments by people is taken from Reddit Website, and other news related to the particular firm/company is taken from that firm/company's Website. The data collected from the Websites is analyzed, and a polarity is assigned to the data of the stocks for a particular day. These news data along with the data of the market indexes (economic indicators) is directly considered as a feature in the hybrid model including LSTM output.

5.2 *Pre-processing of the Data and Polarity Assignment*

The data for the prediction of a firm/company's future stocks are taken from the news Website Reddit and news of the economic indicators regarding the market health. Now, the data which is collected from news Website for a firm/company has to be processed, and a polarity is assigned to the data. A starting date is defined from which the news data is to be considered. In our case, the starting date is from year 2011 to year 2017.

5.3 *Methodology Used for Polarity Assignment*

The news headlines which are shown on the Website are split into each word, and a polarity is assigned to each word and stored in a data structure. The polarity which is to be assigned is of three types: positive polarity, negative polarity, and a neutral polarity. Those words which indicate the falling values for a stock are assigned a negative polarity, and those words which indicates the rising values for a stock are assigned a positive polarity. Other than this, those words which are unable to indicate the falling or rising value of a stock are assigned a neutral polarity. Now, the mean positive polarity and the negative polarity are calculated by taking the average of each polarity, respectively. Now, an average polarity for a particular day is calculated by taken into consideration the mean positive, negative, and neutral polarities for that day. Thus, average polarity of each day is calculated in this way from the news Website. In the proposed work, we have taken data for five companies: Apple, Amazon, Facebook, Google, and Tesla.

Now, the data of the economic indicators is selected for the given period of time. The economic indicators show the stats about any economic activity. The economic indicators are generally used by economists and analysts to predict future economic performance.

6 Results

Figure 5 shows the polarity assignment of the data obtained from the news analysis. If the average polarity for the particular day is positive, it shows that the price will increase as per the prediction, or if the average polarity for the day is negative, it shows that the price of the stock will decrease as per the prediction. The fourth column in the table shows the actual scenario where 0 represents that the price of the decreased that day and 1 represents that the price of the increased that particular day. The data about the stock's values for a particular firm/company is taken from Google or Yahoo finance (depending on availability). Now, all these data about the

Date	Polarity	Prediction	Actual
01-07-2011	-0.049007937	0	1
7/15/11	0.038047138	1	1
7/29/11	0.031026171	1	1
8/26/11	0.104386457	1	1
9/23/11	0.040737564	1	1
10/21/11	0.0675	1	0
11/18/11	0.021616161	1	0
12/16/11	-0.043576	0	1
12/30/11	0.096296296	1	1
1/13/12	0.033333333	1	1

Fig. 5 Polarity assignment

values, data of the news sentiment analysis, and data of market sentiment are merged and taken as an input in the long short-term memory (LSTM) algorithm. The LSTM neural networks is used for the prediction of the stocks based on these combined data. In our case, we have done the analysis for Apple company stock value. The prediction is divided into two parts: (1) One day prediction of stocks and (2) 30 days prediction of the stocks. Now, for both the predictions, the LSTM methodology gives different outputs.

Figure 6a shows the prediction of stock's values of Apple firm/company for one day only. The prediction is done through LSTM based on the combined data, i.e., data from sentiment analysis of news and market and the past data values of the stocks. From Fig. 6b, we can see the prediction of future stock's values of Apple firm/company for one day only. This prediction is also done through LSTM approach, but the data input to the LSTM include only the past data of the stocks and not the data from sentiment analysis of the news and market.

Figure 7a displays the prediction done through LSTM which includes the combined data input of data of the stocks as well as the data of sentiment analysis. From Fig. 7b, we see the prediction done through LSTM, but the data input given to the LSTM includes only the past data of the stock values of APPLE firm/company and does not include the data from sentiment analysis of news and market.

7 Discussion

In the proposed work, we have done the prediction of stock values for a particular firm/company based on the data obtained from the past stock values of the firm/company and from the sentiment analysis of the data obtained from the news Website and data obtained from the market. The results of the experimental analysis show that the prediction of future stocks for one day prediction using the combined data has less accuracy than the one day prediction using only the past stock values. On the other hand, the prediction of the future stocks for 30 days time period has more accuracy when the LSTM approach uses the combined data, and the accuracy is comparatively less when the LSTM approach uses only the stock value data. The reason behind such observation is because, for one day curve prediction, the sentiment analysis cannot work properly due to the less time period. The news sentiments prove to be a disadvantage here. On the other side, the prediction for 30 days time period has enough time for the news sentiment. So, the sentiment analysis used is found to be a disadvantage here. Although, the accuracy for 30 days prediction with sentiment analysis is not highly satisfying, but it can correctly measure the increments or decrements or even the consistency in future stock prices with perfect accuracy. However, the long prediction without sentiment analysis can not measure the change in the stock prices.

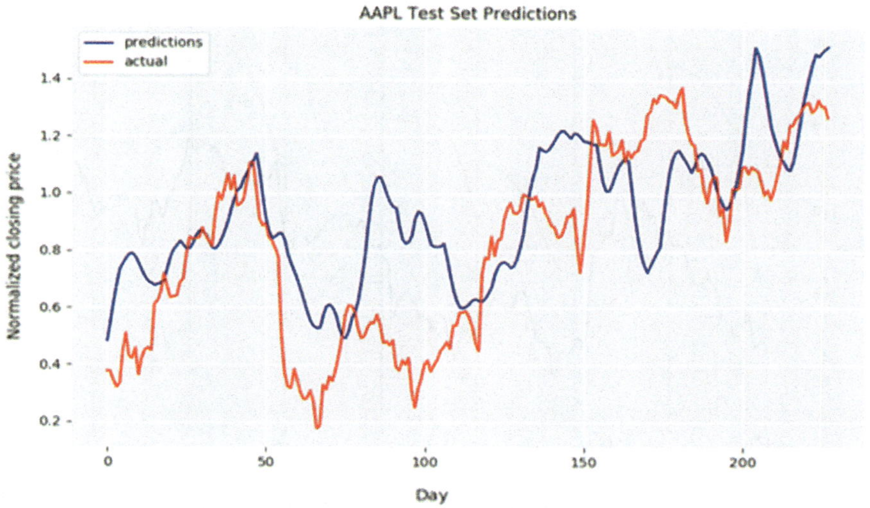

(a) Prediction of stocks with sentiment

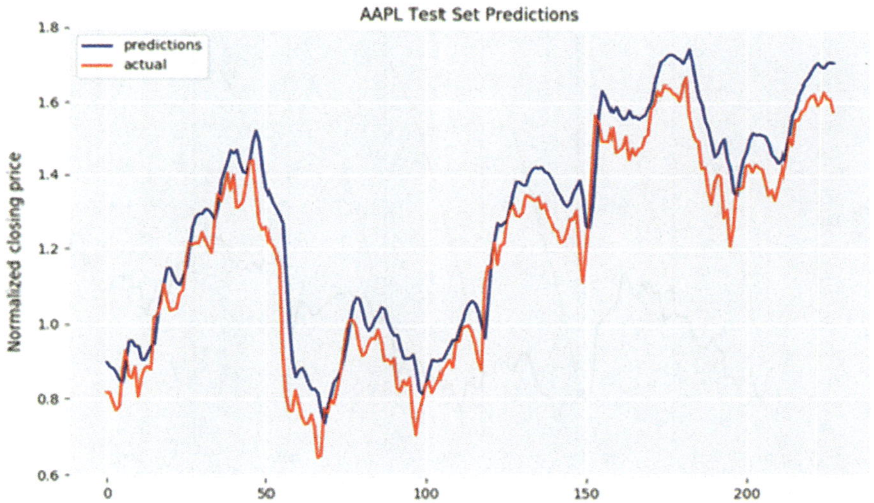

(b) Prediction of the share price without sentiment

Fig. 6 AAPL test set prediction for one day

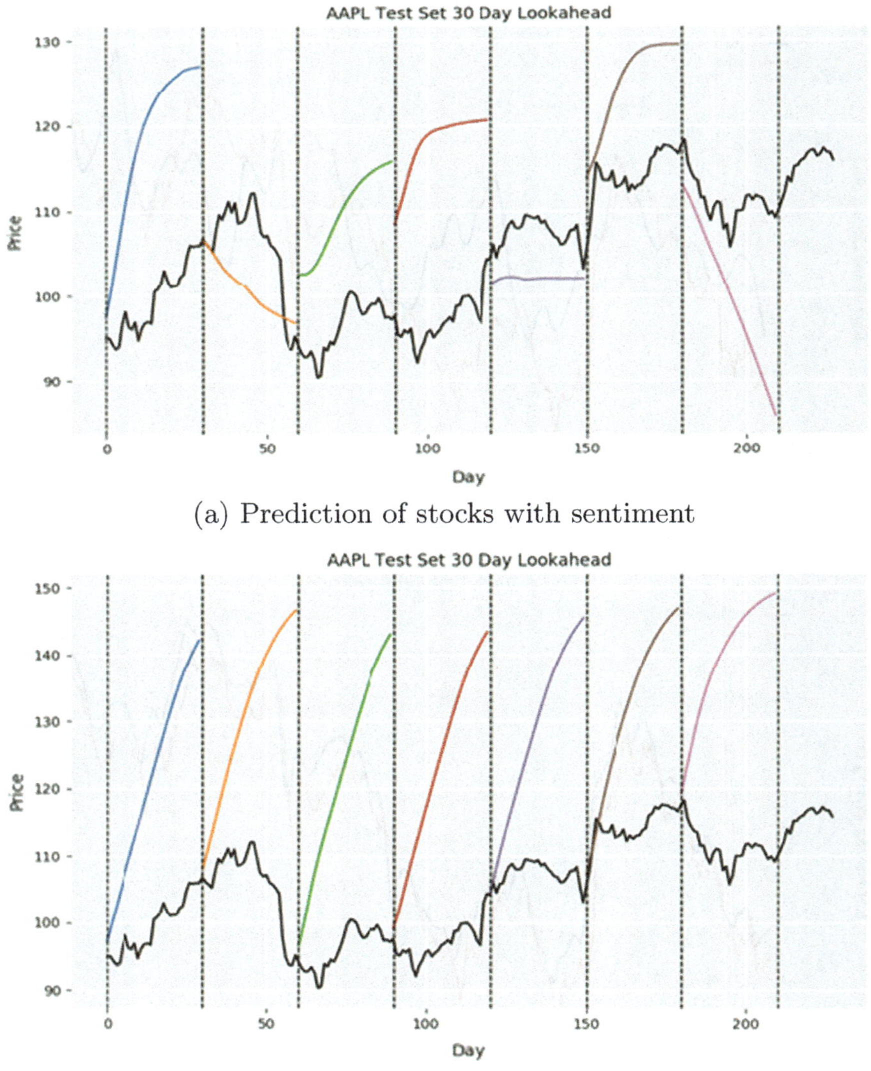

(a) Prediction of stocks with sentiment

(b) Prediction of stocks without sentiment

Fig. 7 AAPL test set 30 days prediction

8 Conclusion and Future Work

The proposed work includes the prediction of future stock's values of a firm based on the long short-term memory (LSTM) methodology. In this, we first done the sentiment analysis on the news for the company and then include the output of this as the features in LSTM module with other features like past stock values and market indices. The accuracy shown by this model is comparatively much more than the accuracy obtained by other methodologies used for the prediction. From the experimental analysis, we conclude that the LSTM approach which uses combined data including the data of stock values and the sentiment analysis output data of news shows very high accuracy when we predict the price for long duration. For small data, or for a small period of time like one day prediction, the LSTM approach without the sentiment analysis of the news proves to be more efficient. So, by this, we can predict the stock values for very long duration by using the news sentiment analysis with nice accuracy. It is very beneficial to estimate the long duration stock prices for the investors for preventing them to invest in wrong companies.

References

1. Hajizadeh E, Seifi A, Zarandi MHF, Turksen IB (2012) A hybrid modeling approach for forecasting the volatility of s&p 500 index return. Expert Syst Appl 39(1):431–436
2. Nelson DMQ, Pereira ACM, de Oliveira RA (2017) Stock market's price movement prediction with lstm neural networks. In: 2017 international joint conference on neural networks (IJCNN), pp 1419–1426
3. Verma JP, Mankad SH, Garg S (2018) Big data analytics: performance evaluation for high availability and fault tolerance using mapreduce framework with hdfs. In: 2018 fifth international conference on parallel, distributed and grid computing (PDGC), pp 770–775
4. Wei B, Jun Y, Yulei R (2017) A deep learning framework for financial time series using stacked autoencoders and long-short term memory. PLOS ONE 12(7):1–24
5. Tsai CF, Wang SP (2009) Stock price forecasting by hybrid machine learning techniques. Lect Notes Eng Comput Sci 1
6. Akita R, Yoshihara A, Matsubara T, Uehara K (2016) Deep learning for stock prediction using numerical and textual information. In: 2016 IEEE/ACIS 15th international conference on computer and information science (ICIS), pp 1–6
7. Verma JP, Patel A (2017) Evaluation of unsupervised learning based extractive text summarization technique for large scale review and feedback data. Ind J Sci Technol 10(17)
8. Fama EF (1965) The behavior of stock-market prices. J Bus 38(1):34–105
9. Verma JP, Tanwar S, Garg S, Gandhi I, Bachani NH (2019) Evaluation of pattern based customized approach for stock market trend prediction with big data and machine learning techniques. Int J Bus Analytics (IJBAN) 6:1–15
10. Singh PK (2019) Futuristic trends in network and communication technologies: first international conference, FTNCT 2018, Solan, India, February 9–10, 2018: revised selected papers. Springer
11. Kaushal NC, Paprzycki M, Bhargava BK, Singh PK, Hong WC (2020) Ditzinger: handbook of wireless sensor networks: issues and challenges in current scenarios. Springer International Publishing
12. Singh PK, Kar AK, Singh Y, Kolekar MH, Tanwar S (2020) Proceedings of ICRIC 2019: recent innovations in computing. Springer

13. Abu-Mostafa YS, Atiya AF (1996) Introduction to financial forecasting. Appl Intell 6(3):205–213
14. Praekhaow P (2010) Determination of trading points using the moving average methods
15. Huang W, Nakamori Y, Wang SY (2005) Forecasting stock market movement direction with support vector machine. Comput Oper Res 32(10):2513–2522. https://doi.org/10.1016/j.cor.2004.03.016, http://www.sciencedirect.com/science/article/pii/S0305054804000681, applications of Neural Networks
16. Verma JP, Mankad SH, Garg S (2019) A graph based analysis of user mobility for a smart city project. In: Next generation computing technologies on computational intelligence. Springer Singapore, Singapore, pp 140–151
17. Rather AM, Agarwal A, Sastry V (2015) Recurrent neural network and a hybrid model for prediction of stock returns. Expert Syst Appl 42(6):3234–3241
18. Wang J-H, Leu J-Y Booktitle=Proceedings of international conference on neural networks (ICNN'96), t.y.v.n.p.k.I.m.:
19. Li J, Bu H, Wu J (2017) Sentiment-aware stock market prediction: a deep learning method. In: 2017 International conference on service systems and service management, pp 1–6
20. Göçken M, Özçalıcı M, Boru A, Dosdoğru AT (2016) Integrating metaheuristics and artificial neural networks for improved stock price prediction. Expert Syst Appl 44:320–331

A Review: Efficient Transportation—Future Aspects of IoV

Ajay Dureja and Suman Sangwan

Abstract In recent years, the term Internet of Vehicles (IoV) has gained popularity among researchers. We are in the era of smart things, smart cities, smart homes, smart transport and smart home appliances, which has been realized due to Internet of Things (IoT). With the invent of smart cars and emerging communication technologies among vehicles, IoV has become the field of research and thereby attracted several vehicle industries and researchers. Internet of Vehicles is the amalgamation of VANET with IoT. Several research challenges need to be addressed for the development of efficient transportation system using IoV. This paper presents the architecture of the IoV with five layers and simple network model of IoV. Recent trends of IoV have been identified in this paper and are discussed to present state-of-the-art advancements and future trends in IoV.

Keywords Internet transportation system · Internet of Vehicles · Internet of Things · Network model · VANET

1 Introduction

Internet of Vehicles (IoV) [1] is an extension of vehicular ad hoc networks (VANET) which is also a branch of Internet of Things (IoT).

In IoV, vehicles communicate with each other through the use of IoT devices. These vehicles nodes share information with each other as well as with roadside units using Internet. IoV vehicles or nodes collect and disseminate the environmental information like speed of vehicles, air pressure, tyre air, breaking oil level of

A. Dureja (✉) · S. Sangwan
Department of CSE, Deenbandhu Chhotu Ram University of Science and Technology, Murthal, Sonipat, Haryana, India
e-mail: ajaydureja@gmail.com

S. Sangwan
e-mail: suman.cse@dcrustm.org

P. K. Singh et al. (eds.), *Evolving Technologies for Computing, Communication and Smart World*, Lecture Notes in Electrical Engineering 694,
https://doi.org/10.1007/978-981-15-7804-5_8

vehicle, number of vehicles running in a region of environment, accidental information in a nearer region, etc., using the embedded vehicular sensors. Different types of sensing devices are used to collect the information from the environment like vibration sensors, spill detector, etc.

In IoV, vehicles can also communicate with roadside units popularly called RSUs and transmit the information or data onto the cloud system.

Although IoV is a new concept, it would surely emerge as an established research and development area in the near future. The concept and use of IoV technology [2] have gained traction started in many countries like USA and Japan. Smart chips are incorporated in vehicles to define their vehicle identities on Internet. In India, almost all registered vehicles like autos, government buses and metro rails are having preinstalled GPS and Wi-Fi.

This incorporation of smart chips facilitates the use of IoT [3]. In IoT, heterogeneous networks are integrated seamlessly. The major aim of IoT is interoperability among these devices which are heterogeneous in nature. IoT enables integration of things which are directly connected to Internet. Smart homes, smart industries, smart appliances, smart cities and smart vehicles are the most sought examples of applications of IoT [4]. IoV is different from VANET in the sense that VANET does not address traffic safety and efficiency. VANETs include the issues related to network, which is ad hoc in nature, unreliable web service, show incompatibility with devices used personally, lacking the feature of cloud computing and lesser accuracy of the services; whereas, IoV can provide security, reduce traffic congestion, prior information of any accident information to vehicles, etc.

VANET architecture [1] also lacks in the management and handling of global information because it has no capability to investigate, process and evaluate information is gathered from different vehicles. But IoV is intelligent network as all the vehicles are equipped with smart IoT devices which collect information from the environment and transmit them on the cloud server in a secure and efficient manner.

IoV also consists of objects such as people in the surrounding vehicles in the region, sensor nodes/devices and networks with high capabilities of services like global traffic analysis, management services such as pollution levels, road conditions and congestion traffic level for large cities or country.

In this paper, overview of IoV has been given, and then the layered architecture has been described and networking model of IoV has been discussed. Arrangement of this paper is as follows: Sect. 1 describes the introduction of IoV in which major aim of IoV is described. In Sect. 1.1, the five-layered architecture of IoV is presented. Simple network model of IoV is described in Sect. 1.2. Related work on IoV that has been proposed by many authors is outlined in Sect. 2. Section 3 presents important findings from these studies. Last Sect. 4 presents the conclusion and future work of this paper.

1.1 Layered Architecture and Network Model of IoV

Various architectures have been proposed by researchers in the field of IoV. Some proposed simple architecture, some proposed layered architecture with protocol stack, and some proposed CISCO-based architecture [5] of IoV. Figure 1 shows layered architecture of IoV in which there are five layers, namely perception layer, coordination layer, artificial intelligence layer and business layer [1]. The functionality of these layers depends upon each others. Each layer supports upper layer for functionality to provide various services.

I. **Perception Layer**
   The most important and first layer of this architecture is perception layer. This layer is responsible for gathering the information from environment in which vehicles are moving. Vehicle's sensors get the information from the environment and send the processed information onto the cloud. This information may be speed of the vehicle, direction of moving, position, acceleration, fuel level, density of vehicles in a region, on road condition, weather condition and information about the people who are in the environment.
   Two main dedicated works of this layer are perception and gathering. A different type of sensors on the vehicle gathers or collects the information at a centre place and sends it onto the cloud server which is highly dedicated for storing and processing of large data. This layer primarily concerned with the different types

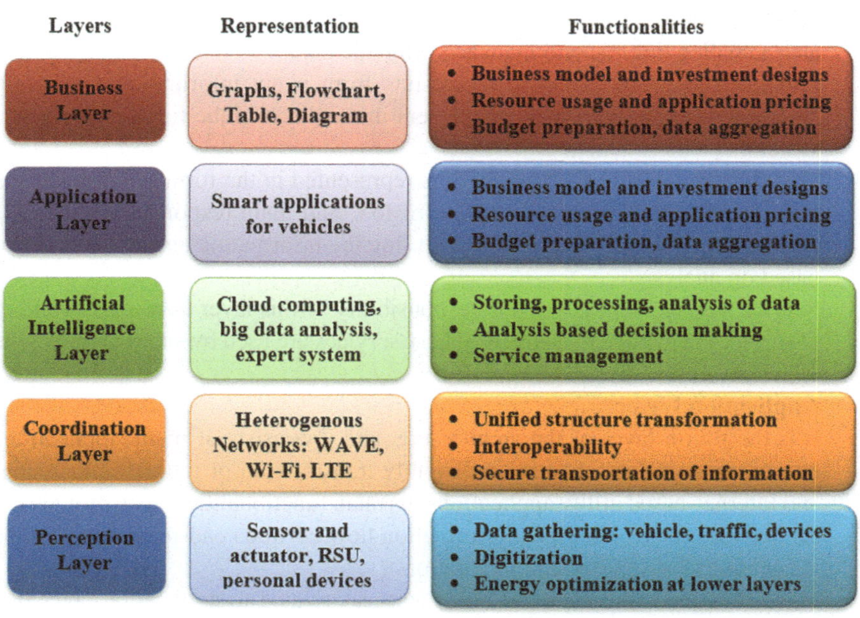

Fig. 1 Five-layered architecture of IoV

of motes and actuators which are connected to the vehicles, RSUs, multimedia devices and other devices.

This layer also provides energy optimization functionality at lower layers. More efficiently sensors get the information from the environment, more optimization of resources will be. One difficulty or issue may arise when vehicles are moving at very high speed. The more the speed of vehicle will be, the more it would be difficult to sense the information from the environment efficiently.

II. **Coordination Layer**

Coordination layer is the next layer to the perception layer which is the second layer of the layered architecture. This layer is concerned with the module which is responsible for connecting the heterogeneous networks like WiMAX, Wi-Fi, E-UTRAN and satellite networks. These networks are also responsible for the transmission of secure data to the next layer, i.e. artificial intelligence layer.

This layer provides interoperability provisions among the vehicles in IoV. Interoperability in IoV is the major issue as there is lacking of standards for communication among moving vehicles through Internet. But this layer provides number of methods of connecting or cooperating between vehicles.

The prime responsibility of coordination layer is to collect different types of information from heterogeneous networks and join together the information in uniformity structure.

III. **Artificial Intelligence Layer**

It represents cloud computing where all the data are gathered from vehicles and processed. It requires managing large data as many vehicles store the data and gather at one location. So, there is requirement of big data analysis which is also a challenging issue to manage such large data.

For processing and inferring the results from many information, deduction system is also required which is present in this layer in the form of the expert system.

This third layer of IoV architecture is represented in the form of virtual cloud infrastructure. It is also called brain of IoV, and main responsibilities of this layer are storing, processing and analysing the information and make decision based on analysis.

Service management is also the responsibility of this layer as it offers number of services in cloud environment. These services are used by smart applications which are embedded in the vehicles.

IV. **Application Layer**

It is the fourth layer of the architecture, and it consists of many applications ranging from safety of traffic, efficiently transmission of multimedia content and Internet-based utility applications. Its main functionalities are to provide the service like smart, intelligent information flow between end-users and provide the service of discovery and integration.

VANET architecture also had many services related to application layer like safety and efficient applications but main focus was not on smart and intelligent applications as there was no provision of gathering of data, communication through Internet and cloud services.

Discovery services are also provided by this layer for combining smart applications for customers. It is the fifth layer of the architecture, and it provides end-user application usages data to the fifth layer of the architecture which is business layer. Uses of many smart applications of this layer are force to think and motivate in research to develop new techniques and applications for end-users.

V. **Business Layer**
This is the fifth and last layer of the architecture which represents operational management of IoV. The different types of representation tools are used to show the statistical analysis of data which was gathered during the processing of above layers. Flow charts, graphs, comparison tables and use case diagrams tools are used in this layer. Usages of resources, overall budget preparation for operation and management are major responsibilities of this layer.

1.2 Network Model of IoV

Below diagram shows the network model [6] of Internet of Vehicle (IoV). It consists of two basic models: swarm model and individual model. This network model contains integration of people, things, vehicles and environment.

The people are those who use the services and provide the services or applications of IoV. In IoV network model, not only people who are driver involves in integration, other people who are in the environment also play a main role in network model of IoV.

In this model, vehicles are the components that share or consume the services. The terminology 'thing' used here may be element other than people and vehicles. Things can be inside or outside of the vehicles like access point, road side unit, IoT sensors, etc. Environment terminology used here is human, vehicles and things.

The individual model is related to only one vehicle, but swarm model is related to many vehicles. As shown in Fig. 2, there is intra-vehicle network in which there is interaction between human and vehicle and in between vehicle and thing. In inter-vehicle network, there is an interaction between people and environment, vehicle and environment and thing and environment.

Swarm model scenario involves interaction between different networks in the form of multi-vehicles, multi-things, multi-users and multi-networks, so-called integrated network. In this model, human, thing and vehicle are connected with integrated network and interact with the environment.

2 Related Work

Previously, many attempts have been made on the studies on the Internet of Vehicle's challenges and future aspects. But these surveys were only based on the concept of

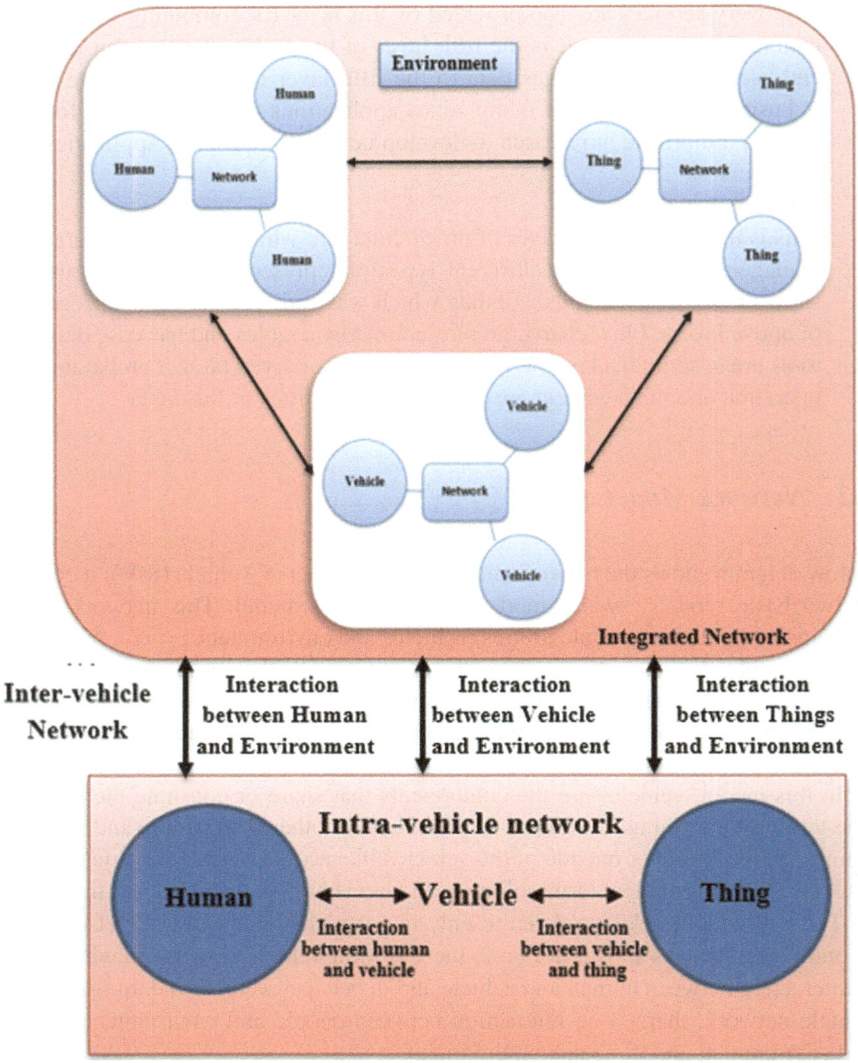

Fig. 2 Simple network model of IoV

term IoV, and there was an issue of defining them. The issue is that these terms were not discussed precisely. In this article, we have reviewed many papers and tried to find out the challenges and future aspects of the IoV.

George [7] surveyed and gave many challenges on the intelligent transportation systems in Internet-connected vehicles. Author firstly proposed theoretical concept of IoV and presented a case study on an area of Internet-connected vehicles like intelligent parking management. Parking management "i-park-wireless" solution was proposed with the following features: formation of networks with sensors among vehicles, combination of infrastructure information with data generated from onboard fused sensor data, aggregation and interpretation of in-vehicle sensor measurements and pattern matching and context recognition.

Lu et al. [8] focused on wireless techniques and challenges to provide vehicle-to-x connectivity. Authors have also mentioned about the challenges and existing solutions to the intra-vehicle connectivity and inter-vehicle connectivity. The authors also discussed the V2I and V2R connectivity solutions. In this paper, authors have presented the challenge of efficient and robust wireless connections to combat the harsh communication environment inside and/or outside the vehicles.

Yang et al. [6] proposed a model of the IoV. Author also discussed the techniques which are required to create the IoV and discussed various applications related to IoV. Authors have divided network model in two sub-models: individual model and swarm model. Individual model concerns with the one vehicle and shows the connection between people and vehicle and vehicle and thing. Swarm model scenario involves interaction between different networks in the form of multi-vehicles, multi-things, multi-users and multi-networks so-called integrated network.

Authors also concluded with following future aspects: building of a cognitive learning model, creation of intelligent technologies to enhance the communication ability and to reduce the redundant traffic and swarm intelligence computing at service providing stage.

Sherly and et al. [9] have focused on to an urban IoT system which is used to build intelligent transportation system. Authors have also presented traffic monitoring system which was based on real-time system to give the solution to the problem of public traffic controlling and monitoring in an efficient manner.

Dhananjay and Madhusudan [10] have given an idea to utilize dashboard camera (Smart-Eye) to prevent the accident and monitoring the services.

Some authors focused on proposing of layered architectures, network model, challenges and future aspects. For instance, Omprakash et al. [1] have presented the concept of five-layered architecture with protocol stack of IoV. Various protocols are defined in the protocol stack for VANET and IoV. Authors also presented a network model of IoV in terms of three major network elements like client, connection and cloud. Many challenges are posed by the authors like location accuracy, location privacy, location verification and operational management of vehicles.

Usha and Rukmini [3] reviewed many challenges and issues related to Internet of Vehicles. But there is an issue author has not discussed, i.e. the challenges and issues related to the efficient transportation management system using IoV. Authors suggested the challenges like integration of non-homogeneous elements on existing IoT architectures in connected vehicles. Another challenge is the collection of data from sensing devices, and this data should synchronize. There is also a need of cloud platform which is also a challenge given by authors.

Yang et al. [11] surveyed on key technologies and introduced various architectures. Authors discussed the first architecture based on current academia and industry standard in which architecture is divided into four main layers that are vehicle network environment sensing and control layer, network access and transport layer, coordinative computing control layer and finally application layer. Another architecture proposed by the authors is based on virtual vehicle which are having two layers, namely perception and execution layer which is in physical space and service layer in cyber space. The proposed architecture focused on intelligent transportation system research which is focused on intelligence of vehicle driving and safety as well as on operation to guide and control traffic flow based on the information of environment traffic and current state of traffic. Realization of VV is the future work suggested by the authors.

As large data are gathered and processed at the cloud server which is sent by many vehicles connected in the IoV network, so there is a need of high security mechanism to protect the data from unauthorized access and to prevent of sending false information. This is also a big challenge in respect to Internet of Vehicle concept. So in this context, Guo et al. [12] proposed a mechanism which is secure for big data collection in large-scale IoV. Authors also proposed two different secure protocols for business data and confidential data collection. In addition to it, simulation result-based performance evaluation has also been shown by the authors to compare how the data are more secure by these algorithms in the proposed approach versus simple approach. In respect to future work, authors suggested the future research on practical demonstration which is required to check the proposed scheme, real-time practical with growing number of vehicles in the IoV. The most important future research work is the development of routing protocol of IoV to make proposed security scheme in more optimized manner.

In 2017, Ramkumar and Linesh [13] focused on the use of adaptive Internet of Things and suggested a framework which solves the issue which was not discussed by the authors Y. Usha and M. S. S. Rukmini. Authors' framework carried out a new IoT-based traffic management (IoT-TM). This system can help to take a decision about the traffic management, and it can help to find the efficient and traffic clearance. Authors also have shown the performance evaluation simulation using MATLAB on the basis of traffic management.

Li et al. [14] proposed an architecture of IoV named content-centric networking architecture which is more enhanced than others. It is novel architecture and different from previously proposed architecture in the sense that it cares about the content itself rather than its source. In traditional architecture vehicles communicate with each other and other things using IP-based networking, but this is different in terms of data exchanges with the help of interest and data. Both of them are two types of CCN packets. CCN always uses a structure for name. This architecture is also divided into five parts: physical layer, data link layer, network layer, perception layer and application layer. The main features of this architecture are short transmission delay, low power consumption and high reliability. The author also analysed and overcome the challenges like relatively high communication delay, large transmission distance and low reliability under high-speed mobile environment.

Contreras-Castillo et al. [5] presented recently proposed protocols based on communication. These protocols enable the seamless integration and operations of the IoV. Authors also discussed the CISCO-based IoV architecture which is based on four layers, namely end point layer, infrastructure layer, operation layer and service layer. Authors also focused on future research directions and challenges related to

IoV. Efficient and scalable coordination and communication among devices and the lack of standards to enable robust V2V communication are the two main challenges were suggested by the authors. Apart from challenges, authors identified following future work: energy efficiency in traffic jams, connected devices, i.e. combination between devices and applications, security features in IoV and various types of safety features.

Jain et al. [15] proposed vehicular social networks based on VIoT and vehicular social network protocol (VSNP) for VIoT based on WSN. Authors discussed about the challenge of reliable and flexible traffic control management. Authors also presented the results and shown the analysis based on VSNP protocol. After analysing on the basis of other parameters, authors also presented future works like improvement of end to end delay, congestion control techniques and throughput parameters to achieve a more efficient IoV.

Many challenges in respect of Internet of Vehicles were suggested and discussed [16]. But one of the major challenges which must be focused is security challenge. Many safety algorithms and secure mechanisms for the avoidance of vehicle collision and avoidance in respect of the concept of IoV were proposed by researchers from time to time. A new VANET environment and an algorithm for collision detection were proposed by Anadu et al. [17]. Authors also proposed a physical model based on IoT devices Arduino UNO, RGB LCD shield and other hardware. The proposed model is highly efficient for prior sharing of accident messages between the sensor nodes of vehicles which avoids collision of vehicles.

Nowadays, the most and current topic of research in case of IoV becomes vehicles' big data and issues like privacy of data for interconnected vehicles in intelligent transportation system. As large scale of vehicles connectivity tends to gather large amount of data onto the cloud servers, so it becomes the need to store this large amount of data with the use of big data applications and secure the data from unwanted users. Internet architecture is not much scalable and not quite efficient to handle extensive amount of data. Similarly, transferring of data is expensive. The processing of big data is time consuming and very expensive.

In context of these challenges, Mahmood et al. [18] proposed a model to secure big data collection in IoV. Authors also suggested many future aspects in context of big data which are creation of vehicular cloud formation schemes, techniques for sharing of IoV data among varied providers of cloud services and development of infrastructure-based application for management of cloud server.

Singh et al. [19] also proposed many issues and challenges related to current scenarios of development in wireless connectivity among vehicles. Author also suggested recent innovations [20] in the field of Internet of Vehicles. Many techniques and technologies related to advancement in IoV have been proposed.

Chen et al. [21] proposed an innovative paradigm to address the challenges which mainly focuses on transportation safety, communication technologies and network security called Cognitive Internet of Vehicles. It uses the mining to find effective information from network and physical data space.

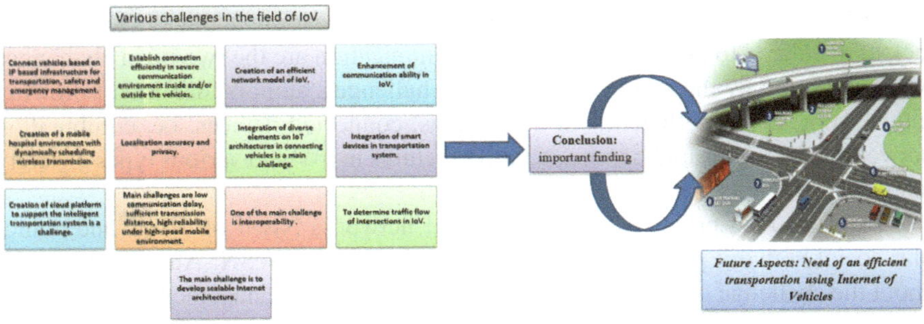

Fig. 3 Important finding: need of an efficient transportation

3 Important Findings

After going through the work of researchers, several challenges have been identified for realization for efficient transportation system using IoV. Some challenges are based on the safety; some are in the area of location-based. These challenges may help researchers to think and implement new ideas to improve the transportation system.

In the era of new technologies development, various smart devices are developing like IoT-enabled smart terminal which may be used for intelligent transportation system.

Some of the challenges are connection of vehicles through an IP-based infrastructure for transportation, reliable and flexible traffic control management, maintaining quality of services for video streaming applications in IoV and many more.

During observations of various issues and challenges in Internet of Vehicles, we have made various important findings. One of the important findings is the efficient transportation. There are many issues in existing transport systems (ITS) [22] using IoV like it do not fully consider and resolve accuracy, do not instantly response, not resolve traffic congestion and do not compete the existing challenges in Internet of Vehicles (IoV) environments.

After review and analysis of many review and research papers, another important finding is that there is a need of efficient transportation management system using IoV. Many researchers have given ideas to make the current IoV better and enhanced. But there is requirement of intelligent and efficient architecture and model of IoV which can make transportation system better. Figure 3 shows an important finding in terms of need of an efficient transportation.

4 Conclusion and Future Work

Due to rapid and fast growth of Internet technologies, there is a need of connecting vehicles for the exchange of important information and data. In this regard, IoV plays an important role. Several challenges and future aspects have been identified in this paper. Based upon these challenges, this field opens several prospects for the researchers who are willing to work in the field of IoV.

A future research area in this field can be improvement of performance in connecting nearby vehicles through wireless technologies. An enhanced model can also be developed for vehicular environment to measure the accuracy of accident and reliability of communication.

In future, we can implement a new transportation system with improved efficiency of traffic safety and travelling costs.

Development of an algorithm to control the speed of the vehicle based on traffic updates can also be research area of this field.

References

1. Kaiwartya O, Abdullah AH, Cao Y, Altameem A, Prasad M, Lin C-T, Liu X (2016) Internet of vehicles: motivation layered architecture network model challenges and future aspects. IEEE Access J 1–17
2. Jadaana K, Zeaterb S, Abukhalil Y (2017) Connected vehicles: an innovative transport technology. Procedia Eng 187:641–648
3. Devi YU, Rukmini MSS (2016) IoT in connected vehicles: challenges and issues—a review. In: International conference on signal processing, communication, power and embedded system (SCOPES), pp 1864–1867
4. Agrawal S, Ahlawat P (2020) Key management schemes in Internet of things: a matrix approach. In: Handbook of wireless sensor networks: issues and challenges in current scenario's, AISC 1132, pp 381–400
5. Contreras-Castillo J, Zeadally S, Guerrero-Ibañez J (2017) Internet of vehicles: architecture, protocols, and security. IEEE Internet Things J 5(5):3701–3709
6. Yang F, Wang S, Li J, Liu Z, Sun Q (2014) An overview of Internet of vehicles. China Commun 11:1–15. https://doi.org/10.1109/CC.2014.6969789
7. Dimitrakopoulos G (2011) Intelligent transportation systems based on Internet-connected vehicles: fundamental research areas and challenges. In: International conference on ITS telecommunications, pp 145–151
8. Lu N, Cheng N, Zhang N, Shen X, Mark JW (2014) Connected vehicles: solutions and challenges. IEEE Internet Things J 1(4):289–299
9. Sherly J, Somasundareswari D (2015) Internet of Things based Smart Transportation Systems. International Research Journal of Engineering and Technology (IRJET) 2(07):1207–1210
10. Singh D, Singh M (2015) Internet of vehicles for smart and safe driving https://doi.org/10.1109/iccve.2015.93
11. Yang F, Li J, Lei T et al (2017) Architecture and key technologies for Internet of vehicles: a survey. J Commun Inf Netw 2:1–17. https://doi.org/10.1007/s41650-017-0018-6
12. Guo L, Dong M, Ota K, Li Q, Ye T, Jun W, Li J (2017) A secure mechanism for Big Data collection in large scale Internet of vehicle. IEEE Internet Things J 4(2):601–610
13. Eswaraprasad R, Raja L (2017) Improved intelligent transport system for reliable traffic control management by adapting internet of things. In: International conference on Infocom technologies and unmanned systems (Trends and Future Directions) (ICTUS) pp 597–601
14. Li Z, Chen Y, Liu D et al (2017) Performance analysis for an enhanced architecture of IoV via content-centric networking. J Wirel Com Netw 2017:124. https://doi.org/10.1186/s13638-017-0905-4

15. Jain B, Brar G, Malhotra J, Rani S, Ahmed SH (2018) A cross layer protocol for traffic management in social Internet of vehicles. Future Gener Comput Syst 82:707–714
16. Singh P, Paprzycki M, Bhargava B, Chhabra J, Kaushal N, Kumar Y (2018) Futuristic trends in network and communication technologies. In: FTNCT 2018. communications in computer and information science, vol 958, pp 3–509
17. Anadu D, Mushagalusa C, Alsbou N, Abuabed ASA (2018), Internet of things: vehicle collision detection and avoidance in a VANET environment. In: IEEE international instrumentation and measurement technology conference (I2MTC), pp 1–6
18. Mahmood A, Zen H, Hilles SM (2018) Big Data and privacy issues for connected vehicles in intelligent transportation systems. https://doi.org/10.1007/978-3-319-63962-8_234-1
19. Singh PK, Bhargava BK, Paprzycki M, Kaushal NC, Hong WC (2020) Handbook of wireless sensor networks: issues and challenges in current scenario's. In: Advances in intelligent systems and computing, vol 1132. Springer, Cham, Switzerland, pp 155–437
20. Singh PK, Kar AK, Singh Y, Kolekar MH, Tanwar S (2020) Proceedings of ICRIC 2019, recent innovations in computing, 2020, Lecture Notes in Electrical Engineering, volume 597. Springer, Cham, Switzerland, pp. 3–920
21. Chen M, Tian Y, Fortino G, Zhang J, Humar I (2018) Cognitive Internet of vehicles. Comput Commun 120:58–70
22. Swarnamugi M, Chinnaiyan R (2020) Context—aware smart reliable service model for intelligent transportation system based on ontology. In: Singh P, Kar A, Singh Y, Kolekar M, Tanwar S (eds) Proceedings of ICRIC 2019. Lecture Notes in Electrical Engineering, vol 597. Springer, Cham

Compact High Gain 28 GHz Concentric Circular Director Low-Cost Antenna for 5G Millimeter-Wave Communication

Raqeebur Rehman and Javaid A. Sheikh

Abstract A novel high-gain concentric circular millimeter-wave patch antenna is presented. The antenna finds its application for high-speed 5G mmWave communication utilizing the licensed band of 28 GHz. The antenna has a better directivity and achieves a peak gain of 10 dBi for both E and H planes. The antenna is designed on 0.787 mm Rogers RT/duroid square substrate having a side dimension of 20 mm. The full dimension of the proposed antenna is $20 \times 20 \times 0.787$ mm^3. The antenna presented resonates at 28 GHz with a return loss value of −24.69 dB and has an impedance bandwidth of 29%. The presented antenna is a low-cost planar structure, and its peak gain is independent for horizontal and vertical polarization. The concentric circular geometry of the proposed antenna makes the vector E-field of antenna to propagate in such a way that the antenna achieves a better directive gain, impedance matching, and a dual polarization. The parameters such as surface current of the radiating structure, polar gain plot, normalized gain are also discussed. The better performance of the presented antenna in connection with return loss, peak gain, impedance matching, radiation pattern, directivity, and impedance bandwidth makes the proposed antenna a novel candidate to be used for 28 GHz mmWave communication.

Keywords Millimeter-wave · Impedance bandwidth · Radiation pattern · Peak gain · Dual polarization

---

R. Rehman (✉) · J. A. Sheikh
Department of Electronics and Instrumentation Technology, University of Kashmir, Srinagar, India
e-mail: raqeeb.scholar@kashmiruniversity.net

J. A. Sheikh
e-mail: sheikhjavaid@uok.edu.in

P. K. Singh et al. (eds.), *Evolving Technologies for Computing, Communication and Smart World*, Lecture Notes in Electrical Engineering 694,
https://doi.org/10.1007/978-981-15-7804-5_9

1 Introduction

The millimeter-wave proves out to be a promising technology for fifth-generation mobile communication due to availability of wide usable spectrum [1, 2]. Also the path loss and signal attenuation at high-millimeter-wave frequencies due to the atmospheric gases and the water molecules can't be neglected. So in order to combat with this problem, the need is to incorporate highly directive antennas with narrow signal beams to make the high-speed and high-capacity 5G millimeter-wave communication link possible [3, 4]. Also with the increase in frequency at millimeter-wave spectrum, the size of the antenna reduces to a greater extent of the order of a few millimeters which lead to the challenges for the incorporation of conventional antennas like dipole arrays, Yagi, horns, and sector antennas inspite of their high gains at some modified geometries [5–10]. In addition to this, the cost regarding their utilization may also increase up to a greater extent. For a simple Omni antenna, the peak gain is of the order of 2.2 dBi where for its collinear array, it is 5.8 dBi. Similarly, for a simple patch antenna, the maximum gain can be 9 dBi or lesser [11]. So the need is to introduce highly compact antennas with high directive gains to be used for 5G millimeter-wave communication. The compact sizes of antennas at millimeter-wave frequencies allow us to design different planar antennas that can felicitate higher gain, lesser interference, higher spectral efficiency with inclusion of better signal coverage. In addition to this, the concept of antenna arrays with modified feeding techniques and specific inter-element spacing of about half wavelength can be introduced to have much higher gains and directivity. But the same geometries have to be designed carefully to achieve a high front to back ratio of antenna radiation patterns and to avoid high side-lobe formation. This can be achieved by regulating the inter-element spacing in an array to about half wavelength. Also the need will be for low cost, wideband, compact size, low complexity, and well integration ability of millimeter-wave antennas. Horn and Yagi antennas at millimeter-wave frequency band can provide much higher gain and bandwidth but their bulky shape and complexity in feeding make them hard to integrate with the micro-strip planar structures. Some novel antennas at millimeter-wave band with various miniaturization techniques have been developed recently, but most of them lag behind to achieve the higher gain requirement at millimeter-wave frequency band. In addition to this, a number of millimeter-wave antennas have been reported in the literature. Kuikui Fan et al. [12] have proposed a wideband horizontally polarized conical-beam omni-directional antenna for millimeter-wave applications. The impedance bandwidth achievement of the design is 22.9% with a better radiation pattern for 39–49.3 GHz band. The gain proceeds from 4.6 to 6 dBi in the prescribed frequency band. It proves as a good prototype for millimeter-wave communication with minimum losses due to incorporation of a dual-mode substrate integrated waveguide (SIW) radial waveguide power divider, conical reflector, and an SIW feed. But still the design of this antenna proves to be a little bit cumbersome in comparison with the design of conventional planar millimeter-wave antennas. Wani et al. [13] have proposed mmWave MIMO antenna having wide-scan angle radiation characteristic. The antenna is designed using three-element quasi Yagi-uda reflectors printed on a metamaterial surface and is generated from an array of unit cells. The presented antenna has wide-radiation scan angle characteristics and is designed at the millimeter-wave band of 28 GHz. The antenna provides a better gain and property to direct the E-field in three separate directions. In [14], a tapered slotted antenna array has been proposed for massive MIMO mmWave communication with multi-beam characteristic. The complexity regarding the computation of the antenna geometry is very low. The integration of the antenna with the planar circuits is very easy due to the incorporation of the SIW feeding structure. The half-wavelength spacing of antenna elements in the H-plane leads to the achievement of better performance. Within the operating frequency ranging from 24 to 32 GHz, the overall gain ranges from 8.2 to 9.6 dBi for each antenna element. The presented antenna array allows a good incorporation in millimeter-wave massive MIMO systems. Kumar et al. [15] have proposed a CPW-fed strip compact square and loaded slots dual-wideband antenna with circular polarization for satellite communications. In the concerned antenna, a grounded L-strip, a rectangular slot in lower left CPW ground plane, and a pair of grounded spiral slots are responsible for the achievement of dual CP. A CPW structure on inverted patch C-shaped square slot antenna and a specific perturbation in ground plane also adds to the enhancement of dual CP. The antenna is a full coplanar structure, and the effect of various geometries on the performance of the antenna has been recorded by following the specific parametric procedure, e.g., effect of spiral M position variation and slot cut outs like rectangular on the impedance bandwidth and AR (axial ratio) bandwidth. The antenna has a dual-band response when subjected to a frequency sweep of 3–14 GHz attains a peak of 6.36 dBic and 3-dB AR bandwidths in both the dual bands. This antenna is a suitable candidate to be used in the Ku-band downlink frequency and wideband wireless but it can't be used for higher Ku-band frequencies of satellite communication because of the FR-4 substrate's high dielectric value which will add to the larger losses at these higher frequencies and naturally much lesser gains. Though all the reported antennas [16–22] have some special and specific characteristics, most of them fail to achieve the required gain and radiation characteristics like the better directivity and polarization diversity. Moreover, the compactness of these antenna structures and their implementation proves out to be a little cumbersome task, but have achieved a good milestone, and the authors have tried their best to overcome most of the limitations of millimeter-wave communication.

Keeping the above circumstances in view, a compact low-cost concentric circular patch antenna for 5G millimeter-wave communication is proposed. The proposed antenna is a square structure having two concentric rings with two hook-shaped structures cut out from the ground plane, and the top radiating patch is encircled by directors as in Yagi-uda antenna. This geometry of the antenna proposed makes it to have a peak gain of 10 dBi for both azimuth and an elevation plane which is much higher than a conventional patch and makes it to have dual polarization with higher directivity. The antenna proposed also achieves an impedance bandwidth of 29%. The antenna can serve as an excellent candidate for portable 5G gadgets and can also be incorporated in a MIMO transceiver to have high speed directional P2P (point to point) millimeter-wave communication.

2 Antenna Geometry and Analysis

2.1 Antenna Structure

The full profile of the antenna is depicted in Fig. 1. The proposed antenna has got a ground plane (defected) comprising of two hook-shaped cut outs at top and bottom and a concentric circular geometry to modify the surface distribution current of ground plane. Similarly, the front face consists of a radiating mini patch bordered by three-concentric circular rings to make the *E*-field to propagate uniformly toward the outer areas of the structure. The antenna proposed has been designed on 0.787 mm Rogers RT/duroid square dielectric of value 2.2 and tan $\delta = 0.0009$. The full structural profile of the antenna is $20 \times 20 \times 0.787$ mm^3. The proposed antenna structure makes it to have a high gain of 10 dBi for both *E* and *H* planes and to achieve dual polarization. The design variables of the proposed antenna are explained below and brought up in Table 1.

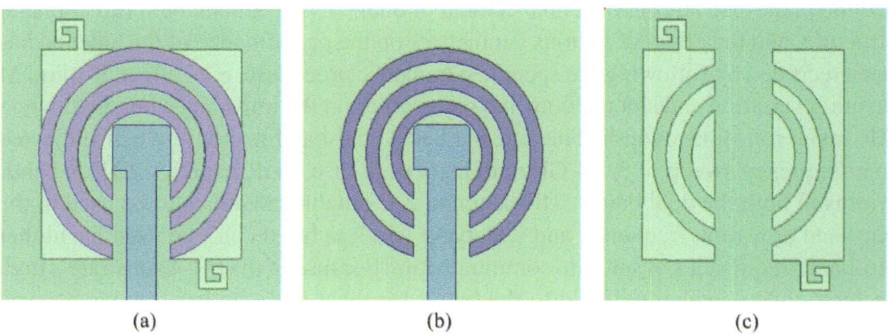

Fig. 1 Structure of the millimeter-wave antenna proposed. **a** Complete view. **b** Top view. **c** Bottom view (structure designed in Ansys EM simulator HFSS V.15)

Table 1 Design variables of the proposed antenna

Parameter	Value (mm)	Parameter	Value (mm)
L_{mp}	4	L_f	9
W_{mp}	3	W_f	2
ro_1	3.6	$S_a = S_g$	20
ro_2	5.38	S_{ico}	14
ro_3	7.2	rgo_1	4.44
ri_1	2.82	rgo_2	6.32
ri_2	4.47	rgi_1	3.6
ri_3	6.32	rgi_2	5.37

A full-fledged parametric analysis of the presented antenna has been followed in the HFSS simulator for the optimization of the respective length and width of the substrate. The detailed analysis of the presented antenna structure has been done in order to select the perfect dimensions of the antenna particularly of the substrate which is consisting of the material Rogers RT/duroid (5880) having a thickness of 0.787 mm and 0.0009 as dielectric loss tangent. The respective return loss value in dB and relative VSWR of the presented antenna for the different values of substrate dimensions are shown in Fig. 2a, b, respectively. The full parametric study has been followed and the finalized values are calculated, and the substrate dimension

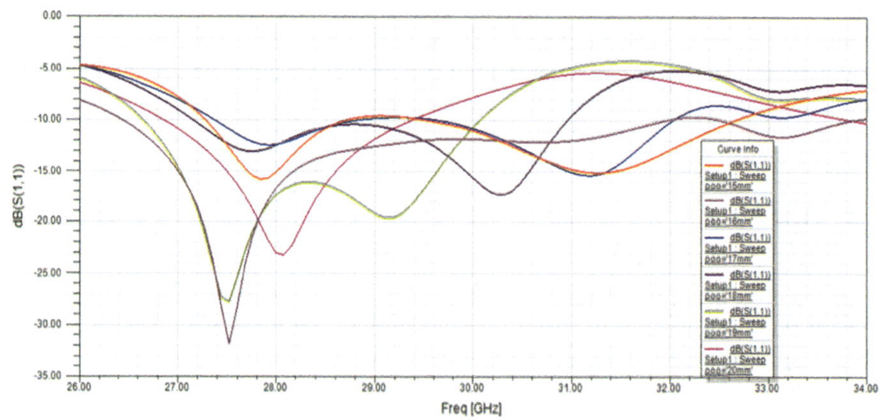

(a)

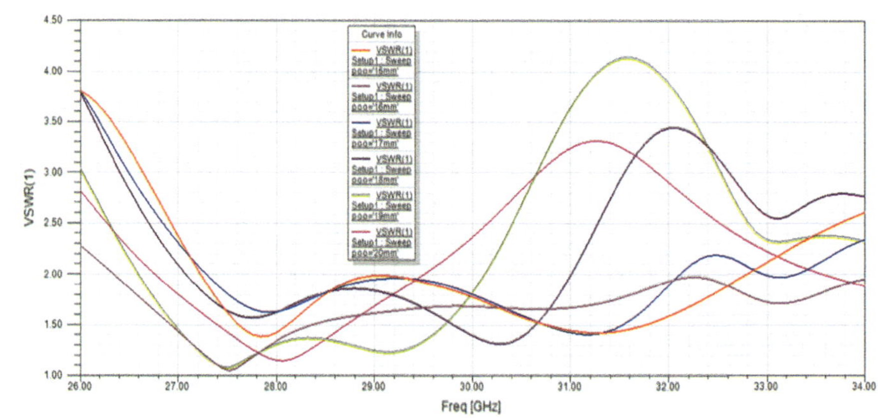

(b)

Fig. 2 Parametric study of the antenna for varied substrate dimensions for optimized values. (Results derived from the simulations carried in HFSS V.15). **a** Parametric study of return loss (dB) for varied substrate dimensions. (Results derived from the simulations carried in HFSS V.15). **b** Parametric study of VSWR for varied substrate dimensions. (Results derived from the simulations carried in HFSS V.15)

is optimized to 20 mm. Thus, it can be inferred from the respective figures that the antenna presented attains the best substrate dimensions at 20 mm due to which the antenna successes in attaining the resonating frequency of 28 GHz with perfect impedance matching, thereby attaining a gain of about 10 dBi. In addition to this, the antenna presented attains a dual-polarization ability which proves very effective in the millimeter-wave band for 5G mobile communication links.

Here, L_{mp} and W_{mp} denote the length and width of the radiating mini patch. The ro_1, ro_2, and ro_3 and ri_1, ri_2, and ri_3 denote the outer and inner radii of the concentric circular directors with center of the mini patch as origin, respectively. The length and width of the corresponding feed of mini patch are denoted by L_{f} and W_{f}. The side of defected square ground which is also the side dimension of full antenna is denoted by $S_{\mathrm{a}} = S_{\mathrm{g}}$. The S_{ico} represents the side dimension of the inner cut out of ground plane. The outer and inner radii of the circular directors on ground plane are denoted by rgo_1, rgo_2 and rgi_1, rgi_2, respectively.

2.2 Analysis

There are numerous plan methods for micro-strip antenna analysis. The cavity model, method of moments, transmission line model, and the method of FDTD are the frequently used models for the antenna system design. The transmission line model is the easiest method and provides a betterment of physical insight and has been applied for the proposed design as well. The equations from (1) to (5) are employed to carry on the designing of the proposed antenna.

$$\varepsilon_{r_{\mathrm{eff}}} = \frac{\varepsilon_r + 1}{2} + \frac{\varepsilon_r - 1}{2}\left[\frac{1}{\sqrt{1 + 12\frac{h}{w}}}\right] \tag{1}$$

$$L_{\mathrm{eff}} = L + 2\Delta L$$

$$\text{where } \Delta L = 0.412h\frac{\left(\varepsilon_{r_{\mathrm{eff}}} + 0.3\right)\left(\frac{w}{h} + 0.264\right)}{\left(\varepsilon_{r_{\mathrm{eff}}} - 0.258\right)\left(\frac{w}{h} + 0.8\right)} \tag{2}$$

For dominant TM_{010} mode f_r is

$$(f_r)_{010} = \frac{1}{2L\sqrt{\varepsilon_r}} * \frac{1}{\sqrt{\mu_0\varepsilon_0}} = \frac{\vartheta_0}{2L\sqrt{\varepsilon_r}} \tag{3}$$

By modifying Eq. (3), the fringing effect may be included and can be derived as

$$(f_{rc})_{010} = \frac{1}{2L_{\mathrm{eff}}\sqrt{\varepsilon_{r_{\mathrm{eff}}}}} * \frac{1}{\sqrt{\mu_0\varepsilon_0}} = \frac{1}{2(L + \Delta L)\sqrt{\varepsilon_{r_{\mathrm{eff}}}}} * \frac{1}{\sqrt{\mu_0\varepsilon_0}} = q\frac{\vartheta_0}{2L\sqrt{\varepsilon_r}}$$

$$q = \frac{(f_{rc})_{010}}{(f_r)_{010}} \tag{4}$$

Equation (4) denotes the length reduction factor and is called as fringe factor

$$W = \frac{1}{2f_r\sqrt{\mu_0\varepsilon_0}}\sqrt{\frac{2}{\varepsilon_r+1}} = \frac{\vartheta_0}{2f_r}\sqrt{\frac{2}{\varepsilon_r+1}} \tag{5}$$

Equation (5) corresponds to the width of the mini patch and obviously depends upon the resonant frequency and the relative permittivity of the material (Rogers RT/duroid in our case) employed for the construct profile of the antenna presented.

3 Antenna Simulation Results and Discussion

The design process and the simulation of the proposed millimeter-wave antenna have been accomplished with the help of electromagnetic simulator Ansys HFSS v.15. A full-fledged parametric analysis has been performed during the design steps for the optimized selection of particular values regarding the dimensions of the antenna and especially for the wideness of the respective circular directors and their corresponding inner and outer radii for the achievement of high directive gains, dual polarization, better radiation performance, and impedance bandwidth. When the antenna is subjected to an excitation with a frequency sweep of 26–34 GHz with 28 GHz as the central frequency, a response of the antenna is depicted for return loss and VSWR as demonstrated in Fig. 3a, b.

3.1 Return Loss, VSWR, and Impedance Bandwidth

Figure 3 shows the simulated S_{11} and VSWR values of the proposed antenna. It can be seen that the return loss value is lower than -10 dB and has achieved the level of -24.69 dB with a good impedance matching corresponding to the VSWR of 1.12. Also it can be depicted from the RL vs. frequency curve that the antenna proposed has a good impedance bandwidth of 29%.

3.2 Radiation Performance

Gain. The antenna proposed is further analyzed for the peak azimuthal and elevation gains which is depicted from Fig. 4a, b. From the respective figures, it can be understood that the presented antenna has almost equal *E* and *H* plane peak gain

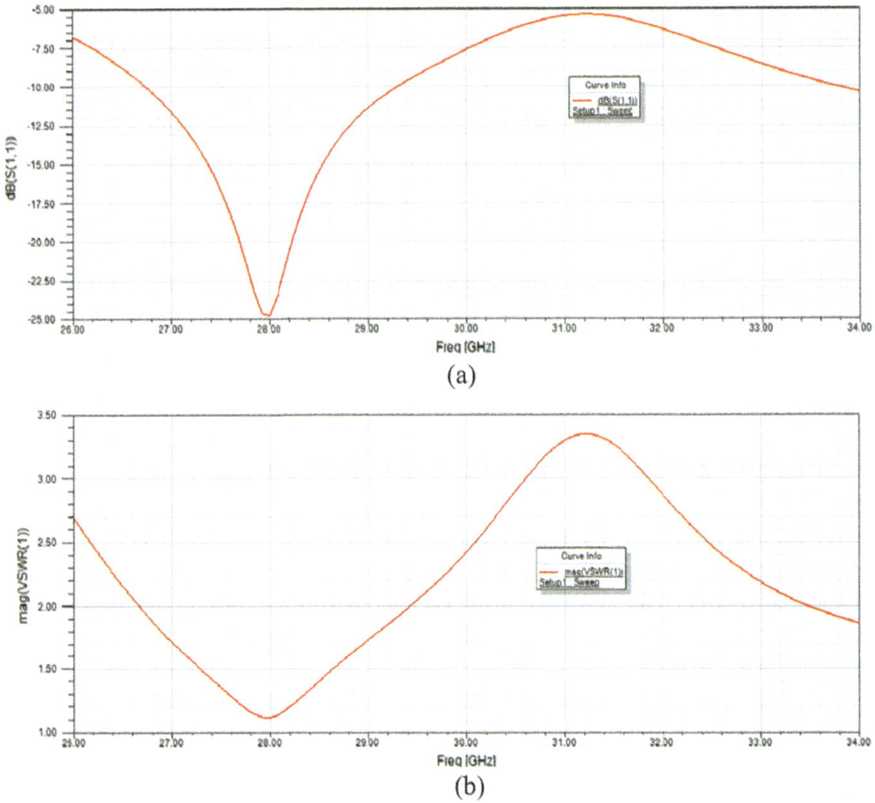

Fig. 3 Return loss and VSWR of the antenna presented (results derived from the simulations carried in HFSS V.15). **a** S_{11} of the antenna presented (results derived from the simulations carried in HFSS V.15). **b** VSWR of the antenna presented (results derived from the simulations carried in HFSS V.15)

corresponding to the varied phi values (0°, 180°, 201°, and 341°) and theta values (47°, 71°, 138°, and 313°). This particular property of the presented antenna makes it to have a dual-polarization state with a peak gain of almost 10 dBi which is much higher than a conventional patch antenna.

Radiation Pattern. The radiation pattern characteristic related with *E*-plane and *H*-plane of the antenna presented is illustrated in Fig. 5. (Results derived from the simulations carried in HFSS V.15).

From the radiation patterns corresponding to the *E* and *H* planes of the antenna, it can be presumed that the antenna proposed has an *E*-plane half-power beam width of 25° and an *H*-plane half-power beam width of about 32° for the respective phi and theta values. It can also be seen that the antenna has much higher gain and a better directivity as compared to other single element planar patch antennas.

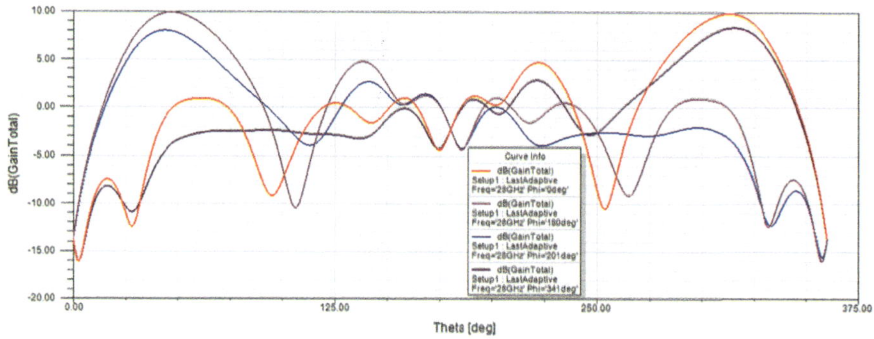

(a)

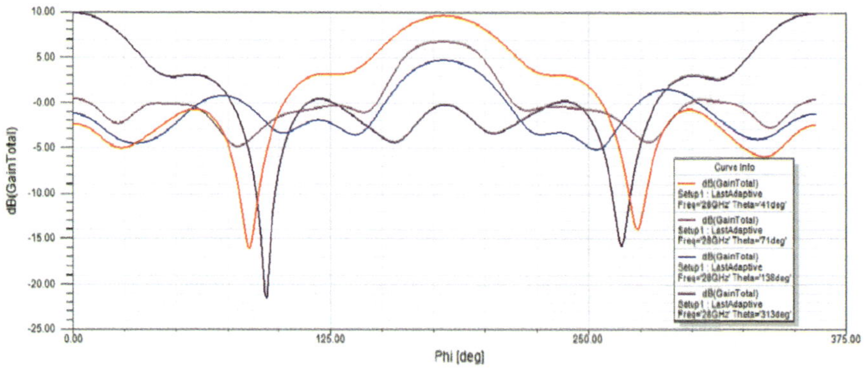

(b)

Fig. 4 Azimuthal and elevation gain of the presented antenna (results derived from the simulations carried in HFSS V.15). **a** Azimuth gain of the antenna presented at 28 GHz for Phi = 0°, 180°, 201°, and 341°. (Results derived from the simulations carried in HFSS V.15). **b** Elevation gain of the antenna presented at 28 GHz for Theta = 47°, 71°, 138°, and 313°. (Results derived from the simulations carried in HFSS V.15)

3D Polar Radiation Pattern and Surface current extent. Figure 6a, b shows the 3D pattern regarding far-field of the presented antenna for the individual sweeps of the phi and theta angles and the surface current distribution of the ground plane and the patch along with the concentric directors, respectively.

From the normalized 3D radiation pattern, it is inferred that the antenna has almost equal beam widths in addition to the respective gains regarding the main-lobe beams in *E* and *H* plane which signifies that the performance of the antenna is independent for its horizontal and vertical orientation. In addition to this, the polar radiation plot signifies that the presented antenna has a broadside radiation characteristic which makes it to have a wide-scan radiation ability. Also from the logarithmic surface current of the defected ground plane and the radiating mini patch surrounded by the circular directors, it is found that the surface current is distributed uniformly over

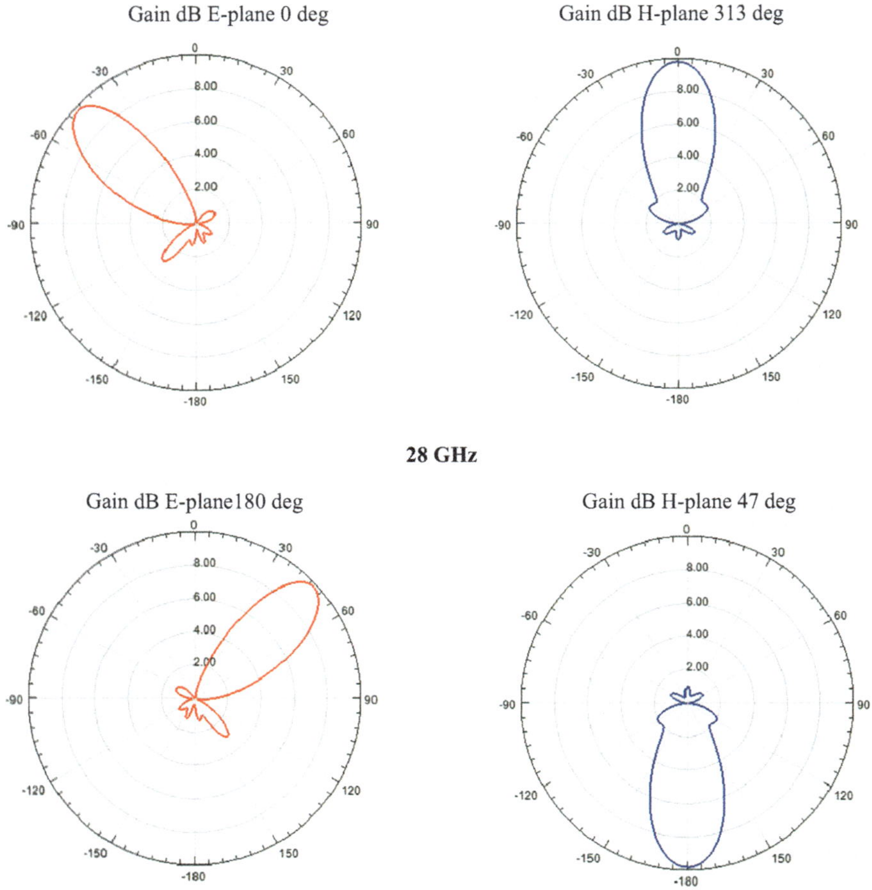

Fig. 5 Radiation patterns of the antenna presented. (Results derived from the simulations carried in HFSS V.15)

the whole micro-strip structure which makes the proposed antenna to have a good sensing capability regarding the reception of EM beams while its use as a receiving antenna.

4 State of Art Comparison

The contribution of this work is supported by a comparison made with previous reported antennas and is shown in Table 2. From the table, it can be inferred that the antenna presented is a low-cost planar geometry with much better impedance bandwidth and peak gain as compared to other reported antennas so far. The antenna proves out to be compact planar structure and easy to implement prototype with the

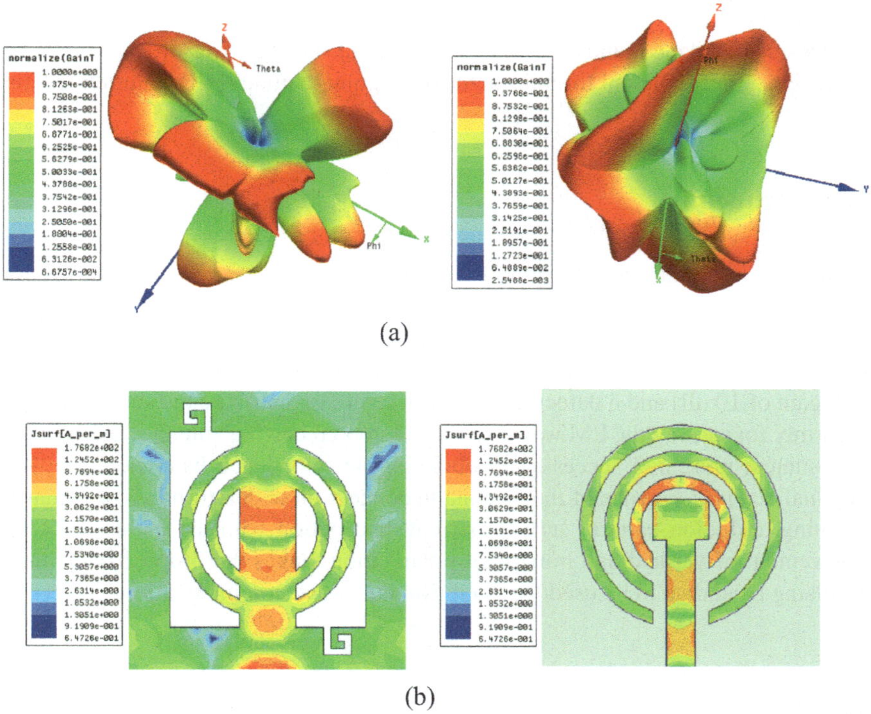

Fig. 6 Normalized 3D radiation pattern and the logarithmic surface current distribution. (Results derived from the simulations carried in HFSS V.15). **a** 3D radiation pattern of the antenna presented for separate sweeps of phi and theta angles. (Results derived from the simulations carried in HFSS V.15). **b** Surface current distribution of ground plane and mini patch along with the concentric circular directors. (Results derived from the simulations carried in HFSS V.15)

Table 2 Proposed antenna compared with the previous reported antennas

References/Year	Freq. band (GHz)	Impedance bandwidth	Isolation (dB)	Peak gain (dBi)	Type
[12]/2018	39–49.3	22.9%	–	5.5	SIW/reflector Omni
[13]/2018	25–30	–	16	11.3	Three-port ML antenna
[14]/2017	24–32	47%	–	9	SIW/tapered slot
[22]/2016	57–67	–	3.2	2.2	Loop micro-strip
This work	26–34	29%	–	10	Planar/concentric circular director

achievement of dual polarization and a high directive gain. The presented antenna achieves an impedance bandwidth of 29% with a gain having a peak of 10 dBi which is much higher and sufficient for millimeter-wave point to point (P2P) communication.

5 Conclusion

A low-cost high-gain concentric circular director antenna with dual polarization operating at the millimeter-wave frequency of 28 GHz was designed and simulated successfully. The antenna has a planar geometry consisting of a mini patch surrounded by the circular directors for the regeneration of E-field to achieve a high peak gain of 10 dBi and a defected ground structure to distribute the surface current uniformly for sensing the EM waves to act as a good receiving antenna, respectively. The antenna presented successes in obtaining an impedance bandwidth of 29%. The finalized performance of the antenna in connection with return loss, impedance matching, radiation pattern, surface current distribution with the inclusion of the achievement of vertical and horizontal polarization makes the proposed antenna a promising candidate to be used for 5G millimeter-wave communication.

References

1. Rappaport TS, Mayzus RH, Zhao S (2013) Millimeter-wave mobile communications for 5G: It will work! IEEE Access 1(1):335–349
2. Ayanoglu E, Swindlehurst AL, Heydari P, Capolino F (2014) Millimeter-wave massive MIMO: the next wireless revolution. IEEE Commun Mag 52(9):56–62
3. Kim Y, Lee H (2016) Feasibility of Mobile Cellular Com. at millimeter wave frequency. IEEE J Sel Top Signal Process 10(3):589–599
4. Roh W, Park J, Park JH, Seol JY (2014) Millimeter-wave beam-forming as an enabling tech. for 5G cellular commun.: theoretical feasibility & prototype results. IEEE Com Mag 52(2):106–113
5. Li M, Luk KM (2015) Wideband 60-GHz magneto-electric dipole ant. for mmWave communications. IEEE Trans Antennas Propa 63(7):3276–3279
6. Wang H, Fang DG, Zhang B, Che WQ (2010) Dielectric loaded substrate integrated waveguide H-plane horn antennas. IEEE Trans Antennas Propag 58(3):640–647
7. Yang TY, Hong W, Zhang Y (2014) Wideband millimeter-wave SIW cavity-backed rectangular patch antenna. IEEE Antennas Wirel Propa Lett 13(13):205–208
8. Zhang Y, Qing X, Chen ZN, Hong W (2011) Wideband mmWave substrate integrated waveguide slotted narrow-wall fed cavity antennas. IEEE Transcs Antennas Propag 59(5):1488–1496
9. Ghiotto A, Parment F, Wu K, Vuong TP (2016) Millimeter-wave air-filled substrate integrated waveguide anti-podal linearly tapered slot antenna. IEEE Anten Wirel Propag Lett 24(5):1–4
10. Wu K, Djerafi T (2012) Corrugated SIW antipodal linearly tapered slot antenna array fed by Quasi-triangular pow. Divider Propag Electrom Res 26(5):139–151
11. Antenna Patterns and Their Meaning, http://www.cisco.com/c/en/us/products/collateral/wireless/aironet-antennasaccessories/prodwhitepaper0900aecd806a1a3e.html. Last accessed 2019-12-21
12. Kuikui F (2018) Wideband horizontally polarized omni-directional antenna with a conical beam for millimeter-wave applications. IEEE Transa Antennas Propag 66(9):4437–4448

13. Wani Z, Mahesh P, Koul K (2019) Millimeter-wave antenna with wide-scan angle radiation characteristics for MIMO applications. Int J Radio Freq Microw Comput-Aided Eng 29(5):2–6
14. Yang B (2017) Compact tapered slot millimeter-wave antenna array for 5G massive MIMO systems. IEEE Trans Antennas Propag 65(12):6721–6727
15. Kumar A, Mahendra MS, Rajendra PY (2019) Dual wideband circular polarized CPW-fed strip and slots loaded compact square slot antenna for wireless and satellite applications. AEU-Int J Electron Commun 108:181–188
16. Tiwari RN, Singh P, Kanaujia BK, Barman PB (2019) Wideband monopole planar antenna with stepped ground plane for WLAN/WiMAX applications. In: Singh P, Paprzycki M, Bhargava B, Chhabra J, Kaushal N, Kumar Y (eds) FTNCT 2018, communications in computer and information science, vol 958. Springer, Singapore, pp 253–264
17. Zhou Z, Wei Z, Tang Z, Yin Y (2019) Design and analysis of a wideband multiple-microstrip dipole antenna with high isolation. IEEE Antennas Wirel Propag Lett 18(4):722–726
18. Tang MC, Li D, Chen X, Wang Y, Hu K, Ziolkowski RW (2019) Compact, wideband, planar filtenna with reconfigurable tri-polarization diversity. IEEE Trans Antennas Propag 67(8):5689–5694
19. Wang J, Lu WB, Liu ZG, Zhang AQ, Chen H (2019) Graphene-based microwave antennas with reconfigurable pattern. IEEE Trans Antennas Propag 68(4):2504–2510
20. Hussain S, Qu SW, Zhou WL, Zhang P, Yang S (2020) Design and fabrication of wideband dual-polarized dipole array for 5G wireless systems. IEEE Access 8:65155–65163
21. Wu GB, Zeng YS, Chan KF, Chen BJ, Qu SW, Chan CH (2020) High-gain filtering reflectarray antenna for millimeter-wave applications. IEEE Trans Antennas Propag 68(2):805–812
22. Ghazizadeh MH, Fakharzadeh M (2016) 60 GHz omni-directional segmented loop antenna. IEEE Int Symp Antennas Propag 1653–1654

Questionnaire-Based Prediction of Hypertension Using Machine Learning

Abhijat Chaturvedi, Siddharth Srivastava, Astha Rai, A. S. Cheema, Desham Chelimela, and Rajeev Aravindakshan

Abstract Machine learning has proven its ability in health care as an assisting technology for health care providers either by saving precious time or timely alerts or vitals monitoring. However, their application in real world is limited by availability of data. In this paper, we show that simple machine learning algorithms especially neural networks, if designed carefully, are extremely effective even with limited amount of data. Specifically with exhaustive experiments on standard Modified National Institute of Standards and Technology database (MNIST) dataset, we analyse the impact of various parameters for effective performance. Further, on a custom data set collected at a tertiary care hospital for hypertension analysis, we apply these design considerations to achieve better performance as compared to competitive baselines. On a real-world dataset of only a few hundred patients, we show the effectiveness of these design choices and report an accuracy of 75% in determining whether a patient suffers from hypertension.

Keyword Machine learning in health care neural network support vector machine random forest

A. Chaturvedi (✉) · S. Srivastava · A. Rai · A. S. Cheema
Centre for Development of Advanced Computing, Noida, India
e-mail: abhijatchaturvedi@cdac.in

S. Srivastava
e-mail: siddharthsrivastava@cdac.in

A. Rai
e-mail: asthar@cdac.in

A. S. Cheema
e-mail: ascheema@cdac.in

D. Chelimela · R. Aravindakshan
All India Institute of Medical Sciences, Mangalagiri, India
e-mail: c.desham@gmail.com

R. Aravindakshan
e-mail: rajeev.a@aiimsmangalagiri.edu.in

P. K. Singh et al. (eds.), *Evolving Technologies for Computing, Communication and Smart World*, Lecture Notes in Electrical Engineering 694,
https://doi.org/10.1007/978-981-15-7804-5_10

1 Introduction

Hypertension is one of the most common ailment in the Indian patients and the severe threat that it poses makes it important to identify it as early as possible in order to stop it from becoming incurable. Fourth National Family Health Survey reports more than 13% of men and more than 8% of women in early to middle ages are suffering from hypertension [1]. The scarcity of health care experts makes the situation even more grim.

In today's world, machine learning becomes an obvious choice to solve such problems. It has already influenced every aspect of health care with major progress reported in for diseases prediction [2], cancer classification [3], drug discovery [4], etc. However, with respect to the current work, these solutions have the following limitations (i) They work on extremely large data with approximately 40; 000 and above instances in the data,set [5] (ii) The data is mostly of non-Indian patients.

Studies indicate that there is disparity in health care with respect to race and ethnicity [6] and hence, it is yet to be established if the same set of techniques can generalize across them.

In this paper, we target the problem of prediction of hypertension. To diagnose hypertension in practice, a health care professional asks a series of questions to the patient and on the basis of answers, it is identified if a patient is suffering from hypertension or not. We follow a similar approach, where a predefined set of questions were asked to several patients and their response was recorded. In view of this, following are the contributions of this paper.

- We evaluate the impact of hyperparameter choices on the performance of neural networks. With exhaustive experiment on MNIST, we evaluate them and provide insights on their impact.
- We collect a data set for hypertension analysis and use the insights from the analysis to design a simple neural network providing better results than competitive baselines.

The rest of the paper is organized as follows. Section 2 contains related works, Sect. 3 discusses background of neural network, compared techniques and notations used in the paper. Section 4 contains designing of neural networks, Sect. 5 explains the experiments done with the models and data set. Section 6 contains results and Sect. 7 contains conclusion.

2 Related Works

Machine learning algorithms are used in predicting diseases [7]. Diabetes prediction is an example of diseases prediction using machine learning [8]. Feras Hatib et al. [9] used machine learning model for predicting hypertension using high-fidelity arterial waveforms. X Li et al. [10] in their paper used feature extraction from sequential data and contextual data using hidden layers of recurrent structure and predicting the blood pressure value. They have used root mean square error as the evaluation metrics for their model. The image-based prediction in health care machine learning is helping in predicting diseases like cancer [11, 12]. This work is not intended in image-based diseases classification, but for the sake of completeness and future progress in this work, image-based studies are also referred.

Chengyin Ye1 et al. [13] uses XGBoost for model creation in hypertension prediction. The data set they used contains 1,504,437 instances from 2013 and 2014. However, they classified the instances into five classes ranging from lowest risk to very high risk. They have used area under curve as the performance metrics. Mevlut Ture et. al. [14] presented a comparative study of machine learning algorithms for predicting hypertension. They have evaluated decision trees, statistical model, and neural network on a data set of 694 instances.

3 Background

In this work, we evaluate various machine learning algorithms for the task of hypertension prediction. Specifically, we design neural networks with varying hyperparameters and structure and provide a comprehensive analysis of impact of these parameters on the performance. Further, we also compare to other machine learning methods to compare and contrast the effectiveness of various paradigms for such prediction. Prior to dwelling in the core designs, we provide a brief background of the techniques used.

3.1 Neural Networks

Neural network is the machine learning algorithm that is designed to recognize patterns in the input data, learn from it, and create a weight matrix on the basis of the relationship among the input parameters. Neural network algorithm is inspired from the human brain and mimics the working of neurons. It has an input layer, some hidden layers and output layer. This architecture is shown in Fig. 1. A neuron in human brains is replaced by a node in neural network; these nodes are interconnected with each other with some weights. These weights are computed by the algorithm itself during the training phase.

The output in neural network is represented as the summation of product of all input values and their corresponding weights along with a bias passed through an activation function, as shown in the following equations:

$$z_{\text{intermediate}} = x_1 w_1 + x_2 w_2 + x_3 w_3 \;:::\; + x_n w_n + b \tag{1}$$

which can be written as,

$$z_{\text{intermediate}} = \sum_{i=1}^{n} (x_i w_i) + b \tag{2}$$

where $z_{\text{intermediate}}$ is the output..., n is the size of input vector, x_i represents the ith input, w_i is the weight of the ith layer, b is the bias.

Here, bias is always 1 and has its own weight value. It is introduced so as to make sure that neuron is activated even when all the inputs are 0. This intermediate output is then passed through an activation function to introduce non linearity in the network. The model has two phases during the training, viz. forward propagation and backward propagation.

During the forward propagation, the inputs are provided in the input layer of the network and an output is acquired after passing it through hidden layer(s) and activation function. This output value is the predicted value. The forward propagation ends here and backward propagation starts. To verify the correctness of the predicted value, we move backwards in the network, updating the weights so as to minimize the error in the predicted value. The minimization of error is ensured by loss function. One forward propagation and one backward propagation are combinedly called an as epoch.

REctified Linear Unit (relu). It is the activation function which gives output as the maximum value between 0 and the input value. It discards the negative input value and changes it to 0. Mathematically,

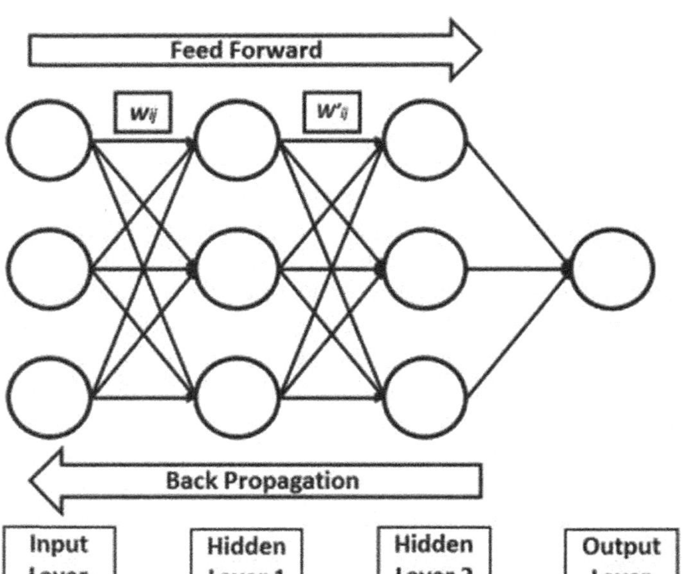

Fig. 1 A simple neural network

$$z_{out} = \max(0; z_{intermediate}) \tag{3}$$

It is mathematically simple; therefore, it is less resource intensive in implementation. Graphically, it is represented in Fig. 2.

Softmax. It is the activation function which is used in the output layer because of its ability to turn numbers into probabilities between 0 and 1. It is used in cases where more than one inputs are received and we need to check the maximum probability of occurrence among multiple inputs. Mathematically,

$$S(y_i) = \frac{e^{y_i}}{P_{j\, ey_j}} \tag{4}$$

where y_i is the ith input to the function. Graphically, it is represented in Fig. 3.

3.2 *Other Algorithms*

Support vector machine. Support vector machine is a supervised learning algorithm that classifies data on the basis of hyperplane that separates the data set. During training, the algorithm draws a hyperplane that linearly classifies the data into the number of classes. In case of binary classification, this plane is a line.

The SVM takes the labeled input as training data. This data is then used to create a plane that separates the data into labelled classes through the following process in case of binary classification:

- A linearly separable line is drawn between the classes.

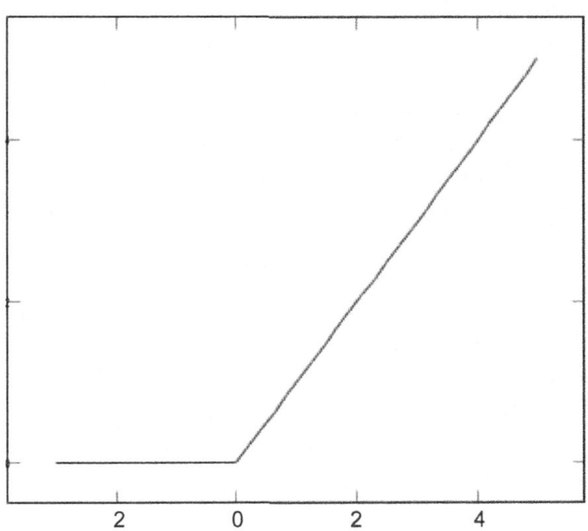

Fig. 2 ReLU activation function

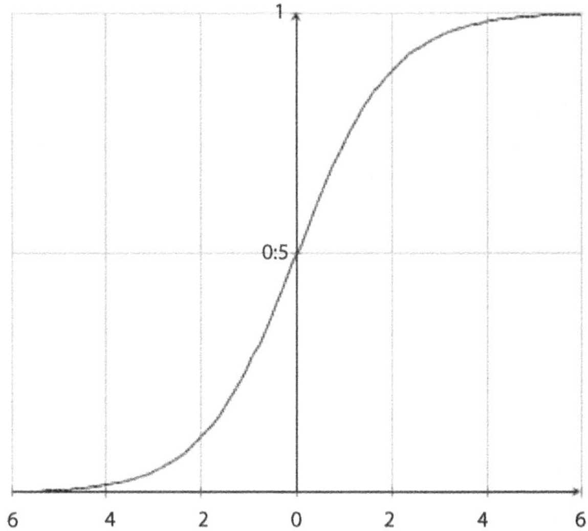

Fig. 3 Softmax activation function

- The nearest two points from the line are selected among both the classes. This is done by calculating distances among all the points and the line. They are called support vectors.
- The distance between support vectors and the plane is called margin. This margin is maximized.

The two hyperparameters in SVM are regularization and gamma. Regularization controls the smoothness of the plane. This translates as the margin of the hyperplane. More value of regularization tells the algorithm to learn every instance in the data set; therefore, smaller hyperplane. However, high value of regularization also risks overfitting. Gamma defines the effect of the points that may cast influence over the hyperplane. Higher value of gamma the points closer to the hyperplane are taken into consideration whereas in low gamma value, the points far from the hyperplane is taken into calculations.

Random forest. Random forest is the classification algorithm that works on the principle of wisdom of crowd. The random forest relays on the fact that unrelated decision trees when combined produce better results. Decision trees are constructed as flowcharts and predict a final outcome. Random forest has two hyperparameters, number of estimators, and maximum depth. Number of estimators is the number of trees that the model is allowed to create. More the number of estimators more is the chance of higher accuracy but also is more computationally expensive and increases the risk of overfitting. Maximum depth represents the depth of each tree in the random forest. More the depth captures more data in the training instances. However it has to be taken into consideration the fact that higher depth also risks overfitting.

4 Designing a Neural Network

Neural network contains multiple hyperparameters, like number of layers, epochs number of neurons in each layer, activation function, etc. The balanced combination of all hyperparameters ensures a good neural network that identifies the patters in the data and predicts the outcomes in the new data smoothly. Hyperparameter tuning is the most challenging task in any neural network design. Masanori Suganuma et al. [15] have demonstrated the use of Cartesian genetic programming (CGP) in automatically designing a convolutional neural network. Dougal Maclaurin et al. [16] use the stochastic gradient descent to derive hyperparameters.

How to choose layers in a neural network? Less number of layers will underfit the model and more number of layers will overfit the model. Increasing the number of layers is also computationally expensive as more number of multiplication and addition will be required to get the output. However, the number of layers and the number of neurons vary greatly from problem to problem as explained by Imran Shafi et al. [17].

What is the impact of neurons? Number of neurons in the input layer is normally equal to the parameters in the training dataset. Neurons in hidden layers varies depending upon the learning required. Whereas output layer contains the number of neurons equal to the class variables.

What is the impact of activation function? Activation function is important in moulding the output in correct shape and induce non-linearity in the network. Without activation functions, the output of the neural network will be linear and thus a liner regression. There are multiple activation functions available such as Sigmoid Function, TanH Function, Rectified Linear Unit (ReLU), Leaky ReLU, and Softmax. Evaluation of multiple activation functions is also done by researchers [18].

Impact of Hyperparameters. Hyperparameter tuning is vital to get the meaningful predictions from the model. The graphs show the relationship between accuracy and batch size, number of layers, and number of neurons.

Figure 4a shows the impact of batch size on accuracy on open-source data set. The graphs show initial increase with decrease in batch size which depicts the learning of the model. 128 batch size makes the model reaching an accuracy of more than 88%, while 16 batch size shows an accuracy of more than 91%. Upon reaching a saturation value, the accuracy flattens and if continued from here the risk of overfitting arises.

Figure 4a shows the relationship between accuracy and number of neurons. More number of neurons increases the accuracy where 32 neurons have an accuracy of more than 93% and 256 neurons reach an accuracy of more than 96%.

Figure 4b shows the impact of number of layers on accuracy. The graphs show a decrease in accuracy with increase in number of layers with some fluctuations. However, if this continues further, the accuracy flattens and condition of overfitting arises.

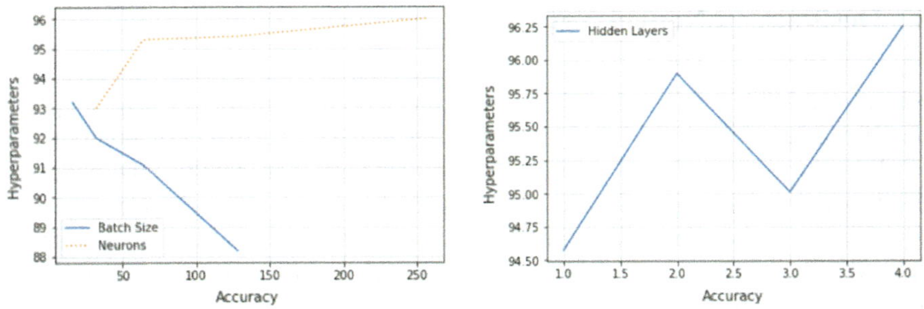

(a) Impact on Accuracy with batch size

(b) Impact on Accuracy with number of neurons

Fig. 4 Impact of **a** batch size and number of neurons in the hidden layers and **b** number of layers on the performance of neural network with MNIST data set

Table 1 Summary of data set for hypertension analysis

	Number	Percentage
Positive instances	134	25.38
Negative instances	394	74.64
Total instances	528	100

5 Experiments

5.1 Data set

The data is collected from a tertiary care hospital in India. A set of questions was prepared and the data was collected from the patients. The data set consists of 528 instances with 141 variables (such as age, religion, education, etc.). The data set contains a mix of continuous and categorical values. Variables such as age and family income are continuous values and variables such as family type and housing type are categorical values. Much of the data is in numeric form except two variables (Table 1).

Exploratory data analysis in the provided data shows the following trends with respect to age, alcohol consumption, and average working hours (Table 2).

5.2 Data Cleaning

- During the analysis of the data, dependent and derived parameters were removed.
- Parameters whose value remain unchanged were also dropped from the training data.

Table 2 Statistics of age, alcohol consumption, and working hours in the hypertension dataset

Age		
	Age > 35	Age < 35
Positive instances (%)	32.02	8.16
Negative instances (%)	67.97	91.83
Alcohol		
	Yes	No
Positive instances (%)	29.96	24.66
Negative instances (%)	70.73	75.33
Working hours		
	WH > 8	WH < 8
Positive instances (%)	23.13	26.14
Negative instances (%)	76.86	73.85

- The string value columns were label encoded.

After data cleaning, 124 variables were left. The class variable is BPgroup which is categorical (0 or 1). The total positive instances are 134, and total negative instances are 394.

5.3 *Implementation Details*

Input representation. The input vector consists of all the parameters along with their valid values. The problem is a binary classification problem where the output from the output layer will be 0 or 1, where 0 represents that patient is not suffering and 1 represents that patient is suffering from hypertension. The input vector is represented by,

$$V = (v_1;\ v_2;\ v_3\ :::\ v_{i=124}) \tag{5}$$

where v_i is the ith parameter of the input vector. The total size of input vector is 124; hence, the last parameter is $v_i = 124$.

Architectures Table 3 shows the different machine learning algorithms that are evaluated viz. Neural network, support vector machine, and random forest along with the parameters used for reporting the results in the paper. Further, we vary the number of hidden layers and neurons in the neural networks while discuss ablation studies on varying other parameters in Sect. 6.

Table 3 Model parameters for various algorithms

Neural network 1 (NN_1)	Batch size = 124
	Epochs = 45
	Layers = 2 with 50 neurons each
	Activation functions = relu
Neural network 2 (NN_2)	Batch size = 124
	Epochs = 45
	Layers = 3 with 50 neurons each
	Activation functions = relu
Neural network 3 (NN_3)	Batch size = 124
	Epochs = 45
	Layers = 4 with 50 neurons each
	Activation functions = relu
Neural network 4 (NN_4)	Batch size = 124
	Epochs = 45
	Layers = 5 with 50 neurons each
	Activation functions = relu
Support vector machine	Train test split ratio = 60:40
Random forest 1 (RF_1)	Number of estimator = 10
Random forest 2 (RF_2)	Number of estimator 20
Random forest 3 (RF_3)	Number of estimator = 30

5.4 Metrics

Accuracy is the fraction of the values predicted by the model that are correct.

$$\text{Accuracy} = \frac{\text{TP} + \text{TN}}{\text{TP} + \text{FP} + \text{FN} + \text{TN}} \tag{6}$$

Precision. The fraction of the correct positive prediction against the overall predictions is called as precision.

$$\text{Precision} = \frac{\text{TP}}{\text{TP} + \text{FP}} \tag{7}$$

Recall (Sensitivity). The fraction of correct positive prediction against overall observation in actual class is termed as recall.

$$\text{Recall} = \frac{\text{TP}}{\text{TP} + \text{FP}} \tag{8}$$

Table 4 Confusion matrix

	Predicted NO	Predicted YES
Actual NO	TN	FP
Actual YES	FN	TP

where TP is true positive, TN is true negative, FP is false positive, and FN is false negative

Confusion matrix. Confusion matrix is a method of evaluation of the performance of a machine learning model. Confusion matrix is the table of predicted values against the actual values (Table 4).

6 Results

The neural network has maximum accuracy of 74.52% (Neural Network 2) followed by random forest at 69.81%. However, recall is maximum of random forest at 26.31. The graphs in Fig. 5 shows the results where accuracy is plotted against epochs under varying hyperparameters. Figure 5a, b shows overfitting after 100 epochs. Figure 5 shows overfitting is a problem to tackle. For this, a regularisation technique dropout [19] is used in the model. Figure 5g, h shows the impact of dropout (Table 5).

As the results shows, neural network has outperformed both the algorithms. Neural network 2, which has an accuracy of more than 75% is the best performing model. However, recall is low in all cases that shows the weak sensitivity of the model. Increasing the recall remains the challenge in this data set. Random forest also performed with more than 70% accuracy. However, random forest outperformed both the algorithms in terms of recall and precision. Random forest 2 reaches a recall of more than 26.

7 Conclusion

We have studied the impact of various hyper-parameters on the performance of neural networks. The analysis was performed on MNIST data set for analysing the impact while the findings have been used further on a custom data set for hypertension prediction. We showed that while medical studies require large data sets, it is possible to provide reasonable performance with simple neural networks which can also generalize well.

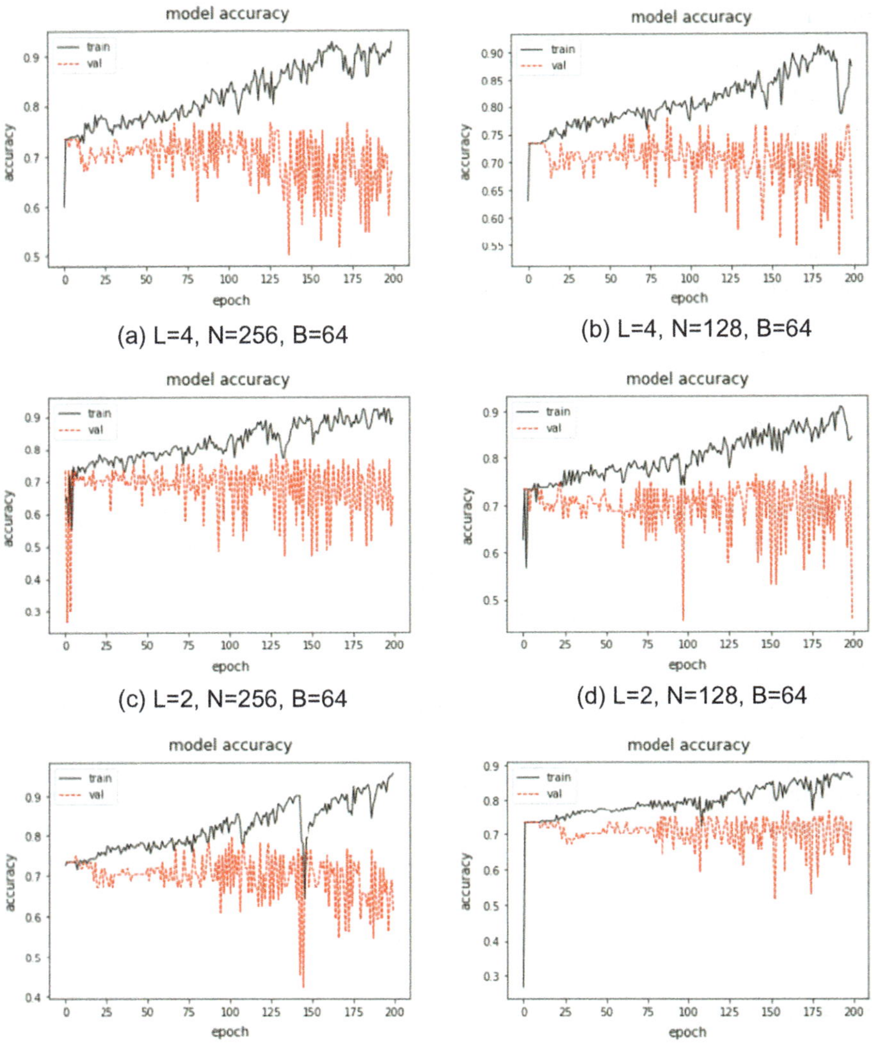

Fig. 5 Ablation study: performance of the network on varying batch size, number of layers, and number of neurons for predicting hypertension. Figure 5h one dropout layer is included and in Fig. 5g, two dropout layers are included in the model

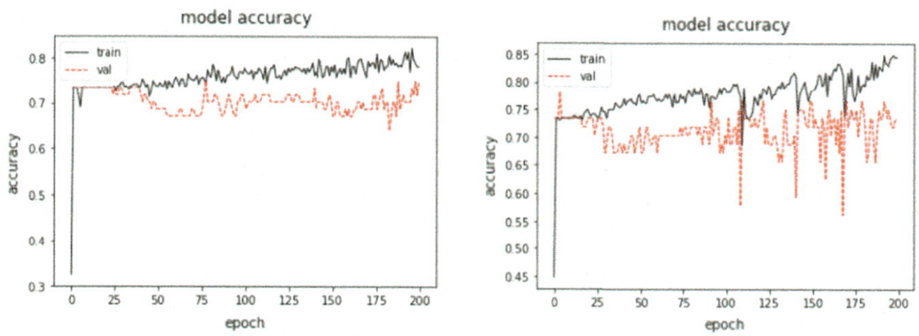

(g) L=4, N=256, B=256, D=2

(h) L=4, N=256, B=256, D=1

Fig. 5 (continued)

Table 5 Results on hypertension data set using neural networks (NN_i), SVM, and random forests (RF_i)

Algorithm	Accuracy	Precision	Recall
NN_1	73.00	55.00	8.7
NN_2	74.52	60.00	15.70
NN_3	72.16	25.00	1.70
NN_4	72.16	25.00	1.70
SVM	73.11	NA	0
RF_1	69.33	38.00	24.00
RF_2	69.81	40.54	26.31
RF_3	68.86	37.83	24.56

References

1. Gupta R, Gaur K, Ram CVS (2019) Emerging trends in hypertension epidemiology in india. J Hum Hypertens 33(8):575–587
2. Chen M, Hao Y, Hwang K, Wang L, Wang L (2017) Disease prediction by machine learning over big data from healthcare communities. Ieee Access 5:8869–8879
3. Tan AC, Gilbert D (2003) Ensemble machine learning on gene expression data for cancer classification
4. Lavecchia A (2015) Machine-learning approaches in drug discovery: methods and applications. Drug Discovery Today 20(3):318–331
5. Miotto R, Li L, Kidd BA, Dudley JT (2016) Deep patient: an unsupervised representation to predict the future of patients from the electronic health records. Sci Rep 6:26094
6. Good MD, James C, Good BJ, Becker AE (2005) The culture of medicine and racial, ethnic, and class disparities in healthcare. In: The Blackwell Companion to social inequalities. Blackwell Publishing, Ltd., Malden, MA, pp 396–423
7. Srivastava S, Soman S, Rai A (2019) An online learning approach for dengue fever classification. arXiv preprint arXiv:1904.08092
8. Kumari VA, Chitra R (2013) Classification of diabetes disease using support vector machine. Int J Eng Res Appl 3(2):1797–1801

9. Hatib F, Jian Z, Buddi S, Lee C, Settels J, Sibert K, Rinehart J, Can-nesson M (2018) Machine-learning algorithm to predict hypotension based on high-fidelity arterial pressure waveform analysis. Anesthesiol: J Am Soc Anesthesiol 129(4):663–674
10. Li X, Wu S, Wang L (2017) Blood pressure prediction via recurrent models with contextual layer. In: Proceedings of the 26th international conference on World Wide Web. International World Wide Web conferences steering committee, 2017, pp 685–693
11. Giger ML (2018) Machine learning in medical imaging. J Am Coll Radiol 15(3):512–520
12. Cruz JA, Wishart DS (2006) Applications of machine learning in cancer prediction and prognosis. Cancer inf 2:117693510600200030
13. Ye C, Fu T, Hao S, Zhang Y, Wang O, Jin B, Xia M, Liu M, Zhou X, Wu Q et al (2018) Prediction of incident hypertension within the next year: prospective study using statewide electronic health records and machine learning. J med Internet Res 20(1):e22
14. Ture M, Kurt I, Kurum AT, Ozdamar K (2005) Comparing classification techniques for predicting essential hypertension. Expert Syst Appl 29(3):583–588
15. Suganuma M, Shirakawa S, Nagao T (2017) A genetic programming approach to designing convolutional neural network architectures. In: Proceedings of the genetic and evolutionary computation conference. ACM, 2017, pp 497–504
16. Maclaurin D, Duvenaud D, Adams R (2015) Gradient-based hyperparameter optimization through reversible learning. In: International conference on machine learning, 2015, pp 2113–2122
17. Shafi I, Ahmad J, Shah SI, Kashif FM (2006) Impact of varying neurons and hidden layers in neural network architecture for a time frequency application. In: 2006 IEEE international multitopic conference. IEEE, 2006, pp 188–193
18. Karlik B, Olgac AV (2011) Performance analysis of various activation functions in generalized mlp architectures of neural networks. Int J Artif Intell Expert Syst 1(4):111–122
19. Srivastava N, Hinton G, Krizhevsky A, Sutskever I, Salakhutdinov R (2014) Dropout: a simple way to prevent neural networks from overfitting. J Mach Learn Res 15(1):1929–1958

Nature-Inspired Computing Behaviour of Cellular Automata

Manisha Ghosh, Pritisha Sarkar, and Mousumi Saha

Abstract Cellular automata are apt to realize a mathematical model for studying and analyzing the behaviour of any complex and sophisticated natural phenomenon. Each element of a natural system generally changes its action or behaviour based on its surroundings and its own state. Cellular automata consist of cells where each cell evolves depending on its neighbouring cells and a predefined logic with discrete time and space. Hence, the cells in cellular automata can be used to model elements of a natural phenomenon, and several physical and biological events of a natural system. John von Neumann's proposed model of self-reproducing cellular automata tried to reduce the gap between man-made automata and natural system. On the other side, the simplified structure of Wolfram's elementary cellular automata described the logic behind the apparently chaotic behaviour of natural systems. The concept of Game of Life developed by John Conway explained the biological processes, like birth, death, population, etc., of various living organisms such as fishes in the sea, bacteria in food or animal body, and growth of infected blood cells in the human body. This close correlation between nature and theory of cellular automata always inspires scientists and researchers to orientate their research interest towards this field. This paper focuses on the nature-inspired complex computing with cellular automata, established by famous scientists. It also shows how modern age scientists effectively utilize those computing methods to bring innovations in science and technology.

Keywords Cellular automata · Game of Life · Elementary cellular automata · Rule mean term

M. Ghosh
NSHM Knowledge Campus, Durgapur, India
e-mail: manishaghosh.uit@gmail.com

P. Sarkar · M. Saha (✉)
National Institute of Technology Durgapur, Durgapur, India
e-mail: msaha.nitd@gmail.com

P. Sarkar
e-mail: pritishapixie@gmail.com

P. K. Singh et al. (eds.), *Evolving Technologies for Computing, Communication and Smart World*, Lecture Notes in Electrical Engineering 694,
https://doi.org/10.1007/978-981-15-7804-5_11

1 Introduction

A natural system can be characterized by its behaviour, and it can be related to any physical, chemical or biological occurrences which mostly have a stochastic or chaotic manifestation. Apparently a particular phenomenon seems to be confusing and random, but inherently there exists a systematic logic which represents a specific pattern or structure. The natural computational models or techniques are inspired by natural systems. The bio-inspired classical models of computations are cellular automata, neural computation and evolutionary computation. Deep relationship between nature and computation stimulates the research on nature-inspired computing from the perspective of information processing. It is rapidly taking its pace of importance as a fundamental research in computing world.

Cellular automaton (CA) is chosen to be one of the oldest models of nature-inspired computing. A cellular automaton represents a lattice assembly of cells. Cells are basic components (units) in a system. The cellular framework can be single to multidimensional. The cells are in a specific neighbourhood relationship with each other. Each cell in a grid is found to be in a specific state at any instant of time. The state can be changed or remains same in the very next instant depending on the present state of its neighbouring cells, and its transition logic is carried out by specific predefined rules. The transition rules can vary for different models. To design a cellular automata (CA)-based model for a complex system, we collect information from the observations that helps to find the transition logic. Dependency of the cells in a system measures the neighbourhood and dimension. Each cell in cellular automata (CA) interacts with other local adjacent cells thereby creating a global distributed environment. The modular and distributed (decentralized) structure of cellular automata (CA) inspires the researchers and scientists to realize several natural phenomenon.

Section 2 describes the related work from von Neumann's self reproduction CA to Wolfram's modern structure of CA. Section 3 introduces the basic concept of cellular automata. Section 4 introduces the Wolfram's elementary cellular automata model with its scope of work. Finally conclusion is given in Sect. 5.

2 Natural Computation and Cellular Automata

In this paper, we would like to discuss some classical and modern natural computational models developed with the help of cellular automata by researchers till date. As the spectrum of different models and its applications are vast, it becomes difficult to include all of them in this work. However, we aim to provide an overview of the topic in the contest of nature-inspired computing and focus an application on fault-tolerant memory test logic design.

2.1 Early Stage of Research—During 1940s

Invention of cellular automata has a significant historical background starting since early 1940s. The first cellular automata model was established by John von Neumann. Von Neumann together worked with Stanislaw Ulam and proposed the architecture of a self-reproducing cellular automata model in 1940s [1]. The invention was totally inspired by the natural process of reproduction of cells in an organ of a living being. Von Neumann and Ulam used two-dimensional cellular automaton to represent their model. He chose four orthogonal nearest cells to be the neighbours of the particular cell to be reproduced. It is called von Neumann's neighbourhood. He defined 29 states at any instant for each cell present in that model. Von Neumann's work was first shown in the University of Illinois in December 1949 [1]. Arthur W. Burks in 1966 wrote the John von Neumann's posthumous book where one can get every conceptual overview regarding that model.

During the period of 1960–1965, we find the modifications of von Neumann's model and different conceptual interpretation in CA by many well-known scientists. Renato Nobili provided additional states in von Neumann's model to represent a memory organ without any input as well as an interference free signal crossing organ of two inputs. Edward Fredkin introduced the reversible computation concept with cellular automata (CA). Moore's invention of finite state machine was extensively used in artificial life and cellular automata (CA). He introduced Moore's neighbourhood in cellular automata (CA) which is an eight-neighbourhood cellular automata [2]. Codd provided some additional and innovative contributions to self-replication of cell using cellular automata (CA) model and universal computing [3]. He simplified the John von Neumann's model by reducing the number of states from 29 to 8 and proposed a modified model of self-reproducing cellular automata (CA) machine.

2.2 Innovation—During 1970s

In late 1960s (1968), we found a mathematically computing CA-based game referred as "Game of Life" [4, 5] by John Conway. It is primarily a combination of cellular automata and computer simulation games. Conway basically studied the phenomenon for both of the rapid growth and decay in population of different living beings and developed some mathematical rules to describe the fundamental logic and pattern behind that changes in population. He developed some logic that can be used to describe different biological processes like growth in fish generation in sea, growth of leaf in tree, etc. Even it can be applied in developing ecological society [6] which is called "symbiopoiesis". He has taken Moore's neighbourhood and a two-dimensional cellular automata (CA) for his mathematical gaming structure and corresponding state transition rules. His application is not limited to living beings. It has been applied in developing logical circuits in digital domain, in developing urban areas, smart cities, etc. Conway's Game of Life rules applied in a two-dimensional grid can generate different patterns. Some of them have the capability to represent the "Turing Machine" simulation which deals with universal computing methods. Later Martin Gardner wrote a column on Game of Life in the "Scientific American".

2.3 *Recent Trends in Computation with Cellular Automata*

Stephen Wolfram in 1983 invented a CA model of one-dimensional cellular automata [7]. In 2002, his published book named A New Kind of Science gives a detail idea of his developed rules and applied models. He used three-neighbourhood model of one-dimensional cellular automata (CA) which was referred as elementary cellular automata (ECA). We found the use of quantum-dot CA in different computational domain [8]. Cellular automata can also be applied in detecting intruder's attack in our secured network systems [9]. Recently, cellular automata concept is being rapidly utilized for pattern recognition and classification [10].

Cellular automata-based models can detect leukaemia-affected blood cells in an organ [11]. Automated instructor can be modelled using cellular automata for drivers to manage driving in busy traffic areas and properly parked cars in a congested parking place [12, 13]. Recent innovations in cellular automata are observed related to very large-scale IC design and malicious function detection, hardware design for high-speed memory testing and its fault diagnosis using built-in-self-testable hardware architecture [14, 15]. We also found a huge application of programmable cellular automata (PCA) [16], used in developing more secured algorithms for cryptography [17].

3 Preliminaries of Cellular Automata (CA)

A cellular automaton consists of a simple array of cells where each cell carries a finite number of states in a cellular space. The next state of a cell is always obtained from the present state of that cell and its neighbouring cells. Specific rules are there for the next state calculation [7, 9, 18]. Each cell in a CA locally interacts with neighbours based on simple rules, but globally, CA shows a complex phenomenon [19].

Figure 1 describes a functional block of cellular automata where the general working principle of a cellular automaton can easily be explained. Let us consider an array of N-cells present in cellular automata module. Consider a cell with all its neighbouring cells. The next state of the cell is calculated using the specific rules applied to the present state of the cell and that of other neighbouring cells. Different logical and mathematical concepts can be used to develop any state transition rule.

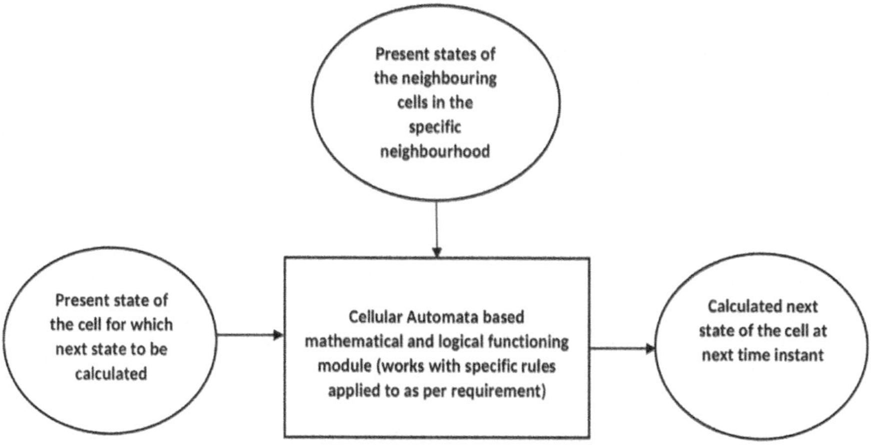

Fig. 1 Functional block diagram of a CA machine

3.1 Characterization of CA

Cellular automata can be characterized based on following features:

1. Dimension: The cellular automata model can be structured in different dimensions. Commonly it is applied with one dimension, two dimensions and three dimensions. Von Neumann [1] and Conway [4] constructed their cell recreating automata models in two-dimensional CA whereas Stephen Wolfram [7] applied his model in one-dimensional system.
2. State: The state of a cell in cellular automata is defined differently by different scientist as per the required properties of their models. But the number of states at which a cell can stay at any instant is always finite. So, cellular automaton is also named as a finite state machine. John von Neumann used 29 states [1] in his CA model to describe all the characteristics of a living organism through his artificial replica. Nobili modified Neumann's CA with 32 states. Conway described four states [4] of a cell in his Game of Life. Wolfram used only two states, i.e., Boolean logic "1" and "0" to de ne his elementary cellular automata.
3. Neighbourhood: The neighbouring cells are always important for a cell to obtain its next generation. The next state of a cell not only depends on its present state but also depends on the present states of its specific neighbours. There are some neighbourhood characteristics defined by many scientists. The neighbours are chosen depending on those neighbourhoods conditions. Von Neumann described five-neighbourhood model [1] where cell interacts with its four immediate orthogonal positioned neighbours. Moore described night neighbourhood mode [2] where cell interacts with eight adjacent neighbours. Wolfram simplified neighbourhood dependence in three-neighbourhood [7] in the same dimension where the cell only depends on its nearest left and right neighbour.

4. Boundary condition: Boundary condition is important because the cells which are always at the boundary positions in a cellular automata model cannot get their right, left, lower or upper neighbours to interact. For this, those cells face difficulties in calculating their next states. To make computation easy in a one-dimensional elementary cellular automata, the left space of a leftmost cell and right space of a rightmost cell are considered to have null value (0 state). This type of cellular automata is called Null Boundary CA [20]. In elementary cellular automata, sometimes it is also found that the leftmost cell becomes the right positioned neighbour of rightmost cell and rightmost cell becomes the left positioned neighbour of the leftmost cell. Together it shows a circular-like structure. This type of cellular automata is called periodic boundary CA [18].

3.2 Scope of CA in Complex Computation

There are some basic properties of cellular automata. Those properties make cellular automaton suitable for designing a complex system based on natural computing methodology. Those properties enhance the scope of utilization of cellular automata in the recent trends of research.

1. Cellular automaton [21] supports the architecture of distributed computing. The cells in a cellular automaton are connected and they cannot be separated, but they can perform their functions independently. If any of the cells becomes damaged and inactive, rest of those can continue to function and ultimately can give output. So, the overall performance of the system remains unaffected. This helps to incorporate the fault tolerance, self-recovery and self-reconstruction features of nature in cellular automata-based system.
2. Parallel computation can be achieved by CA modelling. As the cells in cellular automata work parallel, different processes can be executed simultaneously at the same time in the same cellular automata machine. This feature of cellular automata inspires researchers to design a high speed and an efficient multiprocessing system.
3. The modular and regular organization of cellular automaton enhances the scalability of a system. A small system can easily be modified to large scale. A model of 4-bit cellular automata machine can be enlarged to 64-bit machine by adding 60 cells and applying the transition logic. It may increase the computation time of the system. This can overcome using segmentation of cellular automata approach [15].
4. Cellular automaton has its large variations in dimensions, neighbourhood, states and boundary conditions. This diversity in nature brings a variety in the functionality of a CA-based complex system [18, 20]. Even new features can be implemented to the existing models to make them more sophisticated and efficient.

Cellular automata have significant applications that explain many natural processes. With the applications of cellular automata by simulation in computers, many areas of science and technology start developing. Those domains are like VLSI, complex hardware design and testing, high accuracy memory chip and multiprocessor IC design and testing, network security, cryptography, image processing, pattern recognition and classification, etc.

4 Computing Behaviour of Wolfram's Elementary CA

In the year 1985, scientist Stephen Wolfram simplified the structure of cellular automata (CA). Most of the complexities raised in cellular automata computation are related to dimensions, neighbours and state characteristics. He introduced one-dimensional grid cellular automata with three-neighbourhood and two states. It is well known to us as the elementary cellular automata (ECA) [7, 9]. Here, the next state of one particular cell in the grid only depends on states of the cell itself and that of its left and right neighbours. Rules/logic defined by this simplified structure of cellular automata are also very simple. Two-state cellular automata means at any instant one cell can be either in binary value "1" (logic "1" state) or binary value "0" (logic "0" state).

As Wolfram used binary platform and minimized the state variations, computation became easier. Considering the present states of a cell and its left and right neighbours, the next state of a cell is generated. Using all the possible next state bit combinations, specific Boolean algebraic equation is modelled using Karnaugh map (K-map). Each of such bit patterns represents one rule of elementary cellular automata. So, the rule decides the next state of a particular cell. Rules are specified by a rule number (the number is in decimal form) like Rule 38, Rule 112, etc. Wolfram also demonstrated pictorial models of the rules. He used the binary bit pattern of the rule number to generate Boolean function and applied this to create every next generations of cell and its neighbours. Finally, a geometrical shape is formed using one-dimensional array of cells with their present states and next states. We can see different variations in those designs of rules.

4.1 Elementary CA Modelling

Let us consider one-dimensional three-neighbourhood elementary cellular automata. We know the number of possible states are 1, 0, i.e. two for each cell. Let us assume N number of consecutive cells in a cellular grid and those are taken into consideration for which next state to be generated. Those cells at the same time act as left and right neighbours to each other. We can then get $S = 2^N$ number of binary bit patterns (possible states) in the present states of the set of N cells. Each bit pattern refers to a decimal number which is called rule mean term (RMT). For every present state bit patterns, we can have $T = 2^S$ number of binary patterns in the next state of a cell. The binary bit pattern in T is converted to decimal number. That decimal number represents the state transition rule. All such rules can be applied to a cell and its neighbours to generate their next state [15, 22]. Stephen Wolfram in his Wolfram code has taken one-dimensional three-neighbourhood cellular automata model where he considered a cell only with its very next left and right cells as neighbours [23].

Say one cell is C whose next state to be calculated. For this, the present state of the cell C and its immediate left and right neighbouring cells are to be considered to obtain the next state of the cell C. Let us consider the present state for cell is x_k and for left and right neighbours are x_{k-1} and x_{k+1}, respectively. So, now only three consecutive cells are taken into consideration in the cellular grid. The state variation is binary 1, 0. That means there are $2^3 = 8$ possible bit patterns in the present state (x_{k-1}, x_k, x_{k+1}) of the three cells. Eight such combinations generate $2^8 = 256$ bit patterns in the next state of cell C. Each of these 0–255 bit patterns represents a rule of Wolfram CA. This is called state transition rule. The bit pattern of a rule can be applied to K-map to generate a Boolean function. The Boolean variables are nothing but x_{k-1}, x_k, x_{k+1} (the present state of left neighbour, cell C and right neighbour).

In Table 1, "g" is the function of the three present states of left neighbouring cell, cell itself (C) and right neighbouring cell at time t. The g function can generate the next state at time $t + 1$ of cell (C). So, mathematically we can generally write

$$x_k^{t+1} = g(x_{k-1}^t, x_k^t, x_{k+1}^t)$$

Each next state bit combination forms a decimal number which is the rule number like 11100011 bit pattern in the next state gives the rule as Rule 227. We can see 227 is the decimal conversion of 11100011. A few rules are shown with next state bit patterns in Table 1. We can have such 256 rules in cellular automata. Table 1 is established for three-neighbourhood elementary cellular automata. It can be obtained for five-neighbourhood, seven-neighbourhood and so on. The model of an N-cell one-dimensional cellular automata (CA) is configured depending on a rule vector. Rule vector is a combination of rules applied to each cell. Suppose there is a four-cell CA. Rule applied to each cell is 190. So, rule vector of that cellular automata is written as $\langle 190; 190; 190; 190 \rangle$. So, the cellular automata is modelled as CA $\langle 190; 190; 190; 190 \rangle$. This kind of cellular automata is called uniform cellular automata where same rule is applied to each cell of the cellular automata. If rules applied to the cells in a cellular automata are not same, that cellular automata is referred as non-uniform or hybrid cellular automata. For example, consider a four-cell CA with rule vector $\langle 190; 254; 255; 146 \rangle$. Here, leftmost cell is applied with Rule 190, next is with Rule 254, next to that is with 255 and rightmost is with 146. Every uniform/hybrid CA can be null boundary cellular automata or periodic boundary cellular automata (as mentioned earlier).

Table 1 CA rule and its logic function

Rule mean term	7	6	5	4	3	2	1	0	Transition	Boolean
Present state	111	110	101	100	011	010	001	000	Rule	Function
Next state	0	0	0	0	0	0	0	0	0	$g = 0$
Next state	0	0	0	0	1	0	1	0	10	$g = \overline{x}_{k-1}\ x_{k+1}$
Next state	0	0	1	0	0	0	0	0	32	$g = x_{k-1}\ \overline{x}_k\ x_{k+1}$
Next state	0	0	1	1	0	0	1	0	49	$g = x_{k-1}\ \overline{x}_k + \overline{x}_k\ \overline{x}_{k+1}$
Next state	0	1	0	0	1	1	1	1	79	$g = \overline{x}_{k-1} + x_k\ \overline{x}_{k+1}$
Next state	0	1	1	0	1	1	1	0	110	$g = \overline{x}_{k-1}\ x_{k+1} + x_k\ x_{k+1}$
Next state	0	1	1	1	1	1	1	1	127	$g = \overline{x}_{k-1} + \overline{x}_k + \overline{x}_{k+1}$
Next state	1	0	1	1	1	1	1	0	190	$g = x_{k+1} + x_{k-1}\ x_k$
Next state	1	1	1	0	0	0	1	1	227	$g = \overline{x_{k-1}}\ \overline{x_k} + x_{k-1}\ x_k + x_{k-1}\ x_{k+1}$
Next state	1	1	1	1	1	1	1	0	254	$g = x_{k-1} + x_k + x_{k+1}$
Next state	1	1	1	1	1	1	1	1	255	$g = 1$

4.2 Elementary Cellular Automata (ECA) Properties

Any N-cell cellular automata with a rule vector can be represented with its cell present states and the possible time-dependent next states using the Boolean functions of the rules in the vector. All those possible next states at different discrete time instants can form a diagram showing a graphical path of the next states with time. This graphical path is called state transition graph/diagram (STG/STD) [24].

All the next state bit patterns for all generations of the cellular automata are connected together to form the state transition graph. It is a path which shows pattern of the transition of cellular automata from one generation to another. Using uniform cellular automata, we can obtain different state transition graph for every single rule applied to that. For a four-cell three-neighbourhood uniform elementary cellular automata, there $2^4 = 16$ different bit patterns can be found in 16 generations (from 0000 to 1111) of the elementary cellular automata. If we know the present states of those four cells, we can find next generations using the rules. Each applied rule can generate state transition graph with 16 generations. But the transition path will differ. Let us take two examples of CA ⟨60; 60; 60; 60⟩ (Fig. 2) and CA ⟨143; 143; 143; 143⟩ (Fig. 3).

Figures 2 and 3 show two different transition paths for two different applied rule vectors on a one-dimensional four-cell three-neighbourhood uniform cellular automata. The total number of generation is same for both, i.e. 0000 to 1111, i.e. 16.

Fig. 2 State transition diagram of CA ⟨60, 60, 60, 60⟩

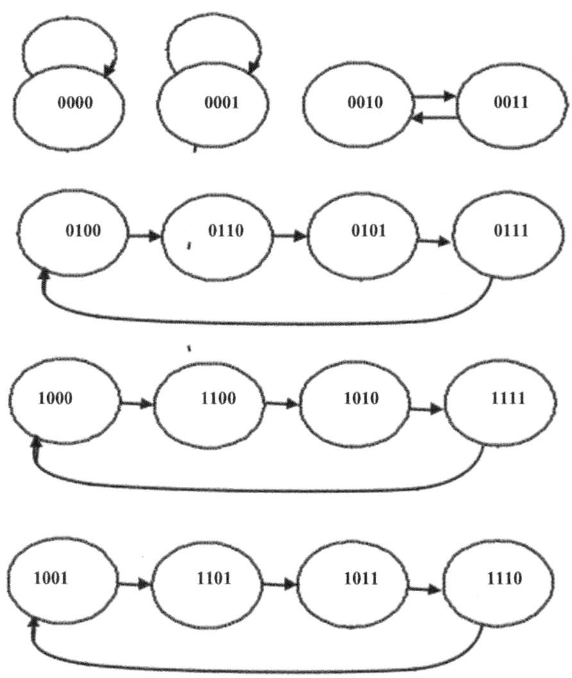

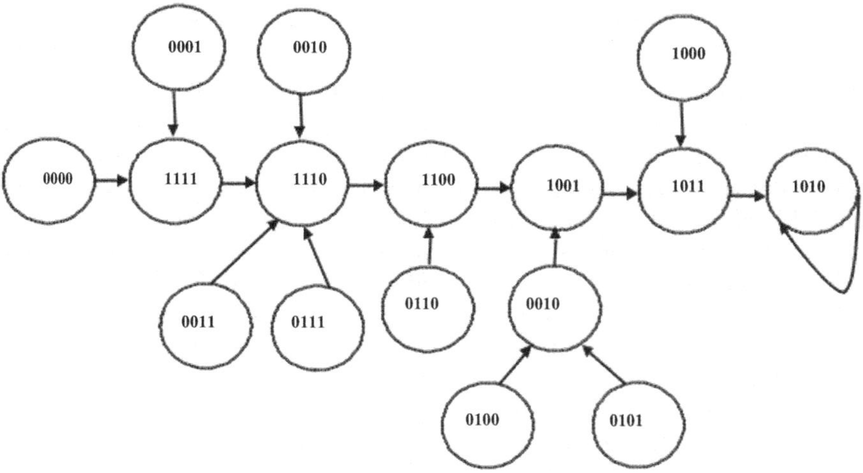

Fig. 3 State transition diagram of CA ⟨143, 143, 143, 143⟩

But state transition diagrams are different. Furthermore, we can observe some other features of CA in the graph. In CA ⟨60; 60; 60; 60⟩ (Fig. 2), there are six cycles, two of them are single length cycle (single node is involved in the cycle) and other four are multilength cycle (more than one node are involved in the cycle). On the other hand, CA ⟨143; 143; 143; 143⟩ (Fig. 3) is having only one single length cycle. CA ⟨60; 60; 60; 60⟩ (Fig. 2) is called single length cycle attractor CA [24], and CA ⟨143; 143; 143; 143⟩ (Fig. 3) is called multiple attractors CA.

From the state transition graph, one of the important properties of cellular automata can be realized, i.e. reversible property. Every reversible CA must have a predecessor (one previous prototype). That means the current state of a reversible cellular automata can determine its previous state. It is analogues to the natural reversible computing. This property can help us to describe an evaluation process from forward to the backward. Reversibility can explain the fundamental law of physics. Any reversible physical system can be observed and logic can be generated based on cellular automata to create artificial reversible computing models.

The elementary cellular automata can be further characterized as additive/linear cellular automata [17, 25] depending on the Boolean functions corresponding to rules. Rules with EXOR/EXNOR Boolean functions if forms a cellular automata, it is referred as additive/linear cellular automata. The additive cellular automata with EXOR functions is called non-complimented cellular automata and with EX-NOR functions is called complimented cellular automata. In Table 1, Rule 110 and Rule 190 can form an additive non-complemented cellular automata, and Rule 227 can form an additive complemented cellular automata.

4.3 *ECA and Natural System Design and Computation*

Elementary cellular automata concept has been extensively used by researches in different domain of modern science and technology like VLSI device testing, data and image encryption, pattern recognition and classification, etc. It can be efficiently used in designing highly precise self-testable hardware and fault-tolerant circuits for high density and high-speed memory devices. A number of design architectures, models and algorithms regarding memory testing have been proposed utilizing elementary cellular automata and its neighbourhood characteristics. March C, a well-known algorithm for high-speed resistive memory testing, is first used by von Neumann 1950s. Later scientists used this algorithm in combination with three- and five-neighbourhood elementary cellular automata to develop innovative self-testable hardware for memory [15, 18, 20].

Researcher has shown innovations with elementary cellular automata (ECA) inspired by the nature. The graphical models of some rules can be compared to some natural phenomenon like design of Rule 30 that can be compared to sedimentation in water. Scientist Matthew Cook in 2004 in a conference at Santa Fe Institute, USA, proved that Rule 110 in elementary cellular automata has the capability of universal computation. He told that it can be simulated in a Turing machine and is considered as Turing complete. Nowadays, elementary cellular automata are randomly applied in the domain of physics. Different physical properties of a material-like diffusion of carriers, fluid flows, lattice space, lattice Boltzmann's modelling, viscosity, etc., can be analysed by elementary cellular automata.

5 Conclusion

The research of nature-inspired computing is expanding so rapidly that it becomes one of the fundamental research domains in computer science. John von Neumann adapted self-reproducing cellular model from biological phenomenon. Conway introduced a cellular automaton model Game of Life to describe the basic nature of death and birth of a living being. Wolfram's defined simplified structure elementary cellular automata have been used to design sophisticated fault detection and tolerance memory devices which are highly efficient in detecting and correcting wrong information in memory just like a human brain works. Through this paper, we have tried to cover proposed cellular automata models by John von Neumann, Conway and Wolfram. From these models, it can be realized that man-made cellular automata models can be compared to natural system. Readers will be familiarized with a good range of applications based on natural computing and cellular automata in the latest technologies of electronics and computer science.

References

1. Von Neumann J (1966) The theory of self-reproducing automata. In: Burks AW (ed) University of Illinois Press, Urbana and London
2. Moore EF (1970) Machine models of self reproduction. University of Illinois Press, Urbana
3. Codd EF (1968) Cellular automata. Academic Press Inc., USA
4. Gardner M (1970) Mathematical games: the fantastic combinations of John Conway's new solitaire game 'life'. Sci Am 223:120–123
5. Gardner M (1971) On cellular automata self-reproduction: the garden of Eden and the game of 'life'. Sci Am 224:112–117
6. Fraile A, Panagiotakis E, Christakis N, Acedo L (2018) Cellular automata and artificial brain dynamics. Math Comput Appl 2018:1–23
7. Wolfram S (1983) Statistical mechanics of cellular automata. Rev Mod Phys 55(3):601–644
8. Qadir F, Ahemed PZ, Bani SJ, Peer MA (2013) Quantum-dot cellular automata: theory and application. In: Proceedings of the IEEE international conference on machine intelligence and research advancement, pp 540–544
9. Wolfram S (1986) Random sequence generation by cellular automata. Adv Appl Math 7(2):123–169
10. Maji P, Shaw C, Ganguly N, Sikder BK, Chaudhury PP (2003) Theory and application of cellular automata for pattern classification. Fundam Inf 58:321–354
11. Ismail W, Hassan R, Swift S (2010) Detecting leukaemia (AML) blood cells using cellular automata and heuristic search. In: International symposium on intelligent data analysis. Springer, Berlin, pp 54–66
12. Małecki K (2017) Graph cellular automata with relation-based neighbourhoods cells for complex systems modelling: a case of traffic simulation. Symmetry 9(12):322
13. Caballero-Gil C, Caballero-Gil P, Molina-Gil J (2016) Cellular automata-based application for driver assistance in indoor parking areas. Sensors 16(11):1921
14. Sikdar BK, Ganguly N, Chaudhuri PP (2005) Fault diagnosis of VLSI circuits with cellular automata based pattern classifier. IEEE TCAD 24(7):1115–1131
15. Saha M, Dalui M, Sikdar BK (2016) A cellular automata based highly accurate memory test hardware realizing March C^-. Microelectron J 52:91–103
16. Anghelescu P (2011) Encryption algorithm using programmable cellular automata. World Congress on Internet Security (WorldCIS), IEEE, pp 233–239
17. Nandi S, Kar BK, Chaudhuri PP (1994) Theory and application of cellular automata in cryptography. IEEE Trans Comput 43(12):1346–1357
18. Saha M, Das B, Sikdar BK (2017) Periodic boundary cellular automata based test structure for memory. EWDTS 2017:1–6
19. Ghosh M, Kumar R, Saha M, Sikdar BK (2018) Cellular automata and its applications. In: Proceedings of the IEEE international conference on automatic control and intelligent systems (I2CACIS 2018), 20 Oct 2018, pp 53–57
20. Saha M, Sikdar BK, Sarkar S (2016) Cellular automata based fault tolerant resistive memory design. In: Proceedings of the IEEE sixth international symposium on embedded computing and system design (ISED), pp 176–180
21. Sarkar P (2000) A brief history of cellular automata. ACM Comput Surv (CSUR) 32(1):80–107
22. Tatashev A, Yashin M (2019) Spectrum of elementary cellular automata and closed chains of contours. Machines 7(2):28
23. Verardo M, de Oliveira PPB (2019) A fully operational framework for handling cellular automata templates. Complexity 2019
24. Saha M, Sikder BK (2016) Test structure for L1 cache in tiled CMPs. IAENG Int J Comput Sci 43(3):392–401
25. Chaudhuri PP, Chowdhury DR, Nandi S, Chattopadhyay S (1997) Additive cellular automata: theory and applications, vol 1. Wiley, Hoboken

A Survey on Transport Layer Protocols for Reliable and Secure Wireless Sensor Networks

Anisha and Sansar Singh Chauhan

Abstract A wireless sensor network consists of a large number of autonomous sensor nodes distributed spatially and working together to observe events in a region to acquire relevant data about the environment. The application areas of WSN include health monitoring, military target tracking, environment exploration, disaster relief operations and other programs which require timely data transport. Therefore, congestion control mechanism to avoid packet loss and ensuring reliable end-to-end communication is vital for these applications of wireless sensor networks. The transport layer is responsible for the reliability, error control, flow control and quality of data being transferred from the source node to sink. Hence, the transport layer protocols play a significant role in maintaining the reliability, handling the packet loss and congestion. In this paper, a comparison of established transport layer protocols has been done on the basis of certain parameters under reliability and congestion control. The present paper represents a detailed review on the transport layer protocols, their functionalities, attributes and research issues associated with it.

Keywords Transport layer protocol · Wireless sensor network · Congestion control · Reliability

1 Introduction

The wireless sensor network is composed of spatially distributed sensor nodes that monitor the environmental and physical phenomena in a region. These sensor nodes in this network communicate through wireless links and consist of a processor, transceivers, memory, power supply and microcontrollers [1]. It is a self-organized network which is efficient in sensing a region with high degree of accuracy, processing

Anisha (✉) · S. S. Chauhan
Galgotias University, Greater Noida, India
e-mail: anishanagpal@outlook.com

S. S. Chauhan
e-mail: sansar@gmail.com

P. K. Singh et al. (eds.), *Evolving Technologies for Computing, Communication and Smart World*, Lecture Notes in Electrical Engineering 694,
https://doi.org/10.1007/978-981-15-7804-5_12

the sensed data and accordingly responding to the specific condition by delivering the information. Therefore, the WSN applications such as military target tracking, health monitoring, event detection, disaster relief operations and environment exploration where the sensed information is delicate and critical need timely data delivery from sensors to the base station. A WSN has its resource constraints that include limited storage, power supplied by batteries, low bandwidth, weak processing and small communication range that can affect the reliable data transmission. As wireless sensor network comprises of larger number of active nodes for transmitting of data, it increases the traffic and load which results in congestion. This congestion in network leads to the degradation of channel quality, packet loss, increase delay and need retransmission [2, 3]. Also, until the detection of congestion takes place and appropriate technique for avoidance is adopted, a considerable amount of loss of packets occurs due to buffer overflow. This causes lower throughput, energy wastage, degraded quality of service and decrease in event detection reliability [4–6]. However, every application of WSN can bear different levels of packet loss, and in order to improve the energy consumption and throughput, it is necessary to detect packet loss and successfully recover the same. This further necessitates the use of transport layer protocols.

Transport layer of WSN ensures the quality of the data and reliable end-to-end data delivery from source to the sink node. It leads to the designing of transport layer protocols in order to have congestion control mechanism and packet loss recovery. These transport layer protocols should be developed independent of the application and must be generic. In WSN, packet recovery and congestion control can be done either end to end or hop by hop [1]. Therefore, in packet recovery, ensuring the reliability means successful delivery of all the packets. The retransmission in hop-by-hop approach is energy efficient as the intermediate sensor node stores the packet information in memory, and the distance for retransmission is shorter, whereas in end-to-end transmission approach, the source node in the network stores all the packet information and retransmits the packet whenever packet loss occurs. Comparatively, the hop-by-hop retransmission has better performance than end-to-end retransmission when high reliability is required. Similarly, congestion control mechanism is used to detect and monitor the congestion which occurs when the sender node overwhelms the receiving node by transmitting data at a higher rate. The congestion detection and avoidance mechanism help in energy conservation by notifying the source node to reduce the rate of data transmission, thereby preventing buffer overflow [7]. The end-to-end approach depends on the nodes deployed at the end for congestion detection, while, in hop-by-hop approach, every sensor node in the network monitors the overflow and changes its behavior whenever congestion is detected, and due to this, it has the faster rate of reducing the congestion. However, the better performance of a particular approach depends on the type of the application [1].

In wireless sensor networks, it is necessary and important to address these issues so that the operation in the network is energy efficient and it must contribute in increasing the lifetime of the network. There exist transport layer protocols which facilitates reliable communication and responsible for the error and flow control. This paper provides a review on the transport layer protocols, its characteristics, functionalities and the approaches needed to acquire reliable data communication in wireless sensor networks.

2 Literature Background

In the literature, there are various transport layer protocols designed for wireless sensor networks. These protocols are responsible to address the reliable transmission of data and congestion control issues.

Sensor Transmission Control Protocol (STCP) [8] is designed by Y. G. Iyer, S. Gandham and S. Venkatesan, which is reliable and generic transport layer protocol for different applications of wireless sensor networks. It contributes to the detection and avoidance of congestion and variable reliability depending upon the diverse application requirement. It executes all the functionalities at the sink node. The sink node that communicates with all the sensor nodes assumes to have high storage capacity, power supply and a processor. In order to adjust data flow, needed reliability and transmission rate, STCP uses session initiation packet consisting of the information regarding number of flows and types of data flow, etc. This session initiation packet must be transmitted to the sink node from the source node before sending any data and have to wait for the acknowledgment from the sink node. The sink node uses NACK-based retransmission of the packets for the applications producing continuous data flows. In case of applications having event-driven data flows, it uses ACK for successfully received packets from source nodes. The sensor nodes use threshold value for setting the congestion bit. The congestion bit is set in all the packets if the buffer reaches to that threshold value. This information lets the sink node to report the source node to lower down the transmission rate.

Yogesh Sankarasubramaniam, Özgür B. Akan and Ian F. Akyildiz proposed event-to-sink reliable transport (ESRT) [9], and it is the upstream transport protocol that offers congestion control mechanism and event reliability. It maintains the desired level of reliability at the sink node using minimum energy for event detection. This protocol defines the term event reliability as the number of packets received within a given interval of time. It also indicates the overall information of the nodes within the radius of an event that are successfully obtained at the base station. ESRT provides reliability by adjusting the reporting frequency of the sensors. The protocol has certain backdrops in terms of processing and power such as if the data is transmitted at a very low rate by the source node, then the reliability will not be achieved. Also, if the data is transmitted at a high rate, this results in the problem of packet loss.

Reliable Multisegment Transport (RMST) protocol proposed by Fred Stann and John Heidemann [10] ensures the reliable transmission of data in the upstream direction. This protocol supports cached and non-cached modes. In the cache mode, all the sensor nodes from the source node to the base station has to maintain the cache and enables hop-by-hop recovery. The non-cached mode is related to the end-to-end transmission of packets where only end nodes are required to maintain the cache. Also, RMST uses the combination of NACK and timer-driven mechanism to facilitate loss detection and notification. If the nodes between the source and sink node fail to detect the lost packet or if it is in non-cached mode, then the NACK is transmitted upstream to the source node. The limitations of RMST lie with the lack of congestion control mechanism, reliability at application level and energy efficiency.

Chieh-Yih Wan, Andrew T. Campbell and Lakshman Krishnamurthy proposed the protocol named as Pump Slowly, Fetch Quickly (PSFQ) [11]. It is a reliable downstream transport protocol which focuses on the scalability and robustness to meet up the resource-related challenges of WSN. It transmits data from sink node to the sensor nodes at a slow speed in comparison with others. However, in case of data loss, it allows the sensor nodes to recover the missing packets from the nearest neighbors. The objective is to attain loose delay bounds by reducing loss recovery through localization of data from immediate neighbors. There are three operations associated with PSFQ, i.e., pump, fetch and report. The pump operation monitors the rate at which packets are injected by the sink node into the network. The fetch operation is used to detect the lost packets from immediate nodes by using gap sequence and recovers it by issuing NACK. The report operation is used to provide feedback status of data delivery in hop-by-hop manner to the users. It has the advantages of being scalable, robust and reliable; however, it has some limitations in terms of large delay due to slow pump. Also, in case of high error rate in wireless channel, it is not able to recover continuing packet loss.

Chieh-Yih Wan, Shane B. Eisenman and Andrew T. Campbell proposed congestion detection and avoidance protocol [12] which can mitigate the congestion. It is an energy-efficient upstream transport control protocol consisting of three components, i.e., congestion detection, hop-by-hop backpressure and end-to-end multi-source regulation. The detection of congestion is facilitated by keeping a check on the buffer occupancy status and rate of load at the wireless channel. Whenever congestion occurs, the corresponding node gives the notification to its immediate upstream neighbor to minimize its transmission rate by using hop-by-hop backpressure technique. At last, it maintains the multisource rate by using closed end-to-end approach. The backdrops in this protocol lie with the unidirectional control from sensor node to the sink node and increase in the response time of multisource control when heavy congestion occurs.

Seung-Jong, Ramanuja, Raghupathy and I. F. Akyildiz have designed a protocol named as GARUDA [13]. It is a downstream transport protocol for reliable data delivery in WSN. Its framework is designed to give attention to the problems of reliable data transmission from sink node to the sensor nodes. The term reliability is expressed into four categories, i.e., ensuring delivery of data to entire region, sub-region of sensors deployed, set of minimum sensors used to cover a particular region and probable subsets of sensors. It uses core infrastructure based on the first packet delivery technique and two-stage NACK for the process of recovery. The method of the first packet delivery ensures the successful delivery of the first packet by using wait for the first packet transmission pulse to the other sensor nodes. The use of NACK message is for detection of loss and notification. The two stages for packet loss recovery process consists core node recovery phase and non-core nodes recovery phase. Therefore, it is a combination of both hop-by-hop and end-to-end transmission schemes. However, apart from such a hybrid scheme, it also has some limitations in terms of lack of data reliability in upstream direction and congestion control mechanism.

V. C. Gungor and O. B. Akan come up with delay sensitive transport protocol [14]. It is used to handle the issues of reliability, timely packet delivery and congestion control. There are two transport mechanisms associated with it, i.e., reliable event transport and real-time event transport. In the reliable event transport mechanism, the focus is on reliability aspects by measuring the deviation between the observed and desired delay constrained event reliability to maintain the level of reliability for event to sink communication, whereas, in real-time event transport mechanism, the focus is on timely delivery of packets. It measures the event transport delay and event process delay which is termed as event-to-sink delay in order to attain the specific objectives of the application. Also, in congestion detection, it computes buffer overflow at each sensor node and evaluates average node delay.

Nurcan Tezcan and Wenye Wang proposed asymmetric and reliable transport protocol [15]. This addresses end-to-end reliability, congestion control mechanism from sensors to sink node and downstream query reliability. The major functions of protocol include distributed congestion control mechanism, reliable event transfer and reliable query transfer. The reliability of protocol is based on the categorization of sensor nodes as essential nodes (E-nodes) and non-essential nodes (N-nodes). It selects subgroup of sensor nodes that efficiently cover the entire area that needs to be sensed. The protocol uses E-nodes as a sub-network which participates in reliable data transmission among sink node and sensor nodes and recovery of lost packets in both upstream and downstream directions. ART adopts ACK and NACK signaling mechanism for reliable end-to-end communication between E-nodes and base station. ART protocol consists of four features. First, there is no overhead of end-to-end communication for Non-essential nodes. Second, congestion control techniques are used to effectively regulate the flow of traffic. Third, only few nodes are required for recovery of lost message. Fourth, it uses distributed congestion control technique for energy efficiency. In ART, two-way reliability and congestion control is not maintained for non-essential nodes; due to this, recovery of any lost packet cannot be ensured.

Price-oriented reliable transport protocol (PORT) [16] is a reliable upstream protocol proposed by Yangfan Zhou. It provides energy efficiency and mechanism for congestion avoidance. The view of reliability in PORT is assuring the sufficient information at sink node about the phenomenon of interest. This reliability is ensured by providing two mechanisms which also help in minimization of energy consumption by avoiding high communication cost. First, sink's application-based optimization mechanism allows sink node to regulate and feedback the reporting rate of every source on the basis of node price. Second, locally optimal routing scheme gives information about the cost of end-to-end communication from source node to sink. This cost is used to dynamically reducing traffic, thereby providing an in-network congestion mechanism. This protocol maintains the needed level of reliability by adapting itself to the network dynamics.

ATP [17] is proposed by K. Sundaresan who provides solution to different problems that traditionally TCP met over wireless network that improves its performance. It uses end-to-end feedback mechanism based on receiver as well as the intermediate nodes in the network to achieve reliability and effective congestion control. It achieves reliability by using selective ACK in order to know the number of packets has been received by receiver and number of packets left to be received in the future. For effective congestion control, the protocol takes feedback from the intermediate nodes in the path to evaluate the network state and gives congestion information to the sink node. It primarily designed to achieve better performance over TCP for network; however, it is not optimized to reduce energy consumption.

Apart from the above-mentioned protocols, there are some protocols developed in recent years for reliability and congestion control in WSN. Some of the most recent protocols are ALORT, RT-CaCC and PSOGSA.

Ahyan Kiraz and Murat Cakiroglu designed a protocol named as reliable and delay-sensitive transport layer protocol (ALORT) [18]. It is an upstream reliable transport protocol having two loss recovery mechanisms (LRM), hop by hop and end to end. These two LRM are used together to get optimized packet latency, reduced energy cost, better reliability and increased packet delivery ratio in different error rates. This protocol delivers packets less than 92% in case of high error rates (i.e., 92 out of 100 successful packet transmissions). A reliable transport with cache-aware congestion control (RT-CaCC) [19] is proposed by Melchizedek I. Alipio and Nestor Michael C. Tiglao. This protocol uses cache management policies for reducing packet loss. These policies are size allocation, cache elimination and cache insertion. The performance of RT-CaCC is evaluated in different scenarios of packet loss, and it shows improvement of 15–38% (average) comparative to baseline protocols. RT-CaCC uses expiration of ETO for congestion detection, implicit NACK for congestion notification and provides congestion avoidance by bounded congestion window. PSOGSA [20] is proposed by Karishma Singh, Karan Singh, Le Hoang Son and Ahmed Aziz. It is a hybrid multi-objective optimization algorithm that decreases arrival data rate from child to parent node for avoiding congestion. The arrival rate depends on the priority, child node's energy and bandwidth. It uses rate adjustment for mitigation of congestion and provides improvement in throughput, delay, queue size, packet loss and congestion level.

3 Comparison of Transport Layer Protocol

The transport layer protocols are classified into three categories [21, 22]. The protocol which mainly considers only reliability, congestion control comes into the categories of reliability only and congestion control respectively. The protocols which support both come into the third category of Hybrid protocols.

The protocols are compared on the basis of certain attributes such as reliability and congestion control. As shown in Table 1, RMST, PSFQ and GARUDA are reliability only protocols and CODA is congestion control only protocol, whereas STCP, ESRT, DST, ART, PORT and ATP support both reliability and congestion control.

In WSN, the term reliability implies successful delivery of each segment which is generated at the source to the target destination. The reliability is achieved by detecting the lost packets and retransmission of these lost packets to the appropriate source. In terms of reliability, the protocols are compared at reliability level, direction, loss detection & notification and error recovery as shown in Fig. 1 [23]. The level is the extent of reliability, a protocol supports, and it can be packet reliability and event reliability. Packet reliability is successful transmission of every packet to its intended destination, and event reliability means successful detection of event which needs reliable transmission of event data from sensing area in a sensor network. In the comparison of both, for event reliability, there is no requirement of retransmission of each dropped packet. As given in the Table 1, STCP, RMST, PSFQ, GARUDA and ATP offer packet-level reliability and ESRT, DST, ART and PORT support event reliability.

In wireless sensor networks, the transfer of data takes place in two directions, i.e., upstream and downstream. Upstream reliability shows the successful flow of data from sensor node to the sink node, STCP, ESRT, RMST, DST, PORT and ATP supports upstream reliability. In downstream reliability, the data flows from sink to the sensor nodes in terms of queries or control packet delivery. PSFQ and GARUDA are the only ones which have downstream reliability. In Table 1, only ART is the protocol which supports bidirectional reliability which means achieving reliability in both upstream and downstream direction.

The reliability in transport layer is attained by ensuring the identification of packet loss at destination, and it must notify the sender for retransmission of the lost packet. The protocols use ACK and NACK kind of feedback for notification of dropped packets. ACK is the positive acknowledgment sent by the receiver for successful reception. Negative acknowledgment (NACK) being a variant of ACK is responsible to notify the sender for missing sequence of the packets and ensure the retransmission of the packets in order. STCP and ART use both ACK and NACK messages for detection of packet loss, whereas RMST, PSFQ, GARUDA and ATP use only NACK messages. Packet recovery mechanism ensures reliability by retransmission using either end-to-end or hop-by-hop approach. End-to-end mechanism works with the storage of packet information at end nodes, and whenever packet loss arises, it retransmits the packets to the destination. Hop by hop uses the approach of storing information at intermediate nodes, and then, detection and retransmission take place. Protocols in Table 1 such as STCP, DST, ART and ATP support end-to-end error recovery. PSFQ offers hop-by-hop error recovery, and RMST supports both end-to-end and hop-by-hop approach for loss. GARUDA is the only one that uses two-stage error recovery process.

In WSN, congestion control mechanism [24] uses three methods in order to deal with congestion, i.e., congestion detection, congestion notification and avoidance as shown in Fig. 2. Congestion detection is the identification of occurrence of congestion

Table 1 Comparison of transport layer protocols in terms of reliability and congestion control

Protocol	Classification (reliability, congestion control and hybrid)	Reliability				Congestion control			Energy efficient	Delay	Caching
		Reliability level	Direction	Loss detection	Error recovery	Congestion detection	Congestion notification	Congestion avoidance			
STCP	H	Packet	Upstream	ACK and NACK	E to E	Buff. occupancy	I	RA and TR	Y	L	Y
ESRT	H	Event	Upstream	–	–	Buff. occupancy	I	RA	Y	L	N
RMST	R	Packet	Upstream	NACK	H by H and E to E	–	–	–	N	M	Option
PSFQ	R	Packet	Downstream	NACK	H by H	–	–	–	N	L	Y
CODA	CC	–	–	–	–	Buff. occupancy and channel status	E	Drop packets and RA	Y	S	–
GARUDA	R	Packet	Downstream	NACK	Two stage recovery	–	–	–	Y	–	Y
DST	H	Event	Upstream	–	E to E	Buff. occupancy and node delay	I	RA	Y	M	N
ART	H	Event	Bidirectional	ACK and NACK	E to E	ACK received at E-nodes	I	Traffic at N-nodes is reduced	Y	S	N
PORT	H	Event	Upstream	–	–	Node price	I	RA and TR	Y	–	N

(continued)

Table 1 (continued)

Protocol	Classification (reliability, congestion control and hybrid)	Reliability				Congestion control			Energy efficient	Delay	Caching
		Reliability level	Direction	Loss detection	Error recovery	Congestion detection	Congestion notification	Congestion avoidance			
ATP	H	Packet	Upstream	NACK	E to E	Buff. occupancy	E	RA	N	M	N

H, R, CC: hybrid (reliability + congestion control), reliability, congestion control; RA and TR: rate adjustment and traffic redirection; S, L and M: small, medium and large; I and E: implicit and explicit; H by H: hop by hop; E to E: end to end; Y and N: yes and no

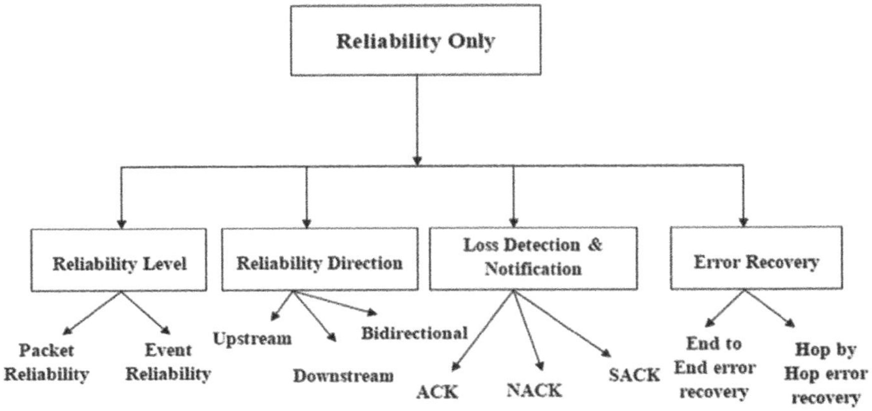

Fig. 1 Comparative attributes in reliability focused protocols

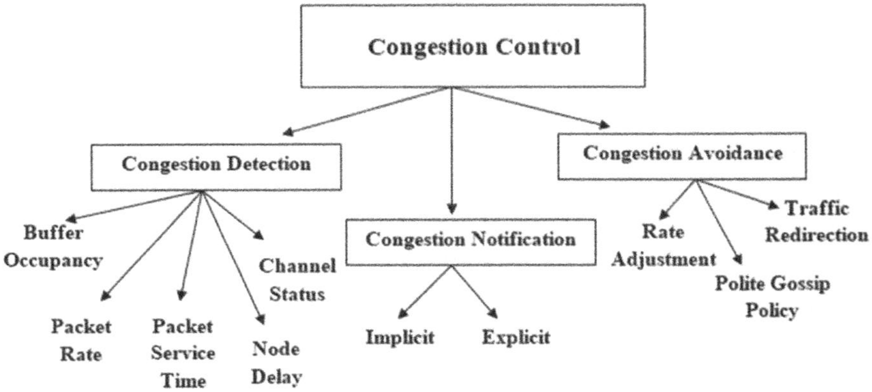

Fig. 2 Methods of congestion control mechanism

in the network. It is a composition of certain attributes such as buffer occupancy, packet rate, packet service time, node delay and channel status. As given in Table 1, STCP, ESRT and ATP use buffer occupancy for congestion detection. CODA uses both buffer occupancy and channel status for detection of congestion, whereas DST uses buffer occupancy and node delay. ART and PORT detect congestion on the basis of ACK received at essential node and node price, respectively. When congestion is detected, it is to be notified to the sender explicitly or implicitly. STCP, ESRT, DST, ART and PORT send implicit notification to the node. CODA and ATP notify explicitly about the congestion occurred. In wireless sensor network, there are three techniques to avoid the congestion, i.e., rate adjustment, polite gossip policy and traffic redirection. ESRT, DST and ATP use rate adjustment technique, and STCP and PORT use both rate adjustment and traffic redirection. CODA follows drop packet and rate adjustment methods for congestion avoidance, whereas, in ATP, traffic at non-essential node is reduced.

Packet loss in a network leads to the wastage of energy, and as in WSN, the sensors are equipped with limited energy sources (batteries) which result in short lifespan of the overall network. Whenever a packet drop occurs, this arises the requirement of packet retransmission which causes delay in delivery and energy loss. There are protocols which are energy efficient as shown in Table 1 such as STCP, ESRT, CODA, GARUDA, DST, ART and PORT. However, RMST, PSFQ and ATP are the protocols which are not energy efficient. There is also significant amount of delay associated with STCP, ESRT and PSFQ which refers that packets are delayed in transmission as expected due to congestion at the sensor node. As delay and packet retransmission affect the energy efficiency, there is a need of optimization in retransmission to make efficient. This leads to the caching of packets at intermediate nodes. STCP, PSFQ and GARUDA support caching of packets at node level, but ESRT DST, ART, PORT and ATP do not support caching, while RMST has the option for it.

4 Research Challenges

The role of transport layer protocols is to have reliable data transfer by checking the state of network for reliability and congestion. The research issues such as cross-layer optimization, real-time queue management, fairness, node prioritization and security need extensive exploration for future researches [25, 26]. As per the review, some of the protocols such as DST and STCP consider change in reliability level and node prioritization. Also, congestion control protocols examine the channel and change the data transfer rate after congestion is detected. There must be a real-time monitoring of queue to check the possibility of congestion prior to its occurrence. Only some protocols such as RT2 facilitate intelligent queue management and can be explored to its full extent [27]. The problem of ensuring fairness in dynamic environment is not fully explored. Another issue to be addressed is achievement of QOS in order to have versatility in sensor-based system. Security is also an important issue that needs attention to ensure correct data being sent from sensor node to sink node or vice versa [28]. Energy conservation can be another challenge in order to have increased network lifetime [29, 30].

5 Conclusion

The transport layer protocol is responsible for the reliability, error control, flow control and quality of data being transferred from the source node to sink. The transport layer protocols play a significant role in maintaining the reliability, handling the packet loss and congestion. In this paper, a review on transport layer protocols is presented which focuses on the characteristics, functionalities and techniques for reliable data transfer. Also, these protocols are compared on the basis of certain parameters related to reliability, congestion control mechanism and performance. These parameters turned out to be crucial for implementation of transport layer protocols.

References

1. Yick J, Mukherjee B, Ghosal D (2008) Wireless sensor network survey. Comput Netw 52(12):2292–2330
2. Kiraza A, Çakıroğlub M (2014) A survey of congestion control protocols providing energy conservation in wireless sensor networks. Turk J Eng 1:12–22
3. Kafi MA, Djenouri D, Ben-Othman J, Badache N (2014) Congestion control protocols in wireless sensor networks: a survey. IEEE Commun Surv Tutorials 16(3):1369–1390
4. Spachos P, Chatzimisios P, Hatzinakos D (2012, December) Energy aware opportunistic routing in wireless sensor networks. In: 2012 IEEE Globecom workshops. IEEE, pp 405–409
5. Mao X, Li XY, Song WZ, Xu P, Moaveni-Nejad K (2009, October) Energy efficient opportunistic routing in wireless networks. In: Proceedings of the 12th ACM international conference on modeling, analysis and simulation of wireless and mobile systems. ACM, pp 253–260
6. Lu T, Zhu J (2012) Genetic algorithm for energy-efficient QoS multicast routing. IEEE Commun Lett 17(1):31–34
7. Jan MA, Jan SRU, Alam M, Akhunzada A, Rahman IU (2018) A comprehensive analysis of congestion control protocols in wireless sensor networks. Mob Netw Appl 23(3):456–468
8. Iyer YG, Gandham S, Venkatesan S (2005, October) STCP: a generic transport layer protocol for wireless sensor networks. In: Proceedings of the 14th international conference on computer communications and networks, 2005. ICCCN 2005. IEEE, pp 449–454
9. Sankarasubramaniam Y, Akan ÖB, Akyildiz IF (2003, June) ESRT: event-to-sink reliable transport in wireless sensor networks. In: Proceedings of the 4th ACM international symposium on mobile ad hoc networking & computing. ACM, pp 177–188
10. Stann F, Heidemann J (2003, May) RMST: reliable data transport in sensor networks. In: Proceedings of the first IEEE international workshop on sensor network protocols and applications, 2003. IEEE, pp 102–112
11. Wan CY, Campbell AT, Krishnamurthy L (2002, September) PSFQ: a reliable transport protocol for wireless sensor networks. In: Proceedings of the 1st ACM international workshop on wireless sensor networks and applications. ACM, pp 1–11
12. Wan CY, Eisenman SB, Campbell AT (2003, November). CODA: congestion detection and avoidance in sensor networks. In: Proceedings of the 1st international conference on Embedded networked sensor systems. ACM, pp 266–279
13. Park SJ, Vedantham R, Sivakumar R, Akyildiz IF (2004, May) A scalable approach for reliable downstream data delivery in wireless sensor networks. In: Proceedings of the 5th ACM international symposium on mobile ad hoc networking and computing. ACM, pp 78–89
14. Gungor VC, Akan OB (2006, June) DST: delay sensitive transport in wireless sensor networks. In: 2006 International symposium on computer networks. IEEE, pp 116–122

15. Tezcan N, Wang W (2007) ART: an asymmetric and reliable transport mechanism for wireless sensor networks. Int J Sens Netw 2(3/4):188
16. Zhou Y, Lyu MR, Liu J, Wang H (2005, November) PORT: a price-oriented reliable transport protocol for wireless sensor networks. In: 16th IEEE International symposium on software reliability engineering (ISSRE'05). IEEE, 10 pp
17. Sundaresan K, Anantharaman V, Hsieh HY, Sivakumar AR (2005) ATP: a reliable transport protocol for ad hoc networks. IEEE Trans Mob Comput 4(6):588–603
18. Kİraz A, Çakiroğlu M (2017) ALORT: a transport layer protocol using adaptive loss recovery method for WSN. Sādhanā 42(7):1091–1102
19. Alipio MI, Tiglao NMC (2018) RT-CaCC: a reliable transport with cache-aware congestion control protocol in wireless sensor networks. IEEE Trans Wireless Commun 17(7):4607–4619
20. Singh K, Singh K, Aziz A (2018) Congestion control in wireless sensor networks by hybrid multi-objective optimization algorithm. Comput Netw 138:90–107
21. Rathnayaka AD, Potdar VM (2013) Wireless sensor network transport protocol: a critical review. J Netw Comput Appl 36(1):134–146
22. Wang C, Sohraby K, Li B, Daneshmand M, Hu Y (2006) A survey of transport protocols for wireless sensor networks. IEEE Netw 20(3):34–40
23. Mahmood MA, Seah WK, Welch I (2015) Reliability in wireless sensor networks: a survey and challenges ahead. Comput Netw 79:166–187
24. Ghaffari A (2015) Congestion control mechanisms in wireless sensor networks: a survey. J Netw Comput Appl 52:101–115
25. Kafi MA, Othman JB, Badache N (2017) A survey on reliability protocols in wireless sensor networks. ACM Comput Surv (CSUR) 50(2):1–47
26. Alipio M, Tiglao NM, Grilo A, Bokhari F, Chaudhry U, Qureshi S (2017) Cache-based transport protocols in wireless sensor networks: a survey and future directions. J Netw Comput Appl 88:29–49
27. Sharma B, Aseri TC (2012) A comparative analysis of reliable and congestion-aware transport layer protocols for wireless sensor networks. ISRN Sens Netw
28. Khan AS, Khan M (2018) A review on security attacks and solution in wireless sensor networks. Am J Comput Sci Inf Technol 7(1):31
29. Sharma P, Chauhan SS, Saxena S (2012, September) Multihop/direct forwarding for 3D wireless sensor networks. In: Proceedings of the CUBE international information technology conference. ACM, pp 344–349
30. Chauhan SS, Gore MM (2015) Balancing energy consumption across network for maximizing lifetime in cluster-based wireless sensor network. CSI Trans ICT 3(2–4):83–90

A Lightweight Exchangeable Encryption Scheme for IoT Devices Based on Vigenere Cipher and MLS Keystream

Yumnam Kirani Singh

Abstract An exchangeable encryption scheme which is lightweight and secure suitable for IoT devices is proposed. The security of the encryption scheme is mainly based on random keystream generated from the Meitei lock sequence (MLS). The encryption scheme is lightweight in the sense that it does not require any memory buffer during its execution and also, it does not involve any complex mathematical operations. It is also size preserving, i.e., plaintext and ciphertext sizes are the same. Also, there is no upper limit for the key size, and hence, breaking it using brute-force attack is theoretically impossible. Moreover, tracking key from the ciphertext is also not possible because of the use of MLS, which produces different random keystreams when there is any slight difference in any of password elements. The encryption scheme is also fast as it does not involve any complex mathematical operation and any rounds of permutation and substitution operations. The encryption scheme has been applied for encryption of text, speech, and image. Experimental results based on correlation coefficient measures show that the proposed encryption is secure.

Keywords Exchangeable encryption scheme · Meitei lock sequence · Random keystream · Size preserving encryption · Lightweight encryption scheme

1 Introduction

Data security is a great concern for IoT and WSN applications since the data from sensor node to gateway are quite often passed over the air, which can be easily accepted by anyone interested. Data security can be provided by cryptographic means. There are many symmetric key cryptography methods for data security which can be broadly classified into three groups—dedicated, exchangeable, and dual functionality. In dedicated cryptosystem, two separate dedicated functions or modules

Y. K. Singh (✉)
Education and Training, C-DAC Silchar, Ground Floor, IIPC Building, NIT Silchar Campus, Silchar, India
e-mail: yumnam.singh@cdac.in

P. K. Singh et al. (eds.), *Evolving Technologies for Computing, Communication and Smart World*, Lecture Notes in Electrical Engineering 694,
https://doi.org/10.1007/978-981-15-7804-5_13

are used for encryption and decryption purpose. That is, encryption function is used only for encryption and decryption function is only for decryption. Examples of the dedicated symmetric key cryptosystem are Data Encryption Standard (DES) and Advanced Encryption Standard (AES). In exchangeable cryptosystem, two functions are used as in dedicated system, but any one of the function can be used for encryption or decryption. In other words, the role of the function can be interchanged. Examples of exchangeable symmetric key cryptosystems are random generalized Vigenere ciphers. In dual functionality cryptosystem, only one function is used for both encryption and decryption. For example, XOR or YOR [1]-based cryptosystem or special cases of generalized Vigenere ciphers. Of these three types of cryptosystem, exchangeable and dual functionality cryptosystem will be the most suitable cryptosystem IoT and WSN; as only one function will be required to be installed in a device, memory requirement will be less and at the same time simpler to implement and operate. The dedicated cryptosystems are mostly block ciphers and are comparatively more complex than the exchangeable and dual functionality cryptosystems which are basically stream ciphers. Even though block ciphers are more complex than the stream ciphers, they are considered more secure than the stream ciphers. Several lightweight cryptosystems based on block ciphers have been proposed for use in IoT and WSN applications in the last decade. Most of these lightweight block ciphers are comparatively more complex than the proposed exchangeable encryption scheme. In [2], the performance analysis of several block ciphers of symmetric key encryption schemes such as Advanced Encryption Standard (AES), Rivest Cipher 6 (RC6), Twofish, SPECK128, LEA, and ChaCha20-Poly1305 algorithms are given and recommended chacha20, LEA, and SPECK128 as suitable for resource constraint devices. Rajesh et al. [3] analyze the avalanche effect of block cipher Tiny Encryption Algorithm (TEA) [4] and proposed a new lightweight encryption scheme for secure transferring of text files between embedded devices. Usman et al. [5] propose a lightweight block cipher based on combination of substitution–permutation and Feistel structure for use in IoT devices.

Also, several lightweight stream ciphers have been proposed in the recent years as they are simpler for implementation on memory and power constrained devices. Most of the stream ciphers are based on XOR function which has several security issues. In [3] performance evaluation of the stream ciphers such as Grain, MICKEY, Trivium etc was conducted in NodeMCU device. These ciphers use fixed length keys and initialization vectors. Using fixed length key has limited keysearch space, and using initialization vectors has the security risk of chosen IV attack. Several other attacks on XOR-based stream ciphers are given in [7, 8].

The proposed exchangeable encryption scheme is a stream cipher which is not based on XOR operation, and hence, the attacks applicable for XOR based stream cipher cannot be used. It is very similar to the polyalphabetic Vigenere cipher [9], which only known effective attack is the differential cryptanalysis. To thwart the differential cryptanalytic attack in the proposed system, it uses a new random keystream known as Meitei lock sequence as in [10, 11]. This keystream is generated from a chosen password string and no initialization vector is used for it generation. Moreover, there is no upper limit of length of the chosen password which makes the keysearch space infinite. As a result, the brute-force attack on the proposed scheme is almost impossible.

The security of any encryption scheme is very much dependent on the tightness of the key used in the encryption system. If the key is easily derivable or traceable from the ciphertext, the encryption scheme is not considered as secure. One of the reasons why original Vigenere cipher is considered insecure is that the keystream used is periodic repetition of any chosen password. Another reason is the use of fixed known table having periodic patterns. The first problem can be solved by using Meitei lock sequence [12] as a keystream. The second problem can be solved by using random tables in generalized Vigenere cipher [11]. Also, another drawback of the original Vigenere cipher is the slow decryption process. This drawback is solved by using separate decryption table in generalized Vigenere cipher [10]. In other words, a generalized Vigenere cipher is a very fast and a secure cipher which can be used for encryption any data type, i.e., text, speech, audio, or image data. The security of the Vigenere cipher is based on the use of random tables and Meitei lock sequence. Generalized Vigenere can be used for developing an exchangeable or a dual functionality cryptosystem. In this paper, a simple lightweight exchangeable symmetric key cryptosystem is based on original Vigenere cipher. In such an encryption scheme, the range of the encryption or decryption tables can be varied easily by appropriately choosing the value of the modulus. For example, for encrypting binary data, the value of modulus can be set to 2. For text only data, the modulus can be set to 26 and so on. However, encrypting low-range data using high-range encryption tables, i.e., higher modulus values make ciphertext more untraceable or unintelligible. So, such an encryption would be able to encrypt any type of data which may range from binary data, textual data, numeric data, audio, image data, or video data acquired by various types of sensors.

2 Exchangeable Encryption Scheme

Exchangeable encryption scheme is a cryptosystem, in which two different functions F_1 and F_2 are used in which any one of which can be used for encryption or decryption. The two functions used in an exchangeable encryption scheme satisfies the following relation.

$$x = F_2(F_1(x, k), k) = F_1(F_2(x, k), k)$$

One such encryption scheme is the Vigenere cipher. In Vigenere cipher, the encryption is done using addition modulo operation and decryption is the performed by inverse addition modulo operation. If F_1 and F_2 denotes the encryption and decryption functions of the Vigenere cipher, they can be expressed as.

$$F_1(x, k) = x + k\%M$$
$$F_2(x, k) = (M + x - k)\%M$$

It can be easily proved that the two functions in Vigenere cipher are exchangeable.

Let x be a data sample to be encrypted with the key sample k using Vigenere cipher.

$$\begin{aligned}
F_2(F_1(x, k), k) &= F_2(x + k\%M, k) \\
&= (M + (x + k)\%M - k)\%M) \\
&= (M - k)\%M + (x + k)\%M \\
&= (M - k + x + k)\%M \\
&= (M + x)\%M \\
&= x\%M
\end{aligned}$$

$$\begin{aligned}
F_1(F_2(x, k), k) &= F_1((M + x - k)\%M, k) \\
&= ((M + x - k)\%M + k)\%M) \\
&= (M + k)\%M + (x - k)\%M \\
&= (M + k + x - k)\%M \\
&= (M + x)\%M \\
&= x\%M
\end{aligned}$$

i.e.,

$$x = F_2(F_1(x, k), k) = F_1(F_2(x, k), k)$$

This shows that F_1 and F_2 are exchangeable. That is, it does not matter which function used first for encryption and which function which letter for decryption.

$X = 15$, $M = 26$, $k = 23$

$$c = F_1(x, k) = F_1(15, 23) = (15 + 23)\%26 = 38\%26 = 12$$
$$d = F_2(c, 23) = (26 + 12 - 23)\%26 = (38 - 23)\%26 = 15 = x$$

Suppose, we apply F_2 first and then apply F_1.

$$d = F_2(x, k) = F_2(15, 23) = (26 + 15 - 23)\%26 = (41 - 23)\%26 = 18$$
$$c = F_1(d, k) = F_1(d, k) = (18 + 23)\%26 = 41\%26 = 15 = x$$

The problem of the Vigenere cipher is that the encryption and decryption based on some tables have regular patterns. Table 1 shows the table generated from function F_1 when $M = 10$ in which off diagonal elements are the same. Similarly, in Table 2 generated from F_2 when $M = 10$, the diagonal elements are the same. Larger tables

Table 1 10 × 10 matrix of F_1

0	1	2	3	4	5	6	7	8	9
1	2	3	4	5	6	7	8	9	0
2	3	4	5	6	7	8	9	0	1
3	4	5	6	7	8	9	0	1	2
4	5	6	7	8	9	0	1	2	3
5	6	7	8	9	0	1	2	3	4
6	7	8	9	0	1	2	3	4	5
7	8	9	0	1	2	3	4	5	6
8	9	0	1	2	3	4	5	6	7
9	0	1	2	3	4	5	6	7	8

Table 2 10 × 10 matrix of F_2

1	0	9	8	7	6	5	4	3	2
2	1	0	9	8	7	6	5	4	3
3	2	1	0	9	8	7	6	5	4
4	3	2	1	0	9	8	7	6	5
5	4	3	2	1	0	9	8	7	6
6	5	4	3	2	1	0	9	8	7
7	6	5	4	3	2	1	0	9	8
8	7	6	5	4	3	2	1	0	9
9	8	7	6	5	4	3	2	1	0

for F_1 and F_2 for $M = 256$ are shown as grayscale images in, respectively, Figs. 1 and 2.

Because of these regular patterns corresponding to encryption and decryption tables, the Vigenere cipher is susceptible to cryptanalytic attacks. To eliminate the regular patterns, use of random tables has been suggested in generalized Vigenere Cipher [13]. But the problem of using generalized Vigenere cipher is that since the tables are random, the random tables cannot be defined in terms of mathematical functions as F_1 and F_2 and hence the tables are to be stored for performing encryption and decryption, which may not be suitable for memory constraint small IoT devices.

Fig. 1 256 × 256 gray image of F_1

Fig. 2 256 × 256 gray image of F_2

2.1 MLS Generation

One weakness of the Vigenere cipher is the use of periodic repetition of a password as keystream. The main reason, why Vigenere cipher is insecure was because of use of the periodic keystream. This problem can be eliminated if we can use a non-periodic sequence as keystream. For this, a random key sequence generator called Meitei lock sequence (MLS) was introduced [12]. In MLS, a random-like sequence is generated from a chosen password string. Use of MLS in Vigenere cipher will enhance its security level significantly.

An MLS is random-like sequence which can be generated from a chosen password string. It has the property that if there is any small change in the chosen password, the generated sequence is drastically changed. This is a desired property for a good keystream as this would make the prediction or tracking of the keystream from the ciphertexts impossibly difficult. The methods for generating MLS keystreams can be found in [10–12]. Also, MLS can be generated from the standard hash functions [14, 15]. For lightweight devices, a simple MLS generation scheme described in [12] will be suitable (Figs. 3 and 4).

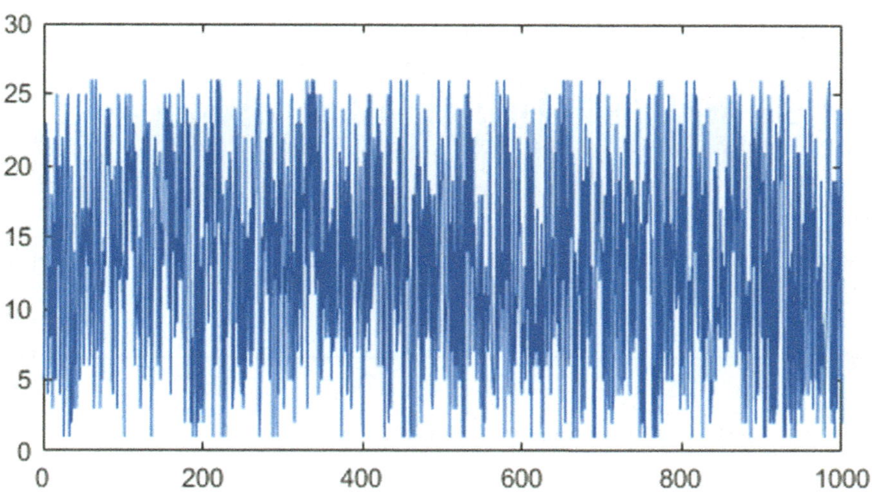

Fig. 3 MLS keystream with range 1–26 generated from '111'

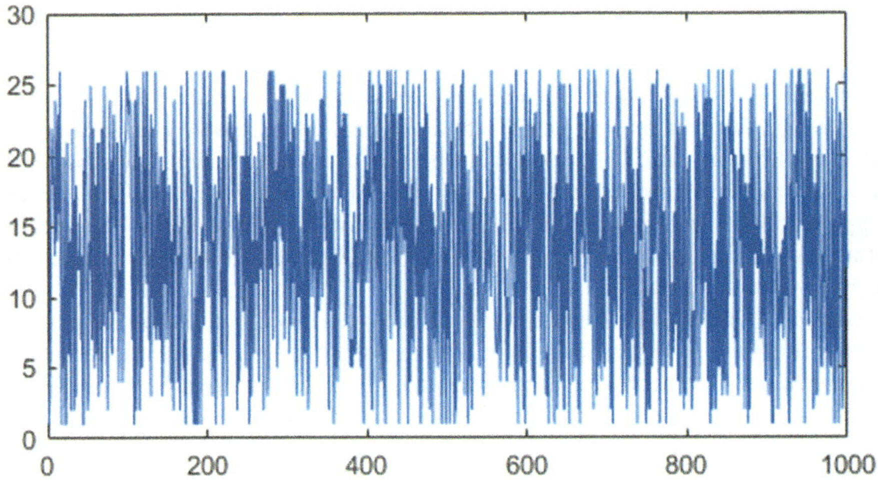

Fig. 4 MLS keystream with range 1–26, generated from '1111'

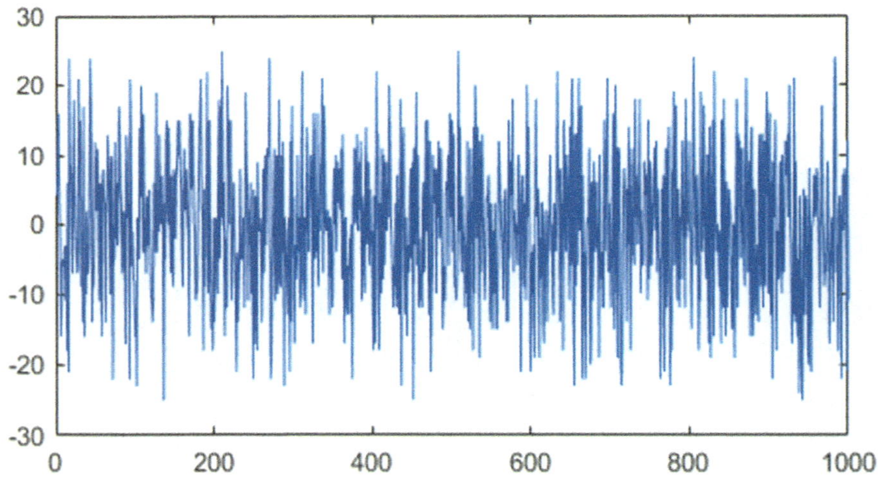

Fig. 5 Difference of MLS keystreams (Fig. 4 from Fig. 3)

Figure 3 shows MLS keystream generated from the password '111' and Fig. 4 shows the MLS keystream generated from '1111'. From these two figures, we can observe that the two keystream are totally different with just a change in length by 1. It can be seen that slight change in either in any elements of the chosen passwords or in the length of the chosen password, the MLS sequences are significantly different. Figure 5 shows the difference of the two MLS whose elements are all ones but the length is differed by 1. This shows that MLS can be used as good keystream to make Vigenere cipher a lightweight secure stream cipher.

3 Proposed Lightweight Cryptosystem

The proposed lightweight cryptosystem is an exchangeable cryptosystem based on the Vigenere cipher and the MLS keystream. In an exchangeable cryptosystem, the role of encryption and decryption algorithms or functions can be exchanged. That is, the encryption algorithm can be used for decryption or the decryption function can be used for encryption. Such an encryption scheme can be used for one to one, one to many, or many to one secure data transfer. For one-to-one data transfer between two devices, one of the devices will be having the encryption algorithm and the other will be having the decryption algorithm. For one-to-many or many-to-one data transfer, e.g., master to slaves or slaves to master, the master device will be having encryption algorithm and the slaves will be having the decryption algorithm. The proposed algorithm is simple being based on Vigenere cipher yet it is secure as it uses MLS as keystream. It can be used for encryption of text, speech, and images. The speech here will be only for telephone quality speech which is coded in 8 bits. The encryption and decryption algorithms are given below.

Encryption algorithm:

Step 1: Read a password string p
Step 2: Choose the appropriate value of M
Step 3: Read a new or next input data sample x
Step 4: Generate a new or next keystream sample k from p
Step 5: Encrypt the data sample x with key sample k as
$c = (x + k) \bmod M$
Step 6: If all data samples encrypted? Stop
Else go to step 3.

Decryption algorithm:

Step 1: Read a password string p
Step 2: Choose the appropriate value of M
Step 3: Read a new or next input data sample c
Step 4: Generate a new or next keystream sample k from p
Step 5: Decrypt the data sample x with key sample k as
$d = (M + x - k) \bmod M$
Step 6: If all data samples decrypted? Stop
Else go to step 3.

It can be seen that the encryption algorithm requires only one addition and one modulus operation for encrypting or decrypting a sample data. Hence, the encryption can be performed very fast. Similar is the case for decryption which involves only two addition operations and one modulus operation for decrypting or encrypting a sample data. The security of the encryption scheme is dependent of the random MLS keystream generated from the secret password p, which also does not involve complex mathematics or time-consuming operations. So, the encryption scheme is lightweight, fast, and secure.

4 Experimental Results

The proposed encryption scheme is applied for encryption of text, speech, and image data. To use the encryption scheme for all the data type, we consider the value of $M = 256$, which is equivalent to generating an encryption and decryption table of Vigenere cipher as shown in Fig. 1. This is sufficient for encryption text, image, and speech signal encoded in 8 bits. There is no restriction in the choosing password. The password can be any alphanumeric character, and its length can be as long as we like. But for the experimental study, we have chosen a simple encryption password of length 4 with all elements equal to 1 to show that the proposed encryption scheme can generate highly uncorrelated data even using such simple password. The decryption passwords are chosen as '11', '111', '1110', 1101', and '1011' to study whether distinguishable trace appear in the decrypted signal when they are closely similar to the encryption password.

4.1 Text Encryption

A text message in English can be expressed by combination of 26 different symbols or characters. It is highly uncorrelated data and can be easily enciphered by simply changing the position of occurrence of characters. We can use value of $M = 26$ for encryption and decryption of the text messages. In that case, the encrypted message will only be the scrambled string of alphabetic characters. The value M can be more than 26 for encryption and decryption of text messages. The more the value of M, the given characters in plaintext can be mapped to different symbols in ciphertext. We consider here $M = 256$, so that characters in the plaintext can be mapped to any symbol which can be represented in a byte. This enhances the security and at the same time this value of M can be used for the encryption of image and 8-bit speech samples.

Plaintext: 'The quick brown fox jumps over the lazy dog'
Password: '1111'
Ciphertext:

'°^§SUPruo~ L$Z¾pðÐæ{ii¡DÎê,»…@DmDê¿þwn° ßbÙ'

Decrypted text with password '111'

Table 3 Correlation coefficients of the original data with the encrypted and decrypted data

Encryption password	Decryption passwords				
1111	11	111	1110	1101	1011
Text 0.08084	0.13596	−0.14024	−0.12933	−0.29262	−0.24195
Image −0.01030	−0.00021	−0.01212	0.02748	0.00621	0.00858
Speech 0.00955	−0.00700	0.00530	−0.02463	0.00983	0.00929

'‹Ð(h‹ <U ì(*µVÌÕP÷! ¾»Ñ±<ÏÞÅÆeËa qv…Â}Æ´Ð:'

It can be observed that the encrypted text with password '1111' is entirely from the original plaintext. The decrypted text using slightly different password '111' is also totally different from the original text. No traceable sequence of characters appears in the decrypted text even though the first three characters in the decrypted password are the same as in the first three characters in the encryption password. The correlation coefficients of the original plaintext with the encrypted text and the decrypted texts with slightly different passwords are shown in Table 3.

4.2 *Image Encryption*

Image is highly correlated data. Standard block ciphers DES and AES which were developed for text encryption are considered to be unsuitable for image encryption. Different chaos-based image encryption schemes have been proposed in the last two decades [13]. In most of the methods, a distorter function is used to decorrelate the pixels in an image. Here, the random keystream generated using MLS is sufficient to encrypt the image without any visible trace. Figure 6, shows the original image popular to the processing community. The encrypted image with keystream generated from the password '1111' is shown in Fig. 7. It may be observed that the encrypted image has no distinguishable feature of the original image. The correlation coefficient between the original image and the encrypted image is −0.01030 which is a negligibly small value indicating that the encrypted image has virtually no correlation with the original image. The encrypted image is then decrypted with slightly different password '111', and the decrypted image is shown in Fig. 8, which is significantly different from the original image. This shows that when there is slight change in the value of the password, the decrypted image has no visible trace of the original image. This prevents the tracing of the key from the encrypted data. The original image can be recovered only when the decryption password is exactly the same as encryption password. The correlation coefficients of the original image with the images decrypted with different passwords are also very small as given in Table 3.

Fig. 6 Original image

Fig. 7 Encrypted image with '1111'

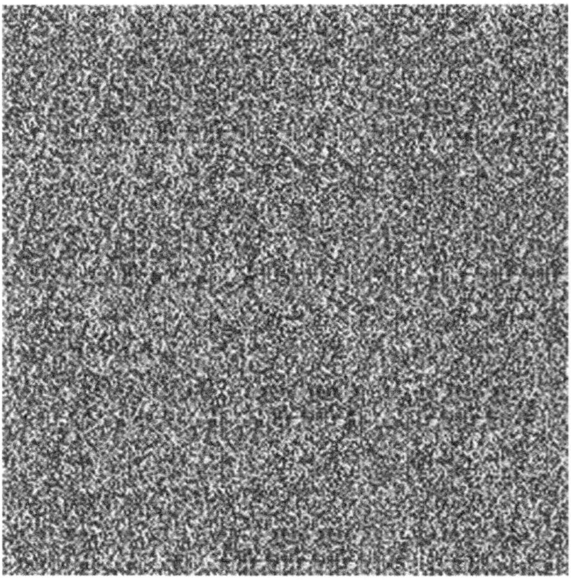

4.3 Speech Encryption

Speech or audio signals are also another highly correlated data which cannot be encrypted using text encryption schemes. Speech data are usually in floating point numbers in the range of −1 to +1. To encrypt such a data, the samples of the speech

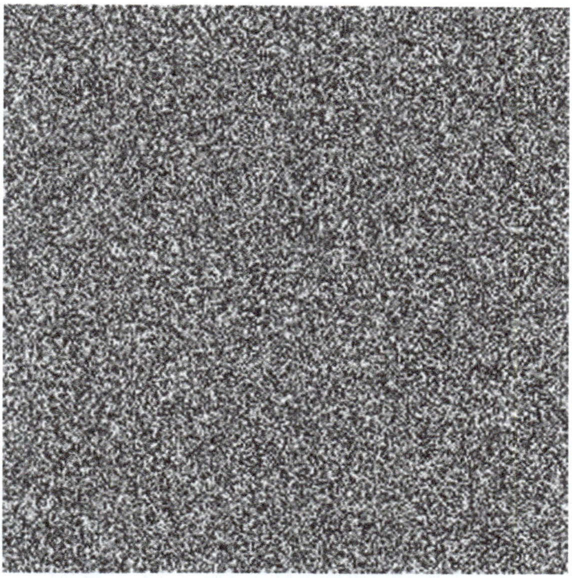

Fig. 8 Decrypted image with '111'

data need to be mapped to suitable integer range. Here, we map the range of the original speech data to the range of 0–255. The original speech signal (i.e., chirp signal) is shown in Fig. 9, and the mapped chirped signal is shown in Fig. 10. It may be seen that the waveforms or the characteristics of original chirp signal remain the same in the mapped signal as well. The mapped signal is then encrypted with keystream generated from the password '1111', and the encrypted signal is shown in Fig. 11. The encrypted signal is decrypted with a slightly different password '111', and the decrypted signal is shown in Fig. 12. It may be seen from Figs. 11 and 12 that the waveforms of the two signals are significantly different from the original signal in Figs. 9 or 10. When such a signal is played, only hissing noise is produced without

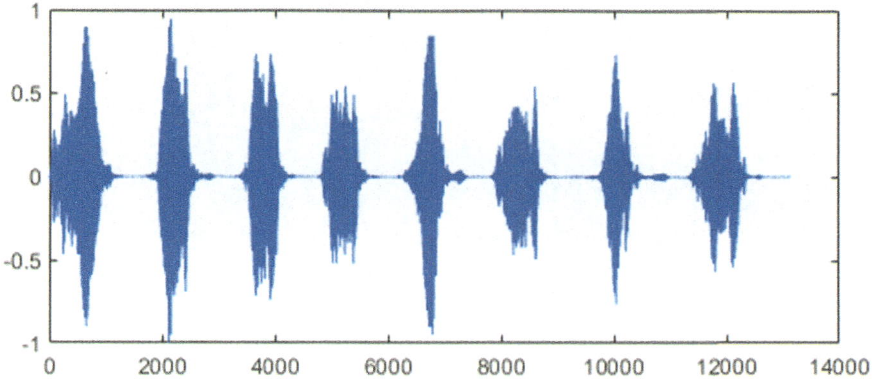

Fig. 9 Original chirp signal

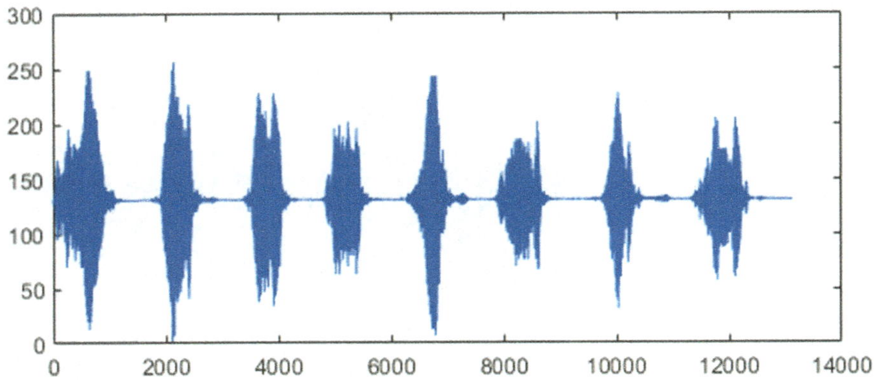

Fig. 10 Chirp signal coded in 8 bits

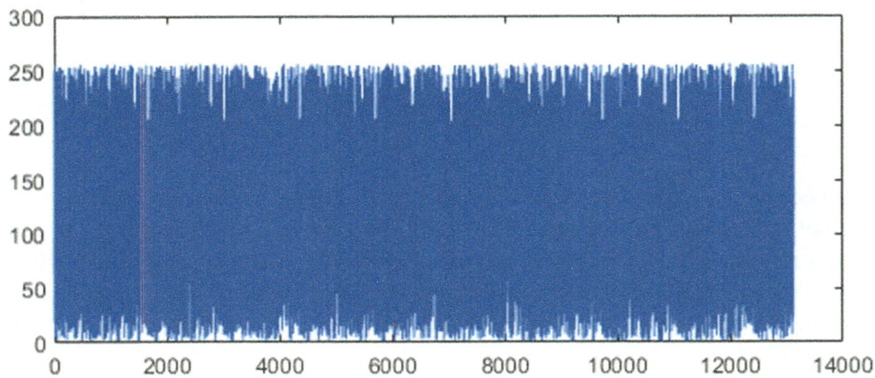

Fig. 11 Encrypted chirp signal with password '1111'

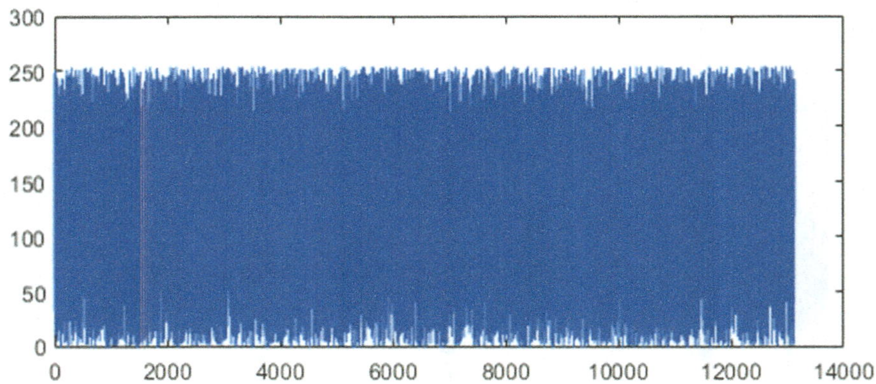

Fig. 12 Decrypted chirp signal with password '111'

any intelligible content of the original signal. It may be noted that for playing the mapped chirp or the decrypted signal, it is necessary to map the range back to -1 to $+1$.

The measure is the similarity between the original signal and the encrypted signal or the decrypted signals, and correlation coefficients are computed and are given in Table 3. The correlation coefficients are found to be negligibly small indicating that there is hardly any similarity between the original signal and the encrypted signal as well as the hardly any similarity between the original signal and the signals decrypted with wrong passwords. This shows that proposed encryption scheme can be used for a secure speech encryption scheme.

5 Conclusions

A lightweight and secure encryption scheme is based on Vigenere cipher and keystream generated from the MLS. It has been applied for encryption of the text, speech, and image data. The correlation coefficients between the original data and the encrypted data have been computed, and they are found to be very small. Also, the correlation is between the original data and the decrypted data using slightly different passwords. It has been found that the decrypted data are very significantly different from the input data if there is any difference in decryption key. This indicates that the proposed encryption scheme is very secure. As the encryption scheme is simple and fast without requiring any complex mathematical operations or multiple rounds of substitution and permutations, it can be conveniently used for low-power and low-memory IoT devices without any compromise to data security. The only concern for the proposed encryption scheme is getting non-repeating lock sequence, i.e., MLS keystream which is difficult to achieve if not properly implemented.

References

1. Singh YK (2019) Performing binary logical arithmetic in decimal arithmetic. ADBU J Eng Technol 8(1)
2. Saraiva DAF et al (2019) PRISEC: comparison of symmetric key algorithms for IoT devices. Sensors 1–23. https://doi.org/10.3390/s19194312
3. Rajesh S et al (2019) A secure and efficient lightweight symmetric encryption scheme for transfer of text files between embedded IoT devices. Symmetry 1–21. https://doi.org/10.3390/sym11020293
4. Wheeler D, Needham R (1995) TEA, a tiny encryption algorithm. In: Proceedings of the 1995 fast software encryption workshop, Leuven, Belgium, 14–16 Dec 1995. Springer, Berlin/Heidelberg, Germany, pp 97–110
5. Usman M et al (2017) SIT: a lightweight encryption algorithm for secure Internet of Things. Int J Adv Comput Sci Appl 8(1)

6. Ertaul L, Woodall A (2017) IoT security: performance evaluation of grain, MICKEY and Triviums—lightweight stream ciphers. In: International conference on security and management, SAM'17
7. Banegas G, Attacks on stream ciphers. https://eprint.iacr.org/2014/677.pdf
8. Verdult E, Introduction to cryptanalysis: attacking stream ciphers. https://www.cs.ru.nl/~rverdult/Introduction_to_Cryptanalysis-Attacking_Stream_Ciphers.pdf
9. C++ code for Vigenere cipher. https://www.geeksforgeeks.org/vigenere-cipher/
10. Singh YK (2011) A simple, fast and secure cipher. ARPN J Eng Appl Sci 6(10):61–69
11. Singh YK (2012) Generalization of Vigenere cipher. ARPN J Eng Appl Sci 7(1):39–44
12. Singh YK, Parui SK (2004) Simplet and its application in signal encryption. Multidimension Syst Signal Process 15(4):375–394
13. Kocarev L (2002) Chaos-based cryptography: a brief overview. IEEE Circ Syst Mag 1(3):6–21
14. Singh YK (2020) Image encryption using Meitei lock sequence generated from hash functions. Submitted to ADBU J Eng Technol
15. Singh YK (2020) Speech encryption using Meitei lock sequence generated from hash functions. Submitted to ADBU J Eng Technol

Improving Steepest Descent Method by Learning Rate Annealing and Momentum in Neural Network

Udai Bhan Trivedi and Priti Mishra

Abstract The backpropagation method of neural network (BPNN) method which is an important algorithm in machine learning has been applied to wide range of problem like pattern recognition, optimization, approximation, classification, and data clustering in real world. BPNN algorithm has been widely used in age estimation, pedestrian gender classification, traffic sign recognition, character recognition, water pollution forecasting models, heart disease classification, breast cancer detection (Keskar et al. in On large-batch training for deep learning: Generalization gap and sharp minima, 2016) [1], remote sensing, and image classification (He et al. in Proceedings of the IEEE conference on computer vision and pattern recognition (CVPR), pp 558–567, 2019) [2]. Algorithm uses a steepest gradient method and suffers with some limitations like convergence to local minima and slow convergence velocity of learning. This research proposed solution for the slow learning convergence velocity by implementing learning rate annealing which implements anneal the learning rate (decline as time progresses) rather than constant learning rate throughout the training. The problem of local minima can be address by momentum, which can be calculated by adding fraction of the past weights updates to the calculation of current weight.

Keywords Machine learning · Backpropagation neural network (BPNN) · Momentum · Learning rate annealing · Steepest descent method · Convergence

U. B. Trivedi (✉) · P. Mishra
PSIT College of Higher Education, Kanpur 209305, India
e-mail: udaibhantrivedi@gmail.com

P. Mishra
e-mail: bca@psit.ac.in

P. K. Singh et al. (eds.), *Evolving Technologies for Computing, Communication and Smart World*, Lecture Notes in Electrical Engineering 694,
https://doi.org/10.1007/978-981-15-7804-5_14

1 Introduction

Neural network is a novel method of programming computers which are constructed and implemented to develop the model of the human brain. Neural networks are exceptionally well at performing pattern recognition, optimization, approximation, classification, and data clustering. Conventional techniques are not sufficient enough to program these tasks. Programs that deploy neural networks are also capable of self-learning and self-adapting to changing conditions [3–5].

Neural network is constituted of a large number of interconnected and interwoven processing elements which work together to solve the particular problem. An artificial neural network draws its information processing mechanism from human brain and inspired by biological nervous systems. ANN's unique collective behavior is composed of special features like—ability to learn, recall, and generalize training patterns or data similar to that of human brains [6, 7].

Figure 1 shows a human brain, wherein a typical neuron collects signals from others through a host of fine tree-like structures called *dendrites*. Nucleus of one cell is connected to another cell through long thin stand known as axon. Nucleus of one cell sends spikes of electrical activity to nucleus of other cells through axon. Synapse connected with dendrites of next neuron receives excitatory input and compared with its inhibitory input. It further prorogates spike of electrical activity down its axon.

Figure 2 represents basic model of artificial neural network. Artificial neural network is composed of set of connecting links, wherein each link is characterized by a weight $W_1, W_2, \ldots, W_m$.

Weighted sum of the inputs is calculated by an adder function

$$v = \sum_{j=1}^{m} w_j x_j$$

Activation function can be used for limiting the amplitude of the output of the neuron.

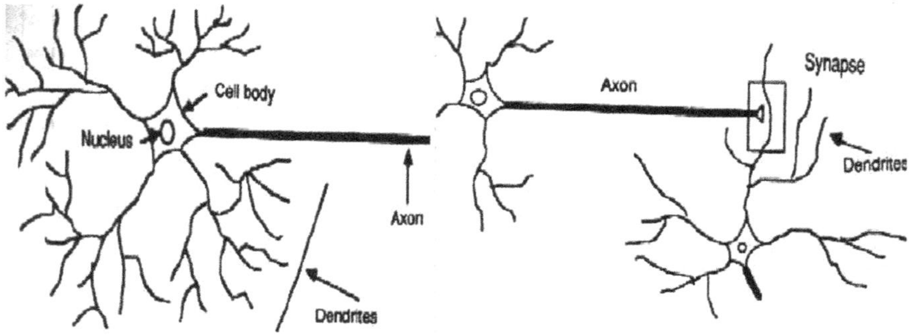

Fig. 1 Nucleus, axon, and dendrites of human brain and its connectivity with other cells

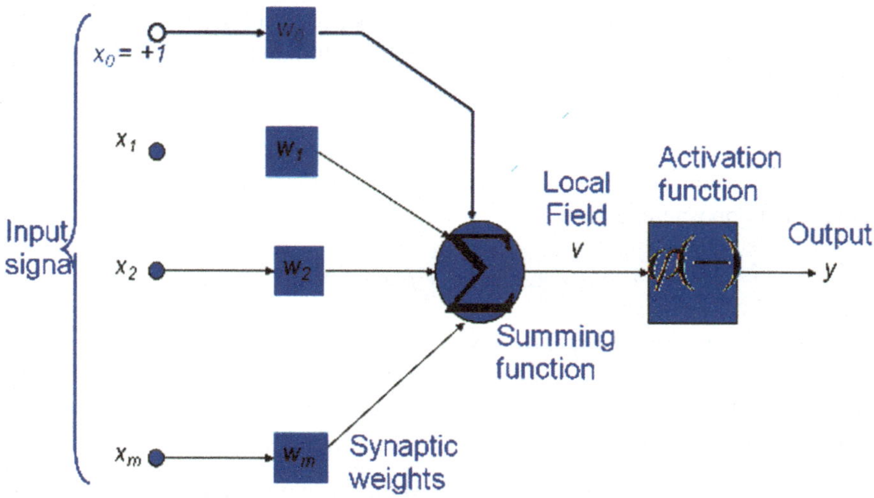

Fig. 2 Basic model of artificial neural network

$$y = \varphi(u + b).$$

Binary sigmoid function and its derivative (Eqs. 1 and 2) is popularly utilized as activation function by neural network [8–10]

$$f(x) = \frac{1}{1 + \mathrm{e}^{-\lambda x}} \tag{1}$$

$$f'(x) = \lambda f(x)[1 - f(x)] \tag{2}$$

The bias is included by adding a component $x_0 = 1$ to input vector $X = 1, x_1, x_2, \ldots, x_n$. Neuron will fire on the basis of threshold which is a set value. The amount of weight adjustment at each step of training is controlled by the learning rate. The value of weight adjustment ranges between 0 and 1.

2 Background and Related Works

The objective of steepest descent method is to find the lowest value of function $f(x)$ from the randomly chosen point $x^{(t)}$. Taylor's expansion formula of $f(x)$ about $x(t)$ in approximation

$$f\left(x^{(t+1)}\right) = f\left(x^{(t)} + \Delta s\right) \approx f\left(x^{(t)} + \left(\nabla f(x^{(t)})\right)\right)^{\mathrm{T}} \Delta s$$

where $\Delta s = x^{(t+1)} - x^{(t)}$ is the increment vector. Since the objective is to find better approximation of the function, the second term of right-hand side should be negative

$$f\left(x^{(t)} + \Delta s\right) - f\left(x^{(t)}\right) = (\nabla f)^{\mathrm{T}} \Delta s < 0.$$

The inner product of $u^{\mathrm{T}}.\, v$ of two vectors u and v will be maximum when they are parallel and opposite in direction

$$\Delta s = -\alpha \nabla f\left(x^{(t)}\right),$$

where $\alpha > 0$ is the learning rate (step size). The choice of α plays a very important role toward the convergence of function to minima. Small value of α means slow movements toward minima and large value sometimes overshoot the value of function from far away from minima.

1. Xu Wang in his research paper discussed the basic phenomenon of the method of steepest descent. In his research paper, he introduced the concept of steepest decent method and also focused on advantages and disadvantages of steepest descent method, and its applications. Author establishes that steepest descent method can start by, from any arbitrary point x_0 which is within a function's range and takes small steps in the direction of greatest slope changes, which is same as direction of the gradient. After much iteration, algorithm can find the minimum of the function. In the literature, the author has also established the fact that for badly scaled system, the convergence of algorithm is very slow [11].
2. Kavita Burse et al. have applied three-term backpropagation to multiplicative neural network learning in their research paper. The researcher has also proposed improved backpropagation algorithm which can solve the local minima and converge faster than traditional backpropagation algorithm. The authors have also done testing on the algorithm for XOR and three-bit parity problem and compared its outcome with the standard backpropagation multiplicative neural network algorithm [12].
3. M. Z. Rehman et al., in their research work, have proposed an algorithm for improving the current working performance of backpropagation algorithm. The proposed method in the research paper quoted as 'Gradient Descent Method with Adaptive Momentum (GDAM)' is compared with 'Gradient Descent Method with Adaptive Gain (GDM-AG)' and 'Gradient Descent with Simple Momentum (GDM).' Simulations on these five classification problems, i.e., breast cancer classification problem, IRIS classification problem, Australian Credit Card Approval Classification Problem, Pima Indians Diabetes problem, heart disease problem, demonstrate the efficiency of the proposed method [13].
4. Kuldip Vora et al. discuss how to solve local minima problem in neural network. The proposed algorithm aims at solving the local minima problem for large number of hidden nodes, with the assumption of having the same mathematical simplicity. The proposed algorithm evolves the input/hidden layers of weights according to the predefined rules whenever the gradient learning algorithm is

being stuck at local minima. However, there is no surety that it will be global minima using this algorithm but it seems to be faster than backproportion algorithm [14].

5. Nazri Mohd. Nawi et al., in their research paper titled 'A New Back-Propagation Neural Network Optimized with Cuckoo Search Algorithm,' have discussed performance of the proposed cuckoo search backpropagation (CSBP). The authors have compared cuckoo search backpropagation with artificial bee colony, using backpropagation algorithm and other hybrid variants with OR and XOR datasets. As per the research paper, the outcome proves enhanced computational performance of backpropagation training process when used with proposed hybrid method [15].
6. S. C. Ng et al., in their research paper, focused on helpfulness of weight evolution between the input and hidden layers in solving the local minima problem. In order to escape from local minimum, we need to have a non-singular hidden output matrix so as to give unique set of solution of the hidden output to have error reduction. The new algorithm evolves the input/hidden layers of weights according to the predefined rules whenever the gradient learning algorithm is being stuck at local minima [16].

3 The Backpropagation for Improvement

Rumelhart, Hinton, and Williams in 1986 proposed supervised learning ANN algorithm known as backpropagation neural network (BPNN). BPNN calculates the error of output layer which will be further used to find errors of hidden layers.

3.1 *Backpropagation Learning*

Figure 3 indicates the connectivity between input layer, hidden layer, and output layer and corresponding weights of various links. These weights have been used for calculation of output at output layer in different pass [17–19].

Dimension of weight matrix between input and hidden layer $[V] = 1 \times m$ and hidden and output layer $[W] = m \times n$.

3.2 *Computation at Input Layer*

Use linear activation function at input layer:

$$\{O\}_{l*1} = \{I\}_{l*1}$$

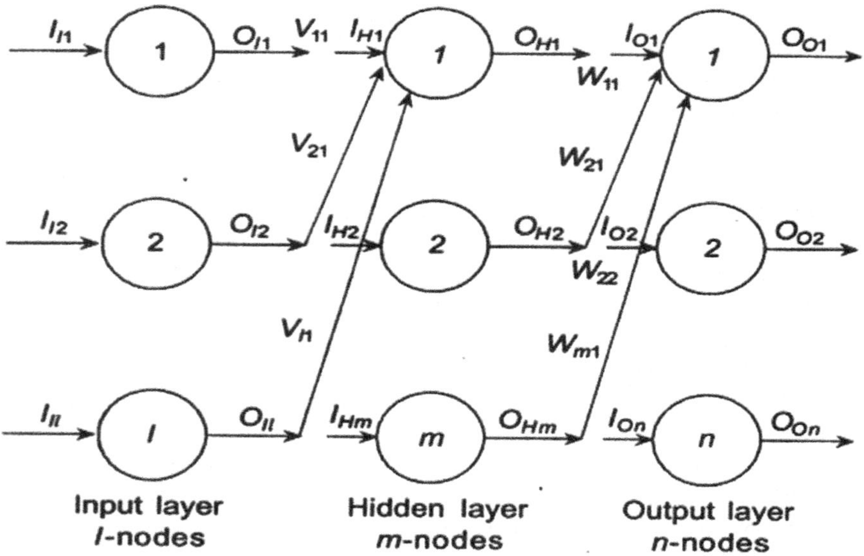

Fig. 3 Weight at different layer

Calculate the weighted sum at hidden layer at input of pth neuron [5, 20]

$$I_{Hp} = W_{1p}O_{I1} + W_{2p}O_{I2} + \cdots + W_{lp}O_{Il} \tag{3}$$

$$I_{Hp} = \sum_{j=1}^{l} V_{jp}O_{Ij} \quad \text{where } p = 1, 2, \ldots, m \tag{4}$$

3.3 Computation at Hidden Layer

Considering sigmoid function, the output of pth hidden neuron is given by (see Fig. 4)

$$O_{Hp} = \frac{1}{1 + e^{-\lambda I_{Hp}}} \quad \text{where } p = 1, 2, \ldots, m \tag{5}$$

Weighted sum at the qth output neuron is:

$$I_{Oq} = W_{1q}O_{H1} + W_{2q}O_{H2} + \cdots + W_{mq}O_{Hm} \quad \text{where } q = 1, 2, \ldots, n \tag{6}$$

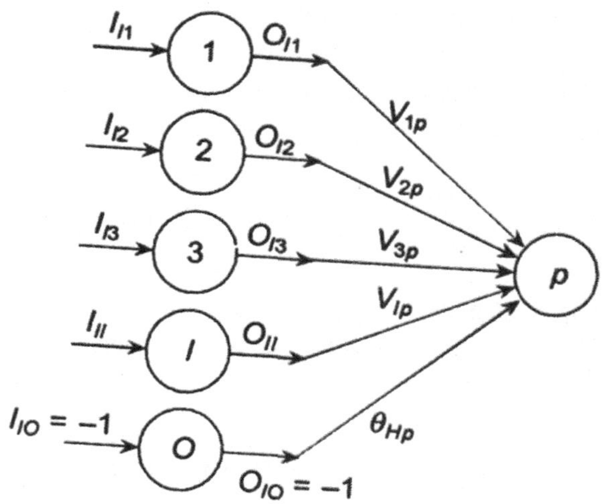

Fig. 4 Weight at hidden layer

$$I_{Oq} = \sum_{j=1}^{m} W_{jq} O_{Hj} \tag{7}$$

3.4 Computation at Output Layer

Considering sigmoid activation function, output of the qth neuron is given by (see Fig. 5).

$$O_{Oq} = \frac{1}{1 + e^{-\lambda I_{Oq}}} \quad \text{Dimension } n * 1 \quad q = 1, 2, \ldots, n \tag{8}$$

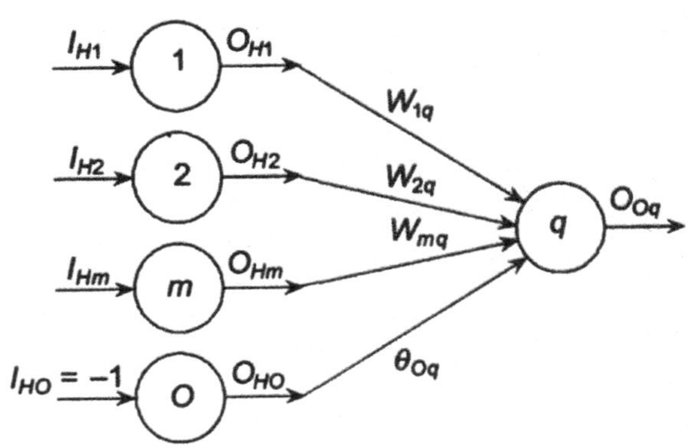

Fig. 5 Weight at output layer

4 Calculation of Error

Multilayer feed-forward networks with nonlinear activation functions have mean squared error

$$E_q^1 = \frac{1}{2}\sum_{q=1}^{n}(T_{Oq} - O_{Oq})^2 \tag{9}$$

where calculated output is O and desired output is T for any qth output neuron. Above error function is for one training pattern. If we use the same techniques for all the training patterns

$$E(V, W) = \sum_{j=1}^{n\text{set}} E^j(V, W, I) \tag{10}$$

5 Method of Steepest Descent

Error surface along the N-dimensional weight space generally consists of many local and global minima and complex to calculate [4, 16, 21].

Figure 6 indicates that how the initial weights will be adjusted with the help of steepest descent algorithm toward the best weight and graph showing relationship between error surface values and weight values [22–24]

$$E(V, W) = \sum_{j=1}^{nset} E^j(V, W, I)$$

$$\overline{AB} = (V_{i+1} - V_i)\bar{i} + (W_{i+1} - W_i)\bar{j} = \Delta V\bar{i} + \Delta W\bar{j} \tag{11}$$

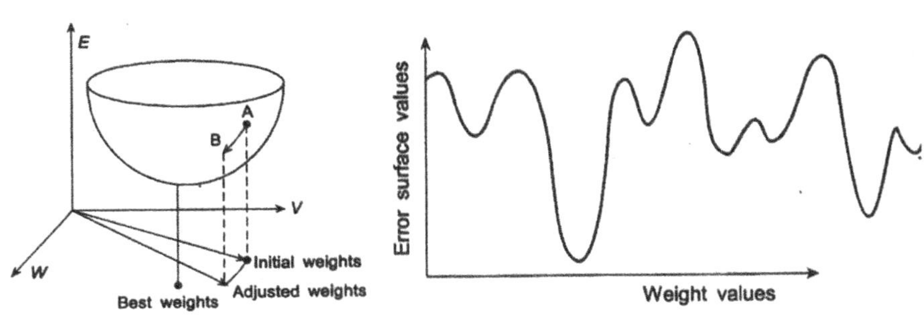

Fig. 6 Direction of steepest descent and weight values

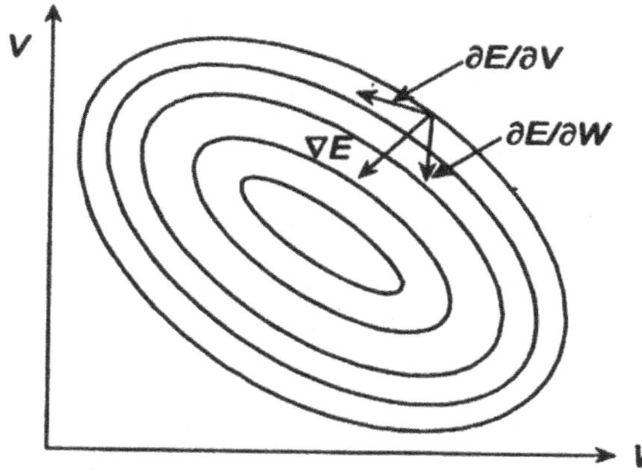

Fig. 7 Direction of steepest descent (two-dimensional view)

$$\overline{G} = \frac{\partial E}{\partial V}\overline{i} + \frac{\partial E}{\partial W}\overline{j} \tag{12}$$

Value of unit vector in gradient direction

$$\overline{AB} = -\eta\left[\frac{\partial E}{\partial V}\overline{i} + \frac{\partial E}{\partial W}\overline{j}\right] \tag{13}$$

By comparing

$$\Delta V = -\eta\frac{\partial E}{\partial V} \tag{14}$$

$$\Delta W = -\eta\frac{\partial E}{\partial W} \tag{15}$$

Figure 7 shows the resultant direction of the gradient which is at 90° of both $\frac{\partial E}{\partial V}$ and $\frac{\partial E}{\partial W}$. Velocity of changes also depends upon learning rate which is constant in simple steepest descent algorithm [25–28].

5.1 Calculation of Weight Update

For the kth neuron, error E_k is given by

$$E_k = \frac{1}{2}(T_k - O_{Ok})^2 \tag{16}$$

To compute $\frac{\partial E_k}{\partial W_{ik}}$ we apply chain rule

$$\frac{\partial E_k}{\partial W_{ik}} = \frac{\partial E_k}{\partial O_{Ok}} \frac{\partial O_{Ok}}{\partial I_{Ok}} \frac{\partial I_{Ok}}{\partial W_{ik}} \tag{17}$$

$$\frac{\partial E_k}{\partial O_{Ok}} = -(T_k - O_{Ok}) \tag{18}$$

$$I_{Ok} = W_{1k}O_{H1} + W_{2k}O_{H2} + \cdots + W_{mk}O_{Hm}$$

$$\begin{cases} O_{Ok} = \frac{1}{1+e^{-\lambda I_{Ok}}} \\ \frac{\partial O_{Ok}}{\partial I_{Ok}} = \lambda O_{Ok}(1 - O_{Ok}) \end{cases} \tag{19}$$

$$\frac{\partial I_{Ok}}{\partial W_{ik}} = O_{Hi} \tag{20}$$

Substitute Eqs. 18, 19, 20 in Eq. 17 and weight change is given by:

$$\Delta W = -\eta \frac{\partial E_k}{\partial W_{ik}} = \eta\lambda(T_k - O_{Ok})O_{Ok}(1 - O_{Ok})O_{Hi}$$

$$[\Delta W] = \eta\{O\}_H\{d\} \tag{21}$$

$$(m \times n) = (m \times 1)(1 \times n) \tag{22}$$

To compute $\frac{\partial E_k}{\partial V_{ij}}$ we apply chain rule of differentiation

$$\frac{\partial E_k}{\partial V_{ij}} = \frac{\partial E_k}{\partial O_{Ok}} \frac{\partial O_{Ok}}{\partial I_{Ok}} \frac{\partial I_{Ok}}{\partial O_{Hi}} \frac{\partial O_{Hi}}{\partial I_{Hj}} \frac{\partial I_{Hj}}{\partial V_{ij}} \tag{23}$$

$$\begin{aligned} \frac{\partial E_k}{\partial O_{Ok}} \frac{\partial O_{Ok}}{\partial I_{Ok}} &= -\{d\} \\ \frac{\partial I_{Ok}}{\partial O_{Hi}} &= W_{ik} \\ \frac{\partial O_{Hi}}{\partial I_{Hj}} &= \lambda(O_{Hi})(1 - O_{Hi}) \\ \frac{\partial I_{Hj}}{\partial V_{ij}} &= O_{Ij} = I_{Ij} \end{aligned} \tag{24}$$

$$\begin{aligned} \frac{\partial E_k}{\partial O_{Ok}} \frac{\partial O_{Ok}}{\partial I_{Ok}} \frac{\partial I_{Ok}}{\partial O_{Hi}} &= -\{d\}W_{ik} = -e_i \\ \frac{\partial E_k}{\partial O_{Ok}} \frac{\partial O_{Ok}}{\partial I_{Ok}} \frac{\partial I_{Ok}}{\partial O_{Hi}} \frac{\partial O_{Hi}}{\partial I_{Hj}} &= -e_i\lambda(O_{Hi})(1 - O_{Hi}) = -\{d^*\} \\ \frac{\partial E_k}{\partial O_{Ok}} \frac{\partial O_{Ok}}{\partial I_{Ok}} \frac{\partial I_{Ok}}{\partial O_{Hi}} \frac{\partial O_{Hi}}{\partial I_{Hj}} \frac{\partial I_{Hj}}{\partial V_{ij}} &= -\{d^*\}I_{ij} \end{aligned}$$

$$\frac{\partial E_k}{\partial V_{ij}} = -\{d^*\}I_{ij}$$

$$[\Delta V] = -\eta \frac{\partial E_k}{\partial V_{ij}} = \eta\{I\}_I\{d^*\} \tag{25}$$

5.2 Backpropagation Algorithm (BP)

1. Initialize b, w_1,w_2, …, w_n with some random value between 0 and 1
2. Repeat until converges
3. $y = b + \sum_{i=1}^{m}(w_i x_i)$
4. Calculate loss $j(b, w) = 1/2\sum_1^n (h_{w,b}(x^i - y^i)^2)$ where x^i is actual value and y^i is predicted value
5. $b^{new} = b^{old} - \eta * \text{gradient } b$
   ${w_1}^{new} = {w_1}^{old} - \eta * \text{gradient } w_1$
   ${w_2}^{new} = {w_2}^{old} - \eta * \text{gradient } w_2$
   -
   -
   ${w_n}^{new} = {w_n}^{old} - \eta * \text{gradient } w_n$
6. Update b, w_1, w_2, …, w_n simultaneously
7. Go to step 2.

6 Proposed Improvement

6.1 Momentum Update

Momentum simply considers fraction of the past weight to the current weight update. This helps backpropagation algorithm to prevent the model from getting stuck in local minima, even if the current gradient is 0.

Figure 8 shows the impact of momentum step on gradient step. The actual change is calculated as gradient step + momentum step, where gradient and momentum step both are vectors [29–31].

With simple momentum, the weight change equation can be rewritten as follows, where α is momentum factor:

$$\text{weight}(t+1) = \text{weight}(t) + \Delta\text{weight}(t+1) + \alpha(\Delta\text{weight}(t)) \tag{26}$$

Fig. 8 Momentum update on gradient step

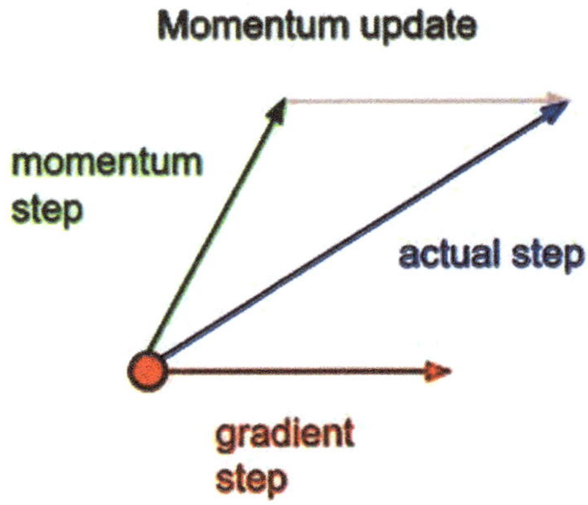

6.2 Learning Rate Annealing

The backpropagation algorithm can be improved by implementing annealing learning rate instead of constant learning rate. This annealing learning rate declines as the time progresses.

The most common relationship which can be defined between the learning rate and annealing learning rate is a **1/t** relationship where **T** and η are provided hyperparameters, and η_t is the current learning rate (Fig. 9) [27, 32]:

$$\eta = \eta_0/(1 + t/T) \tag{27}$$

Generally, in experiment, we find that if learning rate is too small the convergence is too slow and if it is too big the error component will not decrease in every iteration or may not converge.

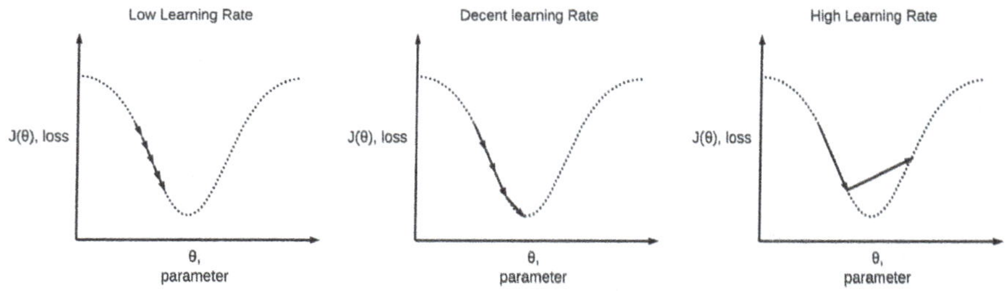

Fig. 9 Impact of learning rate on convergence

7 Conclusions

Backpropagation algorithm is widely implemented by neural network and machine learning in performing pattern recognition, optimization, approximation, classification, and data clustering. Steepest decent methods have been used to find out optimal solution. Paper proposes that the backpropagation algorithm can improve further by dynamic weight adjustments of links of neural network along with momentum coefficient to enhance the learning speed (Eq. 26).

Learning rate annealing is important because if the value of learning rate is set to be the large then the algorithms unable to converge to local minima and gradually decrease losses, which will lead to overshoot the local lowest value and if the value of learning rate is set to be small then algorithm may take too long time to converge.

We have tested the proposed algorithm on online simulator by implementing adaptable learning rate (learning rate annealing) (Eq. 27) and concept of momentum (Eq. 26). We find the proposed algorithm converges faster (i.e., epoch requirement is lower down) and having higher accuracy in comparison than simple backpropagation algorithm. The proposed algorithm also has a drawback that it does not guarantee to find global minima.

References

1. Keskar NS, Mudigere D, Nocedal J, Smelyanskiy M, Tang PTP (2016) On large-batch training for deep learning: generalization gap and sharp minima. arXivpreprint arXiv:1609.04836
2. He T, Zhang Z, Zhang H, Zhang Z, Xie J, Li M (2019) Bag of tricks for image classification with convolutional neural networks. In: Proceedings of the IEEE conference on computer vision and pattern recognition (CVPR), pp 558–567
3. Mutasem KSA, Khairuddin O, Shahrul AN (2009) Back propagation algorithm: the best algorithm among the multi-layer perceptron algorithm. Int J Comput Sci Netw Secur 9(4):378–383
4. Shamsuddin SM et al (2009) Study of cost functions in three term backpropagation for classification problems. In: World congress on nature & biologically inspired computing, NaBIC 2009. IEEE
5. Sharda R, Delen D (2006) Predicting box-office success of motion pictures with neural networks. Expert Syst Appl 30:243–254
6. Whitney TM, Meany RK (2006) Two algorithms related to the method of steepest descent. SIAM
7. Olson RS, Moore JH (2019) TPOT: a tree-based pipeline optimization tool for automating machine learning. In: Hutter F, Kotthoff L, Vanschoren J (eds) Automated machine learning. The springer series on challenges in machine learning. Springer, Cham
8. .
9. Majdi A, Beiki M (2010) Evolving neural network using genetic algorithm for predicting the deformation modulus of rock masses. Int J Rock Mech Min Sci 47(2):246–253
10. Bista R, Thapa A (2020) Handbook of wireless sensor networks: issues and challenges in current scenario's, advances in intelligent systems and computing, vol 1132. Springer, Cham, Switzerland, pp 239–259
11. Wang X (2008) Method of steepest descent and its applications. IEEE Microwave Wirel Compon Lett

12. Burse K, Manoria M, Kirar VPS (2010) Improved back propagation algorithm to avoid local minima in multiplicative neuron model. World Acad Sci Eng Technol Int J Electr Comput Eng 4(12)
13. Jastrzębski S, Kenton Z, Arpit D, Ballas N, Fischer A, Bengio Y, Storkey A (2017) Three factors influencing minima in SGD. arXiv preprint arXiv:1711.04623
14. Vora K, Yagnik S (2014) A new technique to solve local minima problem with large number of hidden nodes on feed forward neural network. IJEDR 2(2). ISSN: 2321-9939
15. Im DJ, Tao M, Branson K (2016) An empirical analysis of deep network loss surfaces. arXiv preprint arXiv:1612.04010
16. Ng SC, Leung SH, Luk A (1999) Fast convergent generalized back propagation algorithm with constant learning rate. Neural Process Lett 9:13–23
17. Liu H, Simonyan K, Yang Y (2019) DARTS: differentiable architecture search. In: International conference on learning representations (ICLR)
18. Singh PK, Bhargava BK, Paprzycki M, Kaushal NC, Hong WC (2020) Handbook of wireless sensor networks: issues and challenges in current scenario's, advances in intelligent systems and computing, vol 1132. Springer, Cham, Switzerland, pp 155–437
19. Singh PK, Kar AK, Singh Y, Kolekar MH, Tanwar S (2020) Proceedings of ICRIC 2019, recent innovations in computing, 2020, vol 597. Lecture notes in electrical engineering. Springer, Cham, Switzerland, pp 3–920
20. Yan H, Jiang Y, Zheng J, Peng C, Li Q (2006) A multilayer perceptron-based medical decision support system for heart disease diagnosis. Expert Syst Appl 30(2):272–281
21. Zou H, Xia G, Yang F, Yang H (2007) A neural network model based on the multi-stage optimization approach for short-term food price forecasting in China. Expert Syst Appl 33(2):347–356
22. He Y, Lin J, Liu Z, Wang H, Li L-J, Han S (2018) AMC: AutoML for model compression and acceleration on mobile devices. In: Proceedings of the European conference on computer vision (ECCV), pp 784–800
23. Howard AG, Zhu M, Chen B, Kalenichenko D, Wang W, Weyand T, Andreetto M, Adam H (2017) Mobilenets: efficient convolutional neural networks for mobile vision applications. arXiv preprint arXiv:1704.04861
24. Singh P, Paprzycki M, Bhargava B, Chhabra J, Kaushal N, Kumar Y (2018) Futuristic trends in network and communication technologies, FTNCT 2018. Communications in computer and information science, vol 958, pp 3–509
25. Li H, Kadav A, Durdanovic I, Samet H, Graf HP (2017) Pruning filters for efficient convents. In: International conference on learning representations (ICLR)
26. Ratner A, Bach SH, Ehrenberg HS et al (2020) Rapid training data creation with weak supervision. VLDB J 29:709–730
27. Nashed MZ (1970) Steepest descent for singular linear operator equations. SIAM J Numer Anal 7(3):358–362
28. Mcinerney JM, Haines KG, Biafore S, HechtNielsen R (1992) Can backpropagation error surfaces have non-global minima? In: Proceedings of the IEEE-IJCNN 8911-627
29. Gori M, Tesi A (1992) On the problem of local minima in backpropagation. IEEE Trans Pattern Anal Mach Intell 14(1):76–86
30. Chang LY (2005) Analysis of freeway accident frequencies: negative binomial regression versus artificial neural network. Saf Sci 43:541–557
31. Huang G, Liu Z, Van Der Maaten L, Weinberger KQ (2017) Densely connected convolutional networks. In: Proceedings of the IEEE conference on computer vision and pattern recognition (CVPR), pp 4700–4708
32. Zhang NM, Wu W, Zheng GF (2006) Deterministic convergence of gradient method with momentum for two-layer feed forward neural networks. IEEE Trans Neural Netw 17(2):522–525
33. Nikolopoulos K, Goodwin P, Patelis A, Assimakopoulos V (2007) Forecasting with cue information: a comparison of multiple regression with alternative forecasting approaches. Eur J Oper Res 180(1):354–368

Fingerprint and Face-Based Secure Biometric Authentication System Using Optimized Robust Features

Vandana and Navdeep Kaur

Abstract Security holds an integral position in every field. Numerous security measures and recognition system have been implemented to enhance the security aspects. In this pape, the authors have proposed two biometric recognition systems to deal with the security and authentication systems. Fingerprint recognition system (FPRS) is based on minutiae feature extraction of fingerprint image and in face recognition system (FRS). Viola–Jones algorithm (VJA) is implemented for face detection from static images. Image features are extracted using speeded up robust features (SURF) that is further optimized using genetic algorithm (GA). Recognition efficiency of the proposed systems is improved by training and classification of the optimized features based on feed-forward back-propagation neural network (FFBPNN). These unimodal biometric recognition systems are evaluated in terms of confusion matrix parameters, precision, TDR, f-measure and detection accuracy. Simulation results have established that the systems demonstrated an average recognition accuracy of 91.25% (FPRS) and 92.28% (FRS). Moreover, it has also been established that by increasing a sample size from 10 to 500 images, the recognition accuracy of the FPRS gets enhanced by 28.8% and FRS by 19%. Additionally, FRS outperformed FPRS by exhibiting 0.84% higher detection accuracy.

Keywords Biometric recognition · Viola–Jones algorithm · SURF · Genetic algorithm · Artificial neural network · Face recognition and fingerprint recognition

---

Vandana (✉)
Chitkara University Institute of Engg. and Technology, Chitkara University, Punjab, India
e-mail: vandana.bajaj@chitkara.edu.in

Vandana · N. Kaur
SGGSW University, Fatehgarh Sahib, Punjab, India
e-mail: Drnavdeep.sggswu@gmail.com

P. K. Singh et al. (eds.), *Evolving Technologies for Computing, Communication and Smart World*, Lecture Notes in Electrical Engineering 694,
https://doi.org/10.1007/978-981-15-7804-5_15

1 Introduction

A biometric system offers a realistic identification technique that takes the advantage of body's measurable characteristics [1]. Biometric recognition has been shown its applications that include restricted entry to buildings, limited access to computer systems and related gadgets that include ATM, mobiles, tablets, laptops, etc. [2]. These biometric systems may be based on soft or primary biometric traits. These systems gather extra information like height, age, gender, eye color, marital status in addition to the primary characteristic like iris, face, fingerprint, voice and hand geometry [3, 4]. However, only primary biometric characteristics hold the authenticated position to identify legitimate individuals. With the rising theft and malicious acts, top-level security systems have become the need of the hour [5]. Hence, the authentication system refers to the positive recognition systems where many individuals are restricted from taking the advantage of a single identity, whereas a negative recognition system identifies the individual as the one who is completely denied the peculiar access. To simplify the recognition systems, positive systems are more popular [6].

1.1 Motivation and Contribution

Privacy and authentication are big issues and need to create a more secure environment for users because in the all era, easy and wide availability of the information which must be secure is a major concern. Privacy and reliability of the information is critically important to users, and it has to be secured from unauthorized users or access. Security refers to prevent the information from unauthorized persons which can access some important data or precious assets [7, 8]. So, precise and secure biometric authentication system is needed in various applications such as home, mobile phones, banking services, ATM and motor vehicles. For this purpose, we designed a model using the fingerprint and face-based secure biometric authentication system using optimized robust features by utilizing the genetic algorithm with feed-forward back-propagation neural network (FFBPNN) as classifier.

The rest of the paper is organized as: Sect. 2 describes the state of the art in the biometric recognition system field and covers papers from 1998 to 2019. The step-by-step procedure of the research work followed for fingerprint and face recognition is discussed in Sect. 3. The results with detail description and the comparison of the proposed work with the existing work are provided in Sect. 4. In Sect. 5, the concluding remarks are drawn followed by references.

2 Related Work

Success of a recognition system is measured in terms of high recognition accuracy, fast execution with least flaws. A number of researchers have proposed various recognition systems and approaches to enhance the potentials of biometric-based recognition systems. In 1998, Nefian and Hayes developed HMM-based face recognition technique. The simulation results had demonstrated considerably lowered computation difficulties of the design while conserving the high recognition efficiency as compared to other recognition approaches [9]. Phillips et al. in 2002 summarized the facial characteristic that forms the core of face detection system. They included facial features such as, shape and location of nose, eyes, eyebrows, chin, lip and their overall spatial relationship. Authors covered the issues of recognition systems that arise due to varied illumination levels and background characteristics. They concluded that same facial image under different illumination levels raised the complexity for effective recognition [10]. Over the decades, interest toward fingerprint-based recognition has also shown a great rise. In 2002 Maio et al. presented a platform to confront merits and issues adjoining fingerprint recognition systems. It was a common platform where institutes and companies could explicitly compare and contrast their performance. In the process four datasets were produced. One was artificially produced while the three were based on high-tech sensors. The work disseminated under FVC2000 offered a standard for cross evaluation in future research [11]. Jain et al. 2004 presented a review to cover the concept of biometric based identifications. They covered the evolution of biometric technology, various traits and related risk management. Enhanced market interaction and application adjoining biometric technology were also covered in addition to its defense, privacy and liability aspects. They concluded their research while discussing the major merits, strengths and limitations of biometric technology [2]. Więcław in 2009 described the minutia-based fingerprint matching technique. He exemplified the minutia extraction followed by matching as a two-step mechanism involved in the recognition system. The major challenging task in this process was extracting a high-quality fingerprint with high-quality minutia visualization. Further, it was established that during matching stage, transformation method and tolerance distance needed a wiser selection because a wrong transformation of minutiae would result in higher false rejection rate [12]. Malhotra and Kumar in 2017 were inspired with nature-based methodologies and implemented Cuckoo search algorithm for designing a face recognition system. They used discrete cosine transform (DCT) and principal component analysis (PCA) for better feature extraction. The proposed design was also evaluated against as particle swarm optimization (PSO) and genetic algorithm (GA). The simulation results had shown that the proposed work exhibited a recognition rate of 88% when 10 features were used and recognition rate gets improved to 96.5% with increasing the PCA features to 34 [13]. Cao and Jain in 2018 studied the available fingerprint techniques and summarized that latent fingerprints exhibited the most widely used technology by forensic and law departments. Inspired with the fact, they proposed an automatic latent fingerprint recognition system based on convolutional neural networks. The method works by recognizing the minutiae from extracted and the template. The scores obtained by comparing latent and the test fingerprint were used to extract a list of candidates from a reference or training dataset. The experimental evaluation had exhibited a accuracy of 64.7% with NIST and 75% with WVU latent datasets [14]. Kalita and Saikia in 2018 were greatly inspired with the astonishing recognition ability of human brain in identifying individuals despite of changes in terms of hairstyle, bearded, mustaches, aging, glasses, etc. In designing a face recognition systems, the authors tried to mimic the strength of human brain. Face recognition system had challenged the fields of image processing and computer vision. The authors had summarized the aspects of face recognition, challenges, hurdles and the applications that were employed to successfully deal the rising issues [15]. Sandar and Oo in 2019 designed a face recognition system to offer secure door lock. The design was implemented using Raspberry Pi and GSM and took the advantage of Haar-like features for detection. The facial recognition was achieved by employing local binary pattern method, and GSM unit was used to raise alert. The proposed design was found to be time saving when compared to the normal systems [16]. In 2019, Niaraki and Shahbahrami were greatly inspired with the potential applications of face recognition systems in banking sector in addition to door access authentication systems. They proposed an improved face recognition system that combined the strengths of local median binary pattern (LMBP) and its co-occurrence matrix. The design was evaluated using images from Yale and ORL databases, and results demonstrated high recognition accuracy [17]. In 2019, Aworinde et al. have examined the performance of feature extraction approaches such as kernel principal component analysis (K-PCA) and kernel linear discriminant analysis (KLDA) by using 1054 population of Nigeria group along with deep learning concept [18]. Alsmirat et al. have conducted research on a large dataset, which have been collected manually in 2019. The raw images are then compressed and have been examined that the images retain their quality up to the compression ratio of (30–40%) of the test images [19].

From the literature analysis, we founded some major limitations faced by biometric systems which are given as follows:

- System performance is low in terms of accuracy due to the presence of noise in sensed biometric data.
- Due to lack of optimization approach, false accept rate (FAR) increases with less unique set of biometric features.
- The spoofing of the existing biometric traits is easily available for an intruder which can create authenticity of biometric model due to lack good training algorithms.
- From the survey, the uses of behavioral biometric can open a massive way for an intruder to attack on the user's privacy and security.

3 Proposed Biometric Designs

The proposed biometric recognition system constitutes of two independent biometric recognition systems, namely fingerprint recognition system (FPRS) and face recognition system (FRS) described one by one.

3.1 Fingerprint Recognition System (FPRS)

Database: The images for fingerprint were obtained from Fingerprint Verification Competition (FVC). Database consists of four databases, namely DB1, DB2, DB3 and DB4. The fingerprint data consists of students enrolled for computer science degree program in Italy. The image data represents high-resolution grayscale images available in *.tif format. The database consists of total 1440 impressions that include 120 fingers with 12 different impressions [20].

FPRS Methodology: FPRS works on the basis minutiae features extracted from the input fingerprint image obtained from the above-mentioned database. The quality of fingerprint image is enhanced using preprocessing technique followed by minutiae feature extraction. These features are further optimized using GA followed by training and classification based on ANN. The framework of employed methodology for FPRS is shown in Fig. 1.

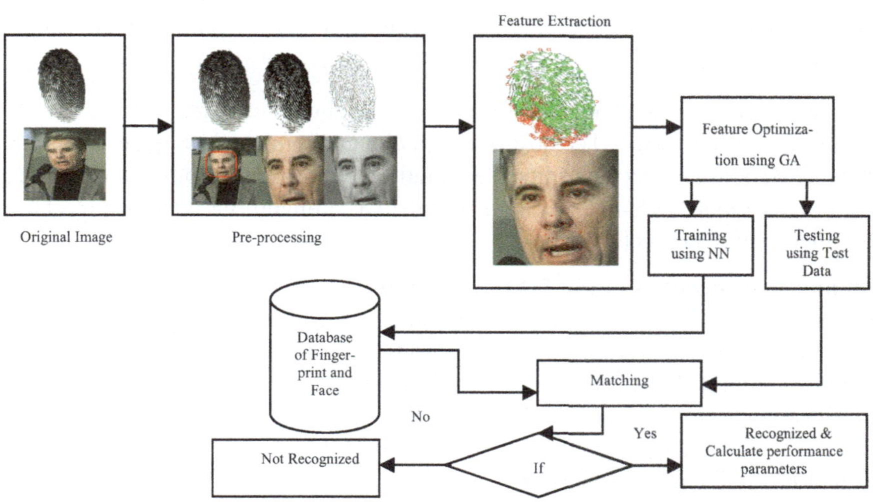

Fig. 1 Block diagram of fingerprint recognition system (FPRS) with face recognition system (FRS)

Preprocessing: In this step, quality of image is enhanced that facilitates feature extraction and image analysis. It can be understood as a process of image normalization commonly implemented in most of the feature extraction techniques. The morphological operations in the preprocessing involve color conversion, binarization and morphological operation [21].

Minutiae Feature Extraction: A fingerprint image has minutiae points as the key features that are employed in the FPRS. These unique features are used to determine the identity of an individual [22, 23]. The improved fingerprint image from the last step is fed to minutiae algorithm for minutiae feature extraction as follows:

Algorithm 1: Minutiae Feature Extraction Algorithm

1. **Input:** $Image_{thin}$**// Input improved image obtained after thinning process**
2. **Determine:** $[R_{value}, C_{value}] = size(Image_{thin})$**// Find number of rows and columns**
3. $For_{each} i_{value} in R_{value}$
4. $For_{each} j_{value} in C_{value}$
5. **Calculate:** $Centroid_{(i,j)} = properties(Image_{thin}, ridge)$
6. **Check:** $if Centroid_{(i,j)} = 1$**// Represents termination**
7. **Assign:** $Term_{(i,j)} = incr_1 + 1$
8. $else Bifur_{(i,j)} = incr_2 + 1$**// Represents bifurcation**
9. End_{if}
10. End_{for}
11. End_{for}
12. $Minutiae_{points} = [Term_{(i,j)}, Bifur_{(i,j)}]$
13. **Output:** $Minutiae_{points}$**// Returns minutiae feature points of the fingerprint image.**

The algorithm is implemented for minutiae feature extraction from the preprocessed fingerprint image. The size of morphologically thinned image is calculated that is used to iteratively determine the minutiae features such as bifurcations, linear and circles. Finally, $Minutiae_{points}$ is returned as a fingerprint image representing minutiae feature points [24, 25].

Feature Optimization Using GA: The minutiae features of the fingerprint image extracted using minutiae algorithm are optimized using genetic algorithm in this step. Genetic algorithm is a heuristic approach that is biologically inspired from the natural evolution. The idea here is to obtain the best solution in the light of defined criteria and constraints [26, 27]. The steps of genetic algorithm are as follows:

Algorithm 2: Genetic Algorithm

1. **Input:** $Minutiae_{points}$**// Minutiae feature points of fingerprint image**
2. **Initialize parameters:**
3. Itr_{value}**// Number of iterations**
4. P_{size}**// Population size**

5. **$F_{crossover}$// Crossover function**
6. **$F_{mutation}$// Mutation function**
7. **F_{fit}// Fitness function**
8. **$F_{selected}$ and $F_{threshold}$// Selection functions**
9. **Calculate: $T_{value} = size(Minutiae_{points})$// Calculate size of image**
10. **$For_{each} i_{value}$ in Itr_{value}**
11. **Calculate: $F_{selected} = \sum_{i=1}^{P_{size}} f(i)$// Determine function based on size**
12. **Calculate: $F_{threshold} = \frac{\sum_{i=1}^{P_{size}} f(i)}{Length_{feature}}$// Determine function based on feature length**
13. **Calculate fitness:**
14. **$F_{fit} = \begin{cases} 1, & F_{selected} < F_{threshold} \\ 0, & F_{selected} \geq F_{threshold} \end{cases}$ // Determine fitness value**
15. **Initialize variable: $N_{variable} = 1$// Number of variable**
16. **Calculate:**
17. **$F_{optimized} = GA(F_{fit}, N_{variable}, P_{size}, F_{crossover}, F_{mutation})$**
18. ***End_{for}***
19. ***While $Itr_{value} \neq max(T_{value})$***
20. ***Assign: $F_{points} = F_{optimized}$***
21. ***End_{while}***
22. ***$Image_{optimized} = optimize(F_{points})$***
23. ***Output: $Image_{optimized}$*// Returns optimized feature points of fingerprint image.**

The algorithm inputs the minutiae feature points of preprocessed fingerprint image. Population size is initialized that represents the number of solutions. Fitness function is calculated that determines the fitness score for individual image. The fitness value corresponds to the quality of resultant solution. Optimization is performed based on features such as population size, fitness function, crossover function and mutation function. Over a number of iterations, the best fit is selected and high-quality optimized solution or image is obtained. This optimized solution or optimized image is further processed in training and classification step.

Training and Classification using FFBPNN: A feed-forward back-propagation neural network is a non-cyclic artificial neural network in which information is passed from input neuron layer to output neuron layer through hidden layer. The classification accuracy of the recognition system is enhanced by training the optimized features using FFBPNN to create a training dataset. This biologically inspired algorithm is further used to enhance the classification accuracy of the recognition system [28]. The steps of FFBPNN algorithm are as follows:

Algorithm 3: FFBPNN Algorithm

1. **Input Parameters: T_{data}// Optimized training data**
   $Target_{data}$// Target data
   N_{num}// Neurons

2. **Initialize FFBPNN parameters:**
   E_{num}// Number of Epochs
   N_{num}// Number of Neurons
   P_{para}// Performance parameters of training
   $Tech_{LM}$// Employed technique as Levenberg Marquardt
3. $For_{each} i_{value}$ *in* T_{data}
4. **Calculate:** $G_i = categories(T_{data})$**// Define groups for various categories of training data**
5. End_{for}
6. $T_{netdata} = Newff(T_{data}, Target_{data}, N_{num})$**// Initialize FFBPNN using training data and groups**
7. $T_{ndata} = Train(T_{netdata}, T_{data}, G_{data})$**// Perform training of the system**
8. **Output:** T_{ndata}**// FFBPNN Trained Structure**

The above algorithm inputs the optimized training dataset along with number of neurons and target data. Based on the training data, images are categorized into various groups. In the process, various parameters of FFBPNN such as epoch number, number of neurons and employed technique are initialized. The output of the neural network is fed back to input neuron layer. In case of any fault, weight is adjusted to reach a required output. The input data is finally trained, and trained structure is returned in T_{ndata}. FFBPNN is used for both training and classification of fingerprint images [29].

3.2 Face Recognition System (FRS)

Database: The images for face recognition framework were obtained from Georgia Tech face database that comprises the images of 50 individuals obtained at the Georgia Institute of Technology. The image size of the faces consists of the square matrix of 150 pixels. Background pruned face images are also available in the database. Fifteen color images exhibit a *.jpeg format [30].

FRS Methodology: In FRS, the quality of face image is also enhanced using preprocessing approaches in a similar fashion as in FPRS. In the next step, face portion is extracted from the preprocessed image using Viola–Jones algorithm. The features of the face image are extracted using SURF algorithm followed by feature optimization implemented with GA. Finally, recognition strength of FRS is enhanced using training and classification performed using ANN.

Face Detection and Preprocessing: Face in itself is a very complex structure. As such, it requires multidimensional and high-level recognition techniques. To extract the face region from the whole image, Viola–Jones approach is used. Viola–Jones is a feature-based face detection method designed by Viola and Jones in 2000 that consists of three major parts, namely Haar feature followed by AdaBoost and cascading [31–33].

Feature Extraction: The extracted face region is further used to extract its corresponding features by applying SURF algorithm. The major steps of SURF algorithm are as follows:

Algorithm 4: SURF Algorithm

1. **Input:** $Image_{pre}$**// Preprocessed face image**
2. $For_{each} i_{value}$ in $Image_{pre}$
3. **Calculate:** $Ext_{det} = Image_{pre}(i)$**// Detect extremes of the** i**th part of preprocessed image**
4. $Local_{keypoint} = Ext_{det}$**// Find key localization point of detected extrema**
5. **Check: if** $Local_{keypoint}$ $need\ reoriented == 'YES'$
6. **Assign:** $Local_{orint} = Local_{keypoint}(i)$ **// Find orientation point in the vicinity of** $Local_{keypoint}$
7. End_{if}
8. $F_{points} = Best F_{points}$**// Stores all the best feature points of individual parts of face image**
9. End_{for}
10. **Output:** F_{points}**// Feature points of face image.**

The enhanced face image obtained as a result of preprocessing and face detection using VJ is used as input for SURF algorithm. Iteratively, extreme points are detected that are further used to evaluate for the occurrence of any other location of the key extreme point in its vicinity. If any other extrema point gets detected, the localization point is reassigned, otherwise assigned to be the key localization point for considered pixel area. Over the number of iterations, the best feature points are stored as F_{points} that are further optimized to enhance the quality of recognition system.

Feature Optimization: The extracted features of the face image are further optimized using genetic algorithm. The steps of the algorithm remain same as discussed in FPRS mentioned under Sect. 3.1.

Training and Classification: This step is also common among the two recognition systems. The ANN is used for training optimized feature points of face image as in Sect. 3.1. In the next step, classification or testing of the uploaded images is performed in order to evaluate the recognition accuracy of the FRS. Overview of the employed methodology for FRS is shown in Fig. 1.

4 Results

The section covers the detailed observation and discussion of results for both FPRS and FRS separately. Starting with experimental evaluation parameters, results of recognition system are discussed and compared with the other literature cited work.

4.1 Experimental Evaluation of FPRS and FRS

The performance of the two biometric systems is separately evaluated in terms of confusion matrix parameters, precision, f-measure, recognition accuracy and True Detection Rate (TDR). The parametric calculations are done using following formulae:

$$\text{Precision} = \frac{\text{True}_{\text{Positive}}}{\text{True}_{\text{Positive}} + \text{False}_{\text{Positive}}} \quad (1)$$

$$\text{Accuracy} = \frac{\text{True}_{\text{Positive}} + \text{True}_{\text{Negative}}}{\text{True}_{\text{Positive}} + \text{True}_{\text{Negative}} + \text{False}_{\text{Positive}} + \text{False}_{\text{Negative}}} \quad (2)$$

$$\text{TDR} = \frac{\text{True}_{\text{Positive}}}{\text{True}_{\text{Positive}} + \text{False}_{\text{Negative}}} \quad (3)$$

$$F_{\text{measure}} = 2 * \frac{\text{Precision} * \text{TDR}}{\text{Precision} + \text{TDR}} \quad (4)$$

4.2 Image Processing Results of FPRS and FRS

Table 1 summarizes the results of three best results of FPRS. The quality of original fingerprint image is enhanced using preprocessing morphological operations. Minutiae features of the fingerprint image are recognized from the binary image that is also referred to region of interest (RoI) of the image. Feature extracted image displays the terminations in red color points and bifurcation points in green color points. These minutiae points optimized and are used to train and test the proposed system.

Table 1 Fingerprint image at different stages of FPRS

Fingerprint image	Original uploaded image	Enhanced image	Binary image	Thinning operation	Feature extraction
FP image 1					
FP image 2					
FP image 3					

Table 2 Face image at different stages of FRS

Face image	Original uploaded image	Face region selection	Face extraction	Preprocessed image	Feature extraction
FI 1					
FI 2					
FI 3					

Similarly, Table 2 summarizes the three best image processing results of FRS. The second column displays the original image that is preprocessed and face is detected using VJ algorithm. Following this, features of face image are extracted using SURF. Image with extracted feature is shown in sixth column that is further optimized using GA to enhance the quality of the recognition system.

4.3 *Performance Parameters of FPRS and FRS*

In the experimentation, 10–500 image samples of fingerprint and face images were evaluated. The parameters of performance matrices for precision, true detection rate (TDR), f-measure and accuracy observed for FPRS are depicted in the following section.

Observed precision, TDR, f-measure and accuracy values are evaluated against number of fingerprint image samples in Fig. 2a. The graph shows that for an image sample of 10 images, precision of 85.8% is observed. With increase in the number of image samples to 20, 50, 100, 200 and 500, precision gradually increases to

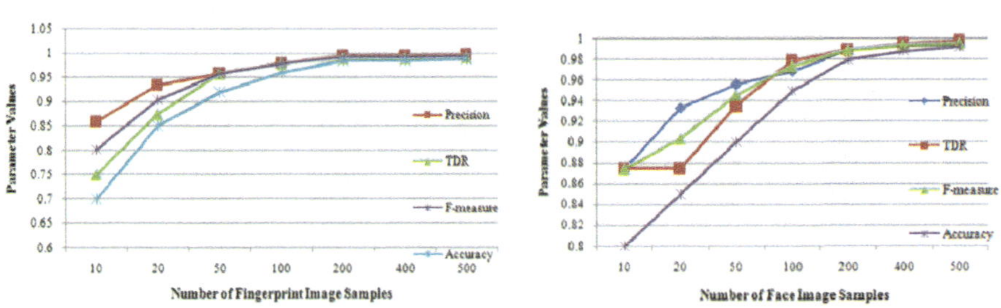

(a) (b)

Fig. 2 Parametric plot of precision, TDR, f-measure and accuracy for **a** FPRS and **b** FRS

93.4%, 95.8%, 97.9% and gets stabilized near 99.5% for rest of image samples. Similar, trend is observed with each of TDR, f-measure and accuracy. \query{Please check the clarity of the sentence 'For a sample size of 10 images, observed TDR, f-measure and accuracy…'.}For a sample size of 10 images, observed TDR, f-measure and accuracy of 75%, 80.1% and 70% is observed that increases to 99.2%, 99.4% and 98.8% for fingerprint image sample of 500 images. Overall, it is observed that with increase in the number of fingerprint image samples, there is a corresponding rise in the individual parameters and accuracy of system gets improved by 28.8%.

The parametric values of precision, TDR, f-measure and accuracy are plotted against number of image sample of 10–500 images in Fig. 2b. Over number of fluctuations, an overall increase in precision, TDR and f-measure have been observed with increase in the sample size. Accuracy of the face recognition system has shown a steep increase from 80 to 98% with rise in the sample size from 10 to 200 images. \query{Please check the edits made in the sentence 'Later, from 200 to 500 image samples, relatively…'.}Later, from 200 to 500 image samples, relatively gradual increase in the accuracy amounting to 98%, 98.8% and 99.2% has been observed. Hence, accuracy of system is improved by 19% by increases the sample size from 10 to 500 face images.

4.4 Comparison of the Proposed Recognition Systems Against Existing Work

The proposed FPRS was also evaluated with the existing fingerprint recognition systems. Figure 3a shows the comparison of the proposed FPRS against two fingerprint-based biometric recognition systems proposed by Kumar and Zhang in 2006 [28] and Galbally et al. in 2012 [29]. Kumar and Zhang had achieved fingerprint recognition accuracy of 89% with SVM, and Galbally et al. had achieved fingerprint recognition accuracy of 90% based on fingerprint parameterization. Our proposed FPRS outperformed both the recognition systems by achieving a recognition accuracy of 91.44%.

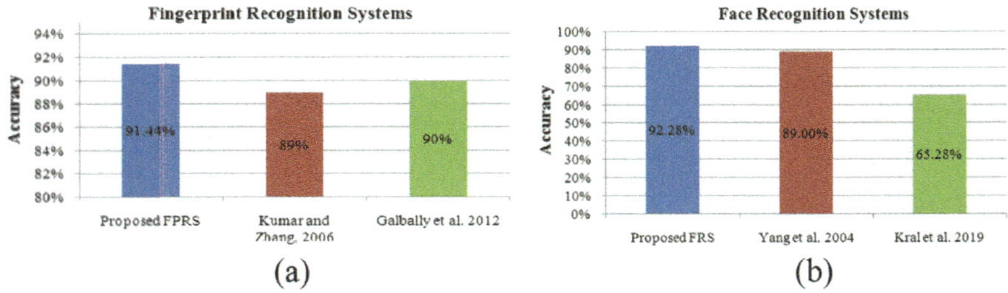

Fig. 3 Comparison of the proposed with the existing work for **a** FPRS and **b** FRS

Similarly, recognition accuracy of the proposed FRS is also evaluated against the literature cited face recognition-based biometric systems. Figure 3b shows the comparison of the proposed FRS against two face recognition systems proposed by Yang et al. in 2004 [30] and Kral et al. in 2019 [31]. Yang et al. had implemented a two-dimensional principal component analysis system to achieve a recognition accuracy of 98%, while Kral et al. work was based on enhanced local binary patterns and exhibited a recognition accuracy of 65%. In contrast to these two works, our proposed face recognition system demonstrated a high recognition accuracy of 92.28%.

5 Conclusion

The proposed work comprises two independent biometric recognition systems; one is based on fingerprint recognition and other is based on face recognition system because of the easy availability of face and fingerprint sensors. The fingerprint recognition system (FPRS) is based on minutiae features, and the face recognition system (FRS) is based on Viola–Jones and SURF-based face feature detection system. The extracted features in both the systems are optimized using genetic algorithm. Performance of the systems is enhanced by implementing FFBPNN for training and classification of the image samples. The simulation results over 10–500 image samples were evaluated in terms of confusion matrix parameters along with precision, TDR, f-measure and accuracy. It has been observed that TP for both FPRS and FRS increases with increase in the sample size with corresponding decrease in TN, FP and FN values. It has been observed that accuracy of FPRS and FRS is improved by 28.8% and 19% by increasing the sample size from 10 to 500 images. Similar trend has been observed in case of precision, TDR and f-measure. The proposed FRS has also outperformed the proposed FPRS by demonstrating 0.84% higher recognition accuracy. In the future, the work can be extended by fusion of face and fingerprint feature to create a multimodal biometric system.

References

1. Gressel CD (2001) U.S. Patent No. 6,311,272. U.S. Patent and Trademark Office, Washington, DC
2. Jain AK, Ross A, Prabhakar S (2004) An introduction to biometric recognition. IEEE Trans Circ Syst Video Technol 14(1)
3. Jain AK, Dass SC, Nandakumar K (2004) Soft biometric traits for personal recognition systems. In: International conference on biometric authentication. Springer, Berlin, Heidelberg, pp 731–738
4. Prabhakar S, Pankanti S, Jain AK (2003) Biometric recognition: security and privacy concerns. IEEE Secur Priv Mag 1(2):33–42
5. Nareshkumar RM, Kamat A, Shinde D (2017) Smart door security control system using Raspberry Pi. Int J Innov Adv Comput Sci 6(11)

6. Wayman L (2001) Fundamentals of biometric authentication technologies. Int J Image Graphics 1(1):93–113
7. Hathaliya JJ, Tanwar S, Tyagi S, Kumar N (2019) Securing electronics healthcare records in Healthcare 4.0: a biometric-based approach. Comput Electr Eng 76:398–410
8. Mehmood R, Selwal A (2020) Fingerprint biometric template security schemes: attacks and countermeasures. In: Proceedings of ICRIC 2019. Springer, Cham, pp 455–467
9. Nefian AV, Hayes MH (1998) Hidden Markov models for face recognition. In: Proceedings of the 1998 IEEE international conference on acoustics, speech and signal processing, ICASSP'98 (Cat. No. 98CH36181). IEEE, pp 2721–2724
10. Phillips PJ, Grother P, Micheals RJ, Blackburn DM, Tabassi E, Bone JM (2003) FRVT 2002. Overview and summary
11. Maio D, Maltoni D, Cappelli R, Wayman JL, Jain AK (2002) FVC2000: fingerprint verification competition. IEEE Trans Pattern Anal Mach Intell 24(3):402–412
12. Więcław Ł (2009) A minutiae-based matching algorithms in fingerprint recognition systems. J Med Inf Technol 13
13. Malhotra P, Kumar D (2019) An optimized face recognition system using cuckoo search. J Intell Syst 28(2):321–332
14. Cao K, Jain AK (2018) Automated latent fingerprint recognition. IEEE Trans Pattern Anal Mach Intell 41(4):788–800
15. Kalita N, Saikia LP (2018) A survey on face recognition based security system and its applications. Int Res J Eng Technol 5(6):1664–1666
16. Sandar S, Oo SAN (2019) Development of a secured door lock system based on face recognition using Raspberry Pi and GSM module. Development 3(5)
17. Niaraki RJ, Shahbahrami A (2019) Accuracy improvement of face recognition system based on co-occurrence matrix of local median binary pattern. In: 2019 4th international conference on pattern recognition and image analysis (IPRIA). IEEE, pp 141–144
18. Aworinde HO, Afolabi AO, Falohun AS, Adedeji OT (2019) Performance evaluation of feature extraction techniques in multi-layer based fingerprint ethnicity recognition system. Asian J Res Comput Sci 1–9
19. Alsmirat MA, Al-Alem F, Al-Ayyoub M, Jararweh Y, Gupta B (2019) Impact of digital fingerprint image quality on the fingerprint recognition accuracy. Multimedia Tools Appl 78(3):3649–3688
20. FVC2004 database. Accessed URL https://bias.csr.unibo.it/fvc2004/databases.asp
21. Anwarul S, Dahiya S (2020) A comprehensive review on face recognition methods and factors affecting facial recognition accuracy. In: Proceedings of ICRIC 2019. Springer, Cham, pp 495–514
22. Wang L, Leedham G, Cho DSY (2008) Minutiae feature analysis for infrared hand vein pattern biometrics. Pattern Recogn 41(3):920–929
23. Prabhakar S, Jain AK, Pankanti S (2003) Learning fingerprint minutiae location and type. Pattern Recogn 36(8):1847–1857
24. Yang W, Zhang X, Li J (2020) A local multiple patterns feature descriptor for face recognition. Neurocomputing 373:109–122
25. Gosavi VR, Sable GS, Deshmane AK (2018) Evaluation of feature extraction techniques using neural network as a classifier: a comparative review for face recognition. Int J Sci Res Sci Technol 4(2):1082–1091
26. Tan X, Bhanu B (2006) Fingerprint matching by genetic algorithms. Pattern Recogn 39(3):465–477
27. Girgis MR, Sewisy AA, Mansour RF (2009) A robust method for partial deformed fingerprints verification using genetic algorithm. Expert Syst Appl 36(2):2008–2016
28. AL-Allaf ONA, AbdAlKader SA, Tamimi AA (2013) Pattern recognition neural network for improving the performance of iris recognition system. J Sci Eng Res 4
29. Thomas T, Vijayaraghavan AP, Emmanuel S (2020) Neural networks and face recognition. In: Machine learning approaches in cyber security analytics. Springer, Singapore, pp 143–155
30. Georgia Tech face database. Accessed URL https://www.anefian.com/research/face_reco.htm

31. Damanik RR, Sitanggang D, Pasaribu H, Siagian H, Gulo F (2018) An application of viola jones method for face recognition for absence process efficiency. J Phys Conf Ser 1007(1):012013 (IOP Publishing)
32. Wang YQ (2014) An analysis of the Viola-Jones face detection algorithm. Image Proc Line 4:128–148
33. Viola P, Jones M (2001) Rapid object detection using a boosted cascade of simple features. CVPR 1(1):511–518 (Accepted Conference on Computer Vision and Pattern Recognition)
34. Kumar A, Zhang D (2006) Personal recognition using hand shape and texture. IEEE Trans Image Process 15(8):2454–2461
35. Galbally J, Alonso-Fernandez F, Fierrez J, Ortega-Garcia J (2012) A high performance fingerprint liveness detection method based on quality related features. Future Gener Comput Syst 28(1):311–321
36. Yang J, Zhang D, Frangi AF, Yang JY (2004) Two-dimensional PCA: a new approach to appearance-based face representation and recognition. IEEE Trans Pattern Anal Mach Intell 26(1):131–137
37. Král P, Vrba A, Lenc L (2019) Enhanced local binary patterns for automatic face recognition. In: International conference on artificial intelligence and soft computing. Springer, Cham, pp 27–36

DevOps, DevSecOps, AIOPS- Paradigms to IT Operations

Abhijit Sen

Abstract DevOps, the widely used buzzword in the software industry, is an umbrella term to describe the technologies and processes used to bring together software development and IT operations to build and deliver software in a repeatable and automated manner. Because of widespread breach of personal and confidential data, the software industry is increasingly taking steps to build security in to the various stages of DevOps lifecycle. This process of integrating security as a key element within the DevOps life cycle is called DevSecOps. Recent advances in artificial intelligence and machine learning techniques can augment DevOps processes by automating the path from development to production, predicting the effects of deployment on production environments, and automatically responding to changes in the environment. This process of incorporating AI in DevOps life cycle is called AIOps. In this paper, the author discusses the recent trends and some challenges in the software industry to incorporate security and artificial intelligence into the various stages of the DevOps life cycle to deliver efficient, reliable, and secured software.

Keywords DevOps · DevSecOps · Aiops

1 Introduction

One of the barriers in traditional software development and release process is the perceived lack of collaboration, coordination, and communication between the development team and the operations team that work in isolation [1]. The result of two teams working in isolation thwarted the pace of the software development and the release of new software applications or features [1]. The lack of collaboration and communication between development and operations teams is more pronounced in organizations where there are many different IT teams and business units.

A. Sen (✉)
Kwantlen Polytechnic University, Surrey, BC V3W 2M8, Canada
e-mail: abhijit.sen@kpu.ca

P. K. Singh et al. (eds.), *Evolving Technologies for Computing, Communication and Smart World*, Lecture Notes in Electrical Engineering 694,
https://doi.org/10.1007/978-981-15-7804-5_16

The DevOps approach to software development tends to address the long time problem by emphasizing collaboration between development and operations by sharing responsibilities and working cooperatively across the entire system development lifecycle. DevOps process attempts to build a culture of collaboration between teams that historically functioned in relative isolation. The development and operations team work together throughout all phases of the software life cycle from software design, development to software release to customer. With this emphasis of cultural change between development and operational teams, the entire application development life cycle can be improved and streamlined. DevOps implements the concept of "shift left" and uses software development practices by shifting responsibilities of delivering quality product to all members of both development and operational teams in all phases of life cycle from the beginning to product deployment.

As a result, overall productivity will be increased, features can be delivered faster, and fixes can be released more frequently to meet changing business needs.

In recent years, several major organizations have been impacted by data breaches, including Yahoo, First American Financial Corp., Facebook, and Marriott international [2]. Data breaches can cause massive losses, resignations/terminations, and overall loss in trust from end-consumers damaging reputation of an organization.

The natural question to ask is how to add security into the DevOps process. The security principles should be integrated into the development process from the beginning, not merely as an afterthought, when security breaches have already affected the business. Inclusion of security into DevOps workflow has resulted in an extension to DevOps, called DevSecOps. The primary purpose of DevSecOps is to identify and application vulnerabilities earlier and take appropriate remedial or preventive actions.

The core impetus behind DevOps adoption is to improve flexibility and agility in the software development process. As Forrester, 2019 report [3] indicates large number of industries have either implemented or planning to implement artificial intelligence(AI) and machine learning (ML) techniques to automate various elements in the development life cycle phases from planning to deployment.

Algorithmic IT operations or AIOps for short embeds AI, ML techniques into DevOps workflow to support various IT tasks. The principal motivation of using AIOps in DevOps lifecycle is to apply tools of AI, and ML to better monitor data, application performance, and IT systems management processes and to quickly anticipate problems, identify the root cause of any issues, and suggest appropriate solutions. With these developments, enterprises are looking and implementing the process by merging DevOps, DevSecOps, and AIOps in system development life cycle.

The rest of the paper is organized as per the following manners.

Section 2 discusses DevOps principles and paradigms. Section 3 discusses DevSecOps principles and paradigms. Section 4 discusses AIOps principles and paradigms. Section 5 discusses sample tools available for each of three paradigms followed by conclusion in Sect. 6.

2 DevOps Paradigm—Principles and Processes

DevOps is a set of practices that merges software development (Dev) and information technology operations (Ops) to improve application delivery process, thereby increasing frequency of production releases of high quality software as shown in Fig. 1.

DevOps principle emphasizes communication, collaboration, and integration between development and operations teams to ensure continuous delivery of applications and services to end-users (Fig. 2).

DevOps is achieved through the principles of:

- Continuous planning
- Collaborating development
- Continuous testing, release, and deployment
- Continuous monitoring.

The major outcome of implementing DevOps is continuous integration (CI) and continuous deployment (CD) pipeline to deliver applications features more frequently. The whole process is supported by introducing automation tools for building, configuration testing, continuous monitoring and release tools. The software development process is shown in Fig. 3 [4].

It consists of eight stages—Plan, Code, Build, Test, Release, Deploy, Operate, and Monitor, where DEV consists of {Plan, Code, Build, Test, Release, and Deploy} segments, and Ops consists of {Operate, and Monitor} segments. As the figure indicates, the DevOps process is not sequential, but iterative in nature. The whole

Fig. 1 Merging developments with IT operations

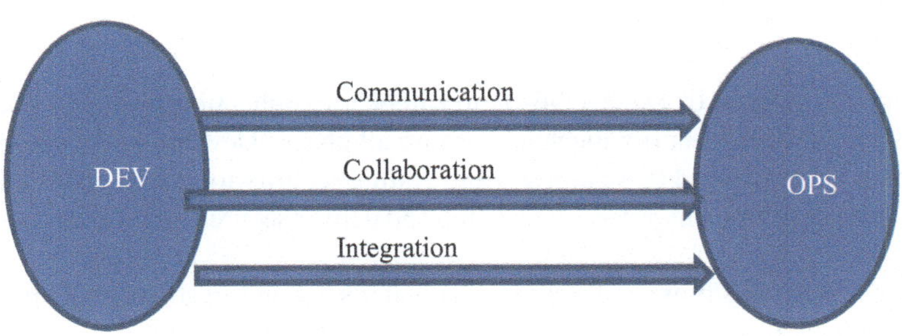

Fig. 2 DevOps principles: Communication, collaboration, integration

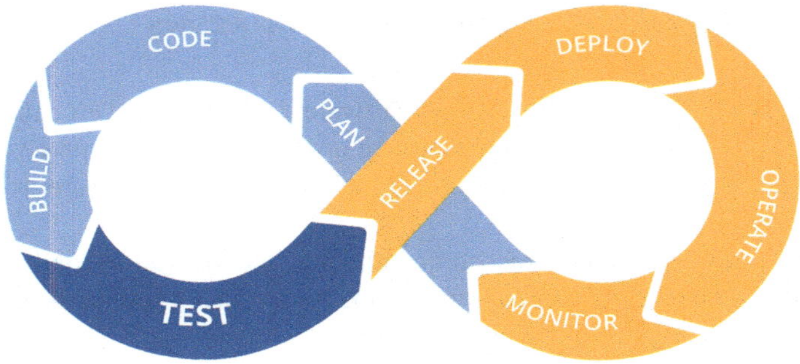

Fig. 3 DevOps life cycle. *Source* http://www.jirehtechconsulting.com/what-is-devops/

process to be successful must include communication between team members across cross-discipline.

1. *Plan*—this phase documents the business vision and requirements. Typical activities include project management, scheduling, release plans, policies/requirements, etc.
2. *Code*—application code is developed using appropriate programming languages and IDEs. Code is usually maintained using version control systems.
3. *Build*—it integrates various software modules to build executable files for product features or fully developed product.
4. *Test*—it ensures that potential errors are eliminated in developed software and reliable product is delivered to production.
5. *Release*—it is a process to deploy whole application or multiple integrated applications to production.
6. *Deploy*—it is a process to promote software components from one environment to next. If it is successfully deployed, the features or product is ready for release.
7. *Monitor*—it ensures that the issues are identified for specific releases that may have impact on end-users, and issues are resolved rapidly.

3 DevSecOps Paradigm—Principles and Processes

In simple term, goal of DevSecOps is to safeguard application from any potential threats by including security mechanisms into all phases DevOps workflow [5].

This is accomplished by seamlessly integrating security tools into entire DevOps application development process to develop and deliver secured application as shown in Fig. 4.

DevSecOps tools provide developers with alerts and notifications about potential security anomalies and defects so they can be investigated and fixed before getting too far along in the process.

Most DevSecOps tools offer some level of automation. The tools in this category automatically scan for, discover and remediate security defects in varying degrees.

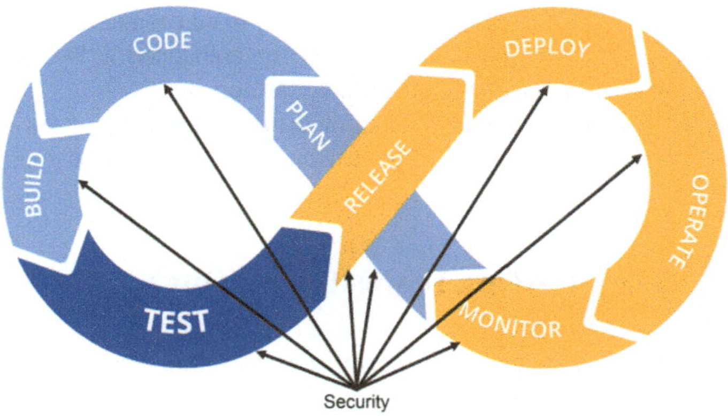

Fig. 4 Embedding security in DevOps life cycle. *Source* https://www.owasp.org/images/7/7e/2017-04-20-AppSecDevops.pdf

4 AIOps Paradigm—Principles and Processes

With recent surge of artificial intelligence (AI) and machine learning (ML) applications, software developers are embracing principles and practices of AI and ML in the various phases of DevOps Lifecycle [6]. AIOps applies AI and ML principles to DevOps lifecycle as shown in Fig. 5 [7, 8]. It analyzes vast amount of data collected from multiple IT operational sources and devices using AI and ML techniques to detect and respond to potential issues in real time.

AI can also be used to learn how problems can be solved by correlating matching problem patterns with corresponding solution patterns over time. AI can then be applied automatically to solve future problems and issues. As a result, problems issues can be handled efficiently and automatically without human intervention and will enhance reliability. In similar fashion, AI can be used to predict security breaches

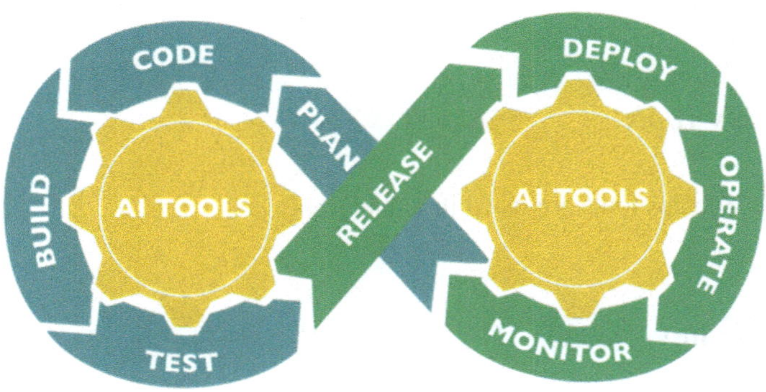

Fig. 5 Embedding AI tools in DevOps life cycle. *Source* https://jamieai.com/artificial-intelligence-for-devops-automation/

or vulnerability of system and respond automatically to take remedial action in anticipation of potential security issues. AI tools can be embedded in DevOps cycle to deliver quality services to end-users in a timely fashion.

In summary, AI can have profound significant impact in the areas of process automation, testing automation, and infrastructure automation of DevOps lifecycle.

5 Sample Tools: DevOps, DevSecOps, AIOps

DevOps tool chain is a collection of tools that assist in the management, development, and delivery of software throughout all DevOps stages. With the proliferation of available and evolving tools, selecting the best DevOps tools for the particular application scenario is a daunting task requires some testing and experimentation. Section 5.1 describes a sample list of tools for DevOps stages. Section 5.2 describes a sample list of DevSecOps tools that can be integrated to achieve security, and Sect. 5.3 provides some AIOps tools that can be embedded within DevOps environment to provide DevOps continuous integration and delivery using AI and ML techniques.

5.1 DevOps Tools

Sample DevOps tools are listed in Table 1.

5.2 DevSecOps Tools

Sample DevSecOps tools are listed in Table 2.

5.3 AIOps Tools

Sample DevSecOps tools are listed in Table 3.

6 Conclusion

DevOps is a business-driven approach to delivering quality software. The goals of DevOps process—continuous integration, continuous delivery and continuous deployment, continuous testing, continuous monitoring and feedback—will be achieved more efficiently by successful integration of DevSecOps, and AIOps with DevOps stages. DevSecOps will have potential to be effective in automatic detection, alerting and correcting security incidents as required. AIOps makes up the technology that integrates intelligence into that system to initiate automatic action when deemed appropriate.

Table 1 DevOps tool chain

DevOps phases: tool functions	Tools examples	Website
Plan: Provides project management, task scheduling support and status	Atlassian Jira Asana Confluence Trello YouTrack	https://www.atlassian.com/software/jira https://asana.com/ https://www.atlassian.com/software/confluence https://trello.com/ https://www.jetbrains.com/youtrack/
Code: provides actual coding and version control	Eclipse or any other IDEs Version Control Tools: Git Github GitLab SVN	https://www.eclipse.org/downloads/ https://git-scm.com/ https://github.com/ https://about.gitlab.com/
Build: Creates executable, integrates different modules; some of these tools may be integrated with other tools to achieve continuous integration, continuous deployment, and continuous delivery	Apache Ants BitBucket Bamboo Gradle Maven Team City	https://ant.apache.org/ https://bitbucket.org/ https://www.atlassian.com/software/bamboo https://gradle.org/ https://maven.apache.org/index.html https://www.jetbrains.com/teamcity/
Test: Tools can perm various tests such as unit, functional, and load tests on mobile web or enterprise applications	Blazemeter IBM Rational Functional Tester UFT 1 Jmeter Junit Katalon Studio Ranorex Studio SoapUI SonarQube Seleniun Test Complete	https://www.blazemeter.com/ https://www.ibm.com/us-en/marketplace/rational-functional-tester https://www.microfocus.com/en-us/products/uft-one/overview https://jmeter.apache.org/ https://junit.org/junit5/ https://www.katalon.com/ https://www.ranorex.com/ https://www.soapui.org/ https://www.sonarqube.org/ https://selenium.dev/ https://smartbear.com/product/testcomplete/overview/

(continued)

Table 1 (continued)

DevOps phases: tool functions	Tools examples	Website
Release: Orchestrate DevOps pipelines for continuous delivery of the product	Ansible BMC Release Life Cycle Management CircleCI RunDeck TeamCity IBM UrbanCode Release XL Release	https://www.ansible.com/ https://docs.bmc.com/docs/ReleaseLifecycleMgt/50 https://circleci.com/ https://www.rundeck.com/ https://www.jetbrains.com/teamcity/ https://www.ibm.com/cloud/urbancode/release https://xebialabs.com/technology/xl-release/
Deploy/Operate: tools automate configuration management t, infrastructure provisioning, and reduce manual and repetitive tasks for infrastructure management	AWS Elastic BeanStalk Bamboo Bitbucket Chef Docker Dynatrace IBMUrbanCode Deploy Jenkins Puppet Terraform Travis-CI	https://aws.amazon.com/elasticbeanstalk/ https://www.atlassian.com/software/bamboo https://bitbucket.org/ https://www.chef.io https://www.docker.com/ https://www.dynatrace.com/ https://www.ibm.com/cloud/urbancode/ https://jenkins.io/ https://www.terraform.io/ https://travis-ci.com/
Monitor: used in various stages for reporting success, provide notification, and find any issues and the root causes	AWS CloudWatch Dynatrace New Relic Prometheus Sensu	https://aws.amazon.com/cloudwatch/ https://www.dynatrace.com/ https://newrelic.com/ https://prometheus.io/ https://sensu.io/

DevSecOps with various security tools provides faster feedback loops to all stakeholders—developer, operation and security personnel, whenever any breach or potential breach of security is detected. To be successful, DevSecOps must address critical issues while mitigating or limiting the impact to the business by integrating appropriate security tools at different DevOps stages.

AIOps when integrated in DevOps can learn the current trends from the running application, analyze the trends, and initiate proactively corrective steps if it anticipates issues or changes that may affect the application. It can not only monitor state of the application, but has the potential to automatically resolve the issues by appropriate effective action.

AIOps using ML, and predictive analytics can forecast various resource requirements such as storage, processing, latency to improve operational aspects of the whole system. AIOps with its technical capability of collecting large volume of diverse data from different sources and learning from these will deliver value to end-users.

Table 2 DevSecOps tools

Tool examples	Tool descriptions	Website
CodeAI	uses AI and machine learning to predict defects in new code and fixes security vulnerabilities in source code to prevent attacks	https://www.qbitlogic.com/codeai/
Veracode Greenlight	provides the ability to scan code within preferred IDE or CI system, during coding, notifies developers, and provides instant insight into any security flaws that are noted	https://www.veracode.com/products/greenlight
IriusRisk	a threat modeling tool which guides developers through the technical architecture, planned features, and security context of an application	https://iriusrisk.com/threat-modeling-tool/
Kiuwan	SAST tool, scans code to identify and remediate vulnerabilities	https://www.kiuwan.com/
SaltStack	Intelligent automation and collaboration software for security operations teams	https://www.saltstack.com/

The adoption of DevOps paradigms is not without challenges [9, 10]. The primary challenges are: to educate various teams in the DevOps process to build expertise, to improve collaborative cultures between Dev, Op, Sec, and AI teams, to select proper tool chains consistent with the skillsets of team members, integrate the tool chains in the work flow, and executive support and proper budgeting for the whole DevOps process adopted.

To integrate security in DevOps, workflow poses other challenges [11–13]: from finding expert security professionals who are knowledgeable in DevOps process, selecting appropriate DevSecOps tools for the adopted platform that can integrate seamlessly with the existing DevOps tools to automate and secure the CI/CD pipeline.

The challenges of merging AIOps within DevOps implementation are no different from challenges posed by DevSecOps [14]: lack of understanding of collaborative DevOps culture, and DevOps processes, difficulties in choosing and integrating right AIOPs tools to meet business and technical goals, and lack of availability of relevant learning data sets. The availability of current tools, and evolving diverse tools of DevOps, DevSecOps, and AIOps will facilitate the process of delivering quality software rapidly to the satisfaction of end-users. The paper provides important trends in IT paradigms and associated tools and will be of interest to practitioners and researches who want to investigate further in specific areas of DevOps landscapes.

Table 3 AIOps tools

Tool examples	Tool descriptions	Website
AIops (from Broadcom)	Correlates data across users, applications, infrastructure and network services, applies machine learning, advanced analytics to provide visibility and actionable insights	https://www.broadcom.com/products/software/aiops#solutions
BigPanda	Detect, investigate, and resolve incidents rapidly	https://www.bigpanda.io/
Dynatrace	Provides performance metrics in real time and detects and diagnoses problems automatically for cloud environment	https://www.dynatrace.com/
SysTrack	Monitors, analyzes, optimizes IT environments, detects anomaly, performs root cause analysis, and analyzes the effect of changes	https://www.lakesidesoftware.com/product
StackState	Provides real-time visibility, impact assessment, root cause analysis, and predictive analytics	https://www.stackstate.com/
Optanix	Monitors performance, provides business impact and service-infrastructure management	https://www.optanix.com/solutions/aiops/
SL1	Automates issue discovery and subsequent remediation across a diverse range of technologies	https://sciencelogic.com/sl1/platform
Splunk	Collects and analyze data to deliver operational intelligence, enables user to analyze, monitor, and report on this data in real time	https://www.splunk.com/
Zenoss Cloud	Provides AIOps analytic, health and performance insights, automate issue remediation	https://www.zenoss.com/

References

1. DevOps Lifecycle I Introduction to DevOps I DevOps Tools. https://www.youtube.com/watch?v=Y-rj4vFc1Q8&t=613s. Last accessed 5 Jan 2020
2. Kiesnoski K, 5 of the biggest data breaches ever. https://www.cnbc.com/2019/07/30/five-of-the-biggest-data-breaches-ever.html. Last accessed 5 Jan 2020
3. The Forrester Wave™, Intelligent Application And Service Monitoring, Q2 2019, https://content.sciencelogic.com/aiops_automation_engine/Forrester_Wave?lx=o7-w7S%3Futm_source%3Dwebsite. Last accessed 5 Jan 2020
4. DevOps, http://www.jirehtechconsulting.com/what-is-devops/. Last accessed 5 Jan 2020
5. AppSec in a DevOps World, https://owasp.org/www-pdf-archive/2017-04-20-AppSecDevops.pdf. Last accessed 5 Jan 2020

6. The Current State of AIOps, https://thenewstack.io/the-current-state-of-aiops/. Last accessed 5 Jan 2020
7. Artificial intelligence for DevOps automation, https://jamieai.com/artificial-intelligence-for devops-automation/. Last accessed 5 Jan 2020
8. AIOps: Using Artificial Intelligence in DevOps, https://opensenselabs.com/blog/articles/aiops. Last accessed 1 Jan 2020
9. A Survey of DevOps Concepts and Challenges, https://arxiv.org/pdf/1909.05409.pdf. Last accessed 5 Jan 2020
10. 30 common challenges to DevOps and how to resolve them, https://techbeacon.com/devops/30-common-challenges-devops-how-resolve-them. Last accessed 20 Jan 2020
11. DevSecOps-A Multivocal Literature Review, https://www.researchgate.net/publication/319633880_DevSecOps_A_Multivocal_Literature_Review. Last accessed Jan 2020
12. The challenges of shifting to DevSecOps, https://www.itproportal.com/features/the-challenges-of-shifting-to-devsecops/. Last accessed 5 Jan 2020
13. DevSecOps Definition, New Challenges, New To-Do's, https://www.itprotoday.com/devops/devsecops-definition-new-challenges-new-dos. Last accessed 5 Jan 2020
14. Essential guide to AIOps: Top tools and implementation tips, https://techbeacon.com/enterprise-it/essential-guide-aiops-top-tools-implementation-tips. Last accessed 20 Jan 2020

Optimized Route Discovery and Node Registration for FANET

Vinay Bhardwaj and Navdeep Kaur

Abstract The capabilities of unmanned aerial vehicle (UAVs) have evolved gradually for their excess use in various domains. The new architecture flying ad-hoc network (FANET) was adopted recently, which took care of the UAV by providing various facilities. The excess use of the FANET due to the high mobility of the objects makes it popular among military and civilian applications to discover the optimum route. FANET is a part of the ad-hoc network which connects the UAV to solve the complex problems in the networking. In this paper, FANET has been used to optimize the problem of route discovery. The route discovery mechanism is designed by implementing algorithms to detect the optimum path. Besides, issues related to node registration and optimal routing method also have been focused. The proposed architecture uses a combination of artificial bee colony, AODV protocol, and Lagrange interpolation (LI). The performance of the proposed architecture is validated using three parameters, namely throughput, packet delivery ratio (PDR), and Jitter in different UAVs. The results are compared with existing techniques to show the effectiveness of the proposed work in terms of throughput rate and Jitter. The throughput rate has been improved by 48%, while for Jitter, it is revamped by 31%.

Keywords FANET · UAV · Route discovery · AODV protocol · ABC algorithm · LI · Throughput · Jitter · PDR

V. Bhardwaj (✉)
Shri Guru Granth Sahib World University, Fatehgarh Sahib, Punjab, India
e-mail: Vinaybhardwaj0708@gmail.com

N. Kaur
SGGSW University, Fatehgarh Sahib, Punjab, India
e-mail: drnavdeep.iitr@gmail.com

P. K. Singh et al. (eds.), *Evolving Technologies for Computing, Communication and Smart World*, Lecture Notes in Electrical Engineering 694,
https://doi.org/10.1007/978-981-15-7804-5_17

1 Introduction

The flying ad-hoc network (FANET) is a new arena of communication (wireless) that takes care of the free travel of unnamed aerial vehicles (UAVs), and it targets UAV-UAV communication [1–4]. If the network looks at each UAV as a router, the routing architecture will be borrowed from MANET or VANET are not adequate for FANET due to the high mobility of flying objects. In addition to the complicated routing architectures, the other tasks related to flight architecture make the structure more complicated [5]. A novel routing architecture is presented in this paper to avoid congestions in the network, considering the MAC protocol and self-adaptive learning structure. Multiple UAV architectures are being addressed by a lot of commercially successful companies and research workers as well [6]. A UAV network or architecture will consist of the UAVs, their associated properties. Definition 1 defines FANET architecture.

Definition 1 FANET F(Nu, Nup, CS) is a structure that contains a Nu number of utility flying objects with Nup associated properties. The associated properties may contain the total associated fuel, the consumption of fuel in the traveling architecture, CS be the communication center.

Lemma 1 *Each UAV is associated with a CS, but the issue arises when the secure connection between UAV and CU fails. The CU can do nothing other than waiting for the UAV to send the signal back to the CU.*

Lemma 1 *It defines the issues raised, which the UAV can face when the establishment is vulnerable. These issues are getting addressed by other research scientists and commercially successful companies like Google, Amazon, and Facebook* [7]. *This paper considers the future aspects of the FANET architecture and hence takes the following things as assumptions.*

- One UAV can talk to another UAV
- If the UAV is not able to communicate with its CS, it can communicate through other UAVs
- The information-sharing requires a nearby connection mode.

Definition 2 FANET has coverage sensors which through which the UAV can calculate its nearby aerial vehicles. The routing architecture is followed by the ad-hoc on-demand vector (AODV) routing protocol. The network also considers trustworthy nodes for secured communication [8]. If a new node aims to enter the network, it has to go through Lagrange's interpolation system.

There are two issues that are addressed in this research.

(a) Registration of a new node in the communication network
(b) Optimal routing method for the existing identities in the network.

The rest of the paper is organized as follows. Section 3 describes the proposed architecture, whereas Sect. 4 represents the evaluation results. Section 5 concludes the paper.

2 Related Work

This section describes the literature review of past studies based on FANET.

Singh [9] demonstrated the challenges and advancements in FANET. Researchers highlighted the functions of UAV-based system to address the problem of security. Alnuami [3] proposed an approach that describes the efficiency of different protocols in FANET. The outcomes showed that the result attained using the AODV protocol are better than the DSR. The specific parameters, such as PDR, throughput, and delay factor, were considered in the proposed approach to evaluate the results. Arafat and Moh [4] proposed a robust approach using the cluster-based routing protocol in the presence of high dynamic topology and rapid mobility challenges. Clustering is essential due to the rapid increase in the number of UAVs. The various performance parameters, such as delay, throughput, and energy efficiency, were considered to validate the results. The qualitative analysis for the clustering protocols has been presented to understand the limitations and competitive advantages of the clustering network of UAV-based routing protocols.

Khan et al. [10] proposed a topology-based routing protocol to solve issues in the FANET. The network efficiency improved by considering the various parameters such as throughput, delay, load on the network. The benefits and limitations of each protocol were discussed in detail to validate the results—besides, the robust and efficient communication between the networks supported by understanding the context of each protocol. Practitioners define FANET as the most suitable technique for secure communication between the networks. Khan et al. [2] proposed a hybrid wireless approach by following specific guidelines considering the quality of service (QoS) and energy efficiency parameters. A hybrid approach of 802.15.1 and 802.11 introduced in wireless technologies. The proposed technique reduces the communication cost and considers the various metrics such as throughput and delay to determine the effectiveness of the proposed approach. Zeng and Zhang [11] proposed a UAV-based energy-efficient approach considering the throughput and energy consumption parameters. The various parameters of the UAV such as flying speed and direction have been configured to validate the proposed approach—the energy maximization technique proposed for the UAV to improve energy consumption and reduce the propulsion energy. The proposed results further compared with other benchmark techniques. Bujari [12] proposed a FANET-based technique using routing algorithms. The paper considered scalability as a significant problem; therefore, UAV-based FANET architecture has been proposed to extend the routing path. The three-dimensional scenario has been taken to develop the algorithms, and state-of-the-art techniques have been elaborated well to determine the efficiency of the proposed work.

The survey resulted in the conclusion that there is a lack of an algorithm for the FANET having efficient energy and throughput rate. Due to the lack of this point in existing work, and increased Jitter rate, previous studies do not achieve better performance because, in the presence of attackers, the discovery of optimal route is the most vital task. More specifically, most of the used FANET architecture based on protocol saving considerable computation time and unsupervised clustering methods, so the detection of cluster centroid is not done correctly in the previous work.

3 Proposed Architecture

The proposed architecture is divided into multiple segments, including the network setup, the route discovery process, enhancement of route discovery, and registration of a new areal object in the network.

3.1 The Setup Phase

The network is first set up with the deployment of the aerial nodes and the initialization of the properties of the nodes.

Pseudo Code 1: Node Setup and Deployment$Network_{Width} = 500\ KMs$
$Network_{Length} = 500\ KMs$
For_each node in the structure
Initialize x and y locations of aerial nodes
Initial_Fuel = Associated_Fuel
Delay_(In_transmission) = New_(Delay_Transmission)
Packet_Dump = Initialize
Deploy (Node)
End_For

The proposed network architecture considers a heterogeneous work environment, and hence, each vehicle is assigned with different values for each parameter. Thus, the network is designed and deployed with some primary fuel, which is different for every vehicle. It is associated with expected delay, which this vehicle is going to consume after a certain period and the packet dump [12]. Each vehicle computes its range of communication, which is evaluated by Algorithm 1.

The coverage prediction algorithm computes the distance of each node to every other node in the architecture. Figure 1 depicts the results of Algorithm 1.

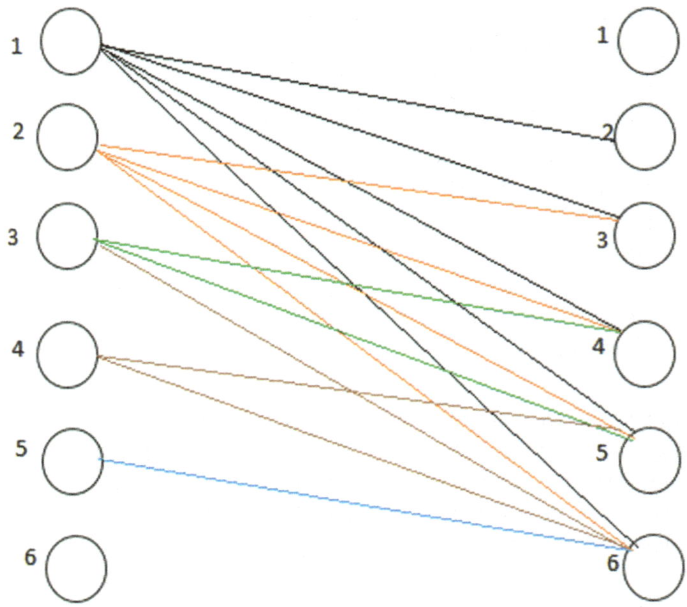

Fig. 1 Coverage evaluation

Algorithm 1: *TheCoverage*$_{Prediction}$ (*Nodes* ,
$X_{loc}, Y_{loc}, Node_{id_{list}}$)
$Coverage_{Prediction} = [\,]$
ForeachnodeinNodes
*Foreachnode*1 *inNodes*
Ifnode ! = *node*1
$dist = \llbracket sqrt\left(\left(X_{loc}(node) - X_{loc}(node1)\right)\right\rrbracket^2 + \left(Y_{loc}(node) - Y_{loc}(node1)\right)^2\right)$
$Coverage\_Prediction\ (node, node1) = Node\_(id\_list\ )\ (node1)$
End_{For}
End_{For}
$End_{Algorithm}$

3.2 The Routing Architecture

The routing architecture is accommodated by the nearest aerial vehicle. Algorithm 2 explains the working of the route discovery.

$$
\begin{aligned}
&\boldsymbol{Algorithm\ 2: RouteDiscoveryandOptimization}\\
&Input: Source\ , Destination_{Aerial}\\
&Distribute_{Msg} = Distribute('Hello');\\
&For_{each} respondentofDistribute_{Msg}\\
&Check_{Route_{Requirements}}\ ();\\
&Select_{Node}(\ Ping);\\
&IfPingsBack\\
&Add_{ToRoute}()\\
&End_{IF}\\
&RepeatUntilDestination\_ArealisNotFound\\
&End_{Algorithm}
\end{aligned}
$$

The discovery mechanism distributes a hello message in the network and waits for the ping. The ping can be sent only to those nodes which come under the coverage range evaluated by Algorithm 1. The routing mechanism based on dynamic architecture evolves the researchers to assess the complex functions [10].

The broadcasted respondents get the update by utilizing artificial bee colony with a new fitness function [13]. The paper presents a solution to the situation when a new aerial node aims to entre in the network. The verification process follows the Lagrange interpolation method. The structure of the interpolation method is as follows.

3.3 *Lagrange Interpolation (LI)*

When a node request the information from a server either directly or through Flying Aerial Side Unit (FASU), the communication server demands three shares from any node in the network and choose two out of them randomly. Three shares overall are considered, including the demanding node. The Lagrange polynomial $S(X)$ containing a degree $\leq (n-1)$ that demands n number of nodes with coordinates $(x_1, y_1 = f(x_1)), (x_2, y_2 = f(x_2)), \ldots (x_n, y_n = f(x_n))$ is given by:

$$S(X) = \sum_{k=0}^{n} P_k(X) \tag{1}$$

where P_k is given by:

$$P_k(X) = y_k \frac{x - x_l}{x_j - x_l} \tag{2}$$

where $l \geq 1$ and $l \leq n$ and $l! = k$.

If written explicitly for $n = 3$ nodes

$$S(X) = \frac{(x - x_2)(x - x_3)}{(x_1 - x_2)(x1 - x3)} y_1 + \frac{(x - x_1)(x - x_3)}{(x_2 - x_1)(x2 - x3)} y_2 + \frac{(x - x_1)(x - x_2)}{(x_3 - x_1)(x3 - x2)} y_3 \tag{3}$$

The separate polynomial follows Szego (1975), which was later called Lagrange's fundamental interpolation.

$$S(X_1) = \frac{x_2 * x_3}{(x - x_2)(x - x_3)} y_1 \tag{4}$$

for the first node

$$S(X_2) = \frac{x_1 * x_3}{(x - x_1)(x - x_3)} y_2 \tag{5}$$

for the second node

$$S(X_1) = \frac{x_1 * x_2}{(x - x_2)(x - x_3)} y_3 \tag{6}$$

for the third node.

The key which is generated by the integration of separate polynomials is represented as

$$G_k = \sum_{k=0}^{n} S(k) \tag{7}$$

If G_k matches the network key only when the node passes any information from the FANET. Secondly, the RSU level security is also applied, which makes the network more secure. Pseudocode is as follows to understand security architecture:

Algorithm 3 for Share Verification

```
Order=2; // Interpolation order
My_VALUE = [ ] // Initializing the key values to be Empty
For i = 1:3   // For 3 nodes
    counter = 1;
      Current_1=Node_ID_I // Taking the first node as an initial reference
    For j=1: Nodes ;
      Current=Vehicles_j; // For each interpolation, there would be 2 Rest Nodes
      if Current1~=Current  // If nodes are not the same
      Rest(counter)=current;
      counter=counter+1;
      End If
    End For
    Deno = 0
    Deno = Current_1 - Rest_1 * Current_1 - Rest_2// The denominator value
    Num=Rest_1 * Rest_2
    My_value[i] = Num / Deno
    Shared_key = Share_Current_1 * My_value[i]
    End For
```

The pseudocode takes the interpolation order 2. This order selects only three nodes for communication. Whether the nodes select for the data communication or not, depends upon the final key result, which is calculated by Lagrange's method. One critical generation method requires a numerator and a denominator. The numerator is calculated by the network id of left nodes for the iteration. As consider 455,361 are the selected nodes for the verification. So numerator value (Num) for 45 is $53 * 61 = 3233$. The denominator (Deno) is calculated by multiplying the difference of network ids of remaining nodes. For 45, the Deno value is $(45 - 53) * (45 - 61) \rightarrow (-8) * (-16) \rightarrow 128$. The verification key would be the product of shared key of 45 to $\frac{\mathrm{Num}}{\mathrm{Deno}}$. In the similar fashion $\mathrm{Shared}_{\mathrm{key}}$ for 53 and 61 is calculated. The final verification key would be the sum of all the generated verification keys.

$$\mathrm{Final}_{\mathrm{key}} = \sum_{k=0}^{i} \mathrm{My}_{\mathrm{value}} \tag{8}$$

If the $\mathrm{Final}_{\mathrm{key}}$ is equal to the network security key, then the nodes are selected for communication. Lagrange's theorem selects a node for the verification in a random manner, although the verification process of Lagrange is good.

4 Results and Discussions

This section describes the results which have two segments, namely result evaluation and comparative analysis. The result analyses describe the evaluation of the proposed approach using different parameters, while comparative analysis elaborates the comparison of the proposed technique with the past approaches [2, 3, 7].

4.1 Results Analysis

This section evaluates the following parameters based on the proposed architecture.

Throughput: It is the ratio of total received packets per unit time.

- PDR: It is the ratio of data packet receiving through the targets to those created by the source nodes. The equivalent used for the packet delivery ratio mathematically
- Total Received Packets/Total Transferred Packets
- Jitter: It is the noise affected the network performance. It is the ratio of total produced delay with the change in the spectrum value.

Table 1 describes the average throughput measurement using AODV protocol, average throughput using AODV and ABC, and average throughput using AODV, ABC, and Lagrange. It is observed that the average throughput using AODV protocol ranges from 11,000 to 14,000, while that of using AODV and ABC lies in the range from 14,000 to 18,000. The throughput rises through the proposed methodology from 20,000 to 22,000. Overall, the proposed methodology provides improved results in contrary to other techniques.

Figure 2 demonstrates the performance of the throughput of the proposed solution in different situations. The proposed solution is compared with AODV architecture. As the proposed algorithm is framed in two sections, the results of both the sections are presented here.

The first section is the enhancement of the route discovery process by ABC, and the second part is the integration of the interpolation architecture by Lagrange's polynomial. The second phase of the proposed architecture exhibits 21,390 after

Table 1 Determination of throughput

Total number of simulations	Average throughput AODV	Average throughput using AODV and ABC	Average throughput using AODV, ABC and Lagrange
100	11,000	14,000	20,000
200	12,000	15,000	22,000
400	12,500	14,500	23,000
500	13,000	17,000	23,500
1000	14,000	18,000	22,000

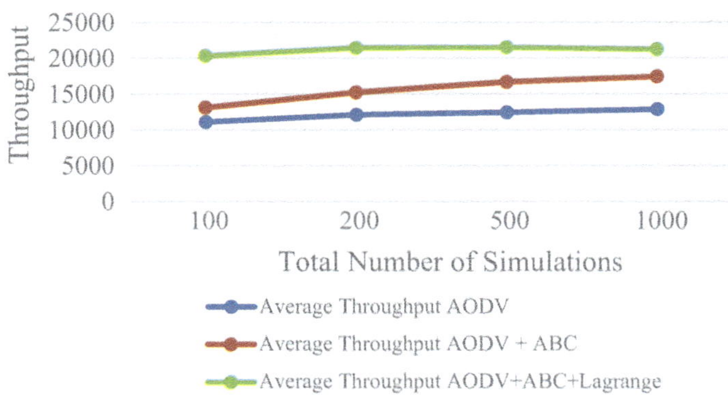

Fig. 2 Throughput rate

1000 simulation rounds, whereas with ABC only, it remains higher than AODV but a little lower than that of the final stage of the proposed work. Similarly, if the throughput is high, PDR will be high, and fuel consumption would below.

Table 2 depicts the PDR rate of the proposed architecture and comparison with other AODV protocol and the ABC algorithm. The results show that the average PDR through the AODV protocol varies from 0.35 to 0.4. While that of ABC and AODV protocol, it lies in the range from 0.5 to 0.54. The average PDR through the proposed approach is approximately 0.7. Overall, it is clear that the average PDR attained through the proposed approach is higher than the other techniques.

Figure 3 demonstrates the performance of the PDR of the proposed solution in different situations. The proposed technique is compared with AODV architecture. The proposed architecture is framed in two sections; the results of both sections are presented. The first section is the enhancement of the route discovery process by ABC, and the second part is the integration of the interpolation architecture by Lagrange's polynomial. The second phase of the proposed architecture exhibits 0.55 after 1000 simulation rounds, whereas with ABC only, it remains higher than AODV but a little lower than that of the final stage of the proposed work.

Table 2 PDR

Total number of simulations	Average PDR through AODV protocol	Average PDR through AODV protocol and ABC	Average PDR using AODV, ABC and Lagrange polynomial
100	0.35	0.5	0.7
200	0.37	0.5	0.69
400	0.39	0.51	0.7
500	0.4	0.52	0.71
1000	0.4	0.54	0.71

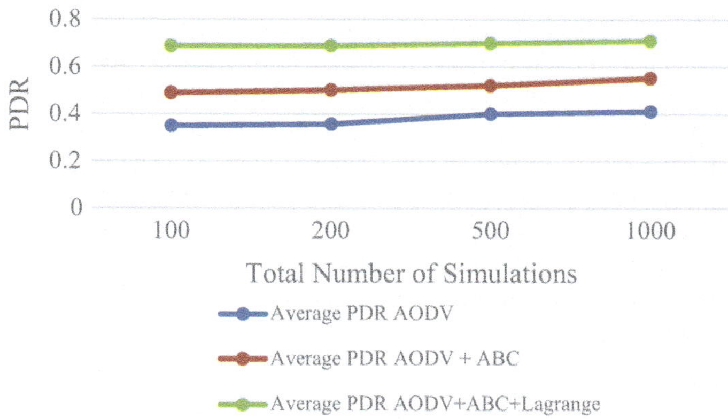

Fig. 3 PDR

Table 3 depicts the occurrence of noise in seconds. The frequent occurrence of Jitter for the proposed methodology for 100 simulations is 0.17 s. The occurrence of Jitter for the proposed methodology lies in the range of 0.17–0.2 s from 100 to 900 simulations, while that of the previous technique lies in the range of 0.24–0.25 s. Overall, it is clear that the common occurrence of Jitter in seconds for the proposed methodology has been better than the previous studies [2].

Table 3 Occurrence of Jitter

Total number of simulations	Average Jitter in seconds for the proposed methodology	Average Jitter in seconds [2]
100	0.17	0.24
200	0.23	0.35
300	0.26	0.36
400	0.25	0.37
500	0.25	0.33

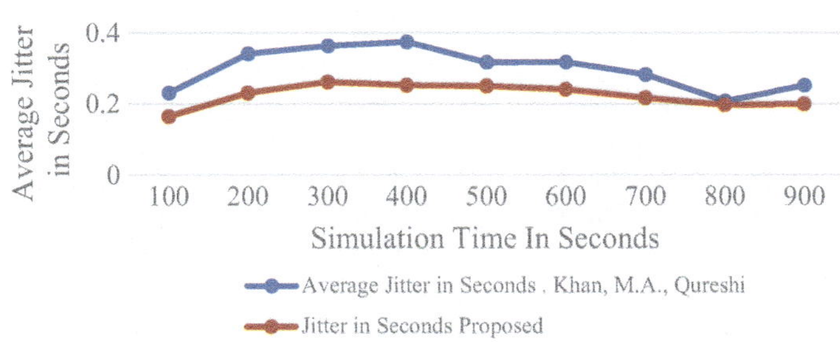

Fig. 4 Jitter in seconds

Figure 4 depicts the occurrence of average Jitter of the proposed work for 800–900 simulation seconds which is noted to be 0.2022 s, whereas [2] demonstrates 0.2544 for the same. The overall average delay for [2] is 0.302267, whereas the proposed work demonstrates 0.22744 s for the same set of simulation values.

Figure 4 represents the Jitter of the proposed network. The change in the data propagation medium produces Jitter in the network as the network takes time to change the architecture of the data as per the propagation medium, and it increases, even more, when the nodes are not trusted worthy. The new node selection method makes the network node more trustworthy, and hence, the delay produces through faulty nodes which is reduced.

4.2 *Comparative Analysis of the Proposed Methodology and Past Studies [2, 3, 7]*

This section describes the comparative analysis of the proposed methodology and past studies [2, 3]. The prominent factors such as throughput and PDR have been considered to determine the effectiveness of the proposed approach.

Table 4 depicts the throughput measurement of the proposed approach and past studies [3, 7]. A comparative analysis is shown between the proposed architecture and previous studies. The efficiency of the proposed approach proves to be better than past results. The average throughput attained through the proposed work approaches to 22,125. The effectiveness of the proposed approach shows better results in comparison with throughput attained through AODV protocol [3], while that of Leonov and Lobachevsky show the average throughput rate 11,500.

Figure 5 depicts the comparative analysis of the throughput measurement rate of the proposed approach and the previous studies [3, 7]. The proposed technique provides effective results in comparison with other techniques. The overall average results are obtained through the proposed work is 22,125, while that of the existing technique is only 300 [3]. The overall accuracy in contrary to [3] is $\frac{22,125-300}{22,125} \times 100 = 98.6\%$. The efficacy of the proposed technique in comparison with [7] is $\frac{22,125-11,500}{11,500} \times 100 = 48\%$.

Table 4 Throughput measurement of the proposed approach and past studies [3]

Number of simulations	Throughput attained through past studies Alnuami (2018) [3]	Throughput attained through past studies Leonov and Lobachevsky [7]	Throughput obtained through the proposed approach
100	350	3000	20,000
200	300	7000	23,000
500	250	16,000	23,500
1000	300	20,000	22,000

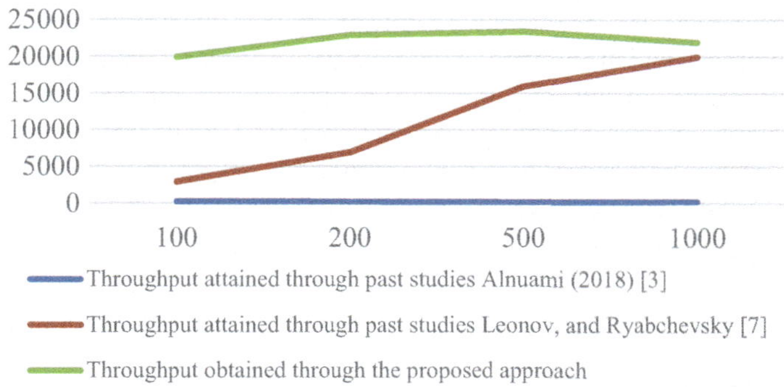

Fig. 5 Comparative analysis for the measurement of throughput

Table 5 depicts the comparative analysis of the Jitter measurement of the proposed approach and past work [2, 7]. The average noise obtained through the proposed architecture is 0.2275, while other techniques include average Jitter 0.33 and 0.2475 for the past work [2, 7], respectively. The effectiveness of the proposed approach shows better results in comparison with other techniques for Jitter measurement [2, 7].

Table 5 Jitter measurement of the proposed approach and past studies

Number of simulations	Jitter obtained through past studies Khan (2019) [2]	Jitter obtained through past studies Leonov and Lobachevsky [7]	Jitter obtained through the proposed approach
100	0.24	0.18	0.17
200	0.35	0.19	0.23
500	0.36	0.16	0.26
1000	0.37	0.45	0.25

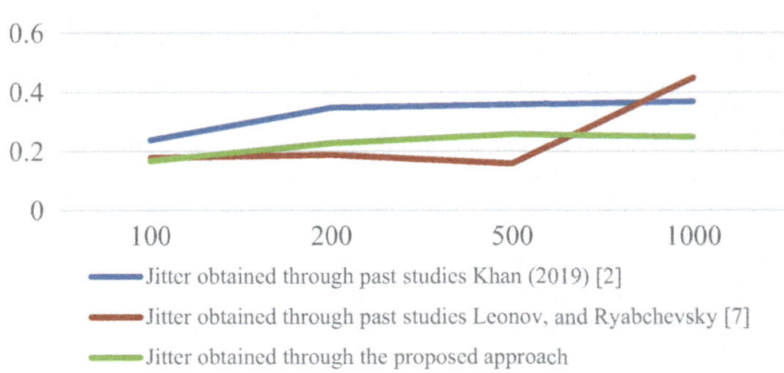

Fig. 6 Comparative analysis for the measurement of throughput

Figure 6 depicts the comparative analysis of the Jitter measurement for the proposed approach and the previous studies [2, 7]. The efficacy of the proposed technique shows effective results in comparison to other techniques. The overall average noise or Jitter obtained through the proposed work is 0.21, while that of past technique is 0.33 [2]. The overall accuracy in contrary to [2] is $\frac{0.33-0.2275}{0.33} \times 100 = 31\%$. The effectiveness of the proposed technique is a comparison to [7] is $\frac{0.2475-0.2275}{0.2475} \times 100 = 8\%$. Thus, overall effective results obtained through the proposed work.

5 Conclusion

The most challenging issue in a communication network is delivering packets through optimum routes. UAVs play a prominent role in the various operational area used by ad-hoc network FANET. The main aim of using UAVs is a reduction in completion time and increasing the reliability of the system. The UAVs have been utilized in FANET to discover the best route. This paper implements the route discovery mechanism using different algorithms. Besides, LI is utilized for node registration and share verification. The throughput using AODV protocol, LI, and ABC is high as compared to traditional approaches. The packet delivery ratio is improved when evaluated through a combination of three protocols.

In comparison with the past techniques, the efficiency of the proposed work shows better results. The overall accuracy obtained through the proposed architecture is 98%. The effectiveness is 48% more as compared to [7] In the case of Jitter, the effectiveness of the proposed technique is improved by 31% in contrary to [2], while that of other technique, the results improved by 8% in comparison with [7].

References

1. Mukherjee A, Keshary V, Pandya K, Dey N, Satapathy SC (2018) Flying ad hoc networks: a comprehensive survey. In: Information and decision sciences. Springer, Singapore, pp 569–580
2. Khan MA, Qureshi IM, Khanzada F (2019) A hybrid communication scheme for efficient and low-cost deployment of future flying ad-hoc network (FANET). Drones 3(1):16–24
3. Alnuami HMT (2018) Comparison between the efficient of routing protocol in flying ad-hoc networks (FANET). J Al-Qadisiyah Comput Sci Math 10(1):9–16
4. Arafat MY, Moh S (2018) A survey on cluster-based routing protocols for unmanned aerial vehicle networks. IEEE Access 7:498–516
5. Oubbati OS, Lakas A, Zhou F, Güneş M, Yagoubi MB (2017) A survey on position-based routing protocols for flying ad hoc networks (FANETs). Veh Commun 10:29–56
6. Sahingoz OK (2014) Networking models in flying ad-hoc networks (FANETs): concepts and challenges. J Intell Rob Syst 74(1–2):513–527
7. Leonov AV, Ryabchevsky VO (2018) Performance evaluation of AODV and OLSR routing protocols in relaying networks in organization in mini-Uavs based FANET: simulation-based study. In: 12th International scientific and technical conference "dynamics of systems, mechanisms and machines" (dynamics). Omsk, Russia, pp 13–15

8. Bekmezci I, Şentürk E, Türker T (2016) Security issues in flying ad-hoc networks (FANETS). J Aeronaut Space Technol 9(2):13–21
9. Singh SK (2015) A comprehensive survey on FANET: challenges and advancements. Int J Comput Sci Inf Technol 6(3):2010–2013
10. Khan MA, Khan IU, Safi A, Quershi IM (2018) Dynamic routing in flying ad-hoc networks using topology-based routing protocols. Drones 2(3):27–37
11. Zeng Y, Zhang R (2017) Energy-efficient UAV communication with trajectory optimization. IEEE Trans Wirel Commun 16(6):3747–3760
12. Bujari A, Palazzi CE, Ronzani D (2018) A comparison of stateless position-based packet routing algorithms for FANETs. IEEE Trans Mob Comput 17(11):2468–2482
13. Khan MA, Safi A, Qureshi IM, Khan IU (2017) Flying ad-hoc networks (FANETs): a review of communication architectures, and routing protocols. In: 1st International conference on latest trends in electrical engineering and computing technologies (INTELLECT). IEEE, pp 1–9
14. Vashisht Sahil, Jain Sushma (2019) Software-defined network-enabled opportunistic offloading and charging scheme in multi-unmanned aerial vehicle ecosystem. Int J Commun Syst 32(8):e3939
15. Singh, PK, Paprzycki M (2020) Introduction on wireless sensor networks issues and challenges in current era. In: Handbook of wireless sensor networks: issues and challenges in current scenario's. Springer, Cham, pp 3–12
16. Vashisht S, Jain S (2019) An energy-efficient and location-aware medium access control for quality of service enhancement in unmanned aerial vehicular networks. Comput Electr Eng 75:202–217
17. Oubbati OS et al (2019) Routing in flying ad hoc networks: survey, constraints, and future challenge perspectives. IEEE Access 7:81057–81105

4.1 GHz Low-Phase Noise Differential XCP LC-VCO with High *Q* and LC Noise Filtering

Akshay Kamboj, Manisha Bharti, and Ashima Sharma

Abstract This paper represents a differential cross-coupled pair (XCP) tuned tank LC voltage-controlled oscillator with optimum noise filtering for low voltage and low-phase noise which can further be utilized in the efficient implementation of phase-locked loop (PLL). The proposed design is characterized under source coupled VCO, wherein a LC filtering method is applied to tail current source to improve phase noise. The design is implemented on 90 nm technology with a supply voltage of 1.4 V and power consumption of 50.08 uW. The novel design has been employed by tuning control voltage in the range of 1.5–1.8 V as this range shows the best linear characteristics. A phase noise of −154.3 dBc/Hz at 1 MHz offset frequency with the tuning range of 5.25–4.0 GHz has been validated for proposed VCO.

Keywords Cross-coupled pair (XCP) · Voltage-controlled oscillator (VCO) · Phase noise

1 Introduction

The reliability of the contact link in a wireless communication system depends on the features of VCO. Higher frequency range in VCO is required in today's wireless communication system. The development of VCO that can achieve low power consumption and low-phase noise is critical, and however, the reduction of noise in analog and RF circuit can enhance the performance [1]. To minimize phase noise, the quality factor of the inductor can be increased [2]. pMOS transistors can be adopted for variable capacitor because their flicker noise is significantly lower than that of the nMOS transistor [3]. The high-frequency harmonics generated by the crosscutting transistors are also eliminated as alternative. The characteristics bit rate (BER) is based on a phase noise of VCO in the RF transceiver [4]. Notwithstanding the desirable role of relaxation and ring oscillator as regards circuit integration, LC oscillators

A. Kamboj (✉) · M. Bharti · A. Sharma
Department of Electronic and Communication Engineering, National Institute of Technology Delhi, New Delhi, India
e-mail: nishukamboj95@gmail.com

P. K. Singh et al. (eds.), *Evolving Technologies for Computing, Communication and Smart World*, Lecture Notes in Electrical Engineering 694,
https://doi.org/10.1007/978-981-15-7804-5_18

remain the only reliable way to achieve such phase noise [4]. With the phase noise, low power consumption and wide tuning range are also significant parameter to be considered. The power consumption increases as the VCO core number increases [5]. Thus, all three parameters should be adjusted well. Two broadly defined VCOs used in PLL are source coupled VCO and current starved VCO [6]. Controlled oscillation may be utilized in local oscillators, which performs demodulation during the coherent detection [7]. The gain of the VCO affects the loop dynamics which determines the stability and settling time of PLL [8]. However, research is highlighting to overcome the shortcomings of individual design.

The organization of the paper is as follows: Sect. 2 addresses the VCO's phase noise and its enhanced architecture development using a completely differential topology, where we examine LC tank phase interference and cross-coupled pair. Section 3 shows the enhanced VCO design and function of filters. Results are discussed in Sect. 4 followed by the conclusion in Sect. 5.

2 Conventional VCO Design and Noise Issue

In phase-locked loop (PLL) cross-coupled (XCP) LC—oscillators are one of the essential components that have been widely utilized in transmitters and receivers. VCO can either be designed using only nMOS, pMOS, or CMOS (pMOS and nMOS). Utilization of XCP has many advantages (Fig. 1).

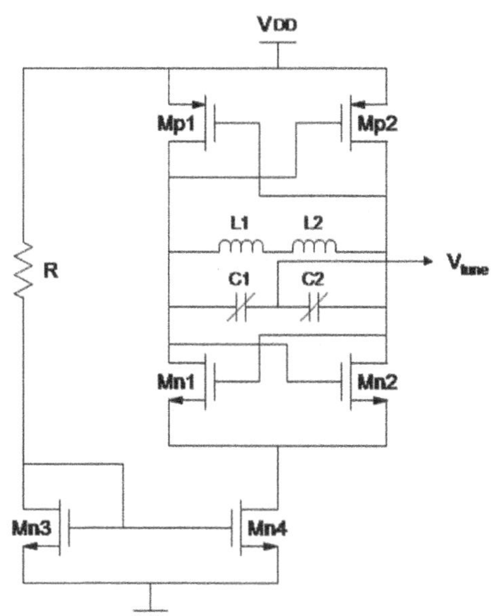

Fig. 1 Conventional CMOS VCO

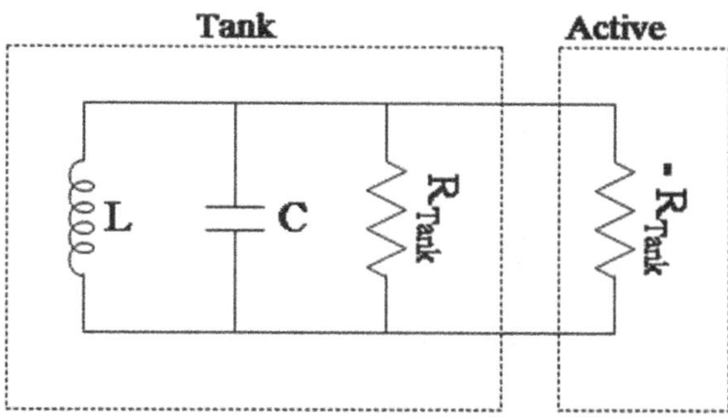

Fig. 2 LC tank equivalent circuit without noise source

At a given current, the complementary configuration offers higher trans-conductance, resulting in XCP differential pair flipping faster [9]. To nullify the effective resistance offered by the tank circuit, active devices (Mn1, Mn2, Mp1, Mp2) are used which provides the negative resistance [10]. As compared to VCO using either nMOS or pMOS, CMOS VCO provides double the trans-conductance due to the utilization of double XCP [11]. Initially, the differential voltage is generated by NMOS transistors across the resonator and the V_{GD} of Mn1 and Mn2 are forced to be equal in magnitude but opposite in direction. The startup of the oscillation is constituted by the small signal conductance caused by the saturation of the transistor Mp1, Mp2, Mn1, Mn2 at zero differential voltage. With the rise of differential voltage more than the threshold voltage of nMOS transistor ($V_{\mathrm{TH},n}$), the gate to drain voltage of Mn1 rises $V_{\mathrm{TH},n}$ because of which V_{GD} of Mn2 falls and gets into cutoff region.

Similarly for falling differential voltage Mp1 and Mp2 get into cutoff and triode region, respectively. The complementary alternative switching of nMOS and pMOS provides oscillation through the resonator [12]. Figure 2 shows the equivalent design of the resonator with R_{Tank} as the induced resistance due to the inductor and capacitor of the tank for which a negative resistance is provided—R_{Tank} using active components.

There are major three sources of phase noise which include XCP, LC tank, and tail current source. Figure 3 shows the contribution of LC tank in noise source. The equivalent resistance of LC tank is given by Eq. 1 [10],

$$R_{\mathrm{tank}} \approx R_{\mathrm{L}} + R_{\mathrm{var}} + \frac{1}{R_{\mathrm{p}}(\omega_0 C_{\mathrm{var}})^2} \tag{1}$$

where R_{var} and R_{L} are the series resistance of the varactor and inductor of LC tank, C_{var} is varactor's capacitance, and R_{p} is the substrate resistance of the inductor. The overall effective resistance of the LC tank contributes to the noise and is given by Eq. 2 [10],

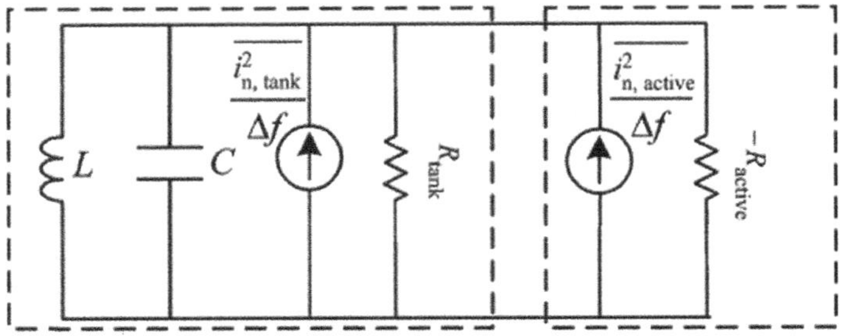

Fig. 3 LC tank equivalent circuit with noise source

$$\frac{\overline{i^2_{\text{tank}}}}{\Delta f} = \frac{4kT}{R_{\text{tank}}} \tag{2}$$

where Δf is the offset frequency from the carrier frequency, k is Boltzmann's constant. Phase noise is generated due to the upconversion or downconversion of multiple noises which includes flicker noise and thermal noise. Due to non-ideal inductor and capacitors only thermal noise is generated in LC tank [13]. LC-VCO provides better phase noise; however, it suffers the challenges of fabrication of large inductors [14]. In XCP, certain parts of flicker noise and thermal noise are transmitted to resonator [13]. Thermal noise generated due to the switching mechanism will also contribute to the phase noise [4]. Among the multiple harmonics mainly the second harmonics is transferred as phase noise [15].

3 Proposed VCO Design

The impact of flicker noise in oscillator increases with decrease in technology node [16]. A major mechanism in upconversion of phase noise includes conversion of amplitude modulation (AM) to phase modulation (PM). This conversion can be minimized by employing smaller varactor for finer tuning [17]. Zero crossing point is more sensitive to phase noise compared to any other point. According to Leeson's phase formula given by Eq. (3) [18],

$$L(\Delta f) = \frac{2kT\text{Req}F}{A^2}\left(\frac{f_0}{2Q\Delta f}\right)\left(1 + \frac{\Delta f_{1/f^3}}{\Delta f}\right) \tag{3}$$

where k is Boltzmann's constant, F is the noise factor, Q is the effective quality factor of the resonator, Δf is the offset frequency from the carrier frequency, $\Delta f_{1/f^3}$ is the flicker noise at corner frequency, f_0 is the frequency of oscillation. From Eq. (1), it can be observed that higher the quality factor (Q) lower will be phase noise. This has been obtained by using high Q inductor in the resonator. Flicker noise is given by Eq. (4),

$$\overline{i_{1/f}^2} = \frac{k_\mathrm{f}}{f C_\mathrm{ox} W L} \Delta f \tag{4}$$

where k_f is the CMOS process constant, L and W are the length and channel width of tail current source transistor, respectively. The flicker noise is inversely proportional to size of the transistors. To reduce flicker noise, the size of the transistor is adjusted in the optimum way as large increase in the size of the transistor will reduce the tuning range due to the effect of parasitic capacitance. However, varying transistor width will also vary its gate resistance which will enhance thermal noise. Figure 4 shows the circuit diagram of the proposed VCO design.

Noise creation in current source is well explained in Rael's analysis [12]. The noise at low frequency is converted into phase noise. In a balanced circuit, even harmonics and odd harmonics flows through common mode path and differential mode path, respectively. The second harmonic among even harmonics is usually dominant. Further harmonics are nearly ignored because their contribution of mean

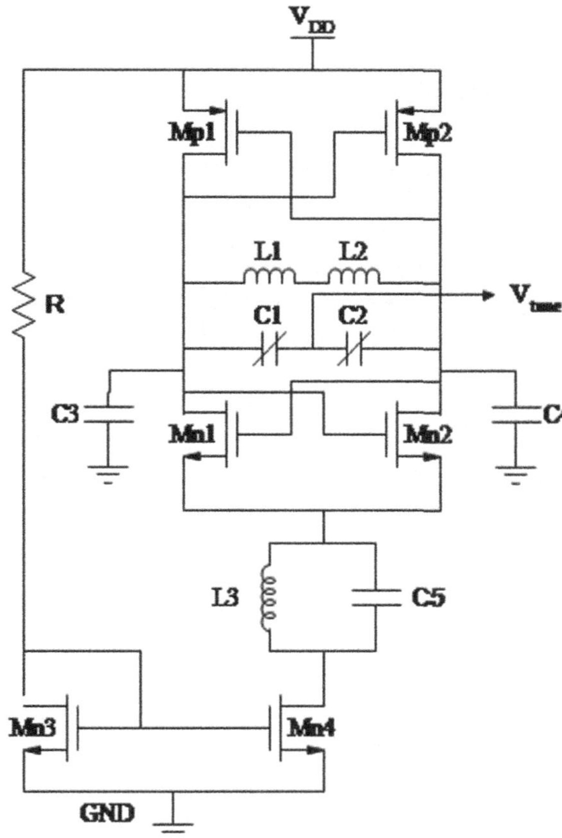

Fig. 4 Proposed differential VCO

square error bows down. Due to the tail current source, the contribution of thermal noise into the tank is large. To prevent this, L3 and C5 are used which block phase noise at frequency $2\omega_0 + \omega_m$ generated due to the thermal noise. It also prevents LC tank to be loaded by transistor Mn1 and Mn2, hence maintaining a high Q which will reduce phase noise as Q is inversely proportional to phase noise.

The phase noise generated due to the thermal noise of the tank and the switching mechanism of the XCP will also affect the performance of the VCO to a larger extent. To reduce such noise, capacitance C3 and C4 are used as filter. Mn3 and Mn4 depict current mirror topology. Utilization of inductor in the tail node decreases the second harmonics by resonating the parasitic capacitor at the same harmonic [19]. Higher current driving capability is achieved using current mirror. 1:3 current mirror source has been utilized. The variable capacitor in the proposed design is obtained using pMOS transistors as pMOS transistors have lower flicker noise [20].

In the proposed design, the varactor designed using pMOS is utilized in the accumulation region in order to have maximum capacitance and hence high-quality factor Q.

4 Results and Discussion

The proposed VCO has been designed for reduced phase noise (Fig. 5). The variation of oscillator output frequency with the tuning voltage is shown in Fig. 6, and the simulated output of the same is shown in Fig. 7. The output frequency is linearly controlled by the very small range of controlled voltage. Figure 8 also shows that the variation of controlled voltage makes an effect of 0.4 dBc/Hz phase noise only.

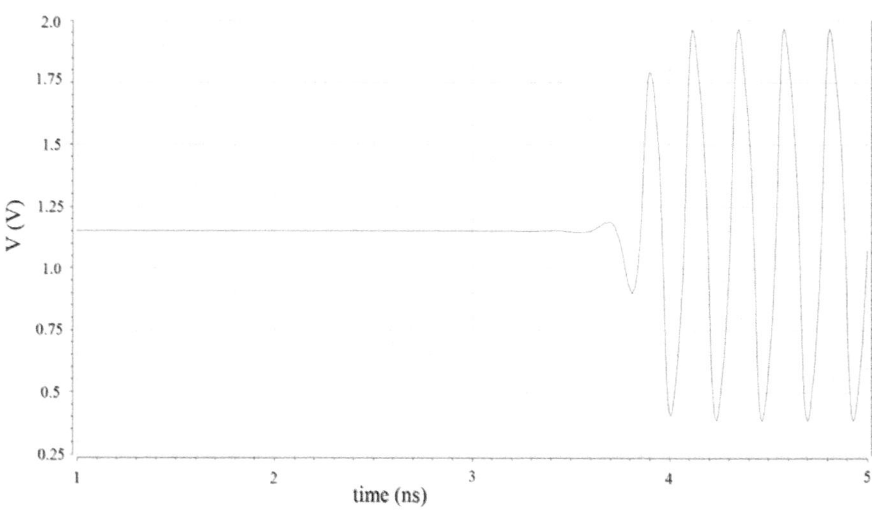

Fig. 5 Output waveform of proposed VCO

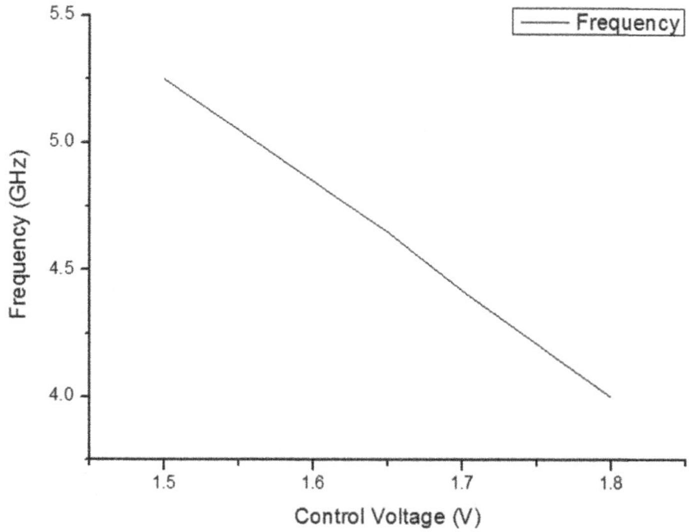

Fig. 6 Observed variation of frequency (GHz) with respect to control voltage (V)

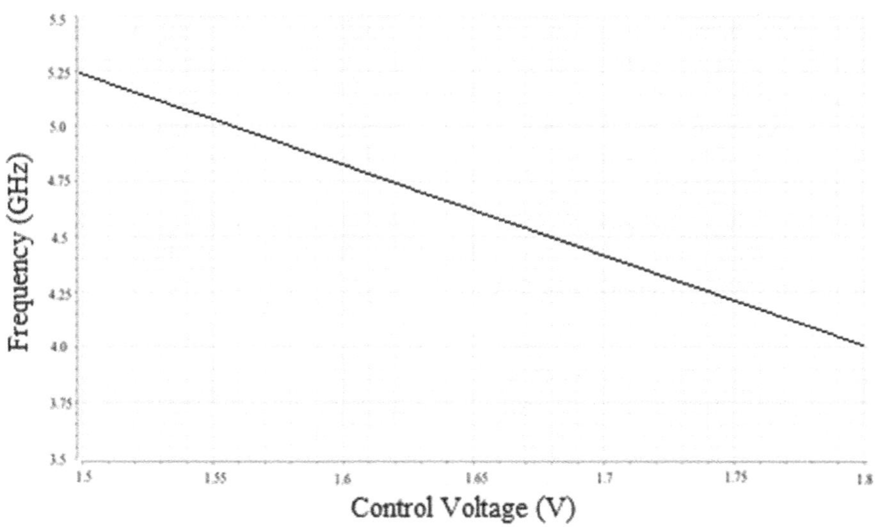

Fig. 7 Simulated results of variation of frequency (GHz) with respect to control voltage (V)

To determine the phase noise and frequency of oscillation, periodic steady-state analysis has been performed. All the transistors have been observed to be operating in saturation region. Figure 5 shows the oscillating output waveform of the proposed VCO. Additionally, the high amplitude can be observed from the waveform which corresponds to the increase in the slope, which significantly contributes to the lower phase noise.

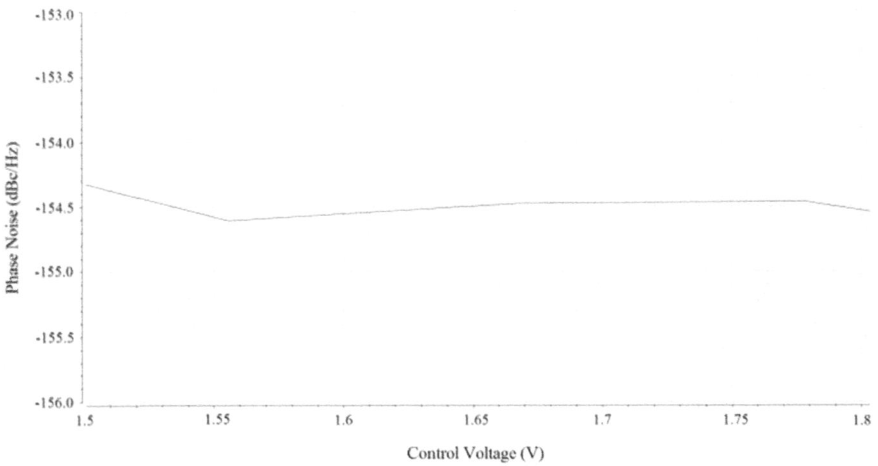

Fig. 8 Variation of phase noise (dBc/Hz) with respect to control voltage (V)

L and *C* in the LC tank form a parallel resonator. For parallel resonator, the resistance of inductor R_L is directly proportional to quality factor *Q*. The resistance *R* in the proposed design is inversely proportional to the bias current. Moreover, the bias current also affects the slope of the output waveform which can improve the phase noise. Quality factor has a direct relation with the energy. The optimum value of R_L and *R* is taken in order to have low-phase noise without impacting the working regions of transistors. The outputs are observed in Cadence Analog Design Environment (ADE).

Figure 4 shows that designed VCO is observed to have a very large tuning range for a very small variation of control voltage of 0.3 V which is one of the exceptional highlights of the performance (Table 1).

Table 1 Performance parameter of designed VCO

References	Node (μm)	f_0 (GHz)	V_{DD} (V)	PC (μW)	PN (dBc/Hz)
[21]	0.09	1.012	1.2	386.64	−138
[22]	0.09	2.44	1.8	53.34	−140
[23]	0.5	1.4	3	300	−107
[24]	0.18	2.25	0.5	323.6	−119.4
[25]	0.18	2.4	1.5	97	−111
[26]	0.09	2.72	1	418	−82.34
This work	0.09	4.1	1.4	50.02	−154.3

5 Conclusion

A low-phase noise differential cross-coupled pair VCO has been demonstrated at 90 nm (0.9-μm) technology. A tuning range of 1.25 GHz from 5.25 to 4.0 GHz with the control voltage varying between 1.5 and 1.8 V has been obtained. With a successful reduction in harmonics near to $2\omega_0$ using multiple techniques and optimum design of high Q resonator, the design is verified with phase noise of −154.3 dBc/Hz at an offset frequency of 1 MHz away from central frequency of 4.10 GHz. On-chip large area requirement of spiral inductors is one of the demerits of design. The observations are made at the supply voltage of 1.4 V. The power consumption of the proposed VCO design is 50.02 μW. Utilization of proposed design can be extended to design and implementation of enhanced PLL or frequency synthesizer.

References

1. Panda M, Patnaik SK, Mal AK, Ghosh S (2019) Fast and optimised design of a differential VCO using symbolic technique and multi objective algorithms
2. Jahan N, Barakat A, Pokharel RK (2019) A −192.7-dBc/Hz FoM K U -band VCO using a DGS resonator with a high-band transmission pole in 0. 18- μm CMOS technology 29(12):814–817
3. Lee SY, Hsieh JY (2008) Analysis and implementation of a 0.9-V voltage-controlled oscillator with low phase noise and low power dissipation. IEEE Trans. Circuits Syst II Express Briefs 55(7):624–627
4. Huang C, De Vreede LCN, Akhnoukh A, Burghartz JN, Low phase noise LC oscillators 31(0):15–18
5. Vco DC, Fu Y, Member S, Li L (2020) A −193 6 dBc/Hz FoM T 28. 6-to-36. 2 GHz. IEEE Access 8:62191–62196
6. Patil RK, Nasre VG (2012) A performance comparison of current Starved VCO and source coupled VCO for PLL in 0.18 μm CMOS process. Int J Eng Innov Technol 1(2):48–52
7. Singh PK, Kar AK, Singh Y, Kolekar MH, Tanwar S of ICRIC (2019)
8. Gui X et al (2020) A voltage-controlled ring oscillator with VCO-gain variation compensation 30(3):288–291
9. Hajimiri A, Lee TH (1999) Design issues in CMOS differential LC oscillators. IEEE J Solid-State Circuits 34(5):717–724
10. Jia L, Ma JG, Yeo KS, Do MA (2004) 9.3–10.4-GHz-band cross-coupled complementary oscillator with low phase-noise performance. IEEE Trans Microw Theory Tech 52(4):1273–1278
11. Hegazi E, Sjöland H, Abidi AA (2001) A filtering technique to lower LC oscillator phase noise. IEEE J Solid-State Circuits 36(12):1921–1930
12. Rael JJ, Abidi AA (2000) Physical processes of phase noise in differential LC oscillators. Proc Cust Integr Circuits Conf 3(c):569–572
13. Yan W, Park CH (2008) Filtering technique to lower phase noise for 2.4 GHz CMOS VCO. In: International conference on solid-state and integrated circuits technology proceedings, ICSICT, pp 1649–1652, 2008
14. Swarnakar J, Sarkar P (2018) Advances in communication, devices and networking, vol 462. Springer Singapore
15. Oh NJ (2014) A phase-noise reduction technique for RF CMOS voltage-controlled oscillator with a series LC resonator. Microelectronics J 45(4):435–440

16. Pepe F, Bonfanti EA, Maffezzoni P, Fiorini C (2013) POLITECNICO DI MILANO DIPARTIMENTO DI ELETTRONICA, INFORMAZIONE E BIOINGEGNERIA Doctoral Programme In Information Technology Analysis And Minimization Of Flicker Noise Up-Conversion In Radio-Frequency LC-Tuned Oscillators The Chair of the Doctoral Program
17. Kral A, Abidi AA (1998) RF-CMOS oscillators with switched tuning, pp 555–558
18. Leeson DB (1998) A simple model of feedback oscillator noise spectrum. Integr Circuits Wirel Commun 429–430
19. Taris T, Rashtian H, Shirazi AHM, Mirabbasi S (2014) A low-power 2.4-GHz combined LNA–VCO structure in 0.13-μm CMOS. Analog Integr Circuits Signal Process 81(3):667–675
20. Bhattacharjee J, Mukherjee D, Gebara E, Nuttinck S, Laskar J (2002) A5.8 GHz fully integrated low power low phase noise CMOS LC VCO for WLAN applications. IEEE Radio Freq Integr Circuits Symp RFIC, Dig Tech Pap, pp 475–478
21. Muddi V, Shinde KD, Shivaprasad BK (2016) Design and implementation of 1 GHz Current Starved Voltage Controlled Oscillator (VCO) for PLL using 90 nm CMOS technology. In: 2015 International conference on control, instrumentation, communication and computational technologies ICCICCT 2015, pp 335–339
22. Bodade P, Meshram MD (2013) Design of differential LC and voltage controlled oscillator for ISM band applications 2(4):1428–1431
23. Scanlon E (2017) A way of being. Prairie Schoon 91(1):52
24. Wang X, Yang X, Xu X, Yoshimasu T, Ic CL Simplified noise filtering in 180-nm CMOS, pp 5–7
25. Lee SH, Jang SL, Chuang YH, Chao JJ, Lee JF, Juang MH (2007) A low power injection locked LC-tank oscillator with current reused topology. IEEE Microw Wirel Components Lett 17(3):220–222
26. Faruqe O, Bulbul AK, Saikat MM, Amin T (2019) A high output power active inductor based voltage controlled oscillator for bluetooth applications in 90 nm process. In: 4th International conference on electrical engineering and information communication technology iCEEiCT 2018, pp 15–20, 2019

Secured Surveillance Storage Model Using Blockchain

D. Soujanya and K. Venkata Ramana

Abstract Nowadays, huge amount of surveillance data is being generated continuously by various resources. Storing and processing of surveillance data are a tedious task for any organization. Distributed environment plays a vital role for analyzing the collected data. However, sharing the data among various distributed devices in different layers raises security concerns where an adversary may capture or tamper with features to mislead the surveillance system. In this paper, we propose a blockchain-based model which makes the data stored in it immutable. Here, the files are saved in a distributed ledger technology called the Inter Planetary File System (IPFS), which works on the principle of cryptography. In IPFS, the data is stored as a Merkle tree by using one-way hash function. Our proposed model includes various policy and maintenance mechanisms, certificate authorities, authentication, and revocation of certificates to the data in IPFS with the help of smart contracts.

Keywords IPFS · Surveillance · Blockchain · Smart contract · Hash function

1 Introduction

In the recent decades, the process of protecting the data from unauthorized access and data corruption throughout its lifecycle has been a major problem in the modern digital world. Especially in an Internet environment, the risk to valuable and sensitive information like digital evidences and personal information is greater than ever before. The enormous sensitive information from the surveillance systems resides in the centralized server which offers better governance but steps back in providing flexible, secure, and computationally efficient way of processing data. Blockchain is an emerging distributed ledger technology which overwhelms all these problems,

---

D. Soujanya (✉) · K. Venkata Ramana
Department of CS and SE, Andhra University College of Engineering (A), Andhra University, Visakhapatnam, India
e-mail: soujanya.297@gmail.com

K. Venkata Ramana
e-mail: kvramana.auce@gmail.com

P. K. Singh et al. (eds.), *Evolving Technologies for Computing, Communication and Smart World*, Lecture Notes in Electrical Engineering 694,
https://doi.org/10.1007/978-981-15-7804-5_19

making it a new decentralized platform. Organizing the information in distributed ledger is done with Inter Planetary File System (IPFS) [1] hash. IPFS and public blockchains mandate user authentication to access files by the protocol called the smart contract. These smart contracts digitally verify and enforce security to the users and stored information.

In this paper, we discussed about the decentralized storage mechanism, integrated with the blockchain and certificate authority. The primary goal of smart contract is to validate the credentials of the user. Later, the Certificate Authority (CA) is embedded with the proposed system to monitor the behavioral pattern of different users of the system by performing various operations like Policy Authentication, Issuing of Certificates, Manufacture and Revocation of Certificates, Maintaining Registration Authority, Authentication of Services, and Repository maintenance.

2 Related Work

Asghar et al. [2] proposed a model and give solution for visual surveillance data protection based on General Data Protection Regulations (GDPR). In the context of GDPR, the roles of machine learning, image processing, cryptography, and blockchain are explored as a way of deploying data protection by design solution for visual solution data. Nikouei et al. [3] suggested a blockchain-enabled scheme to protect the big data generated by numerous surveillance devices, where the data is indexed by encrypting the path between the nodes, thereby reducing the attacks on the small edges and devices. Wong et al. [4] implemented a new video surveillance system with permissioned blockchain, IPFS, CNNs, and edge computing to achieve large scale wireless sensor information acquisition and data processing.

Nagothu et al. [5] developed microservice-enabled architecture for smart surveillance system in which the video analysis algorithms are encapsulated into each microservice. Kerr et al. [6] demonstrated the participation of prototype camera for secured distributed ledger of video streaming. This scheme combines blockchain and digital watermarking for providing trustworthy evidence protection in distributed environments. Jeong et al. [7] created a blockchain network where internal managers play the trusted role. The metadata of the video is recorded in the distributed ledger of blockchain and that further generates a license of the videos. AI-Sahan et al. [8] implemented a national surveillance system to detect intrusion and criminal act to cover all geographical areas and blind spots. Krejci et al. [9] presented and used a middleware architecture that connects to IoT devices for distributing data with guaranteed integrity.

3 Preliminaries

In this section, we present the related knowledge and background for better understanding of our proposed system.

3.1 Blockchain

Blockchain is "an open, distributed ledger that can record transactions between two parties efficiently and in a verifiable and permanent way" [6]. This is a peer-to-peer network of decentralized systems where each and every node holds the replica of the original data in a chained manner. The size of the chain keeps on increasing as more blocks are attached to it. Each block has its own unique ID called the hash pointer which identifies the block uniquely. The initial block is called the genesis block. Any operation performed in the network is stored in the blockchain in the form of transactions. These transactions are stored in the form of cryptographically hashed values. The cryptography function employed in the blockchain is a one-way hash function which prevents the data in the blockchain from being tampered, as the blocks are all arranged in a tree structure called Merkle tree and any changes made to the data will collapse the entire tree, and it has to be built again.

3.2 Ethereum [7]

Ethereum is an open source, public, blockchain based distributed computing platform and operating system featuring smart contract functionality [7]. It provides users with various types of accounts to perform the operations on the blockchain. Any operation performed in Ethereum is calculated in terms gas units. Gas can be purchased using either which is cryptocurrency used in Ethereum or can be mined by the miner. Whenever any transaction is triggered by the user the transaction cost can be calculated as sum of execution cost and transaction cost.

$$\mathrm{PoT} = (\mathrm{EC} + \mathrm{TC})P \tag{1}$$

where PoT is the price of a transaction; EC is the execution cost; TC is the transaction cost; and P is the price of one gas unit.

3.3 IPFS [2]

IPFS is protocol and also a network designed to create a content addressable, peer-to-peer method of storing and sharing hypermedia in a distributed file system [2]. This works on the content it stores in the network. Identifiers are generated depending on the content in the file, i.e., a document, audio, image, video, or any other kind of information that is stored in it. These identifiers are arranged in the form of a Merkle dag. IPFS as a network model uses PKI-based identity. An IPFS Node is a program that can find, publish, and replicate Merkle dag objects. Its identity is defined by a private key.

$$\text{Key}_{\text{Gen}}!\ \text{P UKey};\ \ \text{P RKey} \tag{2}$$

$$\text{NodeId} = \text{MultiHash}(\text{P UKey}) \tag{3}$$

All hashes in IPFS are encoded with multihash, a self-describing hash format. All IPFS nodes support sha2-256, sha-512, sha3 algorithms. The generated hash values should be deterministic, uncorrelated, unique, and one-way. The main functionality of IPFS depends on the content identifiers which posses the following structure <cidv1> :: = <multibase-pre x> <cid-version> <multicode-content-type> <multihash-content-address> where <multibase-pre x> is a code (1 or 2 bytes), to ease encoding CIDs into various blocks. <cid-version> is a varint representing the version of CID for upgradability. <multicode-content-type> is used to represent the content type or format of the data being addressed. <multihash-content-address> represents the cryptographic hash of the content being addressed in the format base58 (<varint hash function code> <varint digest size in bytes> <hash function output>.

3.4 Smart Contract

A smart contract is computer protocol intended to digitally facilitate, verify, or enforce the negotiation or performance of a contract. Smart contracts permit trusted transactions and agreements to be carried out among disparate, anonymous parties without the need for a central authority, legal system, or external enforcement mechanism. They render transactions which are traceable, transparent, and irreversible.

4 Smart Contract for S^3 Model

The entities of our secure surveillance storage model (S^3) are Interrogation Team (IT), Registration Authority (RA), and Certificate Authority (CA) as shown in Fig. 1. The

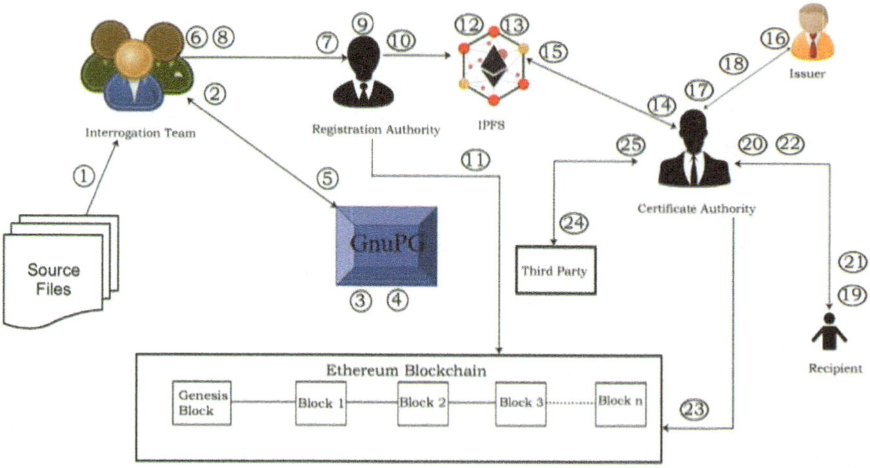

Fig. 1 S^3 architecture

responsibility of the IT is to collect all the valid files from different sources; RA is to validate and store all the files collected by IT. The role of CA is to govern issuer, recipient, third parties and also managing data in IPFS and blockchain. The basic operations performed on proposed system are as follows.

4.1 Process Flow

1. IT acquires all the source files from various sources.
2. The IT sends these source files and user ID to GnuPG for encryption.
3. GnuPG generates unique key pair—public key, private key in specific to users.
4. Using the public key of the user the required file will be encrypted.
5. GnuPG sends back the required encrypted file to IT.
6. The Chief Interrogator(CI) sends his unique ID to RA to validate himself.
7. The RA authenticates the address of CI who will be communicating with the RA.
8. The CI then sends the encrypted documents to RA.
9. The RA verifies the validity of the document hash.
10. Upon successful verification, the RA adds those files to IPFS.
11. RA updates file details as transactions to blockchain.
12. IPFS performs hash operation on the files received from RA and generates the Merkle Dag.
13. Adds valid encrypted files to repository.
14. CA sends a request to IPFS to access encrypted source files.
15. IPFS grants permission to CA if the requested files are available in repository.

16. Certificate issuer places a request to CA to register himself with the system to issue certificates.
17. Upon validating the request, the CA accepts or denies the request.
18. The registered issuer registers the certificates to be issued on further requests.
19. Recipient places a request to CA to register himself with the system to further use the registered certificates.
20. Upon validating the request, the CA accepts or denies the request.
21. Registered recipients place request to CA to issue certificates.
22. CA checks with the repository. If requested certificate is available it issues the certificate to recipient otherwise conveys failure message to recipient.
23. Update all details as transactions to blockchain.
24. Any third party willing to verify the recipients address places request to CA.
25. CA provides the requested details to third party.

The proposed model uses the following algorithms for implementing the above operations.

4.2 Gnu Privacy Guard (GnuPG)

This algorithm generates the unique key pair—public key and private key in specific to user details which are sent to GnuPG along with the input file.

i. Key Gen (name, email, passphrase)! (P_U key; P_R key): The Key Gen algorithm takes the user credentials—name, email address, and passphrase of the user for personal identification and key generation. It generates a key pair—public key which is transmitted to the other user for encrypting the files, the private key is kept personal with the user for the decryption and verification (Table 1).
ii. Share (RP_U key): The recipients generate a request with their public key for a specific file. The request can also be generated by a group of recipients. This algorithm is used to share the recipient key to issuer.

Table 1 Abbreviations used in algorithms

Notation	Description
P_U Key	Public Key of user
P_R Key	Private Key of user
RP_U Key	Recipients public key
F	File
CT_x	Cipher Text
BLOB	Binary Large Object
CID	Content Identifier
Bkey	Secure Hash Key of BLOB
Uid	Unique ID of source file
Pid	Unique peer ID in IPFS
Hid	Unique Certificate Hash
Uaddr	Unique User address
Raddr	Unique Recipient address
Iaddr	Unique Issuer address
CI	Chief Interrogator
Chash	Certificate Hash
CHashIssuer	Hash ID of issuer of certificate

Algorithm 1 GnuPG Algorithm

Input: $User\_Details$
Output: $Cipher\_Text(CT_x)$

```
1: Begin
2: User_Details ← Get_values_from_user(name, email, passphrase)
3: If User_Details are Valid Then
4:     Key_Pair ← Key_Gen(User_Details)
5: Else
6:     Display"error"
7: End If
8: If Key_Pairisgenerated Then
9:     Recipient_key ← Share(RP_U Key)
10:    CT_x ← Encrypt(File, Recipient_key)
11: End If
12: End
```

iii. Encrypt (F,RP_U key)! CT_x: The encryption algorithm takes the input as plain text file or multimedia content and the public key of the recipient. Considering public key encryption method the corresponding input file is converted to the cipher text.

iv. Decrypt (CT_x; $P_R$key)! F: The decryption algorithm takes the private key of the user and cipher text file as the input for the algorithm. If the file has reached the valid recipient, then the decryption process will be succeeded, thereby restricting the unauthenticated users from sniffing the files as the private key will be unknown to him.

4.3 IPFS

Takes the input files and performs various hash operations on the files and stores them in the form of Merkle dag. Receives the requests from CA and RA, and if requested file matches with the IPFS repository the hash value will be send else display an error.

Algorithm 2 Algorithm for Working of IPFS

Input: $BLOB, F$
Output: $Secure\ Hash\ Key\ of\ BLOB, B_{key}$

```
1: Begin
2: Doc ← Get_BLOB_from_user()
3: CID[ ] ← Generate_CID(Doc)
4: While size of blocks!>256kb do
5:     CID[ ] ← Generate_CID(Doc)
6:     blocks[ ] ← make_blocks(Doc)
7:     ID[ ] ← hash(blocks)
8:     hash[ ] ← sha256(blocks)
9:     Append hash_function_size and diges_size to ID[ ]
10:    CID[ ] ← base58(ID)
11:    Construction of merkle tree
12: End While
13: End
```

Our system uses two smart contracts for performing the core functionalities.

4.4 Registration Authority

This smart contract is responsible for performing the following functionalities:

Registering Files: This function is going to take the document type, encrypted documents hash, and peer ID of who is going to store the file in the IPFS.

Algorithm 3 Algorithm for Addition of Source

Input: *Document Details*
Output: *Hash code of the Source added*

```
1: Begin
2: Doc_Details ← Get_values_from_user(DocumentType, DocumentId, UserAccountAddress)
3: If User_account.exists == true Then
4:     addEvidence(DocumentType, DocumentId, UserAccountAddress)
5:     CountCount + 1
6: Else
7:     Display"User not authenticated"
8: End If
9: End
```

Fetching File: This function facilitates the users of the system to easily retrieve the files or view the files by simply entering the unique ID of the document that was already generated in the addition module.

Retrieving Document Hash: Upon taking the files unique ID generated in the registration module, the system is going to retrieve the unique hash key of the document, which can be communicated among the registered users for verification about the case.

Retrieving Peer IDs: This module takes in the unique file ID that is generated in the registration module the system is going to retrieve the unique peer ID to further check whether he is a valid user or not, or if he is the right person to perform the addition task.

4.5 Certificate Authority

This smart contract is responsible for generating certificates and assigns to the valid users. The operation is split up as:

Algorithm 4 Algorithm for Registering Certificate

Input: *Unique Secure Certificate Hash ID*, H_{id}
Output: *Registration ID*

```
1: Begin
2: U_Addr ← fetch_User_Address
3: H_id ← fetch_IPFS_Hash
4: If U_Addr.alreadyregistered != true Then
5:     Display"Issuer not registered to register certificate"
6: Else
7:     Register_user_for_the_requested_hash_code
8: End If
9: End
```

Register Certificate: This function takes in the document unique ID which has already been stored in the registration module, which is the unique certificate that has to be issued for the recipients on their request.

Algorithm 5 Algorithm for Registering Recipient

Input: *Unique Secure Peer* ID, P_{id}
Output: *Registration ID*

Begin
$P_{id} \leftarrow fetch\_user\_address$
If $P_{id}.alreadyregistered != true$ **Then**
  Change registration status of recipient to true
  Increment Count
Else
  Display Recipient already registered
End If
End

Register Issuer: To issue a particular certificate to any of the recipients, the issuer has to register himself with unique ID. Only then he will be allowed to issue certificates to the recipients.

Retrieve Certificate Identifier: Upon entering the unique identification of the certificate, this module will display the identifier of that particular certificate.

Algorithm 6 Algorithm for Registering an issuer

Input: *Unique Identity of the Source file,* U_{id}
Output: *Source Owners ID*

Begin
$U_{Addr} \leftarrow fetch\_User\_Address$
$U_{id} \leftarrow fetch\_document\_unique\_ID$
If $U_{Addr}.alreadyregistered == true$ **Then**
  Display"Issuer already registered"
Else
  $New\_details = retrieve\_Source\_details(U_{id})$
  $Register\_user(new\_details)$
End If
End

Register Recipient: To receive any certificates from the issuers, one has to register himself with the system indicating he is an authenticated user and not any malpractitioner because on performing any malpractice his certificate will immediately be revoked.

Issue Certificate: Upon registering the issuer, the recipient and the certificates to be issued, the certificates will be issued to the recipients on placing a proper request to the certificate authority.

Certificate Counter: This module maintains the count of all the certificates issued till date by all the issuers.

Retrieve All Recipients of a Certificate: Similar kind of certificates can be issued to any number of recipients. This module maintains the list of recipients who are currently being using that particular certificate.

Algorithm 7 Algorithm for Issuing Certificate

Input: *Recipient AddressR*$_{addr}$, *Certificate HashC*$_{hash}$, *Issuer Address I*$_{addr}$
Output: *Status of certificate as transferred from Issuer to Recipient*

Begin
$R_{addr} \leftarrow fetch\_Recipient\_address$
$I_{addr} \leftarrow fetch\_Issuer\_address$
$C_{Hash} \leftarrow fetch\_Certificate\_hash$
$C_{HashIssuer} \leftarrow fetch\_Issuer\_of\_Certificate$
If $R_{addr}.alreadyregistered \mathrel{!=} true$ **Then**
    Display Recipient not registered to be issued a certificate
Else
    If $I_{addr}.alreadyregistered \mathrel{!=} true$ **Then**
        Display "Issuer not registered to register for a certificate"
        $Update(Issuer, Recipient, Certificate, Count\_Of\_Certificates)$
    Else
        Display "Issuer not registered to issue this certificate"
    End If
End If
End

Retrieve Certificate Issuer: When a user has a certificate and to know who has issued that particular certificate, this module is used. This module takes in the certificate identification and retrieves the unique identification of the issuer.

Retrieve All Certificates of an Issuer: A single issuer can issue many varieties of certificates to the recipients. This module retrieves all the certificates that a particular issuer has registered himself to issue.

Retrieve the Recipient of a Certificate: Upon taking the certificates unique id, this module retrieves all the recipients who are currently using that particular certificate.

Retrieve All Certificates of the Recipient: This module retrieves all the certificates of all kinds a recipient possesses.

5 Experimental Evaluation

This application is designed for gathering, validating, and storing the most important files such as audio, video which can include CCTV footage, documents, and images that are very crucial and much prone to tampering. In order to restrict tampering or morphing of data, we implemented this new framework where CCTV footage is linked to blockchain technology which is tamper proof and authentic. In this system, the restriction is imposed on people who are going to perform the operations directly on the system. Users of this system will be provided with their own personal credentials (hash keys); upon validating their details with the system, it grants them permission to use, store, or retrieve the information. These files are stored in IPFS which is a decentralized file storage platform capable of handling larger sized files with great efficiency and less latency by performing multiple hash operations on the blocks of specified size.

The files are identified by the content identifiers which are result of one-way hash functions. All the operations that are performed on the data are validated by multiple smart contracts. We implemented a prototype of the proposed system to study the performance and also to check the attainment of security levels. The prototype is implemented using— node of version 12.7.0, npm of version 6.11.2, go-ethereum as the platform for blockchain, and IPFS version v0.4.13. The solidity framework and IDE for deploying smart contracts true is of version v5.0.30, ganache-cli v6.7.0-beta.0, and the scripting language used to write the contracts solidity is of version v12.7.0. The handling of requests between Ethereum [10] nodes is maintained by web3js is of version v1.2.0. We deployed the blockchain node is four machines, and each possess a 2.30 GHz core Intel processor with 8 GB primary memory. Some nodes were deployed in various environments like— ubuntu 16 OS, Windows 10. All the machines deployed the storage system for handling all kinds of files possible in the project. The complexity of this experiment is calculated in terms of gas as deploying or execution of a smart contract is done with it. This can also be called as space complexity of the algorithm. Table 2 shows the amount of cost incurred for each operation that is being performed in the experiment. The metrics used is Gwei.

Total cost = number of units of gas consumed *Cost of each unit of gas. The execution of the operations discussed above is mandatory for executing the framework. Hence, the total cost for executing this framework is 7.997 USD. The following are the advantages of the proposed system. After gathering and encrypting the required

Table 2 Cost incurred for the operations in the framework

S. No.	Operation being performed	Units of gas consumed		Cost Incurred		Total cost (in GWei)€	Total cost (in USD)
		Transaction (in gas)	Execution (in gas)	Transaction (in GWei)	Execution (in GWei)		
1.	Source details contract deployment	852,719	607,219	4,263,595	3,036,095	7,299,690	1.298
2.	Adding source details	194,209	164,937	971,045	824,685	1,795,730	0.319
3.	Deploying certificate authority	2,840,335	2,110,995	14,201,675	10,554,975	24,756,650	4.403
4.	Registering issuer	707,802	686,402	3,539,010	3,432,010	6,971,020	1.24
5.	Registering certificate	48,888	24,032	244,440	120,160	364,600	0.065
6.	Registering recipient	109,613	84,757	548,065	423,785	971,850	0.173
7.	Issuing certificate	293,937	267,673	1,469,685	1,338,365	2,808,050	0.499

files with their public keys, they are further sent for validation by the smart contract and storage in the IPFS. Upon receiving a file in IPFS, it is stored as (Fig. 2).

The corresponding Merkle tree representation of the file is as follows (Fig. 3).

Further upon receiving requests from issuer, receiver, or third party the certificate authority is responsible for managing the flow of data between them.

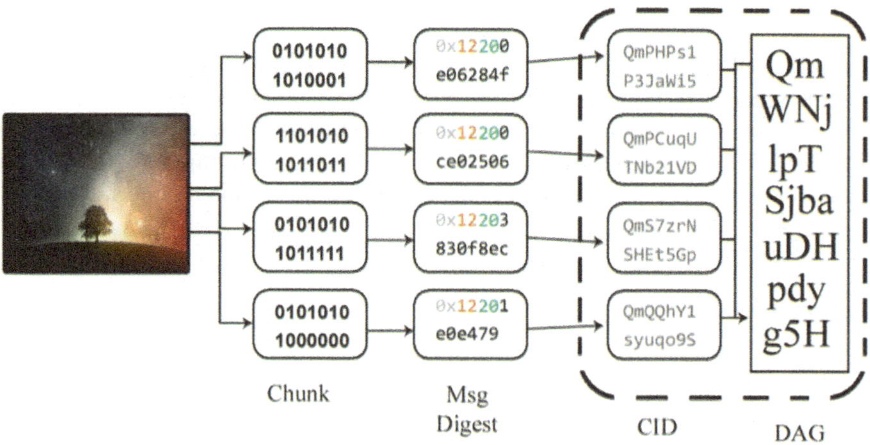

Fig. 2 CID representation of input le

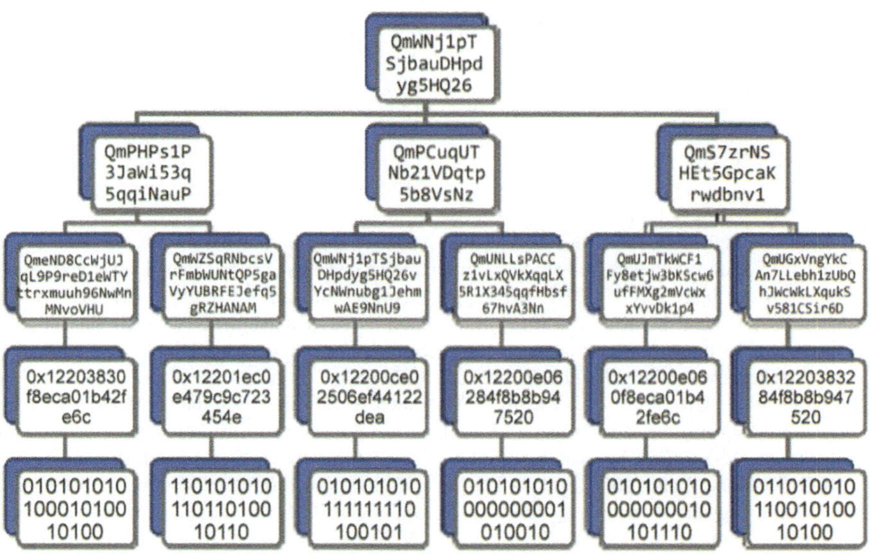

Fig. 3 Merkle tree representation of input file

Modifications are restricted on the blocks as any change to the information in these blocks would collapse the entire tree. All these restrictions and permissions are all governed by the smart contracts registration authority.

6 Conclusion and Future Scope

Our proposed system integrates surveillance system with blockchain technology to provide security to the files stored in it. These are automated and maintained by smart contracts to grant/restrict access of different users. This work can further be combined with neural networks and machine learning algorithms for retrieval of features in faces, i.e., facial recognition, event tracking, session monitoring, etc.

References

1. Benet J IPFS-content addressed, versioned, P 2P File System Retrieved from https://github.com/ipfs/papers/raw/master/ipfs-cap2pfs/ipfs-p2p-le-system.pdf
2. Asghar MN, Kanwal N, Lee B, Fleury M, Herbst M, Qiao Y (2019) Visual surveillance within the EU general data protection regulation: a technology perspective. In: IEEE Access, 2019, vol 7, pp 111709–111726. IEEE, https://doi.org/10.1109/ACCESS.2019.2934226
3. Nikouei SY, Xu R, Nagothu D, Chen Y, Aved A, Blasch E, (2018) Real-time index authentication for event-oriented surveillance video query using blockchain. In: IEEE international smart cities conference (ISC2) 2018, IEEE, pp 1–8. https://doi.org/10.1109/ISC2.2018.8656668
4. Wang R, Tsai W, He J, Liu C, Li Q, Deng E. (2019) A video surveillance system based on permissioned blockchains and edge computing. In: IEEE international conference on big data and smart computing (BigComp) 2019, IEEE, pp 1–6. https://doi.org/10.1109/BIGCOMP.2019.8679354
5. Nagothu D, Xu R, Nikouei SY, Chen Y (2018) A microservice-enabled architecture for smart surveillance using blockchain technology. In: IEEE international smart cities conference (ISC2) 2018, IEEE, pp 1–4. https://doi.org/10.1109/ISC2.2018.8656968
6. Kerr M, Fengling H, van Schyndel R (2018) A blockchain implementation for the cataloguing of CCTV video evidence. In: 15th IEEE international conference on advanced video and signal based surveillance (AVSS) 2018, IEEE, pp 1–6. https://doi.org/10.1109/avss.2018.8639440
7. Jeong Y, Hwang D, Kim K (2019) Blockchain-based management of video surveillance systems. In: International conference on information networking (ICOIN) 2019, IEEE, pp 465–468. https://doi.org/10.1109/ICOIN.2019.8718126
8. Al-Sahan L, Al-Jabiri F, Abdelsalam N, Mohamed A, Elfouly T, Ab-dallah M (2020) Public security surveillance system using blockchain technology and advanced image processing techniques. 2020 IEEE international conference on informatics, IoT, and enabling technologies (ICIoT), Doha, Qatar, 2020, pp 104–111. https://doi.org/10.1109/ICIoT48696.2020.9089523
9. Krejci S, Sigwart M, Schulte S (2020) Blockchain- and IPFS-based data distri-bution for the Internet of things. In: Brogi A, Zimmermann W, Kritikos K (eds) Service-oriented and cloud computing. ESOCC 2020. Lecture Notes in Computer Science, vol 12054. Springer, Cham, pp 177–191 https://doi.org/10.1007/978-3-030-44769-414
10. Wood G (2014) Ethereum: a secure decentralised generalised transaction ledger. In: Ethereum Project Yellow Paper, 2014, vol 151

11. Youssef SBH, Rekhis S, Boudriga N (2019) A blockchain based secure IoT solution for the dam surveillance. In: IEEE wireless communications and networking conference (WCNC) 2019, IEEE, pp 1–6. https://doi.org/10.1109/WCNC.2019.8885479
12. Wang Y, Wang C, Luo X, Zhang K, Li H (2019) A blockchain-based IoT data management system for secure and scalable data sharing. In: Liu J, Huang X (eds) Network and system security. NSS 2019. Lecture Notes in Computer Science, vol 11928. Springer, Cham, pp 167–184. https://doi.org/10.1007/978-3-030-36938-510
13. Tanwar S, Bhatia Q, Patel P, Kumari A, Singh PK, Hong W (2020) Machine learning adoption in blockchain-based smart applications: the challenges, and a way forward. IEEE Access 8:474–488. https://doi.org/10.1109/ACCESS.2019.2961372
14. Mistry I, Tanwar S, Tyagi S, Kumar N (2020) Blockchain for 5G-enabled IoT for industrial automation: A systematic review, solutions, and challenges. Mech Syst Signal Process. 2020 Jan 1;135:106382. https://doi.org/10.1016/j.ymssp.2019.106382
15. Iansiti M, Lakhani KR (2017) The truth about blockchain. Harv Bus Rev. 1–11

Information Retrieval in Financial Documents

M. Kani Sumithra and Rajeswari Sridhar

Abstract The improvement in the computation power and decreasing cost of storage has resulted in the exponential growth of day-to-day data which can be processed and information can be retrieved to gain insights and knowledge. There are some information retrieval models like Boolean, vector, and probabilistic models that help achieve this target. Using these models leads to problems such as documentary silence and documentary noise due to approximate, poor, and partial representation of the semantic content of documents. In this paper, we built a system that constructs knowledge graphs from the unstructured data and later queries the graph to retrieve information. A knowledge graph is the type of knowledge representation consisting of a collection of entities, events, real-world objects, or it can be any abstract concepts and they are linked together using some relations. They are preferred as they contain large volumes of factual information with less formal semantics. Our approach is to preprocess the structured data followed by named entity recognition with appropriate finance-related tags. Entity relation formulator extracts entities and matches them to relations forming a triple of a subject, object, and predicate. These set of triples can be queried by a natural language query language and converting it into a knowledge graph query. The evaluation metrics employed in this paper are accuracy, precision, and recall. The system has an accuracy of 0.822, a precision of 0.837, and a recall of 0.9015 for a set of 500 questions and answers.

Keywords Named entity recognition · Entity extraction · Entity relation extraction · Knowledge graph · Fin tech · Query processing

M. K. Sumithra (✉) · R. Sridhar
Department of Computer Science and Engineering, National Institute of Technology, Tiruchirappalli, Tamil Nadu 620015, India
e-mail: kanisumithram96@gmail.com

R. Sridhar
e-mail: srajeswari@nitt.edu

P. K. Singh et al. (eds.), *Evolving Technologies for Computing, Communication and Smart World*, Lecture Notes in Electrical Engineering 694,
https://doi.org/10.1007/978-981-15-7804-5_20

1 Introduction

The process of dealing with storage, organization, retrieval, and evaluation of the information from documents and text is often termed as information retrieval [1]. The domain which deals with information retrieval (IR) is natural language processing (NLP) where we try to process the text to understand the text and extract meaningful relations. The initiation of NLP is to have a deep understanding of the language semantics which will determine the level of understanding of the system being developed. The challenges that one would be facing when developing these systems are unique and interesting that includes ambiguity, synonymy, syntax co-reference, representation, personality, intention, and style [1]. Financial sector uses and generates a vast amount of data like transactional data, customer data, and logging data which can be used to support decision making. Social media data and data from authentic Web sites also help in understanding and retrieving relevant information from financial documents. The processing and finding knowledge from these data is currently done by bank staff manually, but the task is expensive and time-consuming for three main reasons:

1. The documents are in unstructured form.
2. The volume of "live" documents is large, numbering in the millions of documents.
3. Banks are continuously adding new information to the database.

This paper discusses a novel idea to extract relations in the information collected from this data which is used to represent the knowledge in the form of a knowledge graph. A customer would be able to get the required information from this graph in the form of questions and answers.

The organization of the remaining paper is as follows—Sect. 2 highlights the traditionally available approaches of Information Retrieval. Section 3 illustrates the research methodology that this work has adopted. Performance analysis of the proposed scheme is mentioned in Sect. 4. Section 5 concludes the paper by stating the future scope of the current work.

2 Related Work

IR and NLP have been in study for a long time now [2]. There have been decades of research in the ways to improve and update the methodology such that the information obtained is more accurate and having less noise. The text words are to be converted into vectors called embedding which can be further processed. There have been many such embeddings of which bloom embedding [3] seems to be the good fit here. The input is a sparse binary vector, with ones at places where the user has words and zero otherwise. For reducing dimensionality, a smaller dimension is chosen and original IDs are hashed into it. The hash collisions resulting out of this method can be resolved by collision resolution techniques.

The early approaches to named entity recognition (NER) were simple and intuitive. They are mostly rule-based. The foremost paper was done by Moens et al. [4]. In this method of extracting entities, discourse analysis was done to find the structure of the text and linguistic forms, and text grammars are created. Apart from the basic disadvantage of being a static method, the obtained precision is too little to be used as a good measure. As mentioned in the research paper by Farmakiotou et al. [5], the entities are extracted using grammar rules and gazetteers. The major disadvantage of this method was the creation of rules by hand, which is a very time-consuming task and the overall recall is low. A simple improvement over the previous method was done by Sheikh and Conlon [6], where the rules were based on features including exact word match, part-of-speech tags, and domain-specific features.

Although not much research is done in this area of finance domain, some papers can be found which uses machine learning to solve this problem. One of the papers by Mohit and Hwa [7], where they used a semi-supervised method specifically a Naive Bayes classifier with the EM algorithm, applied to features extracted from a parser to train a model to extract entities. The major advantage of this over the previous methods is the fact that it is robust over novel data. Jiang and Zhai [8] used generalized features across the source and target domain in order to reduce the need for annotation and to avoid over-fitting. One of the related works in this domain is the neural link prediction model where the interactions between input entities and relationships are modeled by convolutional and fully connected layers in [9]. The main characteristic of this model is that the score is defined by a convolution over 2D shaped embeddings. Another related work is [10] that uses graph neural networks. The method learns distributions with help of dependencies among a graph's nodes and edges using probability. The main disadvantage of this method is the use of a probabilistic approach which might have some bias toward the input that might lead to incorrect relationships. To address these problems, this work tends to do a domain specific i.e., finance-related information retrieval on the documents and Web articles.

3 Research Methodology

The architectural block diagram of the proposed system is given in Fig. 1.

The data ingestion phase deals with data collection and preprocessing followed by core NLP. The next phase, entity relation formulation phase is the knowledge graph construction phase in which entities and relations are formulated and stored in database. The final phase is the query processing phase that details the processing of a user query to search the graph database for the information. The following subsections detail the processing done in each phase.

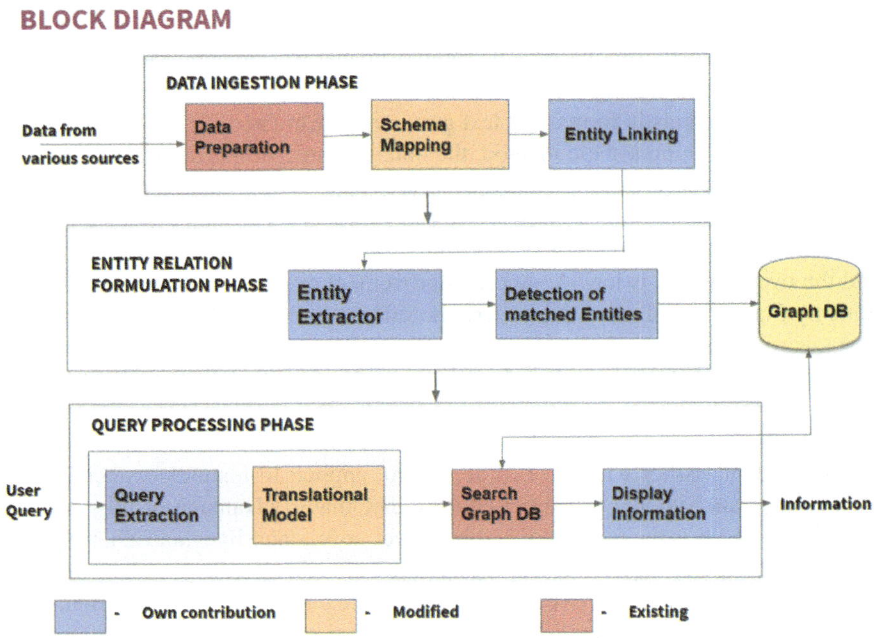

Fig. 1 Block diagram of the system

3.1 Data Ingestion Phase

The first phase of the system consists of a collection of the data, preprocessing it and then extracting the entities from the text which would be used in the further phases. The submodules are detailed as below:

1. Data Preparation
   The data that is collected for this system is from different financial reports of several universities, audits of companies, and Web reports and audits.
2. Schema Mapping
   The collected data is preprocessed at first. The unstructured data is made into sentences and then into words from where the words are categorized. The categories used are novel and finance related such as Person, Organisation, Date, Location, Number, Report, Finance, Positive, Negative and Policy.
3. Entity Linking
   The structure of the data is such that it has large input and output dimensions therefore making it more difficult to train. Therefore, bloom embeddings are used to address these issues. Bloom embeddings use a compression technique which is applied to both the input and output of the neural network (NN) model so that it deals with sparse and high-dimensional cases. Since they are compressed they are efficient, and the accuracy of the model is not compromised till 1/5 compression ratios.

Training and Testing
Convolution neural networks are used for both training and testing the NER model built. This architecture has achieved good NER performance in the general English domain, making it a good choice. Using word embeddings, every word in the input window is mapped to the N-dimension vector where N is the embedding dimension. Then, a convolution layer captures the long-term dependencies (or global features) in the hidden nodes. The local and global features are then concatenated and are given as input to a standard neural network trained using stochastic gradient descending. The tagger, parser, and NER are sharing this convolutional layer. The tagger also predicts a "super tag" for boosting the representation with POS, morphology, and dependency label. This is done by the addition of the softmax layer on the convolutional layer. Ultimately, the convolutional layer is made to learn to give a representation which is one affine transform from informative lexical information. This is good for the parser which back propagates to the convolutions also.

3.2 Entity Relation Formulation Phase

This phase entails the process of finding the entities and their corresponding relations such that they can be represented as nodes and edges of the knowledge graph. The main idea of representing the knowledge as a graph is the benefit of finding the relative answer easily as it models the real world perfectly. Every idea, fact, and information in the real world are connected to something or the other. This is what connects all the information and makes it easy for the retrieval as once the main adjective or verb is identified, the required and related information can be found out easily.

1. Entity Extractor
   The data obtained from the previous phase has entities and its type associated with it. The next step would be to identify these entities individually via identifier so that they can be further fine-tuned and matched to form relations.
2. Detection of Matched Entities
   The core of this approach is a machine learning (ML) classifier that predicts the probability of a possible relationship for a given pair of identified entities in a given sentence. A set of patterns is used to exclude noisy sentences as a filter and then the classifier will extract a set of features from each sentence. We employ context-based features, such as token-level n-grams and patterns. In addition, the classifier also relies on information available from our existing knowledge graph. Past data is also used to automatically detect labeling errors in the training set, which improves the classifier over time. After matching the entities with their appropriate relations, they are stored into a graph database as subject, object, and predicate. The graph has the structure of n-triples which can be structured into subject, object, and predicate.

3.3 *Query Processing Phase*

The customer can provide the query in the English language (natural language) which will be further translated into appropriate query language for searching into a graph database. This phase handles this process.

1. Query Extraction and Translation Model
   The entities and relation of the user provided natural language question is extracted and then—transformed into nodes and entities. This is done by the already trained NLP model which can be further trained for questions and their types. The question now in the for of nodes and entities can be easily converted into any graph query language. The one used here is SPARQL. It is an RDF query language-that is, a semantic query language for databases-able to retrieve and manipulate data stored in resource description framework (RDF) format [11]. The query now in the form of SPARQL is processing in the next submodule.
2. Search GraphDB
   The SPARQL will search into the knowledge graph which is stored as a graph database in the server. The RDF's subject is similar to SQL's entity as the data elements are in many different columns and spread across. There is a unique key by which they are identified. In RDF, the subject has a unique key and predicate that are stored in the same rows as with the object which is the actual data.
3. Display Information
   The answer return from SPARQL will be in the form of JSON dumps. One has to convert it into proper format so that the user can get the correct answer in user readable format.

4 Results and Analysis

The data used in this work is the collection of various financial documents and articles from authentic Web sites. Figs. 2 and 3 details the difference between a generic NER and finance-related NER. One can observe the increase in the precision and number of finance-related tags in the domain-oriented system built.

The stored knowledge base of the entities and their relationships is visualized in a online visualizer Web site, graphcommons.com [12], a snapshot of a portion of the graph database is shown in Fig. 4.

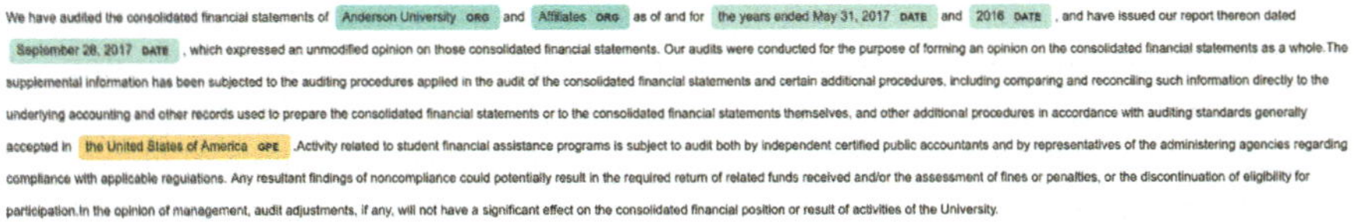

Fig. 2 Generic NER

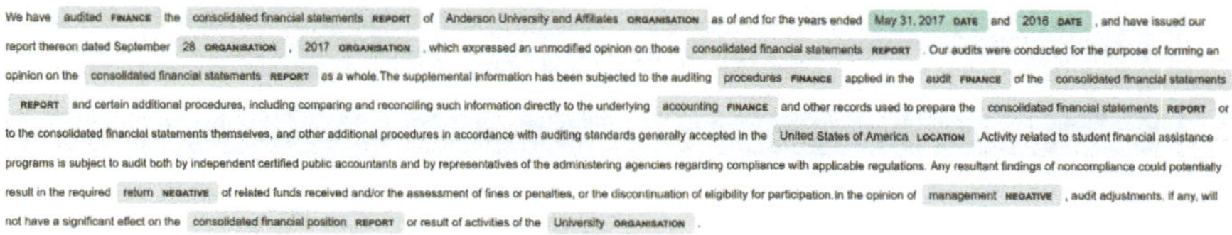

Fig. 3 Financial NER for the same text

Fig. 4 Knowledge graph visualized

Figure 5 is an example of a user query submitted to the system and the consequent response.

The evaluation metrics for this work is based on how many accurate answers one gets for the user query. For 350 questions (where the answer is present in the knowledge base) and 150 questions (where answer is not present or partial relation is present), the following data is obtained. Table 1 gives the confusion matrix for the test data.

From Table 1 it is evident that the accuracy for the system is **0.822**, recall is **0.837** and precision is calculated as **0.9015**.

4.1 Comparison of a Generic System with Proposed System

The proposed system is built with specialized tags which are Person, Organisation, Date, Location, Number, Report, Finance, Policy, Positive, and Negative. For comparison of the proposed system with a generic system, a model is built with just general tags like Person, Organisation, Date, and Location and it's performance is noted as given in Table 2.

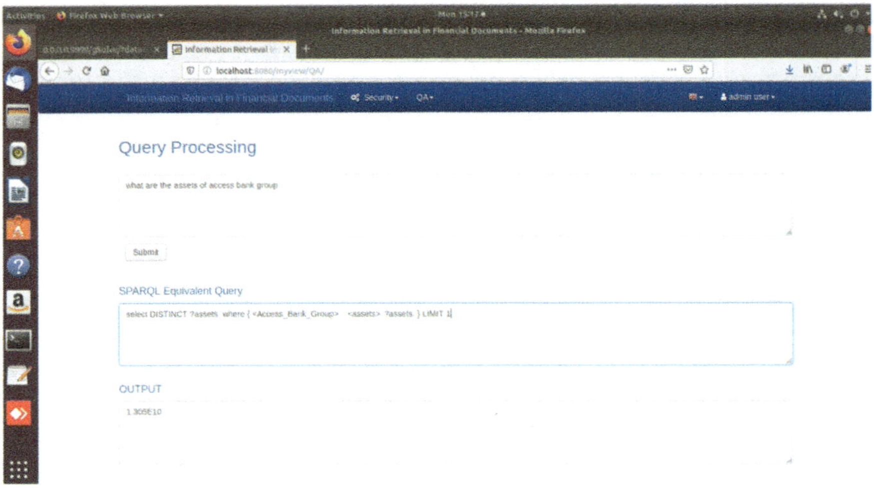

Fig. 5 Customer query window

Table 1 Confusion matrix for proposed system

	Obtained correct answer	Obtained no answer
Actual answer	TP: 293	FN: 57
No answer	FP: 32	TN: 118

Table 2 Confusion matrix for generic system

	Obtained correct answer	Obtained no answer
Actual answer	TP: 272	FN: 78
No answer	FP: 70	TN: 80

From Table 2, it can be observed that accuracy, recall, and precision for this system are 0.704, 0.7771 and 0.7953 respectively.

From Figure 6, it can be observed that the proposed system works better than a generic model of information retrieval. The reasons for the same can be listed as below :

1. The finance domain has a complex way of dealing with terminologies. The data preprocessing in the proposed system has specialized tags for these terms that will help and form more real-world relations in the process of the knowledge graph.
2. The relations in the extraction module are found by parts of speech and the specialized tags. These terms are converted as entities and into subject, object, and predicate which will in turn help in more detailed and accurate relationships in knowledge base that are created.

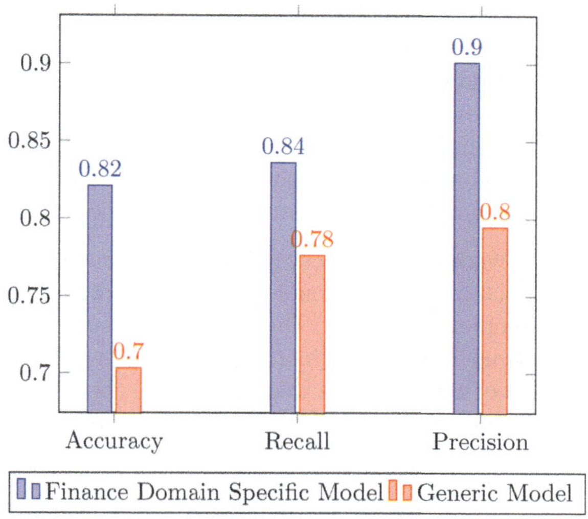

Fig. 6 A comparative study of generic model and proposed system

3. The queries are also converted as a tree (graph) first and then is converted to SPARQL. This helps in understanding the query better and fetching the answer by subgraph matching.

It can be shown from the above analysis that the proposed system is better than the existing models studied according to the dataset used. The improvement in the evaluation metrics are a proof for the system's performance.

5 Conclusion and Future Scope

This goal of this work is to obtain relevant information from the financial documents which are highly unstructured. This is achieved through extracting features of the data using bloom embedding from where the entities are further extracted as per subject, object, and predicate. This is the triple store which is normally stored as entities and relationships so that knowledge graph can be constructed and stored as a database. The reason for choosing graphs for knowledge representation is the fact that it models the real-world facts and relationships well. It will make the process of searching for the answer much easier and more accurate. The query which is in natural language is converted into SPARQL which is query language for graph databases. It will query into the database and fetch the results. Although from the previous section, it can be argued that the proposed system gives us better results, there are some limitations of this work which are listed as below :

1. To add new data into the system, it is required to go through the whole process again. Rebuilding the entire knowledge base takes up huge amount of time, effort, and money in a real-world scenario. This can be mitigated by the introduction of

a new phase which would be able to fuse new knowledge into the already created graph database.
2. The data which is added to the system must be checked and verified to be true as its removal is hard and wrong information might be transferred to the users which might lead to business losses.
3. More complex queries cannot be resolved by the system as finding distant relations in the graph is very difficult and can be incorrect at times.

The work can be further expanded by the inclusion of many different subcategories like stock market terms and government policies which will make the knowledge base more sophisticated. This expansion would allow much wanted and preferred classification in the unstructured data. This work can be used as a basis for development of expert system where another agent can query into the knowledge base to get information that can be formulated as good basis to help others make important decisions.

References

1. StanfordNLP Textbook homepage. https://nlp.stanford.edu/IR-book/html/htmledition/irbook.html. Last Accessed 20 Dec 2019
2. Mang'are Fridah Nyamisa (2017) Waweru Mwangi. Wilson cheruiyot, a survey of information retrieval techniques, advances in networks 5(2):40–46 https://doi.org/10.11648/j.net.20170502.12
3. Serrà J, Karatzoglou A (2017) Getting deep recommenders fit: Bloom embeddings for sparse binary input/output networks. In: ACM RecSys. arXiv:1706.03993
4. Jiang J, Zhai CX (2006) Exploiting domain structure for named entity recognition. In: Proceedings of the human language technology conference of the NAACL (HLT-NAACL). New York City, USA, pp 74–81. https://doi.org/10.3115/1220835.1220845
5. Behrang M, Rebecca H (2005) Financial named entity recognition based on conditional random fields and information entropy. In: Proceedings—International conference on machine learning and cybernetics. https://doi.org/10.1109/ICMLC.2014.7009718
6. Sazali SS, Rahman NA, Bakar ZA (2016) Information extraction: evaluating named entity recognition from classical Malay documents. In: 2016 third international conference on information retrieval and knowledge management (CAMP), Bandar Hilir, pp 48–53. https://doi.org/10.1109/INFRKM.2016.7806333
7. Adak C, Chaudhuri BB, Blumenstein M (2016) Recognition named entity, from unstructured handwritten document images. In: 12th IAPR workshop on document analysis systems (DAS). Santorini 2016, pp 375–380. https://doi.org/10.1109/DAS.2016.15
8. Xu J, He H, Sun X, Ren X, Li S (2018) Cross-domain and semisupervised named entity recognition in chinese social media: a unified model. IEEE/ACM Trans Audio, Speech, Lang Proc 26(11):2142–2152. https://doi.org/10.1109/TASLP.2018.2856625
9. Dettmers T, Minervini P, Stene-torp P, Riedel S (2018) Convolutional 2D knowledge graph embeddings. In: AAAI 2018. arXiv:1707.01476
10. Li Y., Vinyals O, Dyer C, Pascanu R, Battaglia P (2018) Learning deep generative models of graphs, arXiv preprint arXiv:1803.03324
11. https://graphcommons.com/ . Last Accessed 20 Dec 2019
12. https://wikimili.com/en/SemanticWeb . Last Accessed 20 Dec 2019

VNF Security in Telco Environment

Mukesh Kumar Bansal, Aswathy SV, and B. Krishnaswami

Abstract Over the years, Telecommunication service providers are moving towards Network Function Virtualisation (NFV) to greatly reduce and manage Capex and Opex costs. Moving towards virtualisation of services and coupled with software defined networking, service providers are preparing to deliver complex services in the evolving landscape of next generation mobile communications and digital transformation.Thus, moving towards the virtualisation has brought lot of benefits, however the security threat vectors also increases with virtualisation. By sharing common hardware and virtualisation layer additional security boundaries must be created. Defence-in-depth approach is always recommended and in virtualisation it is more relevant to have layers of security at hardware level, virtualisation level, network level and VNF level. This paper describes the ways in which the VNF can be secured as well as our approach to handle the VNF in Telco environment.

Keywords VNF · Telecommunication · Security · Virtulaization · NFV · Network · Radio access network

1 Introduction

Over the years, Telecommunication service providers are moving towards Network Function Virtualisation (NFV) to greatly reduce and manage Capex and Opex costs. Moving towards virtualisation of services and coupled with software defined networking, service providers are preparing to deliver complex services in the evolving landscape of next generation mobile communications and digital transformation. Vir-

M. Kumar Bansal (✉) · A. SV · B. Krishnaswami
Altran, Bengaluru, India
e-mail: Mukesh.bansal@altran.com

A. SV
e-mail: aswathy.sv@altran.com

B. Krishnaswami
e-mail: krishnaswami.b@altran.com

P. K. Singh et al. (eds.), *Evolving Technologies for Computing, Communication and Smart World*, Lecture Notes in Electrical Engineering 694,
https://doi.org/10.1007/978-981-15-7804-5_21

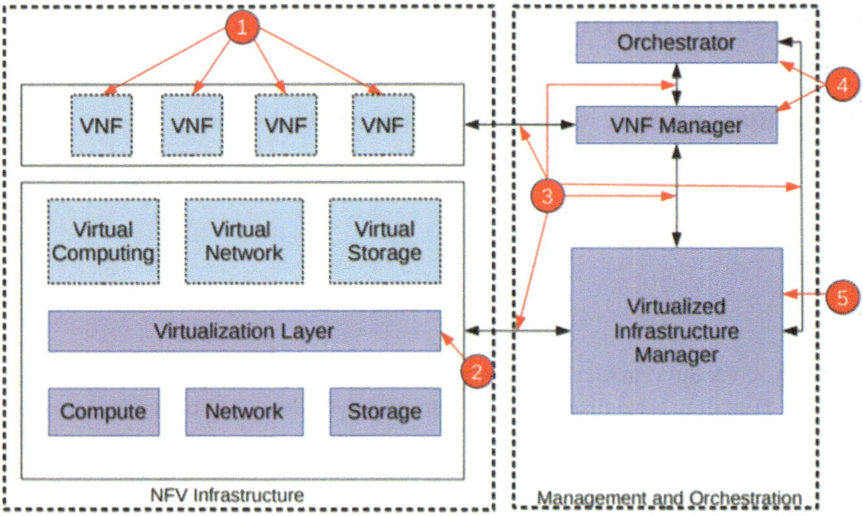

Fig. 1 NFVI architecture

tualisation of telco physical network functions (PNF) and components and deploying them on COTS (commercial of the shelf) hardware provides Telco Service Providers (TSPs) a greater flexibility in managing their network. Managing and orchestrating virtual functions and legacy hardware devices for quick provisioning of services has also evolved along in driving the future of mobile services. Figure 1 shows the NFVI architecture.

1.1 NFV Infrastructure and Its Attacks

The NFVI is the combination of hardware and software components which build up the environment in which the Virtualised Network Functions (VNFs) are deployed.

NFV MANO (Management and Orchestration) is in charge of managing the life cycle and chaining of the VNFs to provide the needed Network Services.

VNF (Virtual Network Function) is a software component provided by a vendor independent of the infrastructure provider. This component is prone to vulnerabilities so malwares can be designed for this layer and attacks are possible.

Virtualization Layer also known as hypervisor, is the middle layer that is responsible for logically partitioning the physical resource and activating the software so that VNF utilizes the virtualization infrastructure and this can be utilized later on different physical resources.

Communications with and within NFV MANO is responsible for the transit between the NFVI and the NFV MANO, as well as traffic within the NFV MANO.

Orchestrator is in charge of the orchestration and management of NFV infrastructure and software resources and realizing network services.

VNF Manager role includes installation, updates, query, scale up/down and termination. A VNF manager may be deployed for each VNF or a single VNF manager may be deployed to serve multiple VNFs.

Virtualized Infrastructure Manager (VIM) is responsible for managing the NFVI resources used by the VNFs (compute, network and storage).

Possible Attacks in NFVI environment

In the above section, the components associated with the NFV infrastructure are explained and the possible attacks are listed down as follows.

Virtualized Network Functions (VNF)

1. DoS attacks is frequent in cloud computing or NFV.
2. Software components in VNF vulnerable to software flaws and it may lead to bypass firewalls, buffer overflow.

Virtualization Layer and communication

1. Code execution on the physical host - code execution on the host from a compromised or malicious Virtual Machine.
2. Buffer overflow.
3. Read or Modify the memory.
4. VM snapshot execution without the original one to bypass security. Orchestrator/VNF manager.
5. Ephemeral storage area can be hacked—eavesdropping, data modification and impersonation.

Virtual Infrastructure Manager

1. Privilege escalation

The above section gives a detailed description about the components in the NFV infrastructure and the attack vectors associated with each module.

1.2 Adoption Towards Virtualisation

In Fig. 2, the stack of the telecom network components is shown as well as the shift to VNFs and CNFs is showcased. The NFV component separates the network functions from the hardware so that virtualized network can run on a commodity hardware so that it can be pliable and cost effective. The PNF component is within the NFV component, as it is a physical node it has not undergone any virtualization. Both the VNF as well as the PNF component formulates the entire network service. The VNF component handles the firewall or load balancing. Individual VNFs can be connected

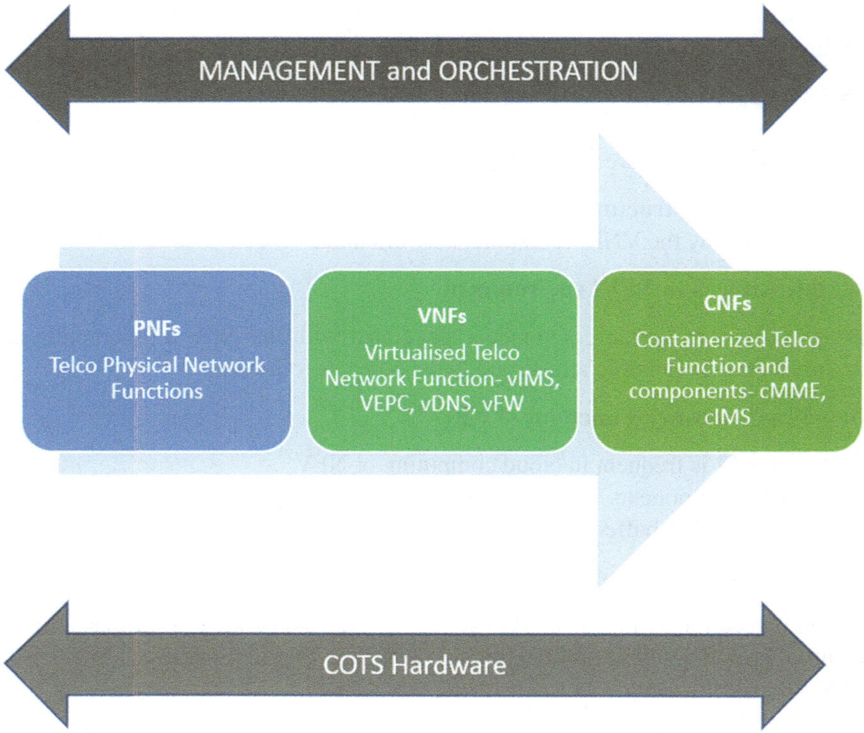

Fig. 2 Telco network components shift to VNFs and CNFs

or combined together as building blocks to create a fully virtualized environment. VNFs run on virtual machines on top of the hardware networking infrastructure. There can be multiple VMs on one hardware box using all of the box's resources. The CNF allows software to be distributed across a network or data-center in small packages and share operating system resources, rather than requiring regular software updates or a virtual machine for each instance of the software. This can run on a shared OS without the need of a virtual machine. They gain access to the upper layer and can be used for dynamic software upgrades and real-time data exchange.

Service providers' adoption towards Virtualisation is on the rise as major telecommunication providers are virtualising their network functions. AT&T has more than 40% of its core telecom network functions virtualised. Verizon has tested a fully virtualised RAN with Intel hardware and Nokia Air Scale cloud base station architecture. Virtualisation has been the need of the hour to manage increasing data demands.

According to Global Market Insights, the NFV market shares to surpass USD 70 billion by 2024, with the growth attributing towards the network service providers adopting virtualisation to deliver next-generation mobile services. According to Fig. 3, telecommunication and enterprise transition towards NFV by 2024 in the UK will be the highest with Healthcare and other industries are predicted to be followed.

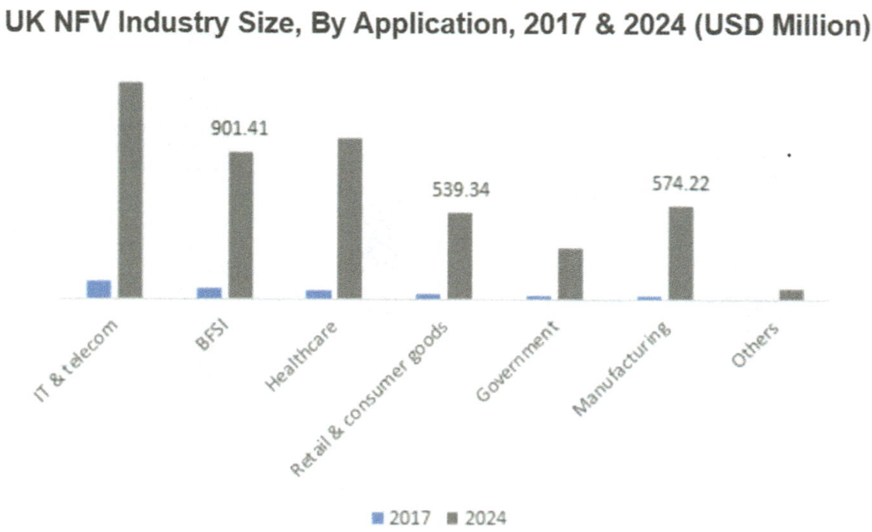

Fig. 3 NFV growth prediction graph [7]

1.3 Evolved Packet Core as Virtualization Network Function

In this section, let us take the example of an Evolved Packet Core (EPC) component like a network function to be virtualized. As the components are virtualized, it is therefore referred here as a Virtual EPC or vEPC. The EPC components include:

- Mobility Management Entity (MME).
- Home Subscriber Server (HSS).
- Serving Gateway (SGW).
- Packet Data Network Gateway (PGW).
- Policy and Charging Rules Function (PCRF).

The components included in the EPC have unique functionalities, MME is the processing element and it performs end to end connectivity between the core components in terms of signalling as well as provideing security services. HSS allows communication service provider to manage customers in real time as well as in a cost effective manner. The S-GW is responsible for the handovers in the network communication, monitors and maintains the context information related to the user equipment whereas the P-GW manages policy enforcement, packet filtration for users, charging support. PCRF is the software node used for real time to determine policy rules in a multimedia network. It plays a central role in next-generation networks.

By virtualizing the core EPC network functions, virtual EPC allows that shift towards a more highly-available, scalable core architecture by leveraging common and shared hardware resources pool to provide high performance and speedy instantiation of vEPC services. Virtual EPC components can then be distributed across

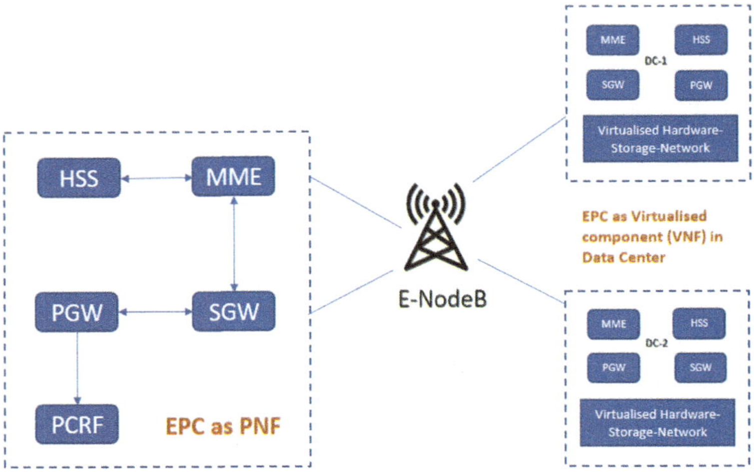

Fig. 4 EPC as VNF

geographical location in different Data Centers (DCs), these DCs may reside in proximity to the end users. Service providers are also looking towards far-edge computing and Mobile Edge Core (MEC) to provide low-latency services by optimizing the hardware footprint [2].

Today, there are many virtual EPC providers that bundle both control and data plane EPC components in one package. EPC VNF package shall be Virtual Machine (VM) for MME, HSS, SGW, PGW, PCRF and any additional VM included as an Element manager for management operations as shown in Fig. 4.

2 Telco VNF Security

Moving towards the virtualisation has brought lot of benefits, however the security threat vectors also increases with virtualisation. By sharing common hardware and virtualisation layer additional security boundaries need to be created. Defense-in-depth approach is always recommended and in virtualisation it is more relevant to have layers of security at hardware level, virtualisation level, network level and VNF level [1].

A Virtual Network Function (VNF) is ultimately a software that is built on an operating system. ETSI definition of a VNF- a network function that can be hosted on a Network Function Virtualisation Infrastructure (NFV-I). In addition to the vEPC other VNFs include, virtual firewall, virtual load-balancer, virtual IMS, virtual DNS etc. [3].

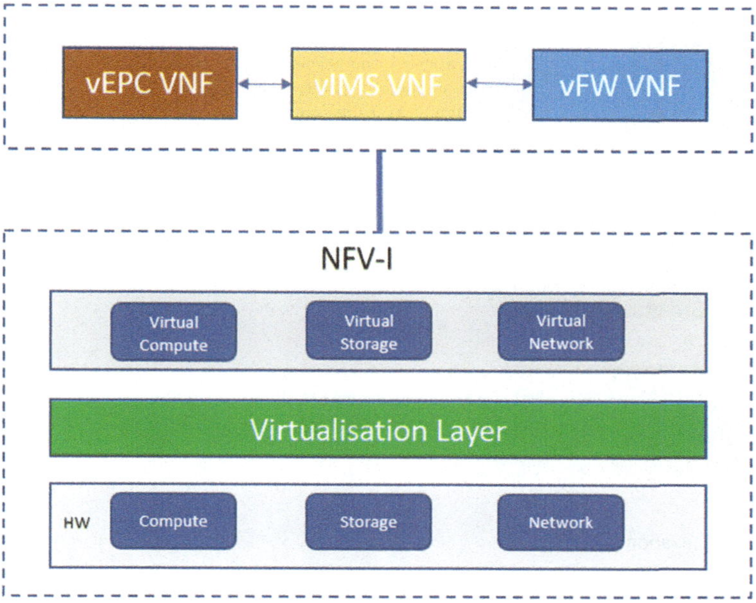

Fig. 5 VNFs hosted on NFV-I

Exploring Fig. 4 further, Fig. 5 below represents a VNFs hosted in NFV-I, as indicated the security threats to VNFs can originate from any of the underlying layers, either from hardware (NFV-I) or from virtualisation layer.

2.1 *Virtualisation Threat Vectors*

As outlined in the previous sections, Virtualised Network Functions (VNFs) are hosted on shared infrastructure in data centres. Virtualisation Infrastructure Manager (VIM) manages the hardware resources, such as memory, CPU and storage to provide adequate resource requirements for the VNFs to be instantiated. In a cloud environment, security threats can be looked at from cloud infrastructure, underlying network level and VNF level. Figure 6 outlines security threats associated with each level.

In the infrastructure layer the possible attacks are DDOS, the hypervisor vulnerabilirties, physical security issues, malware attacks. From the infrastructure perspective the hypervisor attack is most common the attacks and mitigations are listed as follows.

Hyperjacking involves installation of a malicious or fake hypervisor to manage the complete server. The standard security measures are not effective as the operating system is not inteeligent enough to recogonise the compromise. Hypervisor operates

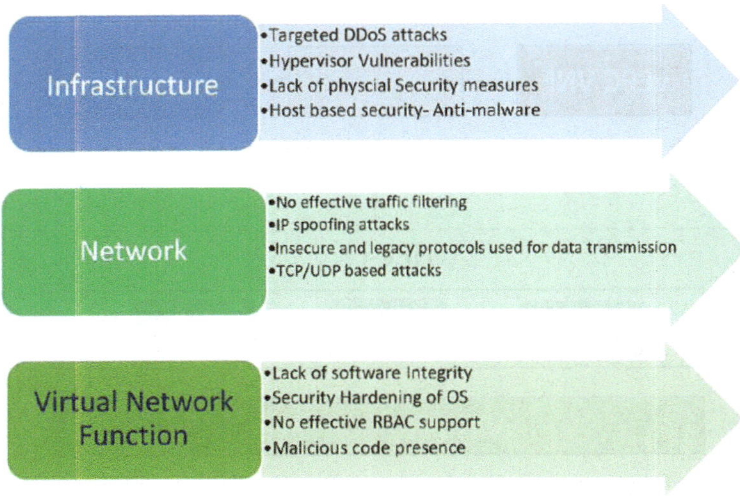

Fig. 6 Virtualization threat vectors

in stealth mode and runs beneath the machine, so it is difficult to detect and an attacker gain access to computer servers where it can shutdown the entire organization. If the access to the hypervisoris gained then anything connected to that server can be manipulated. The hypervisor represents a single point of failure when it comes to the security and protection of sensitive information. To succeed in this attack, the attacker take control of the hypervisor by the following methods

1. Injecting a rogue hypervisor beneath the original hypervisor.
2. Directly obtaining control of the original hypervisor.
3. Running a rogue hypervisor on top of an existing hypervisor.

Mitigation for the hypervisor is as follows:

1. Security management of the hypervisor must be kept separate from regular traffic. This is a more network related measure than hypervisor itself related.
2. Guest operating systems should never have access to the hypervisor. Management tools should not be installed or used from guest OS.
3. Regularly patching the hypervisor.

From the network layer attacks, the perimeter level security has to be provided. The measures for the IP spoofing, TCP/UDP are already in place, so the measures needs to applied efficiently and continuous monitoring has to be performed to eliminate all the attacks related to the network. Continuous monitoring and immediately fixing the vulnerabilities in the network helps to eliminate the crucial attacks.

This paper mainly focuses on the Virtual Network Functions [4]. The VNF layer is prone to few attack such as eavesdropping, data modification, Impersonation, Privilege escalation, memory modification and buffer overflow.

3 Our Approach to Handle VNF Security

VNF security plays an important role in securing the cloud environment. VNFs depend on the underlying infrastructure and network for optimal performance and functioning. Any security threat that affects the VNF can effectively affect the underlying cloud infrastructure and vice-versa.

Telco VNFs such as EPC core or IMS core follows 3GPP technical specification for the packet core components, so it is vital to wrap-around security on top as it is the responsibility of the service provider to ensure adequate security is followed. Some of the security measures to keep in mind specific to Telco VNFs are (Fig. 7),

Security Hardening—It is essential to implement hardening at the VNF level, telco VNFs, such as vEPC or vIMS consists of 6 to 7 virtual machines combined, so it is essential to harden the virtual machine operating system. Industry standard hardening measures such as CIS benchmarks, NIST can be considered for VNF hardening.

Role Based Access Control—Managing VNF access using adequate roles and permission is one of the best security practices, appropriate VNF system level access roles must be defined to avoid unauthorised system access.

Fig. 7 VNF security methods

Software Integrity—This mostly depend on the VNF vendor software development and signing process. VNF components and its software should follow code signing practices and methods to verify software integrity.
Malicious code protection—Before the VNFs are installed in the production environment, security testing methods should include scanning VNF software elements against malicious code presence.

4 Future of Telecommunication Service Provider

Next major shift for telecommunication service providers and VNFs is to move towards Containers. Fundamental difference between virtualisation and containers is that in containers virtualisation happens at the operating system level with the applications sharing same host operating system without need for a guest operating system like virtualisation, Fig. 8 below indicates the virtualisation and container representation.

Shift towards the containers also introduce security threats around the container platform and fundamental functioning of the containers. As containers share common OS kernel they induce a wider security risk at the software level compared to hypervisor-based systems. Other risks such software image vulnerabilities, application vulnerabilities and weak access controls calls for comprehensive security around container platform.

While the telecommunication provider services constantly evolve in next generation mobile core platform, security around the underlying infrastructure, application and solution remains to be a paramount concern. By building layers of security at each level and detection of threats from the underlying hardware to the application, complex security threats can be mitigated in the Telco environment.

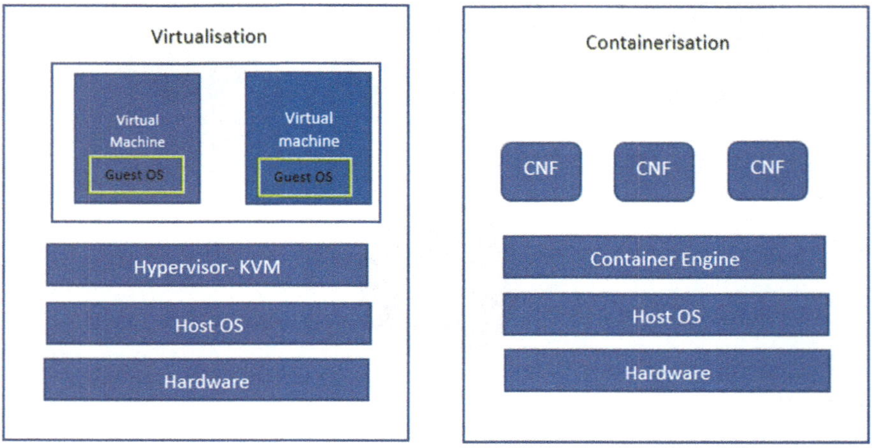

Fig. 8 Virtual machines versus containers

5 Conclusion

In this paper, we concentrate more on the VNF security in the telecommunication environment. The telecommunication plays a major role in our day to day life, everything is virtualised in the current trend. Though virtualization seems to be cost-effective, implementation effective there are security issues associated with them. So, the related security vulnerabilities associated with the VNF and the security measures are explained in depth. If the security measures are handled in appropriate manner VNF in telco environment can be handled in smooth and secure way.

6 Future Work

Future work needs to be done provide container security for heterogeneous cloud environment to make application and VNF secure from malicious attacks, which may disrupt network functionality nay may be whole computer network. Authors are working towards this topic and expected to publish as a separate research paper soon.

References

1. Aljuhani A, Alharbi T (2017) Virtualized Network Functions security attacks and vulnerabilities Journal 2017. In: IEEE 7th annual computing and communication workshop and conference (CCWC)
2. Mijumbi R, Serrat J, Gorricho JL, Bouten N, De Turck F, Boutaba R (2016) Network function virtualization: state-of-the-art and research challenges. IEEE Commun Surv Tutorials 18(1):236–262
3. Lorenz C (2017) An SDN/NFV-enabled enterprise network architecture offering fine-grained security policy enforcement. IEEE Commun Magaz
4. Reynaud F, Aguessy FX, Bettan O, Bouet M, Conan V (2016) Attacks against network functions virtualization and software-defined networking: state-of-the-art. In: 2016 IEEE NetSoft conference work software-defined infrastructure networks Clouds coT service, pp 471–476
5. Mohan AK, Sethumadhavan M (2017) Wireless security auditing: attack vectors and mitigation strategies. Procedia Comput Sci 115(2017):674–682
6. https://www.globenewswire.com/news-release/2018/07/31/1544465/0/en/Network-Function-Virtualization-NFV-Market-worth-70bn-by-2024-Global-Market-Insights-Inc.html
7. https://www.gminsights.com/industry-analysis/network-function-virtualization-nfv-market

Rank-Level Fusion of Random Indexing, Word Embedding, and TF-IDF-Based Rankings for Clinical Document Retrieval

Sanjeev Kumar Sinha and Chiranjeev Kumar

Abstract The amount of clinical data present with medical professionals is growing at a high rate. The data may be collected from clinical records, Internet, social media, and medical books to mention a few. This vast amount of voluminous structured and unstructured clinical data is often tedious to search and analyze. Clinical document retrieval is intended for quick access to the required clinical documents by a staff, patient, doctor, nurse, or any other person in authority. Hence, it becomes paramount to develop a system to search data/document in the medical repositories for efficient and quick analysis of a patient case that may require instant attention. The users of such a system may pose several queries and expect retrieval of relevant clinical documents. The research work proposed in this paper is based on fusion of document retrieval results applied on several novel techniques. Novel technique of rank-level fusion for retrieval is applied on top of other retrieval techniques to obtain a decision of the final rank. The contribution and novelty of the present work are two-fold: firstly, we propose to use two new techniques for clinical document retrieval viz. Random indexing-based retrieval and GLOVE representation-based retrieval. Secondly, we perform the proposed technique of rank-level fusion over the results obtained through various ranking techniques. The fused rank helps in decision making in the scenario that different ranks are produced by different algorithms used for retrieval. The results obtained using this novel approach show improvement over existing techniques.

Keywords Rank-level fusion · Graph · GLOVE · Random indexing · TF-IDF · Ranking · PageRank · Clinical document · Clinical document system retrieval

---

S. K. Sinha (✉) · C. Kumar
Department of Computer Science and Engineering, Indian Institute of Technology (ISM), Dhanbad, India
e-mail: sanjeev.2013dr1029@cse.iitism.ac.in

P. K. Singh et al. (eds.), *Evolving Technologies for Computing, Communication and Smart World*, Lecture Notes in Electrical Engineering 694,
https://doi.org/10.1007/978-981-15-7804-5_22

1 Introduction

The clinical data present with doctors and in the field of medicine on large is growing enormously with large number of documents being added on regular basis. This data is present in medical records, medical journals, books, clinical textual data in form of blogs, reviews, articles, news, etc. This needs to be processed and retrieved for processing by medical experts. Keeping in view the limited time doctors and medical professionals have it becomes paramount for techniques which can retrieve and rank the documents that doctors can process in a short amount of time. This is the task of clinical documents retrieval. It is a special branch of information retrieval which focuses on clinical or medical documents. The area of clinical information retrieval is of immense use considering the amount of textual medical data that is produced at a rapid rate. Handling this voluminous data goes beyond being processing by manual efforts. Information retrieval (IR) [12] algorithms extract and rank documents from a data source based on a user query. Recently, a lot of research [2, 6, 22] has been performed in related areas of retrieving medical content. Some previous works on retrieval lay emphasis on contextual information [15] as well.

Graph-based retrieval [4] is used popularly since its inception in various search-based applications. The most recent application of this technique [4] is combination of graph-based retrieval and co-citation-based weights. The authors [4] state that the Katz algorithm computes the weight between the seed document and another document as the shortest path between them measured as sum of edges. Other graph-based techniques such as maximal flow between the source and sink and random walks are also used as application specific retrieval strategies. A technique of dimensionality reduction is often used on the adjacency matrix-based representation of a graph. LSI is popular technique which performs low-rank approximations of the original matrix. Other retrieval techniques include ontology-based retrieval [17] requiring a pre-defined ontology specific to a particular domain and this can be in particular medical domain as well. Ontology is a descriptive representation of a particular domain which is combined with semantic information.

Supervised techniques such as support vector machines [10] are also used for retrieval. In their work [10], the authors considered the target class computed via clustering the documents given as input. Further, similarity between the query and document is computed. This training data and tagged class are fed into SVM which assigns a relevant cluster to the test data. Some other retrieval algorithms are specific to a particular problem under consideration [3, 9, 13, 16] which include code search as well.

The proposed methodologies combine novel ranking techniques to obtain final ranks of the documents to be searched for a query. The ranks of documents obtained through these algorithms are fused using the methodology of rank-level fusion for computing the aggregate ranks. The motivation behind the current work is that different retrieval algorithms may generate different ranks but there is no strict rule of which algorithm to applied for a particular document set for retrieval. Hence, there is a great need of a system which can combine the ranks generated through retrieval algorithms. The idea is that the final rank so computed is a combination of several ranks and hence represent the actual document rank more preciously and nearly match the expert's view. In this paper, we focus on retrieval of clinical documents and computing their ranks based on several algorithms with subsequent merging of these ranks. It merges the ranks computed through various methods and that the final ranks so obtained must be more accurate that the individually computed ranks.

The techniques proposed for combining retrieval ranks are (i) TF-IDF-based retrieval, (ii) Random indexing-based retrieval and (iii) Word-embedding-based retrieval. The traditional retrieval utilizes TF-IDF scores, while this paper also proposes to explore the textual knowledge captured in random indexing for retrieval. Also, knowledge gathered through pre-trained word-embedding vectors is utilized in GLOVE-based retrieval engine. This is further processed through the rank fusion techniques. The combination of techniques used for retrievals and the use of the techniques for clinical documents are themselves novel.

2 Previous Work

2.1 Information Retrieval

Information retrieval [12] refers to the branch of artificial intelligence which requires retrieval of data from a corpus. The corpus can be present on Internet or locally on a device and may be in form of articles, social media text, or multimedia to mention few. The clinical or medical document retrieval is the process of searching and (or) ranking relevant documents from a collection of medical corpus. The corpus can be medical records, prescriptions, clinical reports, medical blogs, journals, books, and online content related to medicine, treatments, and clinical procedures to mention a few. Clinical retrieval is based on a query posed typically by a person seeking information relevant to data mentioned above. The results of the retrieval changes based on change in query.

Some popular ranking techniques used for retrieval are discussed as follows. Firstly, it is required to represent a document and query in vector space representation. Typically, the document and query are represented in one-hot representation. Each document and query are represented as n-dimensional vector where n is the total number of terms present in the collection of documents which shall be referred to the corpus. The value of vector for a particular dimension shall be one if a term is present in the document/query; otherwise, zero. Other weighing schemes such as term-frequency, inverse-document frequency, and their combinations are also used prevalently. Once the query and the documents are represented in vector space representation, their similarity is computed using several similarity measures.

1. Cosine similarity [12] is a popular and efficient technique to compute the similarity between a document and a query. The formula to compute the similarity between a document $\vec{d_j}$ and a query $\vec{q}$ is given as follows:

$$\text{cosine\_similarity}(d_j, \vec{q}) = \frac{\vec{d_j} \cdot \vec{q}}{\left|\vec{d_j}\right|\left|\vec{q}\right|} = \frac{\sum_{i=1}^{n} w_{i,j} q_i}{\sqrt{\sum_{i=1}^{n} w_{i,j}^2}\sqrt{\sum_{i=1}^{n} q_i^2}}$$

2. Jacard similarity [12] is another popular similarity model for information retrieval wherein collection of terms in document and in query is considered. The documents which have more number of common terms are given higher importance than the one with lesser number of common terms. The formula for computing similarity is computed as follows:

$$\text{Jacard\_ similarity}(A, B) = \frac{|\boldsymbol{A} \cap \boldsymbol{B}|}{|\boldsymbol{A} \cup \boldsymbol{B}|}$$

3. Okapi similarity [7] is one of the efficient techniques of information retrieval. The similarity is mathematically expressed as follows:

$$\text{Okapi\_ similarity}(q, di) = \sum_{t \in q} \boldsymbol{W} \frac{(\boldsymbol{k}1 + 1)\boldsymbol{tf}}{\boldsymbol{K} + \boldsymbol{tf}} \times \frac{(\boldsymbol{k}3 + 1)\boldsymbol{qtf}}{(\boldsymbol{k}1 + \boldsymbol{qtf})}$$

Here, q is the query and di is document under consideration. K and W are parameters dependent on corpus. $k1$ and $k3$ are constant parameters, tf and qtf are term frequencies document and in query, respectively.

Fuzzy inference-based ranking [7] was proposed recently using composite of two fuzzy inference engines. Traditional systems of retrieval using fuzzy sets rely on a single inference engine. The proposed approach superseded the results produced by comparison models. PageRank [18] algorithm was developed to rank documents in a collection based on weights and connection between them. Initially, it was used for Web page analysis. PageRank is highly used in the area of information retrieval.

2.2 Random Indexing-Based Representation

Random indexing [1]-based techniques utilize the contextual knowledge present in the text documents. Initially, suitable random vectors called index vectors satisfying certain constraints are assigned to each of the relevant terms present in the input text. The context vectors of words in text are determined by the neighboring words up to a particular window length. The context vector assimilates this knowledge of contextual information present in document collection by performing weighted addition of index vectors of neighboring words. Random indexing has been successfully used in several NLP applications such as text categorization [5] and information retrieval [20]. The formula for computation of context vectors is given as follows:

$$cv_i = cv_i + \sum_{k=-M}^{k=M} w_k * id_{i+k}$$

Here, M is the window size and cv_i is the context vector of word i which is initialized to id_i, viz the index vector of word i.

2.3 GLOVE-Based Representation

Alternate techniques of representation of words are highly researched lately and have succeeded to some extend in capturing the structural and other relations that are prevalent among words. GLOVE [19] stands for Global vectors which represent words which are trained using log bilinear regression model. It combines the best of matrix factorization and local context window-based methods which are left behind by it in certain NLP tasks. For instance, matrix factorization techniques such as LSA perform well in representing words in a lower dimension but they do not perform well when simple arithmetic operations are performed on vectors. While context window-based methods such as skip-gram methods may perform well on simple arithmetic operations they lack to assimilate the statistical information. GLOVE vectors are pre-trained vectors which maximize the conditional probabilities of occurrence of terms that occur together in a corpus. Traditional methods of vector space representation for words rely typically on distance or angle between the vectors while with GLOVE-based representation, analogies can be established based on simple arithmetic operations.

2.4 Rank Level Fusion

Rank-level fusion [11] has been traditionally used for clubbing results obtained through various modalities in a biometric system. The various modalities in biometric system are face, finger-prints, palm-prints, iris, etc. It has also been used to combine various classifiers results into a consolidated output. This helps to deal with decision making in conflicting situations. The motivation to utilize rank-level fusion for clinical retrieval is that in retrieval as well as biometric systems multiple ranks need to be processed. Both biometric and proposed system re-rank the multiple ranking of the respective objects. These ranks need to be analyzed and cumulative rankings need to be provided in both the applications. Several techniques for rank-level fusion for a biometric system are in use and they are described below:

1. Highest rank method: The highest rank method works on the basis of rankings provided by various modalities. The ranks indicate that a particular classification is relevant to the given input. The highest ranks are the candidate identification for the input to biometric. Further, the ranks for each of the user is computed and

all ranks from all the modalities are fused. The users are then sorted according to the fused ranks.

2. Borda count method: It is a popular unsupervised rank-level fusion technique. The method works on the concept of assigning votes to top N users. Wherein highest vote is assigned to top rank user and next vote to next user rank. All the votes are added and the user that has highest sum of votes is the identified user.
3. Weighted Borda count method: In this method, each modality is assigned a weight based on the perceived importance of the modality. Further, weighted sum of votes is computed.

The following section describes in detail the proposed technique.

3 Proposed Technique

The following are the proposed modification of the rank-level fusion techniques for clinical documents retrieval. The collection of documents retrieved through several techniques viz. (i) TF-IDF-based retrieval, (ii) Random indexing-based retrieval, and (iii) Word-embedding-based retrieval. These three methods are considered as three modalities for fusion of ranks. The ranks in these methods are computed using the PageRank algorithm [18] wherein nodes in graphs are the documents and the links between the nodes are similarity between the documents. The similarity between the documents is computed using the following approach:

a. TF-IDF-based similarity
b. Random indexing-based similarity
c. GLOVE vector-based similarity

The proposed rank-level fusion techniques for clinical document retrieval system are as follows:

i. Highest rank method: The highest rank method utilizes the ranks provided by the three techniques viz

   (a) TF-IDF-based retrieval,
   (b) Random indexing-based retrieval, and
   (c) Word-embeding-based retrieval.

   The highest ranks for each document are selected as the final rank of the document.

ii. Borda count method: The method works on the concept of assigning votes to all the N documents under consideration. The following three techniques viz

   (a) TF-IDF-based retrieval,
   (b) Random indexing-based retrieval, and
   (c) Word-embedding-based retrieval.

provides the vote for each document. The vote is given by the following formula:

$$\text{Vote}(\text{Document}_i) = N - \text{rank}(\text{Document}_i),$$

where N is the total number of documents under consideration.

Further, the votes through the three techniques are combined through the following formula:

$$\text{Combined\_ Vote}(\text{Document}_i) = \sum_{j=1}^{3} \text{Vote}_j(\text{Document}_i)$$

iii. Weighted Borda count method: In this method, each modality is assigned a weight based on the perceived importance of the modality. Further, weighted sum of votes is given as follows:

$$\text{Weighted\_ Combined\_ Vote}(\text{Document}_i) = \sum_{j=1}^{3} w_j * \text{Vote}_j(\text{Document}_i)$$

The proposed technique is presented in Fig. 1. The following section describes the experiments and the results obtained.

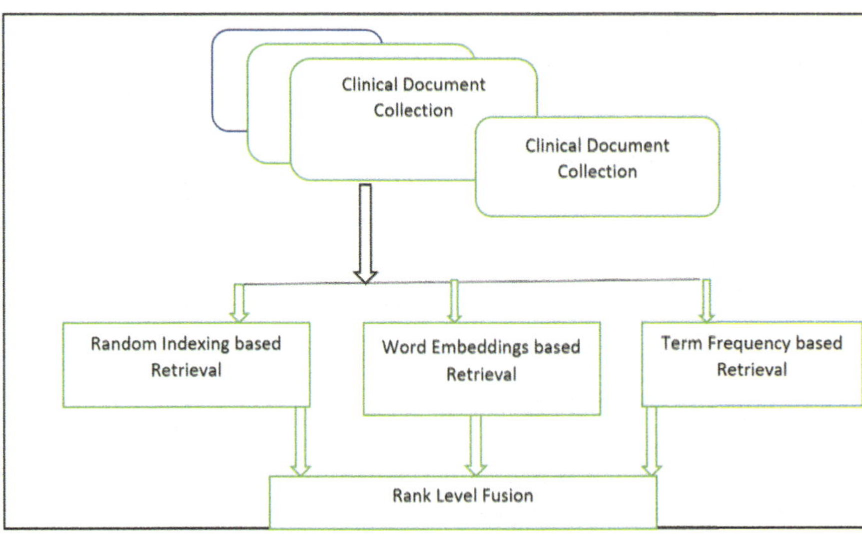

Fig. 1 Overall working of the proposed clinical retrieval system

Table 1 Results of various techniques over eight clinical documents for illustration

S. No.	TF-IDF	Random indexing	GLOVE	Fusion-highest rank	Fusion-Borda count	Fusion-weighted Borda Count
1	7	7	0	7	5	3
2	0	0	7	7	1	3
3	2	5	2	5	3	1
4	1	3	1	3	0	0
5	5	2	5	5	4	5
6	3	1	3	3	1	1
7	6	4	6	6	7	7
8	4	6	4	6	5	5

4 Experiments and Results

We have experimented the system on 50 clinical documents related to cancer which were collected and analyzed with the help of a team of doctors. The data contains clinical documents and the primary purpose of the track is retrieval of clinical documents. The rankings and voting computed for first eight documents are shown in Table 1. It is analyzed that different similarity metrics viz TF-IDF, random indexing, and GLOVE are producing different ranking. The three Rank Level Fusion techniques clubs these retrieval ranks into one using the formulas discussed in Sect. 3. It becomes paramount to assimilate the information processed through various algorithms through fusion techniques. On analyzing the fused results, it was noticed that some documents are ranked at the same rankings, wherein tie is broken based on chronological orderings. In weighted Borda count, the weights are set to 0.3, 0.2, and 0.5 for TF-IDF ranking, RI ranking, and GLOVE ranking, respectively. The graphs obtained by applying Page Ranks on TF-IDF are presented in Fig. 2. The results over 50 documents are given in Table 2. The mean square error (MSE) computed for the techniques is given in Table 3. It is evident that rank-level fusion is decreasing the MSE.

5 Conclusion and Future Work

This paper proposes novel rank-based fusion technique for retrieval of clinical documents which combines three efficient retrieval strategies, namely TF-IDF-based retrieval, random indexing-based retrieval, and word-embedding-based retrieval of clinical documents. No known significant work has been found for random indexing and word-embedding-based retrieval of clinical documents and their subsequent fusion. The proposed method is novel work for clinical document retrieval. The results obtained are encouraging. In the future, appropriate weights for ranking are to

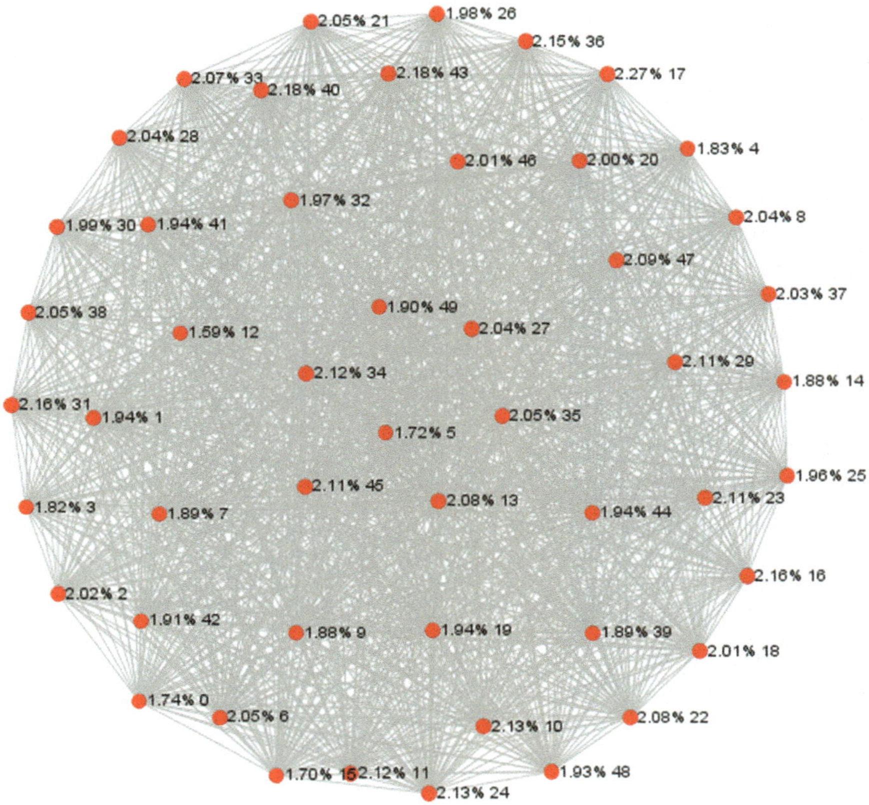

Fig. 2 Graph-based ranking for 50 clinical documents using TF-IDF scores

be determined automatically. It is analyzed that different retrieval techniques provide different rankings; hence, it becomes paramount to perform fusion of these results to come to a conclusion that utilize the best of all the techniques. The method needs to be tested over larger data collection from TREC repository for accuracy computations. In the future, the authors plan to work on extending the model for TREC clinical decision system retrieval [21] track for year 2015, 2016 and 2017. This work can be used for combining ranks over other retrieval models such as neural information retrieval [8] and neural vector spaces for information retrieval [14].

Table 2 Mean square error computed using various proposed techniques over 50 clinical documents

RI	Glove	Fusion-highest rank	Fusion-Borda count	Fusion-weighted Borda Count
0	784	196	784	16
0	36	100	0	100
225	256	16	0	0
441	841	841	0	576
49	169	49	169	16
49	25	169	49	169
81	121	81	121	81
100	529	484	529	400
529	9	4	9	144
1369	25	81	1369	484
441	484	676	0	576
196	9	100	196	196
841	1	1	0	81
0	1849	64	0	1
81	16	1	81	1
81	256	196	256	169
81	9	16	9	1
441	576	121	576	169
225	36	36	225	81
1225	225	361	0	729
4	169	1	0	25
289	729	1	289	121
1	441	324	441	121
36	1089	9	1089	36
49	361	25	49	100
324	961	144	961	64
196	25	1	196	64
1600	1024	576	0	900
225	100	0	225	64
81	121	16	0	36
484	100	400	0	484
196	64	4	64	9
256	196	441	0	361
1681	121	169	1681	576
729	4	4	4	81

(continued)

Table 2 (continued)

RI	Glove	Fusion-highest rank	Fusion-Borda count	Fusion-weighted Borda Count
625	4	16	4	4
900	169	576	900	729
529	576	100	0	49
1225	1764	1089	0	1089
2401	841	625	0	1296
529	784	400	784	400
676	961	529	0	484
36	256	0	256	1
225	121	64	225	1
1156	1	16	1156	196
1444	36	100	1444	400
1521	1521	900	1521	900
1681	1936	1444	1936	1600
9	729	81	729	1
49	256	144	256	81

Table 3 Average MSE results of various techniques over all 50 documents

MSE RI	MSE GLOVE	MSE-fusion-Borda count	MSE-fusion-Highest rank	MSE-weighted Borda count
512.24	434.32	235.84	371.66	285.26

References

1. Chatterjee N, Sahoo PK (2015) Random indexing and modified random indexing based approach for extractive text summarization. Comput Speech Language 29(1):32–44
2. Chou S, Chang W, Cheng CY, Jehng JC, Chang C (2008) An information retrieval system for medical records & documents. In: 2008 30th annual international conference of the IEEE engineering in medicine and biology society, IEEE, pp 1474–1477
3. Chouni Y, Erritali M, Madani Y, Ezzikouri H (2019) Information retrieval system based semantique and big data. Proc Comput Sci 151:1108–1113
4. Eto M (2019) Extended co-citation search: graph-based document retrieval on a co-citation network containing citation context information. Inf Process Manage 56(6):102046
5. Fernández AM, Esuli A, Sebastiani F (2016) Lightweight random indexing for polylingual text classification. J Artif Intell Res 57:151–185
6. Fieschi M (2004) Context-sensitive medical information retrieval. MedInfo 107:282
7. Gupta Y, Saini A, Saxena AK (2015) A new fuzzy logic based ranking function for efficient information retrieval system. Expert Syst Appl 42(3):1223–1234
8. Gysel CV, De Rijke M, Kanoulas E (2018) Neural vector spaces for unsupervised information retrieval. ACM Trans Inf Syst (TOIS) 36(4):38
9. Husain H, Wu HH, Gazit T, Allamanis M, Brockschmidt M (2019) CodeSearchNet challenge: evaluating the state of semantic code search. arXiv preprint arXiv:1909.09436

10. Khalifi H, Elqadi A, Ghanou Y (2018) Support vector machines for a new hybrid information retrieval system. Proc Comput Sci 127:139–145
11. Kumar A (2009) Rank level fusion. Encyclopedia of Biometrics
12. Luhn HP (1953) A new method of recording and searching information. Am Documentation (pre-1986) 4(1): 14
13. Madankar M, Chandak MB, Chavhan N (2016) Information retrieval system and machine translation: a review. Proc Comput Sci 78:845–850
14. Marchesin S, Purpura A, Silvello G (2019) Focal elements of neural information retrieval models. An outlook through a reproducibility study. Inf Process Manage 102109
15. Merrouni ZA, Frikh B, Ouhbi B (2019) Toward contextual information retrieval: a review and trends. Proc Comput Sci 148:191–200
16. Mohamed E, Elmougy S, Aref M (2019) Toward multi-lingual information retrieval system based on internet linguistic diversity measurement. Ain Shams Eng J
17. Munir K, Anjum MS (2018) The use of ontologies for effective knowledge modelling and information retrieval. Appl Comput Inf 14(2):116–126
18. Page L, Brin S, Motwani R, Winograd T (1999) The PageRank citation ranking: bringing order to the web. Stanford InfoLab
19. Pennington J, Socher R, Manning C (2014) Glove: global vectors for word representation. In: Proceedings of the 2014 conference on empirical methods in natural language processing (EMNLP), pp 1532–1543
20. Ponte JM, Croft WB (1998) A language modeling approach to Information Retrieval. In: Proceedings of the 21st annual international ACM SIGIR conference on research and development in information retrieval, ACM, pp 275–281
21. TREC (2014) http://www.trec-cds.org/2014.html
22. Wang H, Zhang Q, Yuan J (2017) Semantically enhanced medical information retrieval system: a tensor factorization based approach. IEEE Access 5:7584–7593

A Novel Feature Engineering Approach for Twitter-Based Text Sentiment Analysis

Hiran Nandy and Rajeswari Sridhar

Abstract With the increasing availability of handheld devices and greater affordability of mobile data, social media has become an inseparable part of the daily life of most of the society. Free availability, diversity, and massiveness of this data have inspired many research work to use it for extracting various insights about a variety of subjects. Text sentiment analysis is the process of mining opinion polarity from a given document in natural language. Social media data is no exception to extract sentiments which could be used for a variety of tasks right from opinion mining to recommendations. In this work, Twitter has been chosen as the source of data corpus required due to the summarized content, ease of availability, and humongous reach among all classes of the society. This work uses traditional machine learning approaches for solving the problem of text sentiment analysis and proposes a novel feature engineering approach for merging the text-based and non-textual features of the dataset by including the predicted output lists also in the final feature set. The result analysis shows significant improvement in the performance of the machine learning model on the test dataset using the proposed approach.

Keywords Sentiment analysis · Feature engineering · Predicted output · Text-based features · Non-textual features

1 Introduction

The expanded action of Microblogging, Tagging, and Podcasting powered by the blast of Web 2.0 has influenced many researchers in mining these enriched information assets for gaining valuable insights [1]. Text sentiment analysis is the process of

H. Nandy (✉) · R. Sridhar
Department of Computer Science and Engineering, National Institute of Technology, Tiruchirappalli, Tamil Nadu 620015, India
e-mail: hiran.tiucse@gmail.com

R. Sridhar
e-mail: srajeswari@nitt.edu

P. K. Singh et al. (eds.), *Evolving Technologies for Computing, Communication and Smart World*, Lecture Notes in Electrical Engineering 694,
https://doi.org/10.1007/978-981-15-7804-5_23

mining the inherent sentiment conveyed by a text document and classifying the emotion to either positive class or negative class. The insights gained from the automation of the sentiment analysis approach can be used in many socioeconomic areas [2] examples including but not limited to the prediction of acceptance of a product launch [3], the stock price prediction [4] and prediction of the success of a political campaign [5]. Twitter [6] is a popular microblogging Web site, that has grown from 5000 tweets per day in 2007 [7] to 500 million tweets per day in 2013 [8]. The summarized content, easy availability, sixth digited growth, and ubiquitous reach of twitter have created a new field in sentiment analysis, namely Twitter sentiment analysis [9–14].

Owing to these reasons, this work has chosen Twitter as its primary data source for sentiment analysis.

The complexities of natural language and the informal format used in Twitter data corpus with limitations of 280 characters per tweet make the task of Twitter sentiment analysis extremely challenging. The unconventional linguistic means like the block letters or repeated letters contained by a word representing an emphasis on that term, frequent misspellings, Hashtags(#), @ mentions, presence of unicode symbols, and URLs make it extremely difficult to preprocess Twitter data.

The features extracted from a dataset have a significant influence on the performances of the machine learning models trained on that dataset. As an enhancement of traditional feature-oriented techniques, this work proposes a novel idea of using the predicted output set of text-based and non-textual features also as individual features in the final feature set. To determine the differences between the predicted outputs of text-based and non-textual features, XOR operation has been performed on these two prediction sets and that XOR operation has also been used as a feature.

The arrangement of the remaining paper is as follows—Sect. 2 highlights the traditionally available approaches of text sentiment analysis. Section 3 illustrates the research methodology that this work has adopted. A performance analysis of the proposed scheme is mentioned in Sect. 4. Section 5 concludes the paper by stating the future scopes of the current work.

2 Related Work

A research work describes sentiment analysis as the intersection of natural language processing and computational machine learning techniques used for extraction and classification of sentiments from polarized documents [15]. The base of the social media-oriented sentiment analysis research works has been established by [16, 17].

Text sentiment analysis task can be performed at many levels [18] namely–word [19], sentence [20], document [21], or aspect [22]. Existing research works [23–26] have taken mainly four popular paths for text sentiment analysis—keyword-based, machine learning-oriented, lexicon-based, and hybrid. The complexities of twitter specific informal linguistic [14, 27], heavy class imbalance [28], and streamed tweet extraction model [29] have often contributed to critical performances of sentiment classification models [30]. User-defined hashtags, punctuation marks, and n-gram word embeddings have been used [31] as features for analyzing twitter sentiments using the nearest neighbor technique. In a work for sentiment analysis, [32] used an n-gram multinomial Naive-Bayes method for classifying sentiments. The emoticon-based tweet extraction technique has been used by the researchers. Unigram Naive-Bayes model was used [33] on a Twitter dataset extracted using three categories—camera, mobile, and movie for classifying obtained sentiment polarities into three categories—positive, negative, and non-opinionated. The useless features have been removed in this research work by using the chi-square method. In the context of text sentiment analysis, features are the attributes of the data corpus, that help in identifying the sentiment polarity express by the text. The feature set and its extraction techniques can broadly be classified into three categories—syntax-oriented based on the occurrence semantics of a term in a document, univariate using chi-square, and information gain feature extraction techniques and multivariate using genetic algorithms [34]. Two of the most popular techniques used for the assessment of the importance of extracted features are—Feature frequency(FF) and feature presence(FP) [35]. Bag of words-based textual feature extraction technique has been used [36], which considered the sentiment polarities of each term, instead of considering the semantics of a word in the context of the whole document. Researchers [37] have got a significant improvement in results, by combining parts-of-speech oriented and word-relation-based features using an ensemble framework. However, no research work has till now focused on using the prediction set of text-based and non-textual features as a feature in the final feature set to improve the accuracy of the sentiment classifier, which has been addressed here. The best performing ensembled model that uses the proposed featurization technique shows an accuracy of 94.75%.

3 Research Methodology

The overall workflow of the proposed system is shown in Fig. 1. The following sub-sections discuss each of the modules in depth.

3.1 Data Collection

Twitter allows the collection of tweets using their API "tweepy". Various methods have been proposed for the collection and labeling of tweets. The first and most naïve approach is to collect the tweets and then labeling them manually. This process is both time-consuming and tedious but ensures the highest possible accuracy in labeling. On the other hand, the second and more practical approach of collection and labeling of tweets is to collect the tweets by querying the Twitter API using positive[), :), :-), :P] and negative[:(, :'(, :-(] emoticons. In this way, automated retrieval of pre-labeled tweet data corpus is achievable. Additionally, the dataset should be both balanced

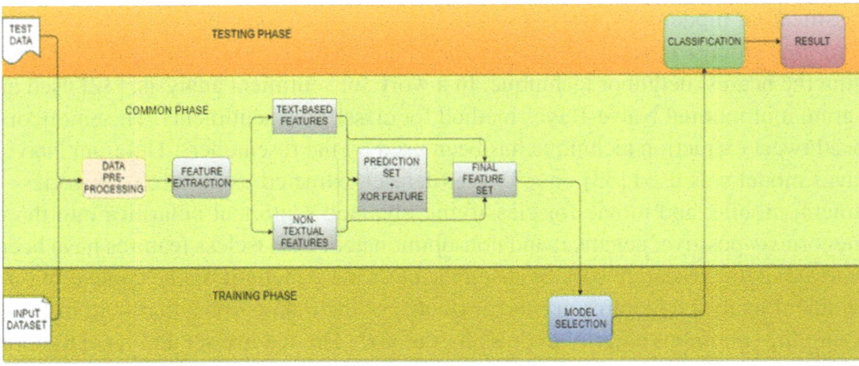

Fig. 1 Overall block diagram

and sufficiently large, as more data means more insights and more insights means more accuracy. This work has used a pre-labeled balanced data corpus of 80,000 tweets from Kaggle [38].

The dataset has three attributes – 1. "id", 2. "label", 3. "tweet". 40,000 of the collected data is labeled as Positive(0) and rest 40,000 as Negative(1).

3.2 *Train-Test Split*

At the beginning of the research work, 20% tweets have been reserved for testing purpose. Machine learning models have been trained on the rest of the 80% dataset.

3.3 *Data Preprocessing*

Following are some of the complexity faced in handling the twitter specific text sentiment analysis problem:

- The use of acronyms is very common on twitter. Now, comprehending these acronyms can sometimes be really tricky. For example, the acronym "Apl"—we are not sure if it means "apple"—the fruit or "Apple" the company or application.
- Sequences of repeated characters are sometimes used to emphasize feeling, for example—"grrrrrrrtttt movie". It becomes very difficult to find out the grammatical roots from these emphasized terms.
- Almost all the tweets contain emoticon symbols like ":(", ":-)", ":O" etc. While these emoticons are very important for analyzing the user's mood, it is really difficult to analyze them in their original form.

- "Urban Grammar" and "spelling mistakes" also finds its presence in social media quite often. For example, "im gunna", "mi"—these kinds of terms are very common. In this case, also finding grammatical root or lemmatization is very hard to achieve.
- Sometimes, people also indicate their moods, emotions or state of mind inside two stars like these—"*sighs*", "*cries*" etc.
- Negative terms also could be present in their original or shortened form. For example, "cannot" and "can't". This ambiguity also needs to be nullified before using the dataset for actual training.

All these practical problems motivated the data preprocessing tasks. Details of the data preprocessing workflow are mentioned below -

Mentions removal. @ mentions are used in twitter for tweeting to someone or just for grabbing attention. These @ mentions do not have much importance in sentiment analysis and that is why they were removed.

HTML entities removal. All the characters that are reserved in HTML are called HTML entities. For getting comprehensible characters from those entities, they need to be decoded. Python library "BeautifulSoup" has been used in this research work for serving the aforementioned purpose.

Case folding. All the characters in the data corpus have been converted to lower-case characters to nullify case-sensitivity.

Stop-words removal. Stop words are the most frequently occurring terms in a particular language. Previous researches have shown these most frequently occurring terms do not help much in sentiment analysis [39]. Python library "NLTK" has a predefined set of stop words for the English language. The negative terms like "no", "not", and "nor" have also been included in that list. In sentiment analysis, those negation words are really important. So, this work has used that predefined stop-words list, by removing all the negative terms from that list.

URL replacement. Uniform resource locator(URL) represents the address of a worldwide web(www) page. These URLs are not at all comprehensible in their raw form and does not help anyway in sentiment analysis. This work has replaced all the URLs present in tweet texts labelled as positive by the phrase "||pos_url||" and the rest as "||neg_url||".

Expansion of negative terms. All the shortened negative phrases were replaced by their full forms. For example, "don't" and "can't" were replaced by "do not" and "can not", respectively.

Emoticon Replacement. All the positive emoticon symbols have been replaced with the phrase "||pos_emotion||" and all the negative emoticon symbols have been replaced with the phrase "||neg_emotion||" using the list [40].

Emphasized term replacement. All the words having three or more repeated occurrences of a character have been replaced either with the phrase "||pos_extreme||" if found in a positively labeled tweet and "||neg_extreme||" alternatively.

Non-alphanumeric characters removal. All the characters except [A–Z], [a–z], [0–9], the emoticon symbols, "||" and "#" have been removed and replaced with whitespace.

Removing small words. All the words of size one were removed as they are almost similar to stop words.

Null tweets removal. After performing all the aforementioned preprocessing tasks, some tweets turned into NULL tweets. These tweets were removed from the data corpus.

Performing lemmatization. Lemmatization is the process of getting the dictionary root term of a word. It is the reverse process of applying morphological changes to a root word. "Wordnet Lemmatizer" [41, 42] has been used in this work for finding the lemmas from the data corpus.

3.4 *Feature Engineering*

The input dataset of a machine learning model comprises of features. Researchers [43] have shown that 80% of the total time spent in a data science project goes in data preparation. Before extracting actual features from the dataset, this work has analyzed the dataset thoroughly to understand which features could be effective. Following are some of the analysis done on the dataset:

Length distribution. From the training dataset, 1000 positively labeled and 1000 negatively labeled tweets were randomly picked and their length distributions have been visualized in Fig. 2. The length distribution of the positive and negative tweets

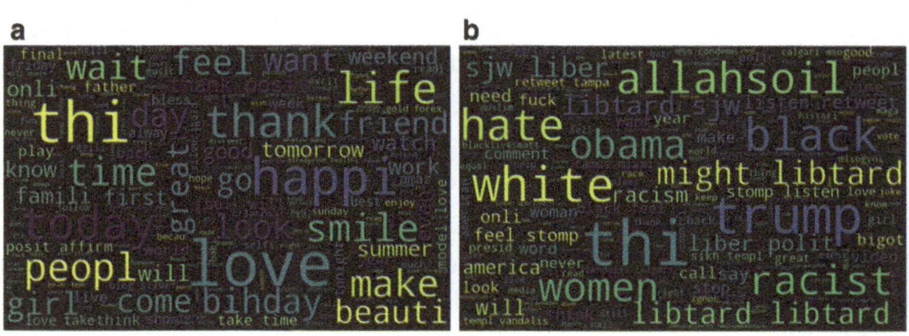

Fig. 2 **a** Positive word cloud, **b** Negative word cloud

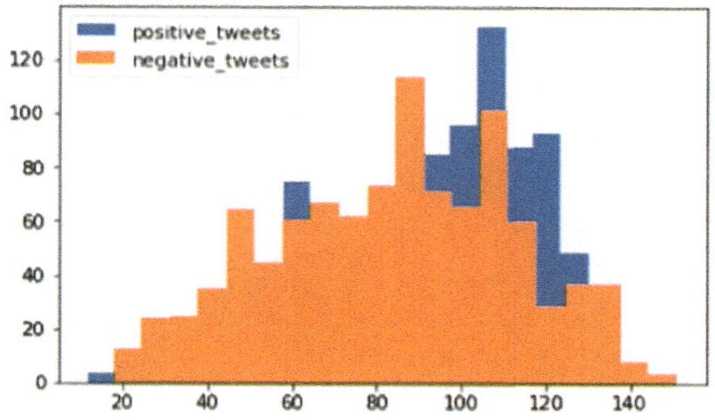

Fig. 3 Length distribution of training dataset

gives an intuition that the length feature can bring some kind of separability between them.

Word cloud visualization. Word cloud is a pictorial representation of words that appeared in a particular data corpus or subject matter where the size of the word appeared in the picture represents the frequency of that term in the corpus. For a sentiment analysis task, it gives a general idea about how important the text feature is for the sentiment classification. Two separate word clouds are visualized for positive and negative training data, respectively, in Fig. 3a, b. It is obvious from Fig. 3a, b that the text appeared in the corpus has a major role to play in this sentiment classification task.

Importance of Hashtags. In twitter, the hashtag is a phrase preceded by a "#" symbol.

If a hashtag is clicked, all the tweets containing that same hashtag are retrieved. Most frequent hashtags in both positive and negative tweets were visually analyzed in Fig. 4a, b.

Need for data imputation checking. Data imputation in feature engineering is the process of filling missing data. Data points having missing tweet texts were already been discarded in the feature engineering stage. In this stage, the label column was checked, and it was found that no entry was missing in that attribute. So, there was no need to impute missing data.

Outlier checking. Two kinds of outlier checking were performed:

- The "label" attribute was checked to find out if it has any value other than 0 or 1. The labels of all the tweets were found to be either 0 or 1.
- The "Tweet" attribute was checked to find out if any tweet has a length of more than 140 characters. In this case, also the tweet lengths were within range.

After all, these analysis features were extracted. First, the text-based features were extracted, then the non-text-based features were extracted. Finally, a prediction set from both these features were extracted and was added to the final feature list.

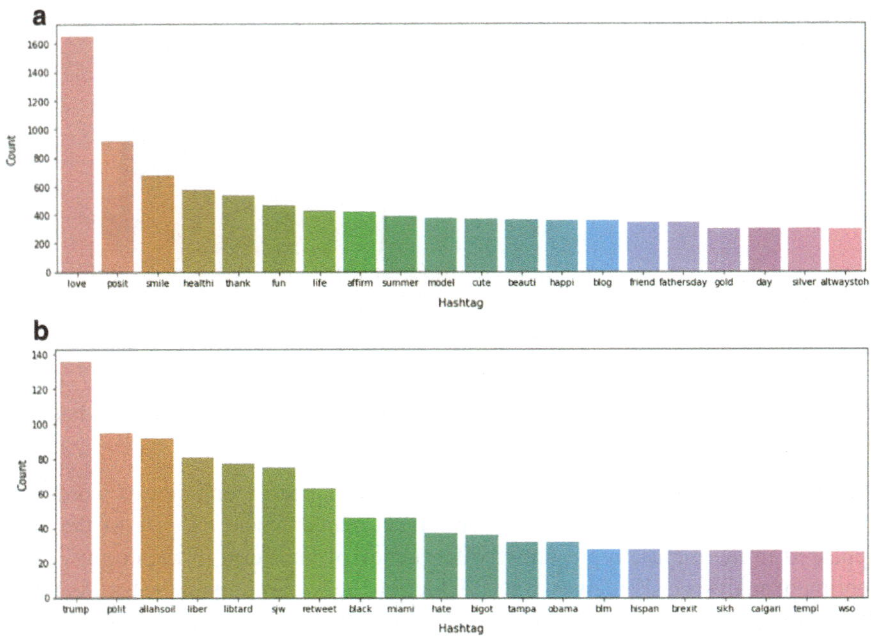

Fig. 4 **a** Frequent hashtags in positive tweets, **b** Frequent hashtags in negative tweets

Text-based feature extraction. Words in a text corpus are discrete and categorical features. They in their raw form and cannot be used in text processing tasks. Some kind of encoding is essential for mapping these words to real-valued features that can be readily used by machine learning algorithms. Two kinds of text-based features were extracted namely—1. Bag of Words and 2. TF-IDF.

Bag of words feature extraction. This is the simplest text feature technique, where the emphasis is only on term occurrence. Grammatical orientation or order of appearances of terms cannot be represented by this technique. In this work, unigram, bigram, and trigram features were extracted limiting the highest number of features to 1000.

TF-IDF feature extraction. Term frequency-inverse document frequency (TF-IDF) is a common technique of text feature which adds term-weighting with the bag of words technique. Instead of just focusing on term frequency, it also considers the current document's topic. The implication of a high TF-IDF score is, the term has appeared rarely in the corpus but finds its high frequency in the current document. This helps in preventing the high frequent terms to overshadow the important rare terms. The formula for calculating TF-IDF is shown in Eq. 1:

$$\text{TF-IDF}(t, d, D) = \text{TF}(t, d) \times \text{IDF}(t, D) \tag{1}$$

Here, d is the whole document corpus and D is the current document. The formula for calculating IDF is shown is Eq. 2

$$\mathrm{IDF}(t, D) = \log \frac{N}{\mathrm{DF}(t)} \tag{2}$$

where $DF(t)$ represents the number of documents in which term "t" appears. This work has again considered all unigram, bigram and trigram features and limited the maximum number of features to 1000.

Non-textual feature extraction. Various analysis of the characteristics of data corpus has indicated the need for non-textual features to be extracted. While extracting non-textual features, this work first extracted the tweet-based numerical features and then according to feature importance chose five best features and performed a few mathematical operations on those features. After those mathematical operations on the features, univariate analysis was performed and all the features that had a significant contribution in the sentiment classification were kept in the final feature list. Following are the details of the non-textual features extracted:

Tweet length. The length of tweets is calculated as the number of characters including white spaces has appeared in it.

Token length. All the space-separated words, lemmas, expressions, or phrases have been considered as separate tokens and the count of these tokens for a particular tweet has been considered as token length.

Stop word count. Before removing the stop words, their counts in each document was calculated.

First-half polarity. Tweet texts were tokenized, and the number of tokens was calculated. For the first half of the tweets, each token's polarity was calculated and the sum of all these polarities has been considered as the first half polarity.

Second-half polarity. Like first-half polarity, the polarity of all the tokens of the second half of a tweet was also calculated and summed. The final sum has been considered as the second-half polarity of the tweet.

Polarity change. In hate speeches or sarcastic speeches, the polarity of the sentence tends to change in the second half of the sentence. So, tracking the change of polarity of a particular tweet is important. This work has proposed the formula of Eq. 3 to track the change of polarity of a particular tweet.

$$\frac{|(First\ Half\ Polarity - Second\ Half\ Polarity)|}{Length\ (Tweet)} \tag{3}$$

Noun, adjective, verb, and adverb count. Variability of parts-of-speech of the same word changes the meaning of the word completely. So, all the tokens of a particular tweet text were parts-of-speech-tagged and the count of these four parts-of-speeches for a particular tweet was documented.

Frequent term count. Top 5000 most frequently occurring terms in the corpus were listed and count of any of these terms in a particular tweet was calculated. These terms work as corpus-specific stop words.

Positive and negative term count. Top 1000 most frequently occurring terms in both the positive and the negatively labeled tweets were listed and their presence in each of the tweets was calculated. For this particular corpus, these terms can be considered as positive and negative terms, respectively.

Positivity ratio. To get an essence of the overall positivity (or negativity) of a particular tweet, the positivity ratio of the tweet was calculated using the formula of Eq. 4:

$$\frac{|(Positive\,Term\,Count - Negative\,Term\,Count)|}{Length\,(Tweet)} \quad (4)$$

As all these extracted non-textual features belong to different numeric ranges, they were standardized using Eq. 5:

$$\frac{(x - Z)}{\delta} \quad (5)$$

where "x" is the feature value, "Z" is the mean of that feature set, and is the standard deviation of that feature set.

After the standardization, a "RandomForest Classifier" was used to obtain the importance of the individual features. Figure 5 is the Barplot representation of the feature importance of the extracted non-textual features.

"Positivity Ratio" was found out to be the most important feature and Fig. 6 represents univariate analysis performed on that feature.

Once the tweet-based non-textual features were extracted and analyzed, the top five most important features were chosen and some mathematical operations like

Fig. 5 Importance of non-textual features

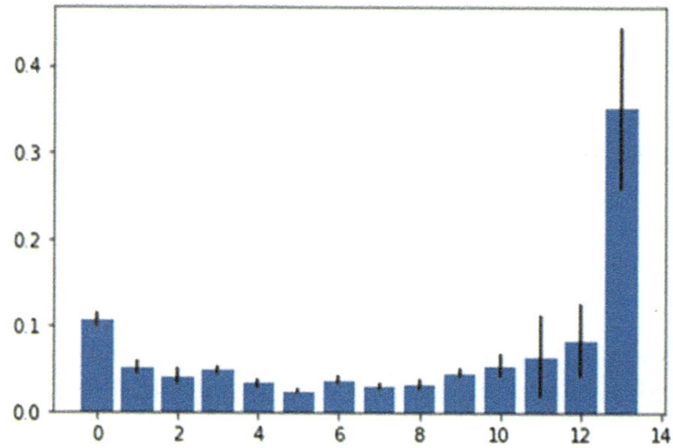

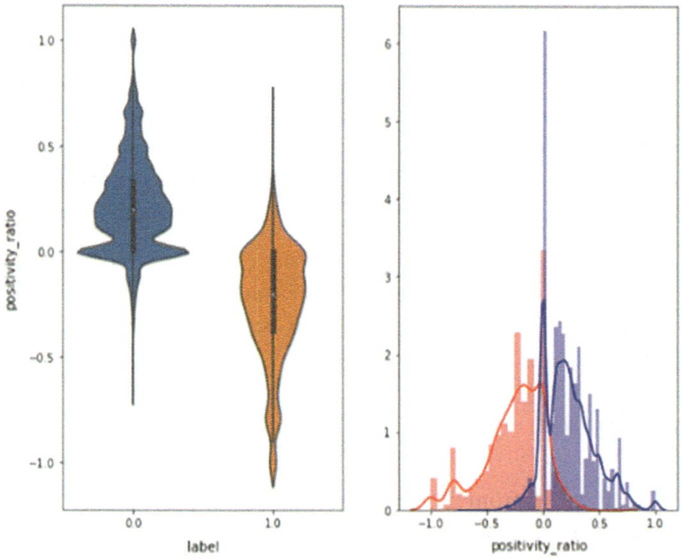

Fig. 6 Univariate analysis of "Positivity Ratio" feature

square root, sin, cos, tan, log were performed on each of those features. After any mathematical operation, the modified feature has been analyzed using the violin plot and it has been kept in the dataset or discarded according to its significance. The final feature set contained 26 non-textual features.

Feature merging. In the classical feature engineering approach, the standardized text-based and non-textual features are merged and a final feature set for training the machine learning model is prepared. This research work has tried to give importance to the predicted output set also and has included them in the final feature set. The prediction set of both the text-based and non-textual features contain some true negative and some false positive values. To minimize that percentage, the prediction set of both these features has been included. Moreover, as X-OR operation can be used as "difference predictor", the X-OR operation of both the predicted output set was also included in the final feature set. For getting the predicted output feature for the training dataset of the text-based features, the whole training dataset has further been divided into nine parts—the first part being of size 20% and all other parts of size 10% of the whole training data. At the beginning, the classifier has been trained using the first part of the training data and output set of the second part, which is 10% of the data has been predicted. In the following stage, the machine learning model was trained using the whole 30% data, without the predicted output feature, and the output list is predicted for the next 10% of data. This process was repeated continued until the training dataset existed and at the last stage, 90% of the training data predicted the output of the last 10% of the training data. Finally, the whole training data (without predicted output) predicted the output set for the test data.

In this stage, the final 80% of the training data and the whole test data have their predicted output feature. Finally, for the first 20% of the training data, the original output set has been kept as its predicted output set. The size of this first chunk of the dataset has been chosen by performing various experiments. For the non-textual features too, the same procedure has been followed for getting the predicted output feature. In this merged-feature model, instead of using bag of words or TF-IDF techniques for text features, TF-IDF weighted Word2Vec model has been used for two reasons:

1. Small-sized dense representation of features and
2. retaining the similarities and dissimilarities between different words.

3.5 Classification

The feature set obtained from the preprocessing step has been used for training the machine learning models for the binary classification task. As the feature set is not too large, the primary intuition was to use decision tree-based classifiers. Decision trees are supervised machine learning techniques, where the dataset is repeatedly split using some parameter. The hierarchical tree like structure consists of two kinds of nodes: decision nodes and leaves. Ensemble classification models like random forest and gradient boosted decision tree also contains decision tree as base models and produce the output using majority voting. From the linear classifiers, SVM with RBF kernel performed impressively well on the dataset. At the end, the classifiers performing well on the dataset have been stacked to create an ensemble of stacking classifier to increase the accuracy. For analyzing the performances of the classifiers, two main parameters have been used—1. Accuracy and 2. F-Score. As the dataset is balanced, "accuracy" has been chosen as a parameter. F-Score, on the other hand, reflects the average precision and recall score of the classifier.

4 Results and Analysis

The performance of the machine learning models has been analyzed based on different parameters for the proposed algorithm. At first, a random classifier has been taken into consideration that does not have any machine learning expertise. This model has been used as a benchmark model for analyzing the performances of the machine learning models that this research work has used. Initially, the performances of the machine learning models have been compared based on textual features. After that, the non-textual features were combined with the textual features, which improved the performances of the classifiers by a considerable amount. The accuracies and F-scores of the classifiers have been compared by training the models with and without the predicted output features and XOR features. Among all the classifiers, KNN ($K = 7$),

SVM(Kernel = RBF) and RandomForest(n_Estimators = 120) classifiers gave best results. The classifiers have been stacked to get improved accuracies.

4.1 Text-Based Features

The text-based features have been obtained using both bag of words and TFIDF techniques. In all the cases, the TF-IDF model beats the performance of the bag of words model by a slight margin (Fig. 7).

As bag of words and TF-IDF models are inherently sparse in nature and the dataset contains only numerical features and no categorical feature, the SVM model outperforms all the other classifiers in this case. As a standalone classifier, the SVM model gave the highest accuracy of 87.1% whereas the stacked Ensemble model has increased the accuracy to 88.16%

4.2 Addition of Non-textual Features

In this step, instead of using bag of words or TF-IDF for text features, TF-IDF weighted Word2Vec technique has been used. This supports two reasons:

- To further reduce the size of the text-based features, so that non-textual features are not overshadowed by the text-based ones.
- To be able to use the whole text-based feature set instead of limiting it to a certain threshold.

The performances of the classifiers have been compared and contrasted by including and discarding the predicted output features and the XOR operation of

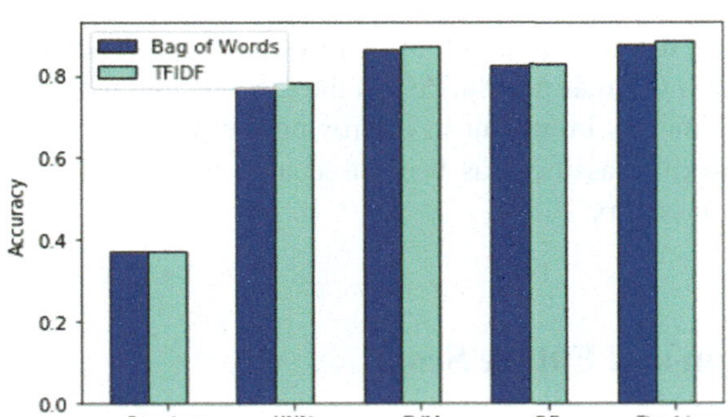

Fig. 7 Comparison of accuracies of different classifiers using bag of words and TFIDF text-based features

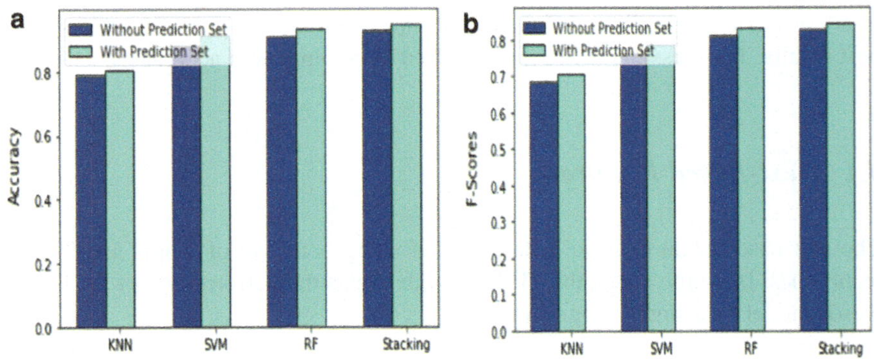

Fig. 8 **a** Comparison of accuracies of different classifiers using and without using prediction set feature, **b**. Comparison of F-Scores of different classifiers using and without using prediction set feature

the predicted output features in Fig. 8a, b. While comparing the performances of the classifiers, both accuracy and F-Score parameters have been taken into consideration.

In all the cases, the addition of the predicted output feature set has helped in increasing the accuracy of the model. Here, the random forest classifier has outperformed all the other classifiers. The reason being, the dense representation of the features, which is a favorable condition for the random forest classifier. Moreover, although there has not been any categorical feature, all the count feature has mostly found its range between 0 and 3, which indirectly helped in creating a proper mix of categorical and numerical features. In this work, no standardization has been performed of the feature set while using the random forest classifier. Here too, the stacking ensemble model has increased the performances of the individual classifiers by a slight margin. Random forest classifier has given 93.3% accuracy for the predicted output included feature set, whereas the stacked model has given the highest accuracy of 94.75% with an F-Score of 84.15%.

The precision–recall curve of the final stacked model has been plotted with and without including the predicted output features in Fig. 9. The AUC score for the first case is 91.2%, whereas it is 88.75% in the second case, which suggests that the prediction set feature is important in improving the performance of the classifier. Here also a no-skill classifier has been used as a benchmark for other machine learning-based classifiers.

5 Conclusion and Future Scope

The ubiquitous reach of social media has influenced many researchers to analyze the sentiment polarities of social media posts. The complexities of natural languages have always made it difficult to automate the process of sentiment polarity detection from human-generated tweets. In this work, a novel feature-oriented technique has

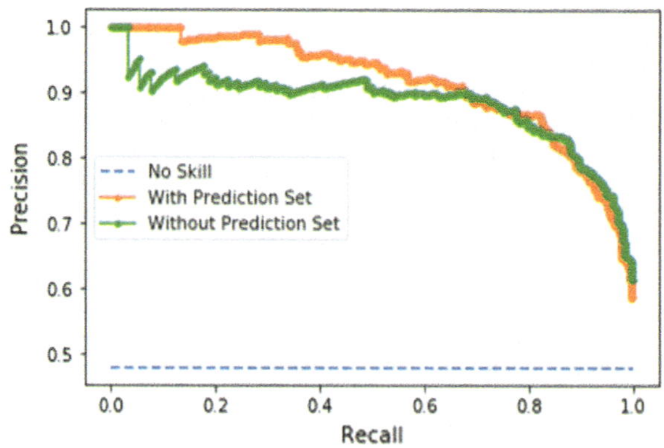

Fig. 9 Precision–recall curve of stacked classifier model with and without prediction set

been proposed that uses the prediction sets as a feature for the sentiment classification task. The result analysis shows significant improvement in the performances of the classifiers while using the proposed feature-oriented technique. The proposed scheme can be applied to any classification problem where the prediction results of two or more classifiers need to be combined and used in the final feature set. The novelty of this approach lies in the fact that this approach can combine the prediction results of different classifiers that uses feature sets of different dimensions. The future work will try to modify the algorithm in such a way that, the need for using actual labels in the first twenty percent of the prediction set feature gets nullified. The future work will also focus on predicting different aspects of sentiment like polarity, sarcasm, negation, etc., for the same dataset and will try to combine them by following the proposed feature-oriented technique to construct an accurate sentiment classification model.

References

1. Kaplan AM, Haenlein M (2010) Users of the world unite! The challenges and opportunities of social media. Science Direct (53): 59–68
2. Jansen J, Zhang M, Sobel K, Chowdury A (2009) Twitter power: tweets as electronic word of mouth. In: JASIST (60), pp 2169–2188
3. Rui H, Liu Y, Whinston A (2013) Whose and what chatter matters? The effect of tweets on movie sales. Decis Support Syst 55: 863–870
4. Bollen J, Mao H, Zeng X-J (2010) Twitter mood predicts the stock market. J Comput Sci 1–8
5. Bermingham A, Smeaton A (2010) Classifying sentiment in microblogs: Is brevity an advantage? In: Proceedings of the ACM Conference on Information and Knowledge Management (CIKM'10), pp 1833–1836
6. Twitter website. www.twitter.com
7. Weil K (2010) (VP of Product for Revenue and former big data engineer, Twitter Inc.).: Measuring Tweets., Twitter Official Blog

8. Krikorian R (2013) (VP, Platform Engineering, Twitter Inc.).: New Tweets per second record, and how! Twitter Official Blog
9. Adedoyin-Olowe M, Gaber M, Stahl F (2013) A methodology for temporal analysis of evolving concepts in Twitter. In: Proceedings of the 2013 ICAISC, International conference on artificial intelligence and soft computing
10. Gomes JB, Adedoyin-Olowe M, Gaber MM, Stahl F (2013) Rule type identification using TRCM for trend analysis in twitter. In: Research and development in intelligent systems. Springer International Publishing, pp 273–278
11. Becker H, Naaman M, Gravano L (2011) Beyond trending topics: real-world event identification on Twitter. ICWSM (11): 438–441
12. Becker H, Chen F, Iter D, Naaman M, Gravano L (2011) Automatic identification and presentation of Twitter content for planned events. In: ICWSM
13. Phuvipadawat S, Murata T (2010) Breaking news detection and tracking in twitter. In: 2010 IEEE/WIC/ACM International Conference Web Intelligence and Intelligent Agent Technology (WI-IAT), IEEE, vol 3, pp 120–123
14. Osborne M, Petrovic S, McCreadie R Macdonald C, Ounis I (2012) Bieber no more: first story detection using twitter and wikipedia. In: Proceedings of the Workshop on Time-aware Information Access. TAIA, vol 12
15. Basant A, Namita M, Pooja B, Sonal Garg (2015) Sentiment analysis using common-sense and context information. In: Computational intelligence and neuroscience, vol 2015, Article ID 715730, 9 pages
16. Das S, Chen M (2001) Yahoo! for Amazon: extracting market sentiment from stock message boards. In: Proceedings of the Asia Pacific Finance Association Annual Conference (APFA)
17. Tong R (2001) An operational system for detecting and tracking opinions in on-line discussion. In: Proceedings of the workshop on Operational Text Classification (OTC)
18. Thomas B (2013) What consumers think about brands on social media, and what businesses need to do about it, Report, Keep Social Honest
19. Nikos E, Angeliki L, Georgios P, Konstantinos C (2011) ELS.: a word-level method for entity-level sentiment analysis. In: WIMS '11 proceedings of the international conference on web intelligence, mining and semantics
20. Noura F, Elie C, Rawad AA, Hazem H (2010) Sentence-level and document-level sentiment mining for arabic texts. In: Proceeding of the IEEE international conference on data mining workshops
21. Ainur Y, Yisong Y, Claire C (2010) Multi-level structured models for document-level sentiment classification. In: Proceedings of the conference on empirical methods in natural language processing. MIT, Massachusetts, Association for Computational Linguistics, USA, pp 1046–1056
22. Haochen Z, Fei S (2015) Aspect-level sentiment analysis based on a generalized probabilistic topic and syntax model. In: Proceedings of the twenty-eighth international Florida artificial intelligence research society conference, Association for the Advancement of Artificial Intelligence
23. Binali H, Wu C, Potdar V (2010) Computational approaches for emotion detection in text. In: 4th IEEE International Conference on Digital Ecosystems and Technologies (DEST), pp 172–177
24. Kao EC-C, Liu C-C, Yang T-H, Hsieh C-T, Soo V-W (2009) Towards text-based emotion detection: a survey and possible improvements. In: International conference on information management and engineering, pp 70–74
25. Shivhare SN, Khethawat S (2012) Emotion detection from text. In: Computer science, engineering and applications (May 2012)
26. Jain VK, Kumar S, Fernandes SL (2017) Extraction of emotions from multilingual text using intelligent text processing and computational linguistics. J Comput Sci 21: 316–326
27. Ghiassi M, Zimbra D, Lee S (2016) Targeted Twitter sentiment analysis for brands using supervised feature engineering and the dynamic architecture for artificial neural networks. J Manage Inf Syst 33(4):1034–1058

28. Hagen M, Potthast M, Büchner M, Stein B (2015) Twitter sentiment detection via ensemble classification using averaged confidence scores. In: Proceedings of European conference on information retrieval
29. Vanzo A, Croce D, Basili R (2014) A context-based model for sentiment analysis in Twitter. In: Proceedings of the COLING conference, pp 2345–2354
30. Hassan A, Abbasi A, Zeng D (2013) Twitter sentiment analysis: a bootstrap ensemble framework. In: Proceedings of the ASE/IEEE international conference on social computing, pp 357–364
31. Davidov D, Rappoport A (2010) Enhanced sentiment learning using Twitter hashtags and smileys. In: Proceedings of the coling conference, Beijing, pp 241–249
32. Pak A, Paroubek P (2010) Twitter as a corpus for sentiment analysis and opinion mining. In: Proceedings of the seventh conference on international language resources and evaluation, pp 1320–1326
33. Liang P-W, Dai B-R (2013) Opinion mining on social media data. In: IEEE 14th international conference on mobile data management, Milan, Italy, pp 91–96, June 2013
34. Abbasi A, France S, Zhang Z, Chen H (2011) Selecting attributes for sentiment classification using feature relation networks, knowledge and data engineering. IEEE Trans 23(3): 447–462
35. Na J-C, Sui H, Khoo C, Chan S, Zhou Y (2004) Effectiveness of simple linguistic processing in automatic sentiment classification of product reviews. In: Conference of the International Society for Knowledge Organization (ISKO), pp 49–54
36. Turney PD (2002) Thumbs up or thumbs down? Semantic orientation applied to unsupervised classification of reviews. In: Proceedings of the 40th annual meeting on association for computational linguistics, Association for Computational Linguistics, pp 417–424
37. Xia R, Zong C, LiS (2011) Ensemble of feature sets and classification algorithms for sentiment classification. Inf Sci Int J 181(6): 1138–1152
38. Kaggle website. www.kaggle.com
39. Pandey AK, Siddiqui TJ (2009) Evaluating effect of stemming and stop-word removal on Hindi text retrieval. In: Proceedings of the first international conference on intelligent human computer interaction. Springer, New Delhi
40. Emojipedia website. www.emojipedia.org/twitter/
41. Miller GA (1995) WordNet: a lexical database for English. Commun ACM 38(11): 39–41
42. Fellbaum C (1998) WordNet: an electronic lexical database. MIT Press, Cambridge, MA
43. Forbes report. https://www.forbes.com/sites/gilpress/2016/03/23/data-preparation-most-time-consuming-least-enjoyable-data-science-task-survey-says/. Last accessed 23 Mar 2016

Survey on DDoS and EDoS Attack in Cloud Environment

Shruti Wadhwa and Vipul Mandhar

Abstract Cloud computing is a heterogeneous distributed environment that provides resources as service through Internet. Cloud consists of various resources like network, memory, computer processing, and user applications provided to the customer on pay-per-use scale. Cloud services are broadly divided as software as a service (SaaS), platform as a service (PaaS), and infrastructure as a service (IaaS). Therefore, data that is stored in cloud needs to be secured from the attackers as it is remotely kept. However, security is one of the major challenges in cloud. EDoS is the latest kind of DDoS attack on the cloud. The purpose is to consume cloud resources although the price of services is pay off by the valid customer. The key intention of DDoS attack brings down the specific service by draining the server's resources whereas EDoS's objective is to create economic unsustainability in the cloud resources for the object and causes financial consequences by exhausting resources and leading to a heavy bill. This paper reviews several DDoS and EDoS modification methods that have been made known in the past years and presents the mechanism which is effective for the mitigation of DDoS and EDoS attack.

Keywords DDoS · EDoS · Cloud computing

S. Wadhwa
SBBS University, Hoshiarpur, India
e-mail: shrutiwadhwa99@gmail.com

V. Mandhar (✉)
NITTTR, Chandigarh, India
e-mail: vipulmandhar130793@gmail.com

P. K. Singh et al. (eds.), *Evolving Technologies for Computing, Communication and Smart World*, Lecture Notes in Electrical Engineering 694,
https://doi.org/10.1007/978-981-15-7804-5_24

1 Introduction

1.1 Cloud Computing

Cloud computing is new IT delivery model, which enables user to store and access data according to their need irrespective of time and place. The idea behind cloud computing is reducing the workload from user's computer to cloud making use of simple Internet connection. Cloud computing and its characteristics are represented by Fig. 1. It allows IT industries to focus on doing what they actually want without spending money on infrastructure and wasting time in arranging them. It gives user the facility of pay-per-use which means provides measured services like networks, servers, storage, and applications as per their demand. Due to cloud providing features like elasticity, pay-per-use, flexibility, scalability, it earns the attraction of big organization and company for hosting their services on the cloud. By means of any latest technology trends, cloud computing is not secured from risk and susceptibilities of security.

Sabahi [1] provides the various issues of security and availability in cloud computing and suggests some obtainable solution for them. Lekkas [2] has defined the requirements of threats, and security is present at the different stages of the cloud execution. Cloud is vulnerable to various attacks being malware injection, metadata spoofing, DNS and DDoS attacks, cross-site scripting, SQL injection, and wrapping attack. And, DDoS is the common type of attack among these attacks that has been performed against cloud infrastructure. The impact of DDoS attacks becomes larger

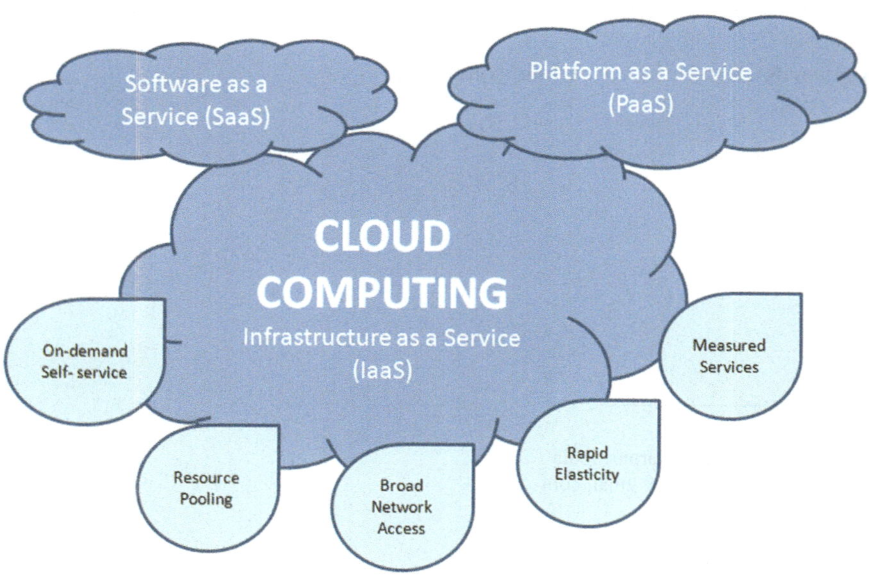

Fig. 1 Cloud computing [5]

and rigid to overlook each year. Though such attacks are rising, various industries have been tried to protect themselves with traditional firewall-based solutions. Alternatively industries had better invested in solutions that provide real protection from unprepared downtime and economic losses.

As with Incapsula Survey, key findings follow [3]:

- 49% of DDoS attacks likely to end amid 6–24 h. It means that with a projected budget of $40,000 per hour, the usual cost of DDoS can be evaluated at about approximately $500,000.
- Budgets are not only constrained to the IT group nonetheless they similarly have a huge impact on risk and security management, sales, and customer service.
- Companies having 500 or more employees are major victim of DDoS attack; experience complex attack costs and involves additional personnel to combat the attack.

Cloud computing is elastic and scalable in nature which allows resources that can be expanded whenever there is demand of more resources. A special kind of DDoS attack is specific to only cloud infrastructure. This is called economic denial of sustainability (EDoS). The main aim of EDoS is to make cloud resources carefully untenable for the victim, whereas DDoS attack focuses on worsen or block cloud services. The time period of DDoS attacks is short while EDoS attacks are more indefinable and performed over a longer time period. EDoS attack takes place just beyond the average movement threshold and beneath the threshold of DDoS attack. Hence, it is tough to be identified by customary systems of intrusion detection and furthermore the procedures practiced to overcome application layer DDoS attacks are not valid to EDoS attack [4]. In this paper, we will evaluate EDoS attacks and several practices to moderate the EDoS attacks.

1.2 Challenges and Issues in Cloud Computing

1. Privacy and Security

The key task to cloud computing is the way it addresses the privacy and security concerns of organizations rational of implementing it. The fact that the crucial enterprise information will exist outside the enterprise firewall increases severe concerns. Hacking and several attacks against cloud infrastructure will probably have an impact on many customers despite the fact that merely one site is subjected to attack.

2. Billing and Service Delivery

Because of the on-demand behavior of the services, it is relatively difficult to measure the costs incurred. Budgeting and valuation of the cost will not be very easy except if the provider proposes comparable and up right benchmarks. The service-level agreements (SLAs) of the supplier are not sufficient to assure the scalability and accessibility. Organizations will be reluctant to shift to cloud without any surety of a high quality of service.

3. Manageability and Interoperability

Organizations must have the control of moving inside and outside of the cloud and swapping providers each time they need, and there should not be any lock-in period. The services of cloud computing should be capable of integrating easily using the on-premise IT.

4. Consistency and Accessibility

Cloud providers still fall short in providing constant service; as a result, there are repeated outages. It is essential to check the service being delivered via internal or third-party tools. It is necessary to have policies to organize usage, service-level agreements, strength performance, and corporate reliance of these services.

5. Bandwidth Cost and Performance

Enterprises can cut back hardware costs but then they need to expend further for the bandwidth. This could be a less cost for the small applications; however, it can be considerably big for the applications that are data-intensive. Appropriate bandwidth is necessary to provide concentrated and composite data across the network. Due to this reason, several organizations are waiting for a lesser cost prior to shifting to the cloud.

1.3 DDoS Attack

Distributed denial of service (DDoS) can be described as an aim to create a machine or network resources unavailable to legitimate users. This attack restrains the availability of resources. It is kind of denial-of-service (DoS) attack where numerous compromise systems usually are contaminated with viruses specially Trojan Horses which are used to aim single system. DDoS attacks are different from that of DoS attacks in such a way that DDoS encompasses multiple systems to attack victim. The widely popular DDoS attacks on Amazon, Yahoo, ebay, and numerous popular Web sites in February 2000 exposed weakness of still fine equipped network and massive Internet users. DDoS has turn out to be a main risk to the entire Internet users. There are various DDoS available tools which can be used with purpose to attack any Internet user. DDoS harms are likely to grow to be more ruthless in future in comparison to other attacks as there may be short of valuable solutions to protect these attacks. Behind major DDoS attacks are botnets and other new emerging DDoS techniques. The botnet makes use of flooding to block the availability of the resources of benign user. Among all prevailing attack weapons, flooding packets are mainly general and efficient DDoS approach. This attack is different from other attacks because it deploys its weapons in "distributed way" across the Internet. The main aim of DDoS is to harm a victim either for individual reasons, for material gain, or to gain popularity. Enormously high-level, "user-friendly" and prevailing DDoS tool kits are accessible to attackers which rise the threat of becoming a sufferer in a DoS or a DDoS attack. The straightforward logic structures and small memory size of DDoS attacking programs make them comparatively simple to employ and hide. There are various detection and mitigation techniques available for preventing DDoS attack. One of the major challenges is the data to be protected from the attacks like DDoS. The data presently is stored in data centers of clouds. Therefore, it is very important to protect data and prevent attacks like DDoS.

DDoS can be categorized into three types [6] and represented by Fig. 2.

I. Attacks targeting network resources
II. Attacks targeting server resources
III. Attacks targeting application resources.

Attacks targeting network resources: The attacks aim for network resources making a struggle to exploit entire bandwidth of a victim's network by applying a vast size of illegal traffic to infuse the corporation's Internet pipe.

Attacks targeting server resources: The attacks aim at server resources making an effort to break down a server's processing proficiency or recollection, which possibly results in denial-of-service state. The scheme of an attacker is to take advantage of an existing exposure or a fault in a communication protocol in a way which aims the target server to turn out to be busy for executing the illegal requests so that it does not have enough resources anymore that it can handle legal request.

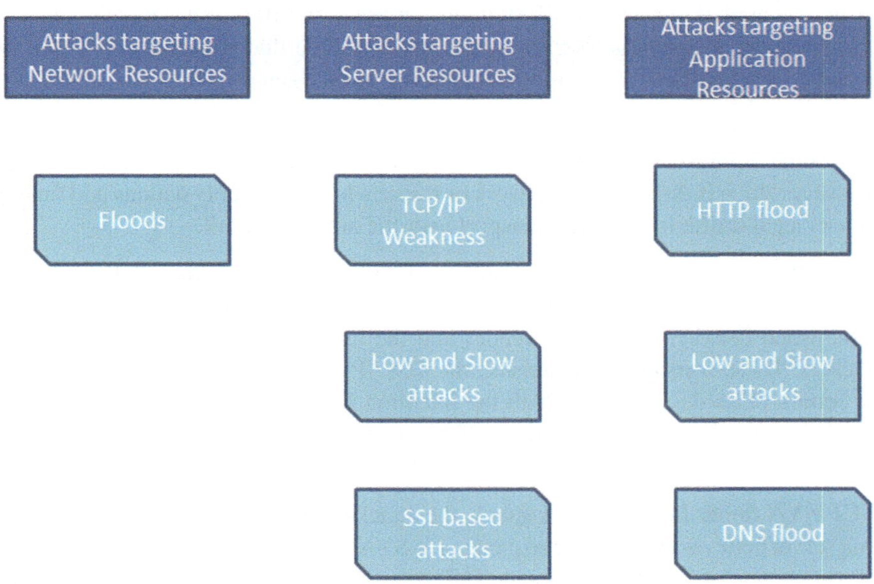

Fig. 2 DDoS attack types [7]

Fig. 3 Types of flood attack [7]

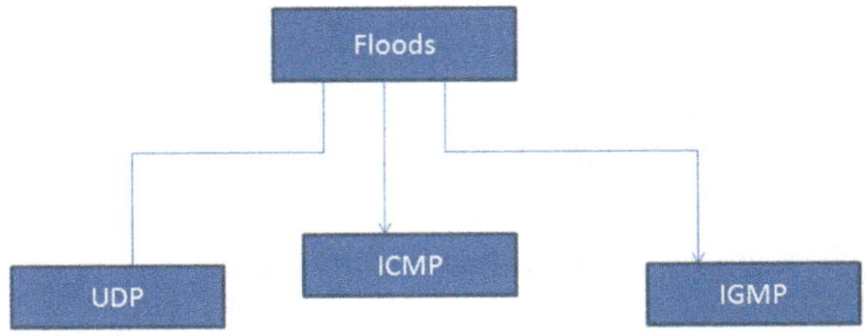

Attacks targeting application resources: The attacks which not only target the Hypertext Transfer Protocol (HTTP), but also other important protocols such as SMTP, HTTPS, FTP, DNS, and VOIP and also the other application protocols which acquire vulnerable weaknesses which can be used for DoS attacks.

Floods: Types of floods are represented by Fig. 3.

UDP: A User Datagram Protocol (UDP) flood attack is that which simply corrupts the normal behavior of victim at a great sufficient level which causes network congestion for the victim network instead of exploiting a specific vulnerability. Attacker sends a large number of UDP packets to random ports on a target server, and the target server is not capable that it processes each request which leads to utilization of its entire bandwidth by attempt to send ICMP "destination unreachable" as a reaction to each spoofed UDP packets to make sure that there was no listening of application on the objected ports.

ICMP Flood: An Internet Control Message Protocol (ICMP) flood is a non-defenselessness-based attack as it does not depend on some certain susceptibility to attain denial of service. An ICMP flood comprises ICMP message of echo request which is sent to the target server as quick as possible that it becomes affected to process all requests which result in a denial-of-service state.

IGMP Flood: An Internet Group Management Protocol (IGMP) deluge is also non-vulnerability-based attack. This flood attack comprises gigantic sum of IGMP message which is directed to a network or router which noticeably detains and finally blocks legal traffic from being transport over the aimed network.

TCP/IP weaknesses: TCP/IP is connection-based protocol unlike UDP and other which are connectionless protocols which means that there should be a full connection established between the packet sender and the intended recipient for sending the packets. These sorts of attacks misuse the TCP/IP procedure by compelling the use of certain of its design flaws. With the intention to dislocate the standard methods of TCP traffic, attacker misuses the TCP/IP protocol's six control bits such as URG, SYN, ACK, RST, and PSH. This is represented by Fig. 4.

TCP SYN flood: In this type of attack, the attacker approached the server in a way that server believes that they are requesting to SYN for legal connections with the help of a sequence of TCP requests with TCP flags set which is in fact appearing from spoofed IP addresses. The victim server opens threads and assigns corresponding

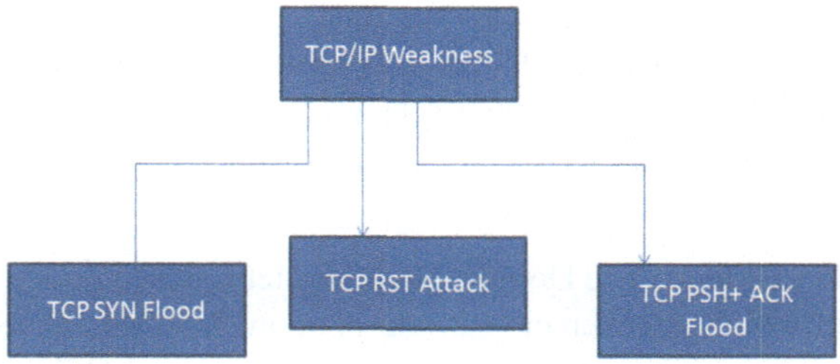

Fig. 4 Types of TCP/IP weakness [7]

buffers so that it can arrange for a connection for handling the each of the SYN requests.

TCP RST attack: In this type of attack, the attacker inhibits amid an active TCP joining among two end points by supposing the present-day system number and forging a TCP RST packet to utilize the IP of client's source which is formerly directed to the server. A botnet is classically utilized to direct thousands of such packets to the server with dissimilar series numbers, which makes it equally tranquil to estimate the exact one. As soon as this happens, the server recognizes the RST packet directed by the attacker, dismissing its association to the client positioned at the forged IP address.

TCP PSH + ACK flood: If a TCP transmitter transmits a packet whose PUSH flag is set to 1, then the outcome pressures the getting server to unoccupied its TCP stack buffer and to refer a byline when this act is comprehensive. The attacker typically uses a botnet to overflow an aimed server with various such requests. This act terminates the TCP stack buffer on the aimed server which causes the server not able to course the legal request or even acknowledge them which eventually roots the denial-of-service condition.

SSL-based attacks: As common services are moving to secure socket layer (SSL) for taming security and address privacy concerns, DDoS events on SSL are also on upswing. SSL is a technique of encryption which is used by many network communication protocols. It is used to offer safeguard to users interconnecting above former protocols by encrypting their interconnections and verifying interconnecting parties. DoS attacks based on SSL can occur in various methods such as harming definite tasks associated to the negotiation process of SSL encryption key, aiming handshake mechanism of the SSL or directing trash data to the SSL server. SSL-based DoS attacks can also be introduced above SSL-encrypted traffic which make it enormously hard to identify. SSL attacks are getting famous because every SSL handshake session utilizes 15 times more server-side resources than the user side. Hence, such attacks are uneven as it takes extensively additional resources of the server to compact with the attack than it does to introduce it.

HTTP flood: An HTTP flood is the DDoS attack which targets the application resources. Attacker exploits the seemingly legal HTTP GET or POST request for attacking the application or Web server. HTTP flood attacks are volumetric attacks and they often use botnet for attack like attack is launched from multiple computers that constantly and repetitively request to download the site pages of the target (HTTP GET flood) which exhaust the resources of application and hence causing a denial-of-service state. They are difficult to detect as it requires less bandwidth to bring down the server than any other attacks.

DNS Flood: The Domain Name System (DNS) floods are symmetrical DDoS attack in which attacker targets one or more than one DNS server. These attacks try to exhaust server-side entity such as memory or CPU with a flood of UDP requests, generated by scripts running on several compromised botnet machines. It is based on the similar impression as former flooding attacks; a DNS flood aims the DNS application procedure by directing a large volume of DNS requests, the DNS server weighed down and incapable to respond to all of its incoming requests, therefore ultimately crashes. The DNS is the procedure utilized to resolve domain names into IP addresses and its fundamental procedure is UDP which takes the benefit of quick request and response intervals without the overhead of having to create connections.

"Low and Slow" attacks: This "low and slow" attack is more related to particularly application resources. These "low and slow" attacks can be launched from a single computer with no other bots as they are not volumetric in nature. They can target specific design flaws or vulnerabilities on a target server with a relatively small amount of malicious traffic, eventually causing it to crash. Additionally, these attacks happen on the layer of application, a TCP handshake is established by this time, effectively making the malevolent traffic appear like regular traffic traveling above a valid connection.

1.4 Economic Denial of Sustainability Attack

The general design of an EDoS attack is to make use of cloud resources without paying for it or to halt the economic drivers of using services of cloud computing. The goal of EDoS attack is to make the cloud cost model unsustainable and therefore making a company no longer capable to affordability use or pay for their cloud-based infrastructure. This is also called cloud-based denial-of-service attacks [8]. The general idea of prevention of EDoS attack is represented by Fig. 5.

Cloud computing follows the model of service where clients are charged on the basis of the practice of cloud's resources. The pricing model has altered the problem of DDoS attack in the cloud to an economic one identified as EDoS attack. The objective of an EDoS attack is to divest the consistent cloud users of their long-term financial capability. An EDoS attack becomes successful when it puts economic liability on the cloud user. For instance, attackers who pretend to be authorized users constantly make requests to a Web site hosting in cloud servers with a motive to consume bandwidth, and the burden of the bill falls on the cloud user who is the owner of the Web site. It appears to the Web server that this traffic does not extent

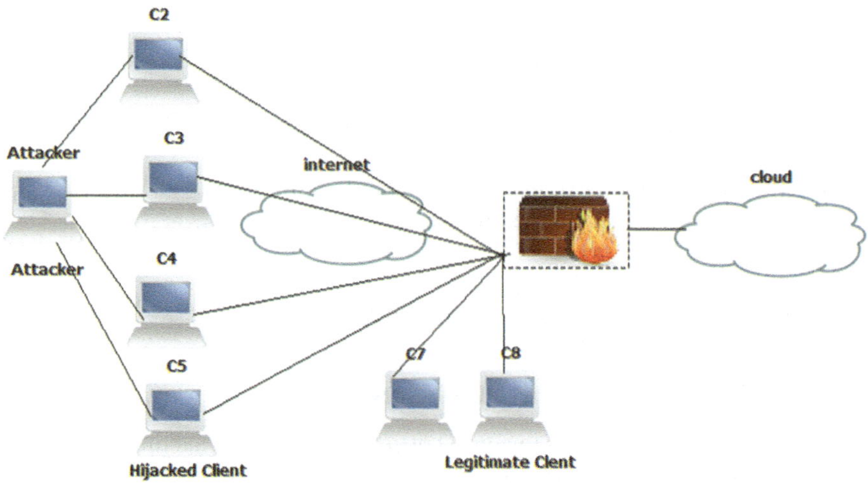

Fig. 5 Prevention of EDoS attack

the service denial level, and it is not easy to differentiate between EDoS attack traffic and legitimate traffic.

If client cloud-based service is intended to upgrade mechanically (such as Amazon EC2), now an attacker can cause financial grief by making large number of automatic requests that seem to be valid externally, however are forged in reality. Client charges will increase as you expand, consuming additional and/or bigger servers (mechanically) to respond to those forged requests. Eventually you will get to a point where your charges go beyond your capability to make payment, i.e., a point where your financial sustainability becomes uncertain.

Many organizations choose cloud infrastructure because of the following reasons:

- Business performance resourcing (compute services)
- Improve employee and partner productivity (Collaboration, QoS)
- Self service and on-demand IT service deliver
- Business Agility (adaptability, simplicity)
- Reduce/optimize cost
- Unlimited capacity (storage).

Service-level agreement (SLA) in cloud works among user and source of the service. When customer instigates request to cloud, then SLA delivers the service conferring to the anticipation of user, i.e., offers the guarantees, service duties, and warranties, and likewise lays down the accessibility and enactment of the service. Client can outspread services that he gains, at whatever stage in cloud structure because of the capability of elasticity.

The cloud service provider's quality and performance can be measured by SLAs in several ways. Certain factors that SLAs could define consist of [9]:

- Accessibility and uptime—the proportion of the time amenities will be accessible

- The amount of synchronized customers that can be assisted
- Specific standards of performance to periodically compare the actual performance
- Response time of application
- The program for notification of network changes in advance that could affect clients
- Response time of help desk for several modules of problems
- The usage statistics that will be offered.

1.5 Difference between DDoS and EDoS.

The difference between EDoS and DDoS attacks are [4].

- The objective of an EDoS is to create cost-effective unsustainability in the cloud resources for the object, while the objective of DDoS attack is to damage or block the facilities of cloud.
- DDoS attacks are capable in a short span of time, and, on the other hand, EDoS attacks are milder and completed in a stretched time span.
- EDoS attack takes place beyond the usual movement edge and beneath the edge of DDoS attack. Thus, it might not be likely to detect it by the help of customary intrusion detection system. Moreover, the approaches employed against application layer and DDoS attacks are not relevant in case of an EDoS attack.

2 DDoS Mitigation Methodology

See Tables 1 and 2.

3 EDoS Mitigation Methodology

See Table 3.

4 Conclusion

Cloud computing allows us to scale our servers up and up in order to provision greater amounts of requests for service. This unlocks a new walk of approach for attackers, known as economic denial of sustainability. DDoS is usually easy to spot given vast upsurges in traffic. EDoS attacks are not essentially easy to detect, because the arrangement and business logic are not present in most applications or masses of applications and infrastructure to provide the connection between requests and

Table 1 Security issues in cloud environment [10–13]

Security issue	Description	Related attack/intrusion/difficulty
Data security	Data stored in cloud database need to be in encrypted format and must not be accessed by other tenants	Data/information breach, unauthorized access to data
Data location	Actual geographical location of storage of data that belongs to cloud client is unknown to the cloud client	Different locations may have different laws and rules, and confliction occurs while determining ownership of the malicious data
Data segregation	Data separation needs to be maintained in host machines that are shared by various clients	Unauthorized access to data of co-resident client
Data integrity	Refers to accurate and consistent database in cloud since the access to the data can be from any device at any time	Inconsistent database
Data confidentiality	Stored data should be compliance with security and privacy policies and terms	Data/information breach
Availability	Services from cloud provider should be available to the client all the time without any downtime	Denial-of-service attacks, flood attacks
Authentication and authorization	Database of user credentials should be kept secure and user access levels must be defined and followed accurately	Insider attack, user to root attack
Privileged user access	Different levels of users require different level of access	Unauthorized user access
Regulatory compliance	To avoid legal issues, cloud user and provider should comply with terms and conditions	Difficulties in legal matters and crime investigations
Recovery and backup	Lost data must be recoverable, and backup is taken for sensitive data	Permanent data loss due to natural disaster or successful attack
Network security	Traffic flowing through network layer should be encrypted with techniques such as TLS or SSL	Packet sniffing
Web application security	Application provided by cloud must not be vulnerable to any security flaw	Service injection attack
Virtualization vulnerabilities	Multi-tenancy may cause troubles since properties of the virtual machines, like isolation, inspection, and interposition may not be followed properly	Blue Pill rootkit, SubVirt, direct kernel structure manipulation (DKSM)

(continued)

Table 1 (continued)

Security issue	Description	Related attack/intrusion/difficulty
Injection vulnerabilities	Design/architectural flaws in the application provided by cloud service may lead to injection attacks	SQL injection, OS injection, LDAP injection
Vulnerabilities in browser APIs	Poor security in handling APIs and vulnerable design of browsers may invite attackers to harm services	SSL certificate spoofing, attacks on browser caches, phishing attacks on mail clients
PaaS-related issues	PaaS providers have to take care of program codes and data related to applications that are being developed in cloud environment	Illegal data transfer, extensive black box testing, attacks on visible code or infrastructure
IaaS-related issues	IaaS provides the computing resources, like storage, RAM, processors to the clients, and security issues related with these resources affect IaaS cloud providers	Reliability of the data stored, trust issues with the provider

Table 2 Comparison of various defense mechanisms of DDoS attack [3, 14, 15]

S. No.	Security mechanism	Benefits	Limitations
1	Filtering of packets (ingress and egress) at edge router of SOURCE	It will perform detection and filtering of packets with spoofed IP addresses at the edge router of source which should be lean on the legal IP address range (used internally in the network)	Spoofed packets might not be discovered if those addresses are covered in the legal IP address range used in the internal network
2	D-WARD	It blocks the attack traffic which is initiated from a network at the boundary of the network's source	More CPU Utilization compared to others
3	MULTOPS	DDOS flooding attacks are detected as well as filtered based on the considerable differentiation between the receiver and transmitter going to and coming from a network node	For observing the packet rates of every IP address, MULTOPS uses a dynamic tree structure which will result in making this a dangerous object of a memory exhaustion attack
4	IP traceback mechanisms	Instead of spoofed IP addresses, it traceback the forged IP packets to their correct sources	These types of mechanisms have heavy computating, network or management overheads. This brings up challenges in the operations and deployments

(continued)

Table 2 (continued)

S. No.	Security mechanism	Benefits	Limitations
5	Packet filtering and marking mechanisms	It marks valid packet at every router alongside with their route to the destination thus the filtering of attack traffic is done by the victim's edge routers	The strength of the attacker is a factor to which this defense mechanism relies on. If the volume grows filters turn out to be ineffective and cannot be installed
6	Increasing backlog	Defends from overflowing a host's backlog of sockets connected	Poor solution for functions, they used linear list traversal. It tries to free the state associated with stale connections
7	SYN Cache	Secret bits in TCP header prevent an attacker from targeting a specific hash value	It is complex in nature
8	Firewalls and proxies	They have policies (inspection) for acting against SYN flooding attacks	attacks can be easily bypassed using advance mechanisms
9	IP-level defense mechanism	It is more devoted to defend SIP servers	Servers are complex to implement and work only at IP level
10	Mitigation on the page access behavior	It is helpful to avoid HTTP-GET flooding attacks	Large False positives

Table 3 Summary of EDoS mitigation techniques [16–18]

Approaches	Methodology	Distributed approach	Learning ability	Limitations
EDoS armor	Packet filtering and authentication	No	Yes	Provide defense only for E-commerce applications
EDoS shield	Virtual firewall and authentication	No	Yes	Does not deal with IP spoofing attacks
Enhanced EDoS shield	Graphical turing test and TTL	No	Yes	–
sPoW	Packet filtering, crypto-puzzle	Yes	Yes	Prevents only network-level EDoS attack

(continued)

Table 3 (continued)

Approaches	Methodology	Distributed approach	Learning ability	Limitations
Cloud traceback	Packet marking and traceback	Yes	Yes	Does not deal with IP spoofing
Cloud watch	Traffic monitoring	Yes	No	Incompetent solution counter to EDoS because user can be still charged for over exploitation of resources
In-cloud scrubber	Authentication through crypto-puzzle	No	Yes	Authentic user is reluctant to resolve such problems; thwarts merely network-level EDoS attacks
DDoS mitigation system	Graphical turing test, crypto-puzzle	No	Yes	Does not covenant with IP packet disintegration, does not covenant with dynamic IP addresses
Digital signatures	Digital signature generation and verification	Yes		Some digital signing processes can be computationally intensive, slowing down business processes and limiting their ability to scale

successful transactions. Current mitigation methodology for DDoS attack and EDoS attack that put forward to address was reviewed in this paper. Machine learning techniques are required for preventing the attack. Therefore, this paper reviews all the aspects of DDoS and EDoS attack [5, 19].

References

1. Sabahi F (2011) Cloud computing security threats and responses. In: 2011 IEEE 3rd international conference on communication software and networks (ICCSN), pp 245–249
2. Zissis D, Lekkas D (2012) Addressing cloud computing security issues. Future Gener Comp Syst 28(3):583–592
3. Incapsula Survey, What DDos attacks really cost businesses

4. Sukhada Bhingarkar A, Deven Shah B (2015) A survey: securing cloud infrastructure against EDoS attack. In: International conference on grid & cloud computing and applications (GCA'15), pp 16–22
5. Abbasi H, Ezzati-Jivan N, Bellaiche M, Talhi C, Dagenais M (2019) Machine learning-based EDoS attack detection technique using execution trace analysis. J Hardware Syst Secur. https://doi.org/10.1007/s41635-018-0061-2
6. Rajkumar MN (2013) A survey on latest DoS attacks: classification and defense mechanisms. Int J Innov Res Comput Commun Eng 1:1847–1860
7. Egress filtering [online]. Available: www.whatis.techtarget.com/definition/egress-filtering
8. EDoS, https://www.elasticvapor.com/2009/01/cloud-attack-economic-denial-of.html
9. Bianco P, Lewis GA, Merson P (2008) Service level agreements in service—Oriented architecture environments, CMU/SEI-2008-TN-021, September 2008
10. Fernandes, Diogo AB, et al (2014) Security issues in Cloud environment: a survey. Int J Inf Secur 13(2):113–170
11. Gartner: Seven Cloud Computing Security Risks, Networkworld, [online]. Available: https://www.networkworld.com/article/2281535/data-center/gartner-seven-Cloud-computing-security-risks.htnl
12. Subashini S, Kavitha V (2011) A survey on security issues in service delivery models of Cloud computing. J Netw Comput Appl 34(1):1–11
13. Modi C, Patel D, Borisaniya B, Patel A, Rajarajan M (2013) A survey on security issues and solutions at different layers of Cloud computing. J Supercomput 63(2):561–592
14. Singh PK, Bhargava BK, Paprzycki M, Kaushal NC, Hong WC (2020) Handbook of wireless sensor networks: issues and challenges in current scenario's. In: Advances in intelligent systems and computing, vol 1132. Springer, Cham, Switzerland, pp 155–437
15. Singh PK, Kar AK, Singh Y, Kolekar MH, Tanwar S (2020) Proceedings of ICRIC 2019, Recent innovations in computing, 2020, Lecture Notes in Electrical Engineering, vol 597. Springer, Cham, Switzerland, pp 3–920.
16. Singh P, Manickam S, Rehman SUI (2014) A survey of mitigation techniques against Economic Denial of Sustainability (EDoS) attack on cloud computing architecture. In: Reliability, Infocom technologies and optimization (ICRITO) (trends and future directions), 2014. IEEE, pp 1–4.
17. Network Ingress Filtering: Defeating Denial of Service Attacks which employ IP Source Address Spoofing [online]. Available: www.tools.ietf.org/html/rfc282
18. Singh P, Paprzycki M, Bhargava B, Chhabra J, Kaushal N, Kumar Y (2018) Futuristic trends in network and communication technologies. In: FTNCT 2018. Communications in computer and information science, vol 958, pp 3–509
19. Nautiyal S, Wadhwa S (2019) A comparative approach to mitigate economic denial of sustainability (EDoS) in a cloud environment. in: 2019 4th international conference on information systems and computer networks (ISCON), Mathura, India, pp 615–619. https://doi.org/10.1109/ISCON47742.2019.9036257
20. Amita (2015) EDoS-shield—a mitigation technique against EDoS attacks in cloud computing. Int J Eng Res Technol 4(05):795–797
21. Rameshbabu J, Balaji B, Daniel RW, Malathi K (2014) A prevention of DDoS attacks in cloud using NEIF techniques. Int J Sci Res Public 4(4):1–4
22. Yu S, Tian Y, Guo S, Wu DO (2014) Can we beat DDoS attacks in clouds. IEEE Trans Parallel Distrib Syst 25(9):2245–2254
23. Nam SY, Djuraev S (2014) Defending HTTP web servers against DDoS attacks through busy period-based attack flow detection. KSII Trans Internet Inf Syst 8(7):2512–2531
24. Darwish M, Ouda A, Capretz LF (2013) Cloud-based DDOS attacks and defenses. In: IEEE international conference, pp 67–71
25. Vashisht S, Kaur M (2015) Study of cloud computing environment, EDoS attack using CloudSim. Int J Adv Res Comput Sci 6(5):181–184
26. Shangytbayeva GA, Karpinski MP, Akhmetov BS (2015) Mathematical model of system of protection of computer networks against attacks DOS/DDOS. Mod Appl Sci 9(8)

27. Strom S (2015) Global information assurance certification paper. As part of GIAC practical repository, December 2015
28. CloudComputing, https://www.google.co.in/search?q=cloud+computing&biw=1366&bih=613&source=lnms&tbm=isch&sa=X&sqi=2&ved=0ahUKEwiI1IKsyb7RAhUhDcAKHWTEChAQ_AUIBygC#imgrc=sOyFCX-Gf5M08M%3A. Accessed on 25 July 2016
29. Ingress filtering [online]. Available: www.whatis.techtarget.com/definition/ingress-filtering
30. Poongodi M, Hamdi M, Sharma A, Ma M, Singh PK (2019) DDoS detection mechanism using trust-based evaluation system in VANET. IEEE Access 7:183532–183544. https://doi.org/10.1109/ACCESS.2019.2960367

Preprocessing Steps for Opinion Mining on Tweets

Arpita, Pardeep Kumar, and Kanwal Garg

Abstract In current scenario, high accessibility to computational facilities encourage generation of large volume electronic data. Expansion of the data persuades researchers toward critical analyzation so as to extract maximum possible patterns for wiser decisiveness. Such analysis requires curtailing of text to a better structured format by preprocessing. This scrutiny focuses on implementing preprocessing in two major steps for textual data generated by dint of Twitter API. A NoSQL, document-based database named as MongoDB is used for accumulating raw data. Thereafter, cleaning followed by data transformation is executed on accumulated tweets. The results for this research demonstrate that even after application of cleaning, text contains some anomalies. Henceforth, data transformation follows the process of cleaning for this research. Ultimately, a preprocessed data suitable for opinion mining is generated.

Keywords Tokenization · Lemmatization · Part-of-speech tagging · Cleaning

1 Introduction

Social media brings people together so that they can generate ideas or share their experiences with each other. The information generated through such sites can be utilized in many ways to discover fruitful patterns. But, accumulation of data via such sources creates a huge unstructured textual data with numerous unwanted formats

Supported by Kurukshetra University, Kurukshetra.

Arpita · P. Kumar · K. Garg (✉)
Department of Computer Science and Applications, Kurukshetra University, Kurukshetra, India
e-mail: gargkanwal@kuk.ac.in

Arpita
e-mail: arpitagrover05@kuk.ac.in

P. Kumar
e-mail: pkmittal@kuk.ac.in

P. K. Singh et al. (eds.), *Evolving Technologies for Computing, Communication and Smart World*, Lecture Notes in Electrical Engineering 694,
https://doi.org/10.1007/978-981-15-7804-5_25

[16]. Henceforth, the first step of text mining involves preprocessing of gathered reviews.

The journey of transforming dataset into a form, an algorithm may digest, takes a complicated road [21]. The task embraces four differentiable phases: cleaning [9], annotation [7], normalization [4], and analysis. The step of cleaning comprehends extrication of worthless text, tackling with capitalization and other similar details. Stop words [2], punctuations marks, URLs, and numbers are some of the instances which can be discarded at this phase. Annotation is a step of applying some scheme over text. In context to natural language processing, this includes part-of-speech tagging. Normalization demonstrates reduction of linguistic. In other words, it is a process that maps terms to a scheme. Basically, standardization of text through lemmatization [20] and stemming [1] are the part of normalization. Finally, text undergoes manipulation, generalization, and statistical probing to interpret features [23].

For this study, preprocessing is accomplished in three major steps, as signified in Fig. 1, keeping process of sentiment analysis in consideration. Foremost step included collection of tweets from Twitter by means of Twitter API. Captured data are then stored in a NoSQL database known to be MongoDB. Thereafter, collected tweets underwent cleaning [25] process. Cleaning phase incorporated removal of user name, URLs, numbers, punctuations, special characters along in addition to lower casing and emoji decoding. Endmost stage involved data transformation comprising of tokenization [17], stop word removal [6], part-of-speech tagging [3], and lemmatization [15].

The remaining paper is organized as follows: Sect. 2 includes discussion of various author's work in concerned arena. Further, entire methodology for preprocessing of data opted for this research is postulated in Sect. 3. Then, the results generated through implementation of code mentioned in Sect. 3 are scrutinized utterly in Sect. 4. Thereafter, Sect. 5 provides conclusion of entire work.

2 Related Work

Many studies centered around the issue of preprocessing for text mining are scrutinized in this section.

Srividhya and Anitha [19] have put forward that preprocessing techniques play a major role in reducing the feature space to bring a considerable rectification to the performance metrics for final text classification. The work is dedicated to mainly three approaches namely stop word removal, stemming, and term frequency-inverse document frequency, whereas the focal points of Hemalatha et al. [8] in their research were removal of URLs, removal of special characters, and removal of questions as these forms of texts do not contribute in any way for determination of polarity. Further, a hybrid algorithm combining TF-IDF and SVD has been introduced by Kadhim et al. [10] for preprocessing the text document in 2014. The ultimate goal

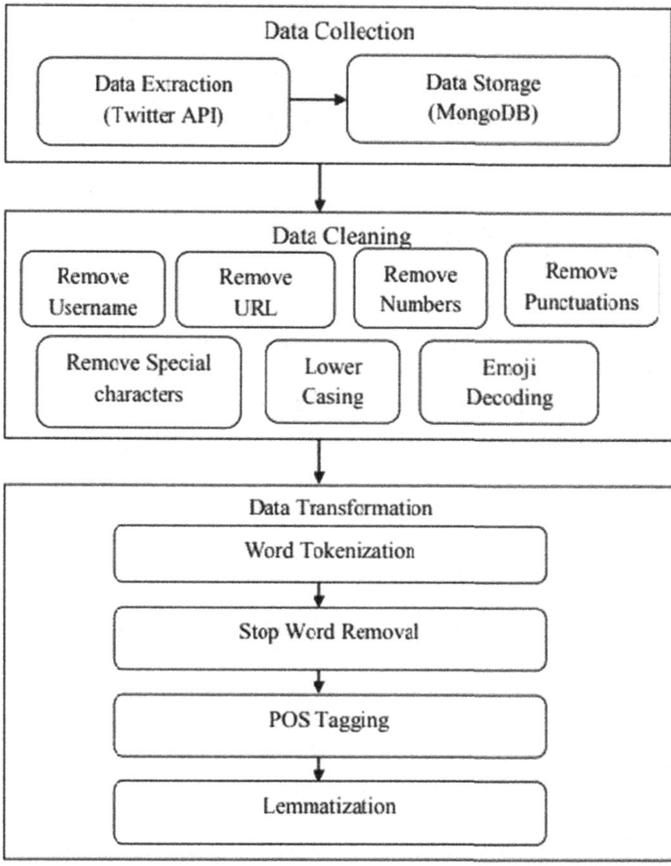

Fig. 1 Preprocessing steps

of their research was dimensionality reduction of feature vector space so as to intensify accuracy in results of clustering. In addition, Kannan and Gurusamy [11] have acknowledged role of preprocessing for efficient information retrieval from text document in their scrutiny. Tokenization, stop word removal, and stemming were the three major techniques which were spotlighted in their research. Later, Vijayarani et al. [22] have given an overview of techniques for preprocessing the text before applying mining approaches to extract useful information so as to reduce the dimensionality of feature space. However, stemming process of preprocessing technique has been centralized by Nayak et al. [18]. MF Porter and Krovetz algorithms of stemming have been analyzed. The survey has put forward some areas of improvement for both algorithms. Moreover, Krouska et al. [13] have empirically visualized the impact of distinct preprocessing approaches over ultimate classification of tweets using four well-known classifiers, viz. Naive Bayes (NB), C4.5, K-nearest neighbor (KNN), and support vector machine (SVM).

All these studies focus a very little on overall implementation of cleaning and transformation steps as whole in context to sentiment mining. Therefore, the focal point of this research is integrated administration of entire procedure for preprocessing of textual data in respect to sentiment analysis.

3 Research Methodology

To carry out present research work, a primary dataset generated through Twitter API is gathered so as to perform an analysis on live tweets. Thereupon, entire coding is done in Python 3.7 on Jupyter notebook. In addition to this, NLTK toolkit for data transformation and entropy model trained on tagset of Penn Treebank for POS tagging were used which are mentioned in Sect. 3.3.

3.1 Data Collection

Data are collected in form of tweets from Twitter using Twitter API in a NoSQL environment, named as MongoDB.

Twitter API One sort of Twitter API called as streaming API [5] is used to collect data for this study as it benefits with retrieval of huge amount of recent data. Moreover, it helps in real-time inspection, such as ongoing social discussion related to a particular entity [12]. The Twitter streaming access has a keyword parameter which restricts domain of collected data. For this work, that keyword parameter was set to Narendra Modi, Hon'ble Prime Minister of India. Furthermore, streaming access has a language parameter whose language code was set to "en" for fetching only English tweets. It uses streaming response of HTTP to accommodate data.

Database Tweets streamed through Twitter API are stored in MongoDB. MongoDB is an open-source NoSQL document database [14]. To capture tweets into MongoDB collections, foremost step is to set up an environment. "Pymongo" module was installed for this intent. Thereupon, MongoClient was instantiated to establish connection with MongoDB. Then ultimately, a database named "Twitter-API" and a collection entitled "Tweet" were created. Data streamed with Twitter Streaming API was stored in this collection.

3.2 Data Cleaning

Social media sites like Twitter generate a huge volume of data. This raw data can be scrutinized to interpret many interesting facts. But, study of such bulky data can prove to be a nasty piece of work without right procedure. Henceforth, for mining propitious patterns out of this huge pile of textual data, foremost obligation is to have an insight into collected data [24]. It is a crucial step so that characteristics of dataset can be explored justly. The concrete understanding of data helps in identification of incompetent content in correspondence to the patterns that need to be mined. In reference to sentiment analysis, this research focuses on emoji decoding, lower casing, removal of user name, URLs, punctuations, special characters, and numbers. Algorithm 1 represents implementation of cleaning process and its output is shown in Fig. 3. The algorithm has encapsulated its entire task in seven functions. "userRemove" is for removal of user names from text, "uRemove" works on removal of URLs, "nJoin" operates on numbers, "pJoin" deals with punctuations in text, "sRemove" abolishes special characetrs, and "cJoin" manages casing while "eJoin" taken care of emoji decoding.

User name People mention their user names with @ symbol while tweeting. The inclusion of user name in text serves no purpose to sentiment analysis. Thus, presence of such data in text does nothing but diminishes the accuracy of classifier. Hence, it is advantageous to get rid of these phrases at preprocessing stage itself. Taking this into consideration, a system was developed to eradicate from mentioned user names in text field.

URLs Persistence of URL till the classification step may result in false prediction. For example, the sentence "Ridhima signed in to www.happy.com" is a neutral sentence but due to the presence of word happy it may leave classifier with positive prophecy. For this reason, to dodge failures of this type, a technique has been employed for weeding out URLs.

Algorithm 1 Clean(text)

```
Input: ìtextî from ìcolî collection
Output: txt
for x in col.find() do
    txt  x[ítextí]
end for
for i in txt do
    txt  userRemove(í@[àns]+í,i)
    txt  uRemove(ìhttp?://[A-Za-z0-9./]+î,i)
    txt  nJoin(i in txt if not i.isdigit())
    txt  pJoin(string.punctuation, i)
    txt  sRemove(rì[àa-zA-Z0-9]+î,i)
    txt  cJoin(i.lower())
    txt  eJoin(emoji.demojize(i))
end for
return txt
```

Numbers As existence of numbers in opinionated text is of no use for polarity determination, it should be removed from the text collected so as to reduce the dimensionality. This way, approach of preprocessing used for elimination of numbers helps with retrieval of accurate results. Hence, numerical terms are abolished from gathered data.

Punctuations English Grammar comprises of commonly fifteen punctuations. These include exclamation mark (!), period (.), question mark (?), comma (,), semicolon (;), colon (:), dash (?), hyphen (-), slash (/), parenthesis (()), brackets ([]), braces (), apostrophe ('), quotation mark (""), and ellipsis (...). All these punctuations increase dimensionality of text inspite of the fact that they are least important for sentiment analysis. Therefore, a method was designed to obliterate punctuations from data.

Special Characters Presence of special characters of sort $%&#' may lead to discrepancy at the time of alloting polarity to sentiments. For example, in phrases like "#bad", many times special characters are taken in unison with words and as a result the word seems unavailable in dictionary to classifier. On account of this problem, a mechanism is applied for removal of special characters.

Lower Casing Textual data usually comprise of diverse capitalization to reflect either starting point of a sentence, or to emphasize the nouns properly. Now, the problem lies in a fact that for computer capital and lowercase letters are two different characters. For example, "A" and "a" are two distinct articles for machine. As its consequence, while stop word removal, words like "The" are treated different than "the". Resultantly, it remains intact in the document. For this reason, it is favorable to translate entire text to either lower or upper case. Thereby, here "lower()" method was implemented to transform overall data into lowercase.

Emoji Decoding Though emoticons are forged with non-alphabets, yet their contribution in sentiment analysis plays a major role. The entire set of emojis if operated appropriately may end up generating better results in process of mining opinions. These minute details may even be beneficial for classification of sarcastic and non-sarcastic text. Consequently, emojis were decoded to their actual meanings in textual form employing "demojize()" function.

3.3 Data Transformation

Even after execution of cleaning process, data are not in a form that can be passed for classification. It is just a lump of characters. Sentiment retrieval from this pile of text requires identification of logical words. For this reason, the process of data transformation is implied next. The phase of transformation was implemented in four sequential parts: tokenization, stop word removal, part-of-speech tagging, and lemmatization. A view of implementation for transformation is presented in Algorithm 2, and its resultant is demonstrated in Fig. 5. The algorithm first takes output generated by implementation of Algorithm 1. Then, transformation of text-to-word tokens is followed by removal of stop words. Subsequently, data undergo part-of-speech tagging. Finally, lemmas are removed from the data generating a preprocessed data suitable for ultimate task of classification.

Tokenization In the beginning phase of data transformation, there is a need of parser for tokenization in document. Henceforth, goal of tokenization steps in preprocessing is to explore existent words in a phrase. Accordingly, it can be termed as a process of working on a streamed text to convert it into worthwhile elements known to be tokens. A token may comprise of a phrase, an idiom, or a symbol. These tokens are then passed on for next level preprocessing. From tokenize class, word tokenize () function is used to carry out this step. Although tokenization is the first essential step of data transformation yet there has to be further scrutiny of resultant text to make it suitable for final analysis.

Stop Word Removal Prepositions, articles, connectors, pronouns, etc., are the most frequently used word forms in a textual document. All such words are considered to be stop words. Abundant occurrence of these words makes a document bulkier and gradually lowers its importance for analysts. Therefore, dictionary of NLTK toolkit is used to rip these words out of the document. Consequently, dimensionality of text is reduced considerably.

Algorithm 2 Transform(text)

```
Input: Clean(text)
Output: txt
txt     Clean(text)
words    tokenize.word_tokenize(txt)
stopWords    set(stopwords.words(íenglishí))
WordsFiltered       []
for w in words do
    if w not in stopWords then
        wordsFiltered.append(w)
    end if
end for
nltktagged    pos_tag(wordsFiltered)
wordnet lemmatizer    WordNetLemmatizer()
for word in wordsFiltered do
    ss   wordnet lemmatizer.lemmatize(word,pos=ìvî)
    if ss.strip() == word.strip then
        sss   wornet lemmatizer.lemmatize(word,pos=ìaî)
        if sss.strip() == word.strip then
            w_n   wordnet lemmatizer.lemmatize(word,pos=ìnî)
            op.append(sss.strip())
        else
            op.append(sss.strip())
        end if
    else
        op.append(ss.strip())
    end if
end for
txt     ì î.join(op)
return txt
```

Part-of-Speech Tagging Part-of-speech tagging is a process of characterizing each word in textual data to its reciprocal PoS. This correspondence of words is established not solely on the basis of its definition, but the context with which it is used in a sentence is also taken into consideration. Part-of-speech tags include verbs, nouns, adjectives, adverbs, etc. The non-generic trait of POS tagging makes it more complex than basic mapping of words to their POS tags. In correspondence to different context, there is fair probability that a word has more than one PoS tags for distinct sentences. For this scrutiny, PerceptronTagger employing maximum entropy model was used. It implements probability model for tagging. Further, the entropy model was trained with a tagset named Penn Treebank.

Lemmatization Lemmatization is a method used for reduction of inflected words to its root. While reducing inflections of words to its lemmas, lemmatization takes into consideration the morphological meaning of text. Therefore, unlike stemming, lemmatization maps inflected words to only those root words which correspond to the language.

4 Result Analysis

Figure 2 represents all the anomalies mentioned in Sect. 3.2 for collected data, with different colors. User name is highlighted with yellow color, green signifies URLs, purple specifies numbers, blue represents punctuations, gray denotes special characters, diversity in capitalization is shown with red color, and emojis are marked with pink color. Subsequently, the code mentioned for cleaning process in Sect. 3.2

S.No.	text
1.	@AkhileshPSingh @NitishKumar If BJP had less than 100 Lok Sabha seats then I'm sure Nitish Kumar would have kicked Modi & NDA . But now out of frustration he's bound to remain shut & wait for the right moment. At least, they have a rebellion in NDA now for the next 5 years which will help our democracy. https://t.co/sG9fd7pEJu
2.	PM Narendra Modi and India's Most Wanted First Week Box Office Collection https://t.co/1MOkqBR8Hh #PMNarendraModi #IndiasMostWanted
3.	In Washington Post I profile the very controversial Amit Shah who is now running India, a man with an equally checkered past on human rights as his mentor Narendra Modi @PostOpinions https://t.co/o5S3pfYyLv
4.	He lives in a small two room mud house, owns a bicycle and nothing else. Salute to this social worker of Odissa who defeated a billionaire and is now a Minister in the Council of Ministers. He is Mr Pratap Sarangi, MP from Balasore! [illegible] Modi sure picks his Ministers well! [illegible] https://t.co/M9SH8ykDbf
5.	Follow recommendation: If there is one voice in the government which gives you such a clear account of things, @ShamikaRavi is the person. Such a great article. So insightful on the unemployment being at its highest issue.
6.	"If Modi can visit a temple we can visit our mosques, if Modi can go sit in a cave, we Muslims can also proudly say our prayers in mosques," @asadowaisi said. https://t.co/l7RsYkRAK3
7.	Farmers and poor have always been a priority for Modi government. As promised, PM @narendramodi has extend the PM-Kisan yojana to all farmers. Cabinet has also approved a new scheme "Pradhan Mantri Kisan Pension Yojana" to provide pension to crores of small & marginal farmers. https://t.co/JNpfMg69eP
8.	TN was, is and will fight against the Hindi imposition at any cost! Modi govt is playing with fire! Federal structure must be respected as per the Constitution! #TNAgainstHindiImposition #StopHindiImposition
9.	While we read Modi Govt Data showing Joblessness at 45 yr high, Here a data shows [illegible]% drop in farmer suicides in Karnataka. Credit to @hd_kumaraswamy lead @INCKarnataka @JanataDal_S govt for giving confidence to the farmers. A lot to be done though. @divyaspandana https://t.co/vauCJIvgOc

Fig. 2 Highlighted anomalies

S.No.	Text	Cleaned Data
1.	@AkhileshPSingh @NitishKumar If BJP had less than [illegible] Lok Sabha seats then I'm sure Nitish Kumar would have kicked Modi & NDA . But now out of frustration he's bound to remain shut & wait for the right moment. At least, they have a rebellion in NDA now for the next 5 years which will help our democracy. https://t.co/sG9fd7pEJu	if bjp had less than lok sabha seats then im sure nitish kumar would have kicked modi amp nda but now out of frustration hes bound to remain shut amp wait for the right moment at least they have a rebellion in nda now for the next years which will help our democracy
2.	PM Narendra Modi and India's Most Wanted First Week Box Office Collection https://t.co/1MOkqBR8Hh #PMNarendraModi #IndiasMostWanted	pm narendra modi and indias most wanted first week box office collection pmnarendramodi indiasmostwanted
3.	In Washington Post I profile the very controversial Amit Shah who is now running India, a man with an equally checkered past on human rights as his mentor Narendra Modi @PostOpinions https://t.co/o5S3pfYyLv	in washington post i profile the very controversial amit shah who is now running india a man with an equally checkered past on human rights as his mentor narendra modi
4.	He lives in a small two room mud house, owns a bicycle and nothing else. Salute to this social worker of Odissa who defeated a billionaire and is now a Minister in the Council of Ministers. He is Mr Pratap Sarangi, MP from Balasore! [illegible] Modi sure picks his Ministers well! [illegible] https://t.co/M9SH8ykDbf	he lives in a small two room mud house owns a bicycle and nothing else salute to this social worker of odissa who defeated a billionaire and is now a minister in the council of ministers he is mr pratap sarangi mp from balasore folded hands medium light skin tone modi sure picks his ministers well raised fist india
5.	Follow recommendation: If there is one voice in the government which gives you such a clear account of things, @ShamikaRavi is the person. Such a great article. So insightful on the unemployment being at its highest issue.	follow recommendation if there is one voice in the government which gives you such a clear account of things is the person such a great article so insightful on the unemployment being at its highest issue
6.	"If Modi can visit a temple we can visit our mosques, if Modi can go sit in a cave, we Muslims can also proudly say our prayers in mosques," @asadowaisi said. https://t.co/I7RsYkRAK3	if modi can visit a temple we can visit our mosques if modi can go sit in a cave we muslims can also proudly say our prayers in mosques said
7.	Farmers and poor have always been a priority for Modi government. As promised, PM @narendramodi has extend the PM-Kisan yojana to all farmers.	farmers and poor have always been a priority for modi government as promised pm has extend the pm kisan yojana to all farmers cabinet has also approved a new scheme pradhan mantri kisan pension yojana to provide pension to crores of small

Fig. 3 Output of cleaning process

puts an impact on text as portrayed in Fig. 3. Second column of output enlightens all errors which were present in text, while third column represents text free from the noises which were highlighted in second column. Further, red highlights in third column denote text which has been lowercased whereas text in magenta comes up with decoded emojis.

Now, it can be clearly seen from Fig. 4 that the cleaned data are still left with some impurities which need to be considered for better results. For this figure, stop words are highlighted in yellow color while text requiring lemmatization is marked with green color. Section 3.3 specifies the code for transformation techniques to further deal with these anomalies. Then ultimately, Fig. 5 delineates output for code of data transformation. Second column of figure corresponds to Fig. 4. Third column demonstrates preprocessed data with lemmas marked in green.

All these results are represented in tabular format for better visualization, though in real data are stored in json format within MongoDB collections.

S.No.	Cleaned Data
1.	if bjp had less than lok sabha seats then im sure nitish kumar would have kicked modi amp nda but now out of frustration hes bound to remain shut amp wait for the right moment at least they have a rebellion in nda now for the next years which will help our democracy
2.	pm narendra modi and indias most wanted first week box office collection pmnarendramodi indiasmostwanted
3.	in washington post i profile the very controversial amit shah who is now running india a man with an equally checkered past on human rights as his mentor narendra modi
4.	he lives in a small two room mud house owns a bicycle and nothing else salute to this social worker of odissa who defeated a billionaire and is now a minister in the council of ministers he is mr pratap sarangi mp from balasore folded hands medium light skin tone modi sure picks his ministers well raised fist india
5.	follow recommendation if there is one voice in the government which gives you such a clear account of things is the person such a great article so insightful on the unemployment being at its highest issue
6.	if modi can visit a temple we can visit our mosques if modi can go sit in a cave we muslims can also proudly say our prayers in mosques said
7.	farmers and poor have always been a priority for modi government as promised pm has extend the pm kisan yojana to all farmers cabinet has also approved a new scheme pradhan mantri kisan pension yojana to provide pension to crores of small amp marginal farmers
8.	tn was is and will fight against the hindi imposition at any cost modi govt is playing with fire federal structure must be respected as per the constitution tnagainsthindiimposition stophindiimposition
9.	while we read modi govt data showing joblessness at yr high here a data shows drop in farmer suicides in karnataka credit to lead govt for giving confidence to the farmers a lot to be done though interesting thread

Fig. 4 Errors left in cleaned data

S.No.	Cleaned Data	Transformed Data
1.	if bjp had less than lok sabha seats then im sure nitish kumar would have kicked modi amp nda but now out of frustration hes bound to remain shut amp wait for the right moment at least they have a rebellion in nda now for the next years which will help our democracy	bjp less lok sabha seat im sure nitish kumar would kick modi amp nda frustration hes bind remain shut amp wait right moment least rebellion nda next years help democracy
2.	pm narendra modi and indias most wanted first week box office collection pmnarendramodi indiasmostwanted	pm narendra modi indias want first week box office collection pmnarendramodi indiasmostwanted
3.	in washington post i profile the very controversial amit shah who is now running india a man with an equally checkered past on human rights as his mentor narendra modi	washington post profile controversial amit shah run india man equally checker past human right mentor narendra modi
4.	he lives in a small two room mud house owns a bicycle and nothing else salute to this social worker of odissa who defeated a billionaire and is now a minister in the council of ministers he is mr pratap sarangi mp from balasore folded hands medium light skin tone modi sure picks his ministers well raised fist india	live small two room mud house own bicycle nothing else salute social worker odissa defeat billionaire minister council minister mr pratap sarangi mp balasore fold hand medium light skin tone modi sure pick minister well raise fist india
5.	follow recommendation if there is one voice in the government which gives you such a clear account of things is the person such a great article so insightful on the unemployment being at its highest issue	follow recommendation one voice government give clear account things person great article insightful unemployment high issue
6.	if modi can visit a temple we can visit our mosques if modi can go sit in a cave we muslims can also proudly say our prayers in mosques said	modi visit temple visit mosques modi go sit cave muslims also proudly say prayers mosques say
7.	farmers and poor have always been a priority for modi government as promised pm has extend the pm kisan yojana to all farmers cabinet has also approved a new scheme pradhan mantri kisan pension yojana to provide pension to crores of small amp marginal farmers	farmers poor always priority modi government promise pm extend pm kisan yojana farmers cabinet also approve new scheme pradhan mantri kisan pension yojana provide pension crores small amp marginal farmers
8.	tn was is and will fight against the hindi imposition at any cost modi govt is playing with fire federal structure must	tn fight hindi imposition cost modi govt play fire federal structure must respect per constitution

Fig. 5 Output of data transformation

5 Conclusion

The foundation or premise for sentiment analysis is preprocessing of textual data. Only the qualitative data can produce accurate and precise results for legitimate decision making. The paper presents cleaning and transformation steps on data collected in Mon-goDB database via Twitter API. Subsequently, results sketch out the impact of cleaning process on different anomalies encountered in assembled data. Further, it is delineated that the step of cleaning still leaves data with many impurities which need attention for accurate results in later stages of sentiment analysis. Consequently, cleaned data are passed for transformation phase. Therefore, for this research, raw data collected through Twitter are filtered with fine sieve of two processes, i.e., cleaning and transformation. In future, generated preprocessed data can be utilized to find many useful patterns although this research was carried out to prepare data for sentiment analysis.

References

1. Alhaj YA, Xiang J, Zhao D, Al-Qaness MA, Elaziz MA, Dahou A (2019) A study of the effects of stemming strategies on Arabic document classification. IEEE Access 7:32664–32671
2. Aro TO, Dada F, Balogun AO, Oluwasogo SA (2019) Stop words removal on textual data classification
3. Belinkov Y, Marquez L, Sajjad H, Durrani N, Dalvi F, Glass J (2018) Evaluating layers of representation in neural machine translation on part-of-speech and semantic tagging tasks. arXiv preprint arXiv:1801.07772
4. Chua M, Van Esch D, Coccaro N, Cho E, Bhandari S, Jia L (2018) Text normalization infrastructure that scales to hundreds of language varieties. In Proceedings of the eleventh international conference on language resources and evaluation (LREC 2018)
5. Das S, Behera RK, Rath SK et al (2018) Real-time sentiment analysis of twitter streaming data for stock prediction. Procedia Comput Sci 132:956–964
6. Effrosynidis D, Symeonidis S, Arampatzis A (2017) A comparison of pre-processing techniques for twitter sentiment analysis. In: International conference on theory and practice of digital libraries. Springer, pp 394–406
7. Guan R, Wang X, Yang MQ, Zhang Y, Zhou F, Yang C, Liang Y (2018) Multi-label deep learning for gene function annotation in cancer pathways. Sci Rep 8(1):267
8. Hemalatha I, Varma GS, Govardhan A (2012) Preprocessing the informal text for efficient sentiment analysis. Int J Emerg Trends Technol Comput Sci (IJETTCS) 1(2):58–61
9. Henry D, Stattner E, Collard M (2018) Filter hashtag context through an original data cleaning method. Procedia Comput Sci 130:464–471
10. Kadhim AI, Cheah Y-N, Ahamed NH (2014) Text document preprocessing and dimension reduction techniques for text document clustering. In: 2014 4th international conference on artificial intelligence with applications in engineering and technology (ICAIET). IEEE, pp 69–73
11. Kannan DS, Gurusamy V (2014) Preprocessing techniques for text mining. Int J Comput Sci Commun Netw 5(1):7–16
12. Kiyavitskaya N, Zeni N, Mich L, Cordy JR, Mylopoulos J (2006) Text mining through semi automatic semantic annotation. In: International conference on practical aspects of knowledge management. Springer, pp 143–154

13. Krouska A, Troussas C, Virvou M (2016) The effect of preprocessing techniques on twitter sentiment analysis. In: 2016 7th international conference on information, intelligence, systems & applications (IISA). IEEE, pp 1–5
14. Kumar P, Kumar P, Zaidi N, Rathore VS (2018) Analysis and comparative exploration of elastic search, mongodb and hadoop big data processing. In: Soft computing: theories and applications. Springer, pp 605–615
15. Liu H, Christiansen T, Baumgartner WA, Verspoor K (2012) Biolemmatizer: a lemmatization tool for morphological processing of biomedical text. J Biomed Semant 3(1):3
16. Miner G, Elder IV J, Fast A, Hill T, Nisbet R, Delen D (2012) Practical text mining and statistical analysis for non-structured text data applications. Academic Press
17. Mullen LA, Benoit K, Keyes O, Selivanov D, Arnold J (2018) Fast, consistent tokenization of natural language text. J Open Source Softw 3:655
18. Nayak AS, Kanive AP et al (2016) Survey on pre-processing techniques for text mining. Int J Eng Comput Sci 5(6)
19. Srividhya V, Anitha R (2010) Evaluating preprocessing techniques in text categorization. Int J Comput Sci Appl 47(11):49–51
20. Straka M, Strakova J, Hajic J (2019). Czech text processing with contextual embeddings: Pos tagging, lemmatization, parsing and ner. In: International conference on text, speech, and dialogue. Springer, pp 137–150
21. Su C-J, Chen Y-A (2018) Risk assessment for global supplier selection using text mining. Comput Electr Eng 68:140–155
22. Vijayarani S, Ilamathi MJ, Nithya M (2015) Preprocessing techniques for text mining—an overview. Int J Comput Sci Commun Netw 5(1):7–16
23. Virmani D, Taneja S (2019) A text preprocessing approach for efficacious information retrieval. In: Smart innovations in communication and computational sciences. Springer, pp 13–22
24. Woo H, Kim K, Cha K, Lee J, Mun H, Cho S, Chung J, Pyo J, Lee K, Kang M et al (2019) Efficient data cleaning using text clustering for semistructured medical reports: application to large-scale stool examination reports. J Med Internet Res 21(1):e10013
25. Zainol Z, Jaymes MT, Nohuddin PN (2018) Visualurtext: a text analytics tool for unstructured textual data. J Phys Conf Ser 1018:012011 (IOP Publishing)

Machine Learning-Based Lightweight Android Malware Detection System with Static Features

Kavita Jain and Mayank Dave

Abstract With the increased popularity and wide adaptation in the embedded system domain, Android has become the main target for malware writers. Due to rapid growth in the number, variants and diversity of malware, fast detection of malware on the Android platform has become a challenging task. Existing malware detection systems built around machine learning algorithms with static, dynamic, or both analysis techniques use a high dimensionality of feature set. Machine learning algorithms are often challenged by such a large number of feature sets and consume a lot of power and time for both training and detection. The model built using such a large number of features may include an irrelevant or negative feature that reduces the overall efficiency of the detection system. In this paper, we present a lightweight Android malware detection system based on machine learning techniques that use fewer static features to distinguish between malicious and benign applications. To reduce the feature dimensions, we have used the feature engineering approach, which utilizes a multilevel feature reduction and elimination process to create a detection model lightweight. Finally, we have built a machine learning-based detection system on the reduced feature set that performs better in comparison to the model build using the original feature set. The proposed detection system achieves accuracy and precision above 98% and drastically reduces the feature set size from 3,73,458 to 105 features.

Keywords Android · Static analysis · Malware · Security · Machine learning

K. Jain (✉) · M. Dave
Department of Computer Engineering, National Institute of Technology, Kurukshetra, India
e-mail: kv1jn7@gmail.com

M. Dave
e-mail: mdave@nitkkr.ac.in

P. K. Singh et al. (eds.), *Evolving Technologies for Computing, Communication and Smart World*, Lecture Notes in Electrical Engineering 694,
https://doi.org/10.1007/978-981-15-7804-5_26

1 Introduction

With the open-source nature and the extensive support of open application space, Android has become the largest shareholder of the global market of the smartphone, with more than 85% unique devices running worldwide [1]. Its open-source nature and simplicity have gained the attention of manufacturers across the world to produce low-cost smartphone devices compare to another platform. Apart from smartphones, Android is gaining popularity in the use of other devices such as tablets, TV, wearable, and recently, IoT devices.

The open-source nature of Android has led to the development of open-source applications that are available for free use. A naive user can also use the Android SDK to develop Apps for Android and contribute them for free use. A developer can distribute Android applications through the official Google Play Store or via a third-party sharing platform. According to the statistics published by Statistica, the number of applications available only at the Play Store is approximately 2.6 million [2], and the count of applications available at another third-party stores is unknown.

Due to the open-source nature and the significant share in the smartphone market, Android has drawn the attention of Android malware writers. In Android, a user can install an application from the official application market (Play Store) or a third party market store assuming it as legitimate. However, it can be a malicious application that has managed to enter into mass-market. A study published by the G DATA shows that around 1.9 million new Android malware were counted in the first half of the year 2019 that indicates that more than 10,000 new Android malware were developed each day during the first six months of 2019 [3]. Google, however, provides a reliable defense system against the malicious apps, but these apps still find new methods to enter inside Play Store and subsequently into an end-users smartphones. Recently, there was in the news that 172 malicious apps were found on Google Play Store, and installation count for them is around 335+ million [4]. If a malicious application gets installed into a smartphone, it can steal sensitive information like contacts, SMSes, GPS location, etc., that are critical to a user.

Existing security techniques to analyze Android malware can broadly be classified into two categories, namely static analysis and dynamic analysis [5]. In static analysis, an application is analyzed without executing it, whereas the execution of an application is required to profile its behavior during the dynamic analysis. However, a detection system can use any technique and detect an application by generating a signature (the traditional approach used by antivirus) or utilize machine learning (modern detection approach) [6]. The detection system developed based on a signature cannot analyze zero-day malwares. To extend the detection method in analyzing a zero-day malware, machine-learning-based detection systems are seeking attention nowadays [7]. As a consequence, many state-of-the-art [8–15] machine-learning-based Android malware detection systems have been developed. Some method uses the high-dimensional feature sets [8], while some other use feature reduction techniques [9, 15] to reduce the feature set size and train the machine learning model to detect malware. Even though feature selection has been used in the past for reducing the dimension of feature space, but they have been applied on a single category of feature, i.e., permission [9].

In this paper, we present a lightweight machine-learning-based Android malware detection model that can detect malware automatically. In our method, we first extract the various static features from the Android manifest file and dex code. The extracted features are then passed to a Feature Engineering and Analysis framework to reduce the feature dimensionality. Feature Engineering and Analysis module uses a multilevel feature selection/reduction methods and reduces feature set size drastically. Feature Engineering and Analysis module is more effective in reducing the feature set size, and it minimizes the feature dimension in each step. To show the effectiveness of our approach, we have collected 20,988 benign samples by crawling the Google Play Store. Malware samples (20,988) have been obtained from the VirusShare. The experimental results show that our detection method can detect malware with high accuracy and precision above 98% while reducing the feature set size drastically from 3,73,458 to 105 features.

In summary, our paper makes the following contribution:

- We design a Feature Engineering and Analysis module, which can reduce the feature set size while maintaining the prediction power.
- The proposed model can give the optimal features of each category of feature set while maintaining the accuracy and precision.
- Finally, we show the efficacy of the reduced feature set against the original features.

The paper is structured into six sections. Section 2 discusses the background and related work. Section 3 introduces the dataset used for this work. Section 4 addresses the architecture of the proposed work and elaborates its various components, along with reducing the feature dimension step by step. In Sect. 5, we show the effectiveness of the proposed system by evaluating the reduced feature set against the original set. The last section, i.e., Section 6 concludes the proposed work.

2 Background and Related Work

In this section, we first discuss necessary background details related to an Android application and its components. This is then followed by the related work.

2.1 Background

2.1.1 Android Application (APK)

Android application is an archive file like JAR, TAR, and ZIP file, and contains multiple components. The description of some most frequently used components of an application is as follows [16]:

- **Activities**: Every visible screen on Android through which a user can interact and provides his/her response is called an activity. Generally, an application contains multiple activities, and communication between them is achieved through a message passing interface called intent.
- **Services**: These are the background processes and used to perform a long-running computation in an Android application. For example, if we want to upload a high volume of data, then the services are the only components that can be used while making our smartphone active to perform other user interactive tasks.
- **Broadcast Receiver**: These components work like the mailbox. If an application wants to respond for a particular event (e.g., when a new SMS arrives), then it registers for that event. Whenever such events occur in a smartphone, the Android system notifies the same to the registered application through the broadcast receiver.
- **Content Providers**: If an application wants to access data of other application or wishes to share its own data to others, then the content providers provide them securely. Through a content provider, we can access databases, files, and other resources that are exposed to public use or within an application.
- **Intents**: Intents are the message passing interfaces in the Android and are useful for establishing communication between multiple applications or different components of a single app.

2.1.2 Permission System

Android security is based on the *permissions* system and on the concept of providing least privileges. Permission is required to access the system functionality and the functionality exposed by other applications. For example, if an app wants to use the Internet, then it should request the Internet permission and register it through the manifest file. Without the proper authorization, an application cannot utilize the system and other apps' functionality available for use [17].

2.1.3 Application Manifest File

It is a core component and the most crucial resource file of an Android application. It contains all the information through which we can convey the requirements of an application. Every executable component that can trigger by an event needs to register in this file. Apart from the executable component, an app developer also guides through the manifest file for the requirement of privileges in terms of permission and hardware features [18].

2.2 Related Work

MADAM [10] uses the in-depth analysis of Android applications performed at the kernel level, application level, and user level and able to detect more than 96% application efficiently. Zhao et al. [15] proposed a detection method based on the permission and API and applied feature selection algorithm by differentiating the frequency of features between malware and benign samples.

Drebin [8] proposed a lightweight detection model for on-device malware detection and achieves an accuracy of 94% while taking an approx 10 sec of average detection time. Drebin used a static analysis technique and extracted more than 5 lakhs features from the malware and benign samples.

Wang et al. [11] proposed a hybrid Android malware detection methods and could only be deployed off-device on clouds for providing malware detection as a service. Some methods, as shown in [13], use ensemble machine learning classifiers to improve the detection power of a detection model. Recently, it has been observed that malware analysis systems are also built on deep learning methods; such methods are shown in [12, 14].

SigPID [9] identifies the significant permissions by mining the permissions data and pruning the same at three levels for correctly extracts the related permission. They have found 22 such permission that is lesser than the dangerous permission list released by Google team and can be utilized to detect Android malware with an accuracy of approx 92%.

Zhu et al. [19] use the random forest classifier on the features extracted from the Android application for malware detection. SAMADroid [20] used both static and dynamic analysis techniques and proposed a client-server architecture for efficient malware detection. In [21], a comparative study on feature extraction and their selection methods has been shown and can be used to increase the prediction power of a machine-learning-based detection model. In [22], such feature reduction techniques have been used on permissions and their associated APIs, and a machine learning model has built for malware classification with high accuracy.

3 Dataset

In malware analysis and classification work, the critical part is to collect the required data to evaluate our model. In this regard, we have acquired 41,976 unique Android applications in which 20,988 samples are malware, and rest belongs to the benign category. The malware samples are provided by the VirusShare [23], and the benign samples are obtained by crawling the Google Play Store. To ensure the benign behavior of the samples collected from the Google Play Store, we have submitted the signature (hash) of each sample on VirusTotal [24]. In the benign category, we have only included those samples that are not detected as malware by any antivirus engine available on VirusTotal. By this analogy, we have acquired 20,988 benign samples in our dataset.

4 Design and Implementation

This section presents an overview of the design of machine-learning-based lightweight malware detection model and discusses the working of its core components.

4.1 An Overview

The main aim of this work is to design a lightweight machine-learning-based Android malware detection model that provides the almost same prediction capability and utilizes fewer resources in terms for computation power and time. Figure 1 shows the architecture of the proposed framework to reduce the feature dimension while maintaining the prediction power. As shown in Fig. 1, there are three main components of proposed work—(i) feature extraction, (ii) Feature Engineering and Analysis, and (iii) learning model.

Feature extraction module work is to extract the static feature from an Android app and saves the result in the JSON file. The JSON file then read by the Feature Engineering and Analysis module, which encodes the extracted feature and further performs various analyses to reduce the feature dimension. At last, the learning model module trains the final detection model on the basis of reduced feature set as selected by the Feature Engineering and Analysis module. Further, we discuss the detailed design and working of each module in the subsequent section.

4.2 Feature Extraction

Feature extraction module is a core component of the proposed system. This module is applied on the collected dataset (see Sect. 3) to extract the static feature and dumping them in a JSON file each for a single application. The architecture of the static feature extraction module is shown in Fig. 2.

Among several things, a typical Android application contains application code, resources, assets, certificates, and manifest file that are packaged into the APK. Using

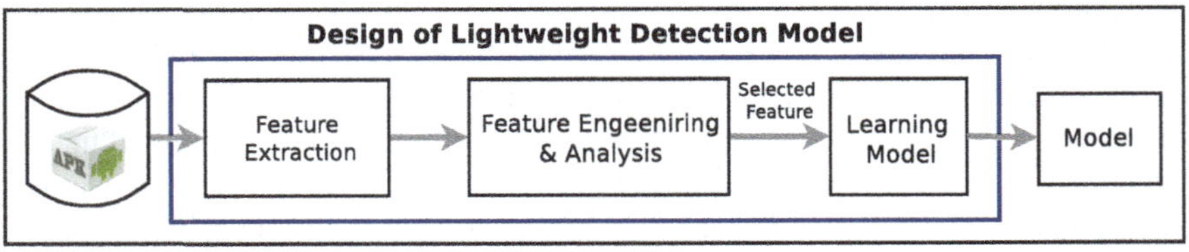

Fig. 1 Architecture of lightweight android malware detection model

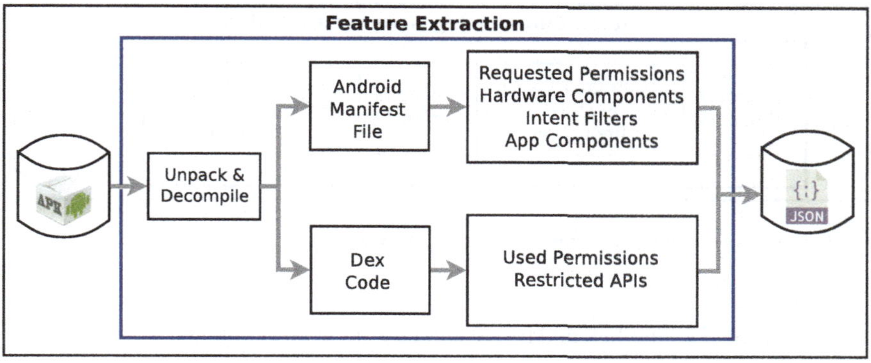

Fig. 2 Static feature extraction process

reverse engineering tools like *Androguard* [25], *apkparser*, etc., there is abundant amount of information that can be extracted and analyzed from these files. The main static features extracted for this work are from the manifest file and the dex code. The proposed framework extracts several categories of information from these two files as shown in Fig. 2.

The first argument to the Feature Extraction module is path to the set of samples. Each sample in the dataset is loaded and passed to *Androguard* for reverse engineering. Using Androguard, first the Android application is unpacked and decompiled and then the manifest file and the dex code are retrieved from the Android application to extract information like activities, services, hardware components, permissions, restricted APIs, etc. from each sample. The extracted information is dumped in the directory provided as the second argument while running this module. Each JSON file contains a dictionary with a key - value pair, where key is the type of information extracted like permission, API, activity, intent, etc., and values are all the relevant information extracted in that category.

4.3 Feature Engineering and Analysis

The machine learning algorithm is always challenged by a large number of features to train a model in order to solve a classification problem. Often machine-learning-based malware detection system contains a huge number of features. In such a large set of features, all features do not contribute to classification results. Even some features are opposite to each other, which may decrease the accuracy of a machine learning model. Feature engineering modules deal with all such issues and reduce the feature set size by including only the relevant features that directly contribute to malware detection while maintaining the prediction power. Figure 3 shows the design of the Feature Engineering and Analysis module. This module contains four main components, namely (i) Feature Encoding, (ii) Category-wise Feature Reduction,

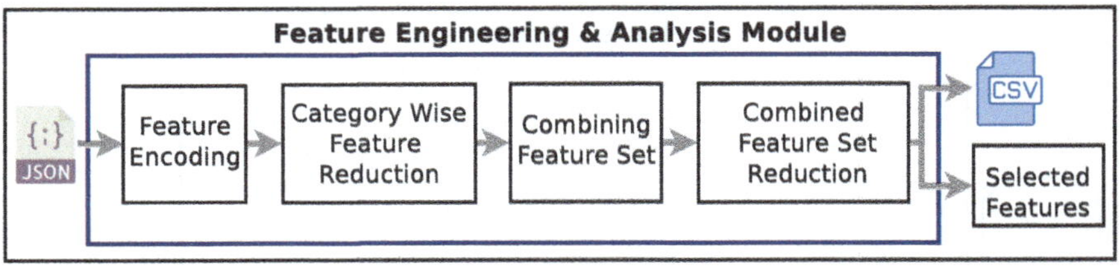

Fig. 3 Design of feature engineering and analysis module

(iii) Combining Feature Set, and (iv) Select Feature from Combined Feature Set. We discuss these components and their effect on the feature set size in the subsequent section

4.3.1 Feature Encoding

The main task of this component is to parse the JSON file obtained from the feature extraction module and produces CSV file one for each category. Initially, all the features in each category of feature sets are of binary type. Taking all the features in the feature set as binary will increase the feature set size because some of the features belong to a user-defined name like activity, services, etc. and can be changed from one sample to another. We can transform such features into one feature by taking their count. Such transformation reduces the feature set size drastically. The features that have a predefined name like APIs (restricted APIs) are considered to be used as a binary feature. Finally, the output of this module is two types of features, i.e., numerical features and binary features for each category. We have generated one CSV file for all numeric features and a separate file for each categorical feature where feature type is binary. In Table 1, column name original shows features count before the transformation, whereas column name encoding corresponds to the features count after the transformation.

4.3.2 Category-Wise Feature Reduction

In Sect. 4.3.1, we have discussed the feature encoding technique to reduce the feature space size where the size is vast due to the humans' defined name to a component. In this section, we discuss the category-wise feature reduction process and utilizing the same to reduce the feature set size of the remaining binary-type categorical feature set. The category-wise feature reduction process is shown in Fig. 4. It comprises two main components—(i) Feature Reduction Based on Correlation and (ii) Optimal Feature Identification, and are discussed as follows:

Feature Reduction Based on Correlation: In correlation-based feature reduction, we have used Pearson's correlation methods to identify the correlated features. As

Table 1 Effect of different feature reduction processes w.r.t feature size

Features category	Original	Encoding	Correlation	Optimal
Activities	280,784	1	1	1
Services	29,417	1	1	1
Content providers	3202	1	1	1
Broadcast receivers	25,850	1	1	1
Intent filters	24,006	1	1	1
Hardware components	159	159	145	96
Requested permissions	9180	506	449	230
Used permissions	59	59	58	30
Restricted APIs	801	801	512	378
Custom permissions	0	1	1	1
Total	3,73,458	1531	1170	740

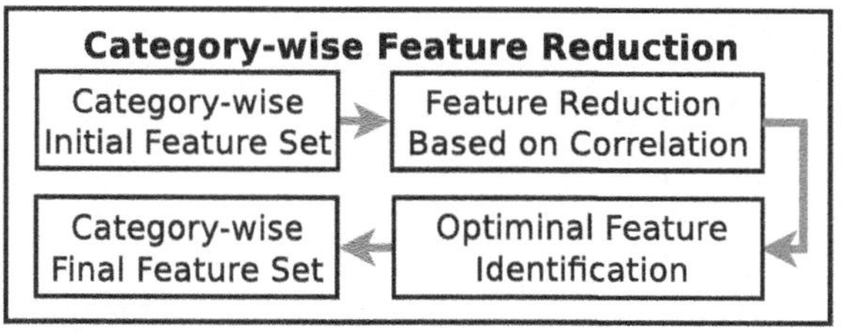

Fig. 4 Category-wise feature reduction process

shown in Fig. 4, the feature reduction process utilizes Pearson's correlation methods to identify the correlated features for the first level of feature reduction. By this approach, we identify the features that are highly correlated to each other and retain only one from them. We have observed the effect of the correlation coefficient with different threshold values, and the results are shown in Fig. 5. Figure 5a shows the effect of threshold accuracy on the accuracy, while Fig. 5b demonstrates the impact of threshold value on the feature set size. We have selected the 0.8 as a threshold value as it results in correct tradeoff between number of feature and accuracy for most of the feature sets. For threshold 0.8, the reduced feature set size is shown in the column correlation of Table 1.

Optimal Feature Identification: After eliminating the feature based on the correlation coefficient, the resulting features are utilized to identify optimal features in each category of binary feature set. To determine optimal features, we have used the Recursive Feature Elimination with Cross Validation (RFECV) algorithm. This algorithm eliminates the features recursively and provides the optimal features based on evaluation criteria defined by a user. We have used accuracy as a metric for evaluation

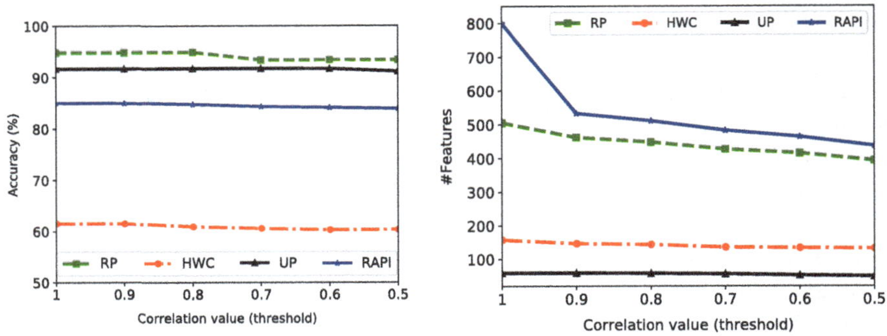

(a) Effect of correlation value on Accuracy. (b) Effect of correlation value on feature size

Fig. 5 Feature reduction based on correlation coefficient for requested permissions (RP), hardware components (HC), used permissions (UP), and restricted APIs (RAPI)

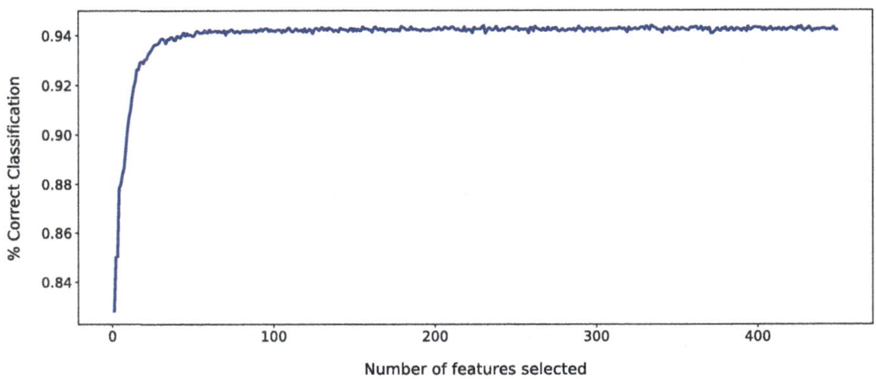

Fig. 6 Optimal features identification for requested permissions using RFECV

criteria in RFECV, and accuracy versus #features graph for requested permissions are shown in Fig. 6. The optimal number of features obtained from RFECV is enlisted in column Optimal of Table 1 for each category of feature set.

4.3.3 Combining Feature Set

The previous section deals with the feature reduction process and reduces the size of the feature set in each category having type binary and generates a reduced feature set. In this phase, we combine all these features with a different set of combinations and generate the combined feature set. As discussed earlier, we have one feature set of type numerical and four categorical features set of type binary, which produces in total 26 different combinations of feature set. Some of the generated combined feature set (out of 26) and their size are projected in Table 2, where column Combination represents the mix of feature set, and Mix Result represents the prediction power along with their size.

4.3.4 Select Features from Combined Feature Set:

This section further optimizes the combined feature set and reduces the correlated features based on the correlation coefficient value followed by the optimal feature identification using RFECV. We have used a similar approach to reduce the feature set size, as used in Sect. 4.3.2 for the correlation-based optimization and optimal feature identification. The results for correlation-based optimization are projected in column Correlation of Table 2, and the column Optimal of Table 2 shows final result after identification of optimal features by utilizing RFECV. As can be seen in Table 2, combination of numeric features (N) with other categorical features requested permissions (R), hardware components (H), used permissions (U), and restricted APIs (RA) are showing highest prediction power with an accuracy of more than 98% while containing only 105 features. So we have **N + R + H + U + RA** as the set of optimal features that can be utilized to train the final detection model.

Note: The optimal feature identification step does not reduce the size of the different combinations of feature set except the mix **N + R + H + U + RA** (same has opted for final model), and the main reason behind not showing the results for other combinations in the Optimal column of Table 2.

Table 2 Feature selection in the combined feature set

Combination	Mix results		Correlation			Optimal	
	#Feat	Acc (%)	#Feat	Threshold	Acc (%)	#Feat	Acc (%)
N + R	236	95.91	228	0.7	95.82	–	–
N + H	102	95.1	96	0.6	95.11	–	–
N + U	36	95.32	36	1	95.32	–	–
R + H	326	95.77	315	0.7	95.8	–	–
R + U	260	96.46	253	0.8	96.5	–	–
H + U	126	95.53	120	0.6	95.67	–	–
N + R + H	342	95.96	320	0.7	96.21	–	–
N + R + U	266	96.79	266	1.0	96.79	–	–
N + H + U	132	95.47	124	0.6	95.77	–	–
R + H + U	356	96.96	349	0.8	96.52	–	–
N + R + H + U	362	96.85	355	0.8	97.19	–	–
N + R + H + U + RA	**740**	**98.28**	**599**	**0.8**	**98.17**	**105**	**98.23**

Note N denotes numeric feature set, R denotes reduced set of requested permissions, H denotes reduced set of hardware components, U denotes reduced set of used permissions, and RA denotes reduced set of restricted APIs

Table 3 Category-wise number of features in final feature set

Category	#Feature
Activities	1
Services	1
Content providers	1
Broadcast receivers	1
Intent filters	1
Hardware components	9
Requested permissions	43
Used permissions	13
Restricted APIs	34
Custom permissions	1
Total	105

4.4 Learning Model

All of the training/testing processes are performed with a 70:30 split ratio of our dataset (see Sect. 3). We use random forest as the base classifier with 100 estimators for each step of the Feature Engineering and Analysis module and utilize the tenfold cross-validation. Random forest is an ensemble of several trees, and prediction is made by averaging the results of each tree. The reason behind choosing random forest as the base classifier is that it can handle unbalanced data, more robust to outliers and nonlinear data, faster training/prediction, and most importantly, it can deal with the high dimensionality of data efficiently. Extra tree classifier also provides the same feature and can be used as the base classifier in place of random forest. For optimal feature identification, we have used RFECV with fivefold cross-validation to avoid over-fitting/under-fitting. The resulting set of features from Sect. 4.3.4 is then used for the final classification process. Table 3 shows the feature set used for the training and evaluating the detection model. For final classifier modeling, we have also used the random forest classifier with 100 estimators and tenfold cross-validation. The final model has been trained on 70% of samples of our dataset, and the remaining 30% for evaluating it.

5 Evaluation

We have evaluated our model against four classifiers, namely extra tree (ET), decision tree (DT), gradient boosting (GB), and random forest (RF). We have used tenfold cross-validation to avoid over-fitting or under-fitting. We have evaluated our reduced feature set against the original feature set without any optimization. To evaluate the model on both sets of features, we have used accuracy (Acc), precision (Pre),

Table 4 Detection results of model trained on original and reduced feature set

Classifier	Original feature set (3,73,458)				Reduced feature set (105)			
	Acc (%)	Pre (%)	Recall (%)	F1 (%)	Acc (%)	Pre (%)	Recall (%)	F1 (%)
RF	98.00	98.02	98.00	98.02	98.23	98.23	98.22	98.23
ET	97.91	97.94	97.91	97.94	98.13	98.14	98.13	98.14
DT	95.47	95.50	95.46	95.46	96.81	96.81	96.81	96.81
GB	96.24	96.25	96.24	96.24	96.24	96.24	96.24	96.24

Note RF denotes random forest, ET denotes extra tree, DT denotes decision tree, and GB denotes gradient boosting classifiers

Table 5 Training and testing time with original and reduced feature set

Classifier	Original feature set (3,73,458)		Reduced feature set (105)	
	Training (s)	Testing (s)	Training (s)	Testing (s)
RF	15.4895	0.103	2.0133	0.0354
ET	22.756	0.1454	0.5441	0.0392
DT	13.5496	0.0058	1.5476	0.0054
GB	47.3529	0.0271	7.6512	0.0301

recall, and F1-score (F1) as evaluation metrics, and results are shown in Table 4. The evaluation results in Table 4 show that the model trained on the reduced feature set outperforms the model trained on the original feature set. The evaluation metric accuracy represents the percentage of samples correctly classified; recall represents the malware detection rate (TPR), precision denotes the portion of the malware detected by our model which is relevant. Generally, there is a race between recall and precision wherein maximizing one of them; the other metric gets penalized. The solution to this problem is the F1-score, which gives an optimal blend of both (recall and precision). All these metrics are obtained through the confusion matrix.

Further, we have also measured the time spent in training/testing processes of different classifiers on the original and reduced feature set. Training is performed on 70% of the samples, and the remaining 30% of samples are used for testing. We have used the same set of classifiers as used for detection, and results are shown in Table 5. As can be seen in Table 5, there is a significant amount of reduction in time for both training/testing when the reduced feature set is considered for the final detection model.

Therefore, we can say that the feature set identified using our approach contains only relevant features, and reduces the feature set size drastically while maintaining/improving the prediction power.

6 Conclusion

With the rapid growth in new Android malware infections and variations in the existing ones, the traditional detection methods often fail to detect them. This work observes the problem and introduces an efficient and computationally in-expensive and lightweight machine-learning-based approach for malware detection and effective identification of zero-day attacks against the signature-based methods. We have designed a Feature Engineering and Analysis module that reduces the feature dimension drastically. With the help of the Feature Engineering and Analysis module, we have reduced the feature dimension size from 3,73,458 to 105 features while achieving the accuracy above 98% with high precision and recall both are more than 98%. Our evaluation dataset contains recent malware and benign apps with the same number of samples for both. This shows that our proposed work is highly efficient in drastically reducing the feature set size while maintaining prediction power.

References

1. Idc - smartphone market share - os. https://www.idc.com/promo/smartphone-market-share/os. Accessed 30 Sep 2019
2. Google play store: number of apps 2019 | statista. https://www.statista.com/statistics/266210/number-of-available-applications-in-the-google-play-store/. Accessed 30 Sep 2019
3. Mobile malware report - no let-up with android malware | g data. https://www.gdatasoftware.com/news/2019/07/35228-mobile-malware-report-no-let-up-with-android-malware. Accessed 6 Dec 2019
4. 172 malicious apps with 335m+ installs found on google play. https://thenextweb.com/apps/2019/10/01/google-play-android-malware-2/. Accessed 5 Dec 2019
5. Faruki P, Bharmal A, Laxmi V, Ganmoor V, Gaur MS, Conti M, Rajarajan M (2015) Android security: a survey of issues, malware penetration, and defenses. IEEE Commun Surveys Tutorials 17(2):998–1022. https://doi.org/10.1109/COMST.2014.2386139
6. Bhat P, Dutta K (2019) A survey on various threats and current state of security in android platform. ACM Comput Surv 52(1). https://doi.org/10.1145/3301285
7. Kreimel P, Eigner O, Tavolato P (2017) Anomaly-based detection and classification of attacks in cyber-physical systems. In: Proceedings of the 12th international conference on availability, reliability and security. ARES'17, association for computing machinery, New York, NY, USA. https://doi.org/10.1145/3098954.3103155
8. Arp D, Spreitzenbarth M, Hübner M, Gascon H, Rieck K (2014) Drebin: effective and explainable detection of android malware in your pocket https://doi.org/10.14722/ndss.2014.23247
9. Li J, Sun L, Yan Q, Li Z, Srisa-an W, Ye H (2018) Significant permission identification for machine-learning-based android malware detection. IEEE Trans Ind Inf 14(7):3216–3225. https://doi.org/10.1109/TII.2017.2789219
10. Saracino A, Sgandurra D, Dini G, Martinelli F (2018) Madam: effective and efficient behavior-based android malware detection and prevention. IEEE Trans Dependable Secure Comput 15(1):83–97. https://doi.org/10.1109/TDSC.2016.2536605
11. Wang X, Yang Y, Zeng Y (2015) Accurate mobile malware detection and classification in the cloud. SpringerPlus 4. https://doi.org/10.1186/s40064-015-1356-1
12. Xu K, Li Y, Deng RH, Chen K (2018) Deeprefiner: multi-layer android malware detection system applying deep neural networks. In: 2018 IEEE European symposium on security and privacy (EuroS P), pp 473–487. https://doi.org/10.1109/EuroSP.2018.00040

13. Yerima SY, Sezer S, Muttik I (2014) Android malware detection using parallel machine learning classifiers. In: 2014 Eighth international conference on next generation mobile apps, services and technologies, pp 37–42. https://doi.org/10.1109/NGMAST.2014.23
14. Yuan Z, Lu Y, Xue Y (2016) Droiddetector: android malware characterization and detection using deep learning. Tsinghua Sci Technol 21(1):114–123. https://doi.org/10.1109/TST.2016.7399288
15. Zhao K, Zhang D, Su X, Li W (2015) Fest: a feature extraction and selection tool for android malware detection. In: 2015 IEEE symposium on computers and communication (ISCC), pp 714–720 https://doi.org/10.1109/ISCC.2015.7405598
16. Android application package - wikipedia. https://en.wikipedia.org/wiki/Android_application_package. Accessed 10 Dec 2019
17. Permission | Android Developers. https://developer.android.com/guide/topics/manifest/permission-element. Accessed 10 Dec 2019
18. App manifest overview | android developers. https://developer.android.com/guide/topics/manifest/manifest-intro. Accessed 10 Dec 2019
19. Zhu HJ, Jiang TH, Ma B, You ZH, Shi WL, Cheng L (2018) Hemd: a highly efficient random forest-based malware detection framework for android. Neural Comput Appl 30(11):3353–3361. https://doi.org/10.1007/s00521-017-2914-y
20. Arshad S, Shah MA, Wahid A, Mehmood A, Song H, Yu H (2018) Samadroid: a novel 3-level hybrid malware detection model for android operating system. IEEE Access 6:4321–4339. https://doi.org/10.1109/ACCESS.2018.2792941
21. Coronado-De-Alba LD, Rodríguez-Mota A, Ambrosio PJE (2016) Feature selection and ensemble of classifiers for android malware detection. In: 2016 8th IEEE Latin-American conference on communications (LATINCOM), pp 1–6. https://doi.org/10.1109/LATINCOM.2016.7811605
22. Qiao M, Sung AH, Liu Q (2016) Merging permission and api features for android malware detection. In: 2016 5th IIAI international congress on advanced applied informatics (IIAI-AAI), pp 566–571. https://doi.org/10.1109/IIAI-AAI.2016.237
23. VirusShare: https://virusshare.com/. Accessed 10 Dec 2018
24. VirusTotal: https://www.virustotal.com/. Accessed 10 Dec 2019
25. Desnos A, Geoffroy Gueguen SB (2019) Welcome to Androguard's documentation!—androguard 3.3.5 documentation. https://androguard.readthedocs.io/en/latest/. Accessed 10 Dec 2019

Area and Power Efficient 2 Bit Multiplier by Using Enhanced Half Adder

Akshay Kamboj, Manisha Bharti, and Ashima Sharma

Abstract In this paper, a novel design of half adder (HA) is proposed using low power and area-efficient XOR and AND gates. Multipliers are essential component mostly used arithmetic unit. Using novel HA, novel design of multiplier and its area utilization is discussed along with conventional design. The performance of proposed circuit is analyzed using Cadence Virtuoso on 90-nm Generic Process Design Kit (gpdk) Complementary Metal Oxide Semiconductor (CMOS) technology at 1 V supply voltage (V_{dd}) at maximum of 50 GHz frequency. Compact high-performance systems require consolidated high-speed multipliers. The simulated result indicates that the proposed multiplier utilizes 33.79% lower area as compared to conventional method. A low power 2 bit multiplier based on the novel full swing half adder (HA) also shows lower power consumption by 12.42% over conventional multiplier.

Keywords Half adder (HA) · Multiplier · Generic process design kit (gpdk) · Complementary metal oxide semiconductor (CMOS)

1 Introduction

Over last few decades, the demand for more functionality and high performance for portable systems has increased drastically which has drawn the consent to increase the compactness of the portable systems. Area and power efficiency are important requirements in high-speed interface circuit [1]. Many research efforts are required to provide enhancement in these parameters.

There are various elementary as well as composite ways to implement multiplier containing multiple digital circuits. The rendition of the multiplier can be enhanced by improvement in any fundamental structural block [2]. A 2 bit conventional multiplier contains 4 NAND gates and 2 half adder. Further, the HA contains XOR and AND

A. Kamboj (✉) · M. Bharti · A. Sharma
Department of Electronic and Communication Engineering, National Institute of Technology Delhi, New Delhi, India
e-mail: nishukamboj95@gmail.com

P. K. Singh et al. (eds.), *Evolving Technologies for Computing, Communication and Smart World*, Lecture Notes in Electrical Engineering 694,
https://doi.org/10.1007/978-981-15-7804-5_27

gate [3]. XOR gates inside half adder can be optimized by precluding the problems which are consequence of reduction of power supply [4]. In this paper, we have evaluated multiple HA using various configured XOR and AND gate. Further, using those HA several multipliers has been discussed and compared with the conventional circuit. This paper represents the implementation of multiplier using a low power and fast XOR gate.

Paper proceeds as follows: Section 2 contains the deliberation about the conventional multiplier. Circuit design of proposed multiplier is described in Sect. 3. The simulated result is analyzed and compared in Sect. 4. Finally, the conclusions are drawn by Sect. 5.

2 Review of Conventional Circuit

2.1 *Half Adder*

Half adder is an illustration of unembellished, operating digital circuit. Conventional HA is made up of two logic gates, i.e., XOR and AND gate. Figure 1 shows the block diagram of conventional HA [5], wherein the XOR and AND gates are conventional CMOS gates. The output of XOR gate responds as sum and the output of AND gate responds as carry. The same can be verified from the truth table given in Table 1 [6].

Conventional CMOS HA comprises of 20 transistors which includes 14 transistors of CMOS XOR gate and 6 transistors of CMOS AND gate.

Fig. 1 Half adder circuit diagram

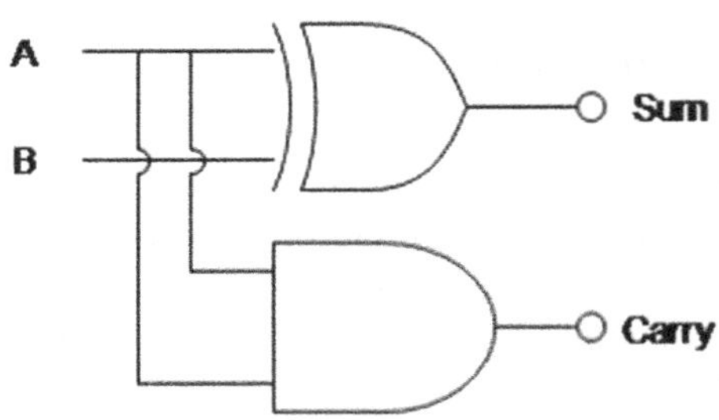

Table 1 Truth table of half adder

Input		Output	
A	B	Sum	Carry
0	0	0	0
0	1	1	0
1	0	1	0
1	1	0	1

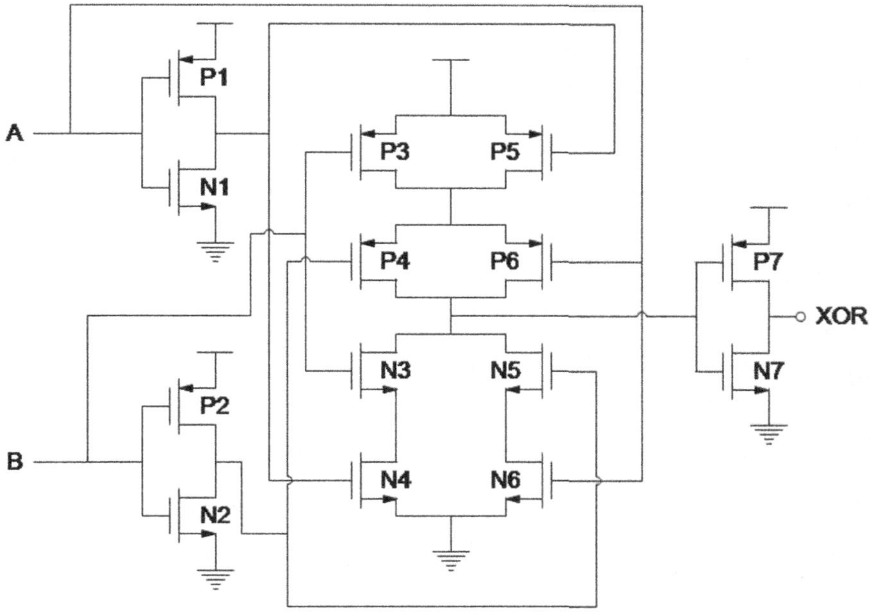

Fig. 2 Circuit diagram of conventional XOR gate

Figure 2 shows the 14 T circuit diagram of conventional CMOS XOR gate which contains 7 PMOS and 7 NMOS transistor [7].

XOR gate within the HA of multiplier is the major source of power consumption. Therefore, the power consumption can be minimized by flawless design of XOR gate [8]. Various XOR gates have been proposed to enhance the performance of XOR gate [9–12] depending upon the application.

A 2 bit multiplier is created using the XOR gate [9] shown in Fig. 3. If the XOR gate can be optimized in some parameters, then it can have a considerable effect on the multiplier created using it.

2.2 *Conventional Multiplier*

Multiplication is a significant function in arithmetic operation. The multiply and mount the inner product are certain repeatedly used calculus acute arithmetic functions prevalently used in numerous digital signal processing applications. Figure 3 shows 2 bit conventional multiplier, made up of 2 HA and 4 AND gates. These basic building block XOR and AND gates are designed through CMOS technique. Here in Fig. 3 of conventional multiplier, the XOR and AND gates used are conventional CMOS XOR gate and AND gate, respectively. The conventional multiplier makes

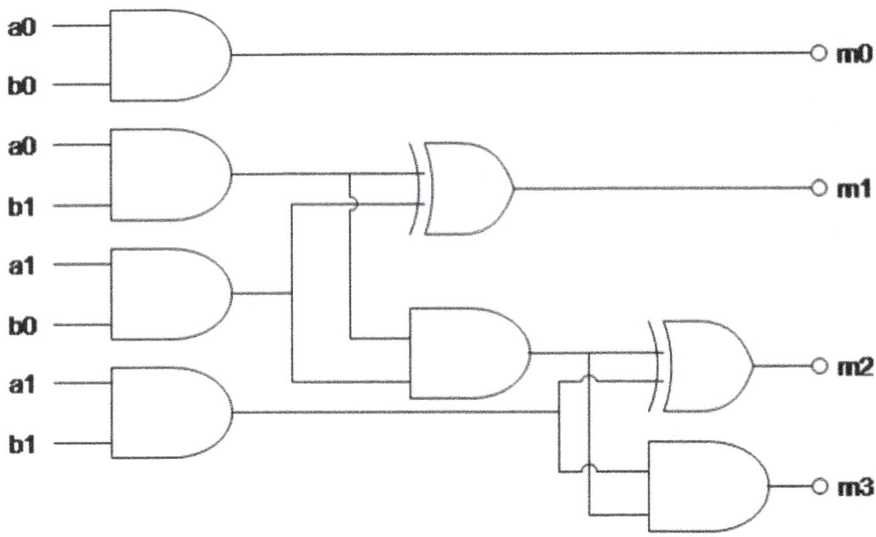

Fig. 3 Conventional multiplier circuit

up of 64 transistor. In order to reduce the power dissipation, the conventional CMOS XOR gate can be replaced by low power full swing XOR gate given by Fig. 2.

This indirectly reduces the power dissipation and area. The only disadvantage with such multipliers is that the delay increases when used in series due to collective capacitance. Figure 4 depicts the multiplication operation of 2 bit multiplier.

Table 2 shows the truth table of 2 bit multiplier. Two 2 bit values are a0 (LSB) and a1 (MSB) and b0 (LSB) and b1 (MSB). 4 bit result are m0 (LSB) to m3 (MSB). The same operation can be seen from Fig. 5.

Fig. 4 Multiplier operations of 2 bit multiplier

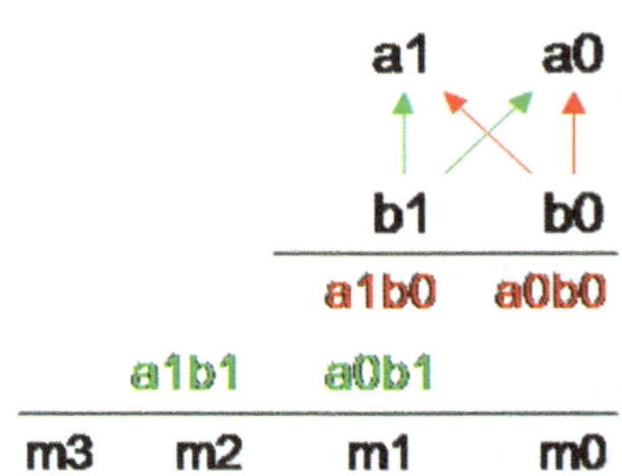

Table 2 Truth table of 2 bit multiplier

Input				Output			
a1	a0	b1	b0	m3	m2	m1	m0
0	0	0	0	0	0	0	0
0	0	0	1	0	0	0	0
0	0	1	0	0	0	0	0
0	0	1	1	0	0	0	0
0	1	0	0	0	0	0	0
0	1	0	1	0	0	0	1
0	1	1	0	0	0	1	0
0	1	1	1	0	0	1	1
1	0	0	0	0	0	0	0
1	0	0	1	0	0	1	0
1	0	1	0	0	1	0	0
1	0	1	1	0	1	1	0
1	1	0	0	0	0	0	0
1	1	0	1	0	0	1	1
1	1	1	0	0	1	1	0
1	1	1	1	1	0	0	1

3 Proposed Circuit Design

In the proposed method, the multiplier is designed using low power consuming XOR gate. If the number of transistor to which inputs A and B are connected to, is same, then the input capacitance of A and B is different. In order to equate it, inputs are connected as displayed in Fig. 5 unlike the conventional multiplier. The critical path of the structure does not contain any NOT gate and the output capacitance is very small. Hence, it consumes less power. The simulation results are observed at gpdk 90 nm technology and 1 V power supply voltage (V_{dd}).

3.1 Proposed Half Adder

Proposed half adder circuits consist of low power XOR and AND gate. In HA or FA circuit, the majority of the power is consumed by XOR gates. Therefore, by prime designing the XOR gate, the average power dissipation of half adder and full adder cell can be reduced.

Figure 6 shows the proposed structure of half adder consisting 10 transistor. This structure is obtained through combination of full swing XOR gate and AND gate.

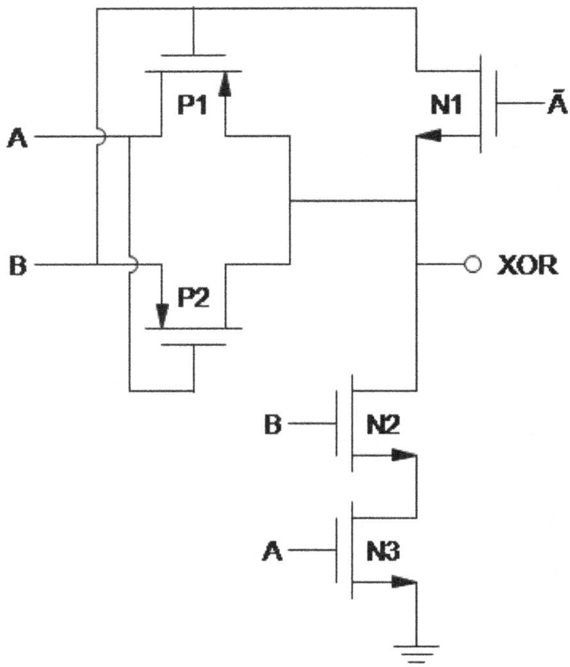

Fig. 5 Full swing XOR gate [6]

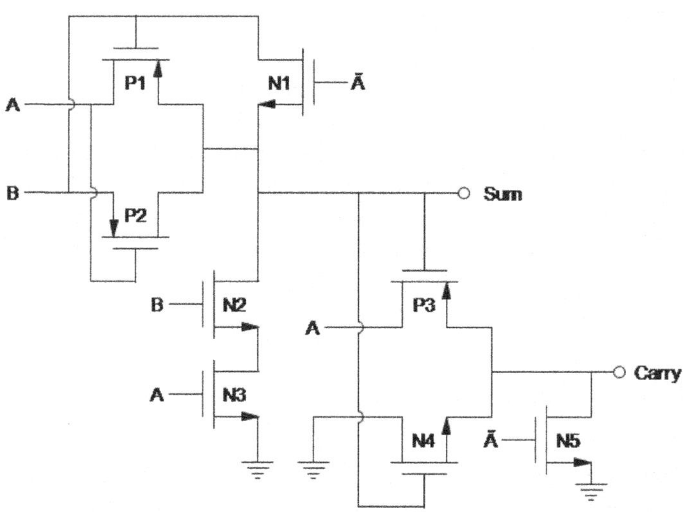

Fig. 6 10 T proposed half adder

Used AND gate is prepared through the output of full swing XOR gate and transmission gate. XOR circuit shown in Fig. 5 has a full swing response for all different combination of input. Its input capacitance is also symmetrical, i.e., the inputs A and B are connected to different transistor count. The proposed HA does not have NOT gate in the critical path. In comparison to conventional method, the proposed HA has good driving capability. Due to utilization of less number of transistor, it demonstrates low power dissipation.

3.2 *Proposed Multiplier*

Proposed 42 T multiplier circuit consists of 4 NAND gate and 2 proposed 10 T half adder. Figure 7 shows the circuit diagram of proposed 42 T multiplier. Due to the reduction in number of transistor, the circuit has lower dynamic power dissipation

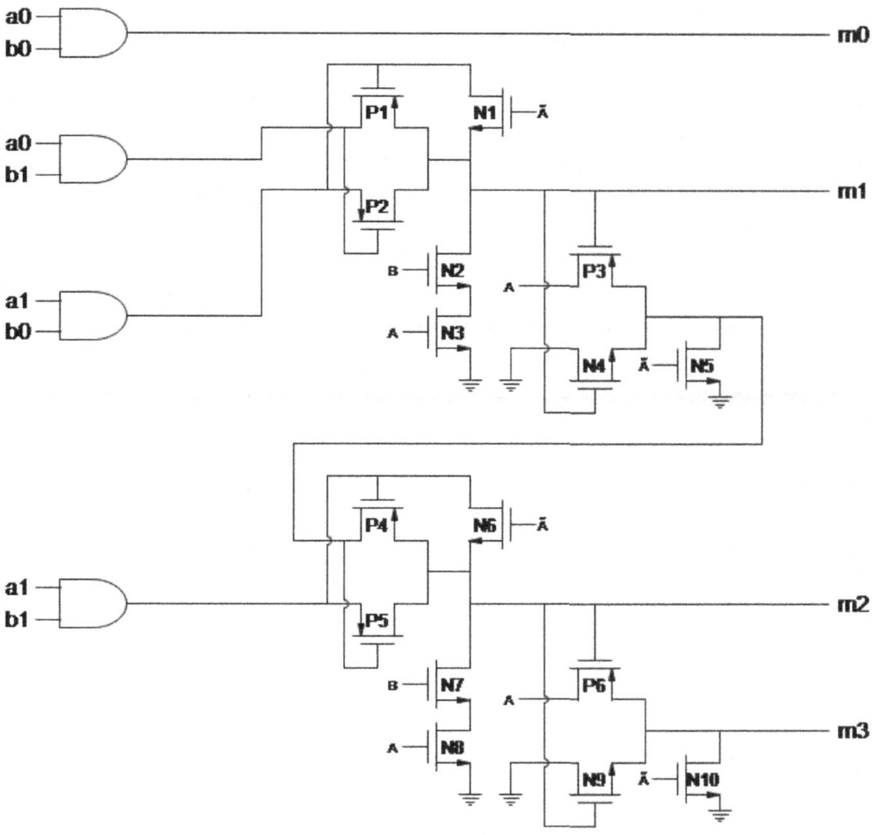

Fig. 7 42 T 2 bit proposed multiplier

than the conventional 64 T conventional multiplier. Also, in order to have some positive effect on speed of operation, the absence of NOT gate in the critical path of proposed multiplier plays an efficient role. Thus, it has good driving capability as compared with conventional method. Here in Fig. 7, the inputs A and B correspond to the input order as given in Fig. 5.

Down conversion of multiple noise which includes flicker noise and thermal noise. Due to non-ideal inductor and capacitors, only thermal noise is generated in LC tank [13]. LC-VCO provides better phase noise; however, it suffers the challenges of fabrication of large inductors [14]. In XCP, certain parts of flicker noise and thermal noise are transmitted to resonator [3]. Thermal noise generated due to switching mechanism will also contribute to the phase noise [15]. Among the multiple harmonics mainly, the second harmonics is transferred as phase noise [16].

4 Results and Discussion

To examine and verify the performance of the proposed 2 bit multiplier, simulations are performed on Cadence Virtuoso tool on 90 nm gpdk CMOS technology at 1 V supply voltage (V_{dd}) at maximum of 50 GHz frequency. Simulations were performed for the proposed multiplier and the conventional multiplier. Figure 11 shows the time domain simulated output of proposed 2 bit multiplier. It depicts that, as inputs a0a1 = 11 and b0b1 = 11 at time 40 μs, the output transits to m3m2m1m0 = 1001 at the same time (Fig. 8).

Conventional multiplier is given by the combination of Figs. 1 and 2. It can be clearly seen from Table 3 that the proposed multiplier has lower power consumption than the conventional multiplier. Structure of half adder shown in Fig. 6 (using which

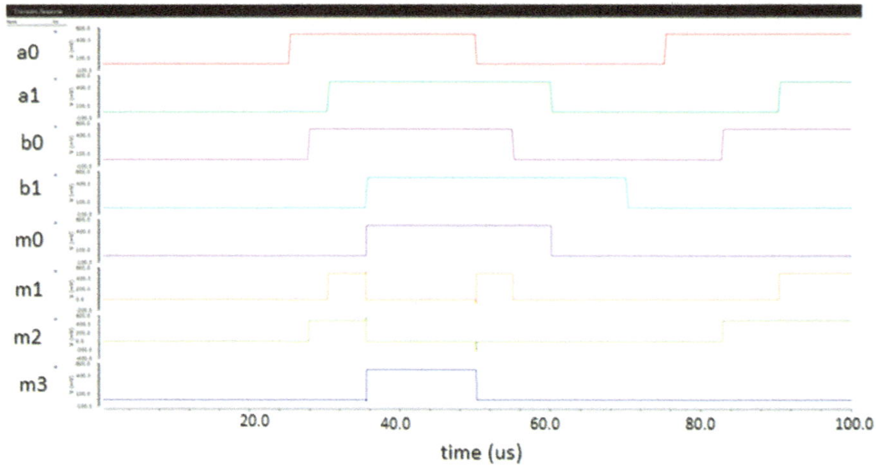

Fig. 8 Output waveform of proposed 2 bit multiplier

Table 3 Simulation results for multiplier circuit at 90 nm technology with 1 V power supply voltage

Design	Power (nW)
64 T [2]	356.5
42 T[a]	312.2

[a]Means proposed design

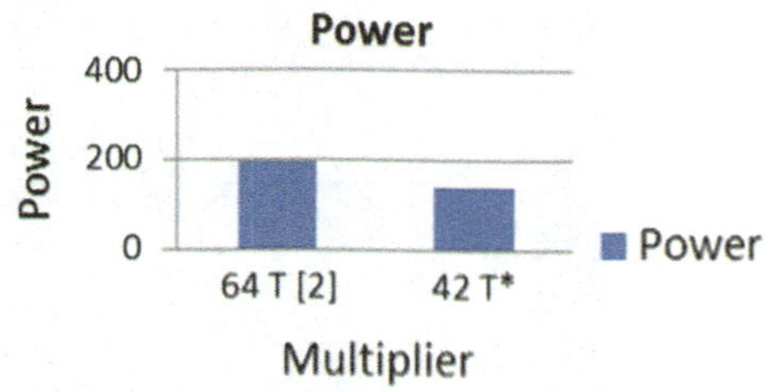

Fig. 9 Power consumption of different multiplier circuit

the proposed multiplier is designed) does not contain any NOT gate in its critical path. It is designed with TG logic style with no short circuit power dissipation. Also, the proposed half adder has very small output capacitance due to which it consumes very low power. The same can be observed through Fig. 9. The results were obtained for optimum transistor size. However, the lowest power consumption is achieved when the width of the transistor is as minimum as possible [9].

Table 4 shows the area utilization of proposed and conventional multiplier. Conventional multiplier utilizes 64 transistors among which 32 of them are pMOS and remaining 32 are nMOS, whereas proposed multiplier utilizes 42 transistors among which 19 are pMOS and remaining of them are nMOS.

By optimizing the size of transistor, the multiplier circuit can be traded for energy [17]. For all the simulations, the size of all the transistors is chosen with the objective to obtain minimum power for the circuit. To improve the performance, downscaling has been a fundamental strategy [14]. Utilization of large number of transistors in conventional half adder stands as one of the reasons for large average power consumption. The XOR gates contained in proposed multiplier do not contain any positive feedback which in result avoids any increase in capacitance and hence dynamic power consumption decreases. The proposed multiplier is able to draw up the power

Table 4 Area utilization of different multiplier design at 90 nm technology

Design	Area (μm^2)
64 T [2]	138.69
42 T[a]	91.826

[a]Means proposed design

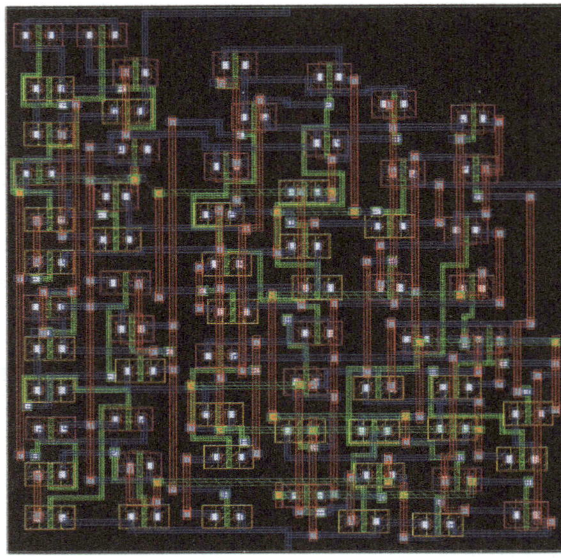

Fig. 10 Layout of 64 T conventional multiplier

consumption better than conventional method by 12.42% and that is evident from Table 3.

Nearly half of the area of multiplier is occupied by half adder. Conventional multiplier contains 2 HA circuits having 20 transistors [2] whereas proposed multiplier contains 2 HA circuits having 10 transistors each with a common input. Figures 10 and 12 show the layout of conventional and proposed multiplier, respectively. Table 4 shows the comparison of layout area among different multiplier circuits. Among both the case, proposed multiplier came out to be best.

The layout area of conventional multiplier is 138.69 μm, whereas the layout area of proposed multiplier is 91.826 μm. Comparing the area of both the multiplier, the layout area of proposed multiplier is 33.79% less than the layout area of conventional multiplier.

Major drawback of conventional method was its NOT gate on the critical path. This drawback was removed in the proposed method along with the optimization of input capacitances.

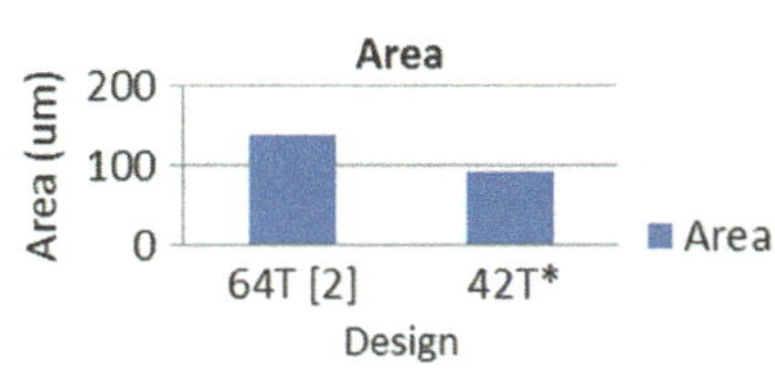

Fig. 11 Area utilization of different multiplier circuit

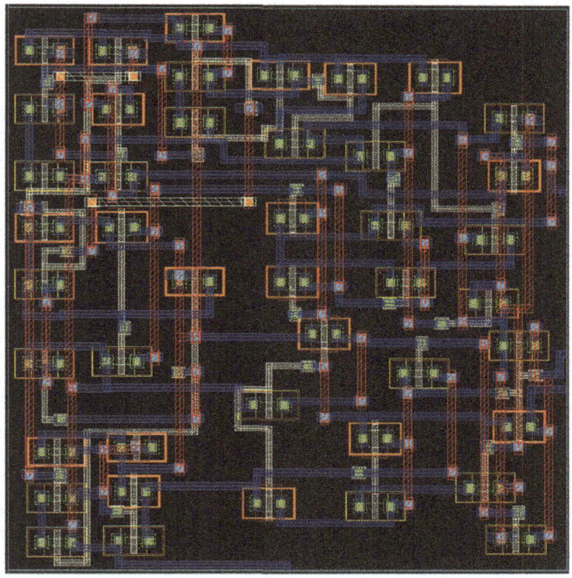

Fig. 12 Layout of 42 T proposed multiplier

The proposed multiplier utilizes the efficient proposed HA and the obtained results show that the average power dissipation of proposed multiplier is lower than conventional multiplier by 12.42%. Figure 9 shows the comparative plot of average power consumption of conventional 64 T, and proposed 42 T multiplier.

5 Conclusion

XOR and XNOR gates are reiteratively involved in high-performance data processing unit such as multiplier, adders, etc. In this work, design and implementation of different HA are carried out. The novel design of half adder utilizes enhanced XOR gates. Proposed half adder is further utilized to implement 2 bit multiplier. The proposed multiplier has low power consumption than the conventional multiplier due to the reduction in the number of transistor and symmetric input capacitance of XOR gate. It also has considerable lower area than conventional multiplier and the same can be observed from the results. The area and power parameters obtained for multiplier conclusively prove that the proposed HA performs considerably better and has lower area than the conventional HA.

References

1. Ahmed I, Shahid MK (2019) Analysis of CMOS full Adder circuits. In: 2019 international conference on advanced science, engineering and technology, pp 1–5
2. Oklobdzija VG, Villeger D (1995) Improving multiplier design by using improved column compression tree and optimized final adder in CMOS technology. IEEE Trans Very Large Scale Integr Syst 3(2):292–301
3. Hasan M, Hossein MJ, Hossain M, Zaman HU, Islam S (2019) Design of a scalable low-power 1-bit hybrid full adder for fast computation. IEEE Trans Circuits Syst II Express Briefs PP(c):1–1
4. Rabaey JM, Chandrakasan A, Nikolic B (2003) Digital integrated circuits a design perspective, 2nd edn
5. Kim JH, Byun YT, Jhon YM, Lee S, Woo DH, Kim SH (2003) All-optical half adder using semiconductor optical amplifier based devices. Opt Commun 218(4–6):345–349
6. Kumar P, Bhandari NS, Bhargav L, Rathi R, Yadav SC (2017) Design of low power and area efficient half adder using pass transistor and comparison of various performance parameters. In: 2017 international conference on computing, communication and automation (ICCCA), vol 2017-January, pp 1477–1482
7. Akashe S, Tiwari NK, Shrivas J, Sharma R (2012) A novel high speed & power efficient half adder design using. In: 2012 IEEE international conference on advanced communication control and Computing technologies (ICACCCT), no 978, pp 157–161
8. Naseri H, Timarchi S (2018) Low-power and fast full adder by exploring new XOR and XNOR gates. IEEE Trans Very Large Scale Integr Syst 26(8):1481–1493
9. Weste NEH, Harris DM (2013) CMOS VLSI design: a circuits and Systems perspective
10. Bhattacharyya P, Kundu B, Ghosh S, Kumar V, Dandapat A (2015) Performance analysis of a low-power high-speed hybrid 1-bit full adder circuit. IEEE Trans Very Large Scale Integr Syst 23(10):2001–2008
11. Yadav A (2017) Novel low power and high speed CMOS based XOR/XNORs using systematic cell design methodology, vol 5, no 3
12. Aguirre-Hernandez M, Linares-Aranda M (2011) CMOS full-adders for energy-efficient arithmetic applications. IEEE Trans Very Large Scale Integr Syst 19(4):718–721
13. Yan W, Park CH (2008) Filtering technique to lower phase noise for 2.4 GHz CMOS VCO. In: 2008 9th International Conference on Solid-State and Integrated-Circuit Technology (ICSICT), pp 1649–1652s
14. Kaur R, Mehra R (2016) Power and Area Efficient CMOS Half Adder using GDI Technique. Int J Eng Trends Technol 36(8):401–405
15. Hegazi E, Sjöland H, Abidi AA (2001) A filtering technique to lower LC oscillator phase noise. IEEE J Solid-State Circuits 36(12):1921–1930
16. Rael JJ, Abidi AA (2000) Physical processes of phase noise in differential LC oscillators. In: Proceedings of the IEEE 2000 Custom Integrated Circuits Conference, vol 3, no c, pp 569–572
17. Sureka N, Porselvi R, Kumuthapriya K (2013) An efficient high speed Wallace tree multiplier. In: 2013 International Conference on Information Communication and Embedded Systems (ICICES), pp 1023–1026

Security Analysis of a Threshold Quantum State Sharing Scheme of an Arbitrary Single-Qutrit Based on Lagrange Interpolation Method

Manoj Kumar, M. K. Gupta, R. K. Mishra, Sudhanshu Shekhar Dubey, Ajay Kumar, and Hardeep

Abstract Nowadays, quantum computing is gaining popularity in managing and distributing keys for data transfer this is because of Heisenberg uncertainty principle of quantum physics, which provides unconditional security compared to conventional cryptosystem. The present paper proposes two quantum secret sharing (QSS) schemes, using Lagrange's interpolation method, for conventional and quantum key distribution system. Security analysis of the proposed schemes exhibits that our schemes are immune to the possible existing attacks, namely intercept-and-resend, photon number splitting (PNS) attack and participants attack. The proposed schemes are studied theoretically; their design and implementation in practical quantum channel indeed may require further research work.

Keywords Quantum cryptography · Threshold QSS · Single qutrit · Lagrange interpolation · Unitary phase shift operation · Three-level mutually unbiased bases

---

M. Kumar · S. S. Dubey (✉) · Hardeep
Department of Mathematics and Statistics, Gurukul Kangri Vishwavidyalaya, Haridwar 249404, India
e-mail: sudhanshusdubey@gmail.com

M. Kumar
e-mail: sdmkg1@gmail.com

Hardeep
e-mail: hardeeppawariya1994@gmail.com

M. K. Gupta
Department of Mathematics, Chaudhary Charan Singh University, Meerut 250004, India
e-mail: mkgupta2002@hotmail.com

R. K. Mishra
Department of Applied Science, G. L. Bajaj Institute of Technology and Management, Greater Noida, 201306, India
e-mail: rk.mishra@glbitm.org

A. Kumar
Defence Scientific Information & Documentation, Defence Research & Development Organization, Delhi 110054, India
e-mail: ajaydesidoc@gmail.com

P. K. Singh et al. (eds.), *Evolving Technologies for Computing, Communication and Smart World*, Lecture Notes in Electrical Engineering 694,
https://doi.org/10.1007/978-981-15-7804-5_28

1 Introduction

With the advancement of the quantum computing, it is unconditionally secure to transfer data between two remote parties. Quantum computing schemes are based on Heisenberg uncertainty principle and do not allow users to create an exact copy of an unknown quantum state. Furthermore, it prevents an attacker to intercept-and-resend the encrypted data. In our daily life, we face a number of problems where we want to share secret information among a group of participants securely. The study on secret sharing scheme is a very fascinating field of research in cryptography since 1979, when Shamir [1] and Blakley [2] have independently proposed the idea of sharing information or data among the group of participants.

In cryptography, QSS schemes [3, 4] are very crucial methods for sharing important data and information among the users of same [5] or different [3] groups. QSS schemes are the extension of traditional secret sharing schemes [1, 6]. In fact, QSS schemes involve shares which are physical particle (known as photons) that store the information to be shared among the participants. It was the Hillery [7] who first time securely shared quantum information in the presence of eavesdroppers. Deng et al. [8] have proposed a multiparty QSS scheme based on entangled state of an arbitrary two qubit states. A QSS scheme in d-dimensional space with adversary structure was suggested by Qin and Dai [9] in 2016. In 2017, Wang et al. [10] have presented a multilayer QSS scheme based on GHZ state [11] and generalized Bell basis measurement in multiparty agents. Recently, Qin et al. [12] have proposed a rational QSS scheme in which they have adopted the concept of game theory to analyze the behavior of rational participants and designed a protocol to clog them from deviating from the protocol. In the mean time, Qin et al. [13] have proposed multi-dimensional QSS scheme using quantum Fourier transform. Very recently, in 2019, Matsumoto [14] has presented a strongly secure quantum ramp secret sharing scheme based on algebraic curves over finite fields. In the mean time, Habibidavijani and Sanders [15] have proposed a continuous variable ramp QSS scheme based on Gaussian states and operations.

As we know that a (t, n)-threshold QSS scheme [16, 17] involves the distribution of a secret data into n-participants such that no participants less than t can reconstruct the secret message. Such types of schemes are helpful where some important data or information is shared among n-participants and the original secret need to be reconstructed by $t(\leq n)$-participants at any time when some of the participants say ψ^{t-1} or less participants are absent from the system at that time. Yang et al. [18] have suggested a verifiable threshold QSS scheme to reduce the much quantum authentication information required in realistic implementation of any threshold QSS scheme. Later on, Kumar [5] has proposed a verifiable threshold QSS scheme using interpolation method. Recently, Lu et al. [19] have proposed a threshold QSS scheme for a single qubit. In this scheme, they have used Shamir's secret sharing technique [1] and unitary phase shift operation on a single qubit to share classical information as well as quantum states. Very recently, Chen et al. [20] have presented quantum homomorphic scheme with flexible number of evaluator using threshold QSS scheme.

In the present paper, authors used the Shamir's method [1] to propose a threshold QSS scheme on a single qutrit. In this work, we have proposed a threshold QSS scheme based on a single qutrit using Lagrange's interpolation method. The present paper consists of seven sections. Section 1 is introductory in nature and it consists of the brief literature review of previous research done on the quantum secret sharing scheme. In Sect. 2, some preliminary results and mathematical background of qutrits are discussed. Section 3 consists of the proposed threshold conventional and quantum state sharing schemes based on single qutrit using Lagrange's interpolation technique. In Sect. 4, we described the correctness of the proposed work. In Sect. 5, we gave a concrete illustration of the proposed scheme followed by the security analysis in Sect. 6. Finally, the last section concludes the present research work.

2 Some Preliminaries

This section involves some basic definitions and results concerning qutrit and their mathematical background which are very gainful to perceive the proposed scheme.

2.1 Qutrit [21]

A qutrit $|A\rangle$ is defined as

$$|A\rangle = \alpha|0\rangle + \beta|1\rangle + \gamma|2\rangle \tag{2.1}$$

where $|\alpha|^2 + |\beta|^2 + |\gamma|^2 = 1$.

2.2 Sequence of Qutrits [22]

A qutrit sequence $\{|A_q\rangle\}$ is defined as

$$\{|A_q\rangle = \alpha_q|0\rangle + \beta_q|1\rangle + \gamma_q|2\rangle\} \tag{2.2}$$

where $|\alpha_q|^2 + |\beta_q|^2 + |\gamma_q|^2 = 1$ and $q = 1, 2, \ldots, m$.

2.3 Unitary Transformation for Qutrits

Unitary transformation denoted by $U(\psi)$ for the qutrit $|A\rangle$ is defined as

$$U(\psi) = |0\rangle\langle 0| + e^{i\psi}|1\rangle\langle 1| + e^{i\psi}|2\rangle\langle 2| = \begin{bmatrix} 1 & 0 & 0 \\ 0 & e^{i\psi} & 0 \\ 0 & 0 & e^{i\psi} \end{bmatrix} \tag{2.3}$$

where $\psi \in (0, 2\pi)$.

2.4 Dealer

It is a trusted party which divides secret information (to be shared among n-participants) into n-parts (each called share) and distributes these shares into n-participants which wish to share the secret information among them.

2.5 Decoy Particles

Decoy particles are some fake states of qutrit which are randomly prepared by the sender participant and inserted in a sequence of qutrits during the sharing of secret information among the participants to maintain the security of the sharing secret information.

2.6 Lagrange's Interpolation Method [23]

For a given set of $(n+1)$ points say (x_j, y_j), $j = 0, 1, 2, \ldots, n$, we can construct the following n-degree Lagrange's interpolation polynomial $f(x)$ as

$$f(x) = \sum_{j=1}^{n} f(x_j) \prod_{r=0, r\neq j}^{n} \frac{x - x_r}{x_j - x_r} = a_0 + a_1 x + a_2 x^2 + \cdots a_n x^n \tag{2.4}$$

where $y_j = f(x_j)$ and $a_0, a_1, a_2, \ldots, a_n$ are some known finite constants over the finite field F_p.

In our proposed scheme, secret information s is constructed by the following assumption that $s = f(0)$. The following lemma describes that on applying a finite number of unitary phase operations on a qutrit $|A\rangle$, it changes into a single unitary phase operation as the sum of their inputs.

Lemma 2.1 [22] *If* $\psi^1, \psi^2, \ldots, \psi^n \in (0, 2\pi)$ *and* $U(\psi)$*is an unitary transformation then for any qutrit* $|A_q\rangle$ *we have*

$$U(\psi^1)U(\psi^2)\ldots U(\psi^n)|A_q\rangle = U(\psi^1 + \psi^2 + \cdots + \psi^n)|A_q\rangle.$$

Proof First, we will show the given result is true for ψ^1 and ψ^2.

For this, we have

$$\begin{aligned} U(\psi^1)U(\psi^2)|\varphi_k\rangle &= \begin{bmatrix} 1 & 0 & 0 \\ 0 & e^{i\psi^1} & 0 \\ 0 & 0 & e^{i\psi^1} \end{bmatrix} \begin{bmatrix} 1 & 0 & 0 \\ 0 & e^{i\psi^2} & 0 \\ 0 & 0 & e^{i\psi^2} \end{bmatrix} |\varphi_k\rangle \\ &= \begin{bmatrix} 1 & 0 & 0 \\ 0 & e^{i(\psi^1+\psi^2)} & 0 \\ 0 & 0 & e^{i(\psi^1+\psi^2)} \end{bmatrix} |\varphi_k\rangle \\ &= U(\psi^1 + \psi^2)|\varphi_k\rangle \end{aligned}$$

This shows that it is true for ψ^1 and ψ^2.

In the same way, the result can be generalized for n-variables $\psi^1, \psi^2, \psi^3, \ldots, \psi^n$.

Lemma 2.2 *If each participant* P_j *in Shamir's secret sharing scheme has a pair* $(x_j, f(x_j))$, $j = 1, 2, \ldots, k$ *with* $t \le q \le n$ *of public information and share, respectively, then the original secret information* s *can be rebuild by using the Lagrange's interpolation polynomial* $f(x) = a_0 + a_1x + a_2x^2 + \cdots + a_{t-1}x^{t-1}$ *of degree* $t-1$ *over the finite field* F_p *for large prime* p *i.e.*

$$s = f(0) = \sum_{j=1}^{q} \delta_j \bmod p, \text{ where } \delta_j = f(x_j) \prod_{r=1, r\ne j}^{t} \frac{x_r}{x_r - x_j} \tag{2.5}$$

Proof We know that Lagrange interpolation formula for q points $(x_j, f(x_j))$, $j = 1, 2, \ldots, q$, under addition modulo p, where is p prime number and is given by

.

$$f(x) = \left(\begin{array}{l} \frac{(x-x_2)(x-x_3)\ldots(x-x_q)}{(x_1-x_2)(x_1-x_3)\ldots(x_1-x_q)} f(x_1) \\ + \frac{(x-x_1)(x-x_3)\ldots(x-x_q)}{(x_2-x_1)(x_2-x_3)\ldots(x_2-x_q)} f(x_2) + \cdots \\ + \frac{(x-x_1)(x-x_2)\ldots(x-x_{q-1})}{(x_q-x_1)(x_q-x_2)\ldots(x_q-x_{q-1})} f(x_q) \end{array} \right) \bmod p$$

$$f(x) = \sum_{j=1}^{q} f(x_j) \prod_{r=1, r\neq j}^{q} \frac{x - x_r}{x_j - x_r} \bmod p \tag{2.6}$$

Put $x = 0$ in Eq. (2.6), we get

$$f(0) = \sum_{j=1}^{q} f(x_j) \prod_{r=1, r\neq j}^{q} \frac{x_r}{x_r - x_j} \bmod p \tag{2.7}$$

Since $s = f(0)$, therefore, Eq. (2.7) implies that

$$s = f(0) = \sum_{j=1}^{q} f(x_j) \prod_{r=1,r\neq j}^{q} \frac{x_r}{x_r - x_j} \bmod p \tag{2.8}$$

Using (2.5) in (2.8), we get the required result

$$s = f(0) = \sum_{j=1}^{q} \delta_j \bmod p \tag{2.9}$$

This completes the proof of the lemma.

Corollary 2.3 *If in Lemma 2.2, the original secret s is reconstructed by the expression* $s = \sum_{j=1}^{t} \delta_j \bmod p$ *with* $\delta_j = f(x_j) \prod_{r=1,r\neq j}^{t} \frac{x_r}{x_r - x_j} \bmod p$, *then* $\sum_{j=1}^{q} \delta_j = Np + s$, *where N is a positive integer.*

Proof From the theory of numbers, we know that

$$a \equiv b \,(\bmod\, p) \text{ if and only if } b - a = Np \text{ or if } b = a + Np \tag{2.10}$$

where N is a positive integer.

Using the above assertion (2.10) to $s = \sum_{j=1}^{t} \delta_j \bmod p$, we immediately follow that $\sum_{j=1}^{q} \delta_j = Np + s$.

3 Proposed Work

The proposed work involves two threshold quantum secret sharing schemes namely conventional information sharing scheme and quantum state sharing scheme. Both schemes ply Shamir's secret sharing method in confluence with phase shift operation on single qutrit to interpolate and rebuild original secret state to share secret information. The following two subsections explain in detail each of the proposed schemes.

3.1 Conventional Information Sharing Scheme

Suppose t-participants, say $P_1, P_2, \ldots, P_t$ with $1 \leq t \leq n$, out of n-participants request a dealer (which has qutrit information $\{|A_q\rangle : 1 \leq q \leq m\}$ defined by (2.2) to rebuild the initial secret information $|A_q\rangle$. This scheme consists of the following two phases, namely share distribution phase and secret reconstruction phase, as explained under:

Phase -I: Share Distribution Phase: This phase (see in Fig. 1) consists of the following steps:

1. First of all dealer chooses a very large prime number p, a finite field F_p and defined a following polynomial $f(x)$ of degree $t-1$ over F_p as

$$f(x) = a_0 + a_1 x + a_2 x^2 + \cdots + a_{t-1} x^{t-1} \tag{3.1}$$

   where $a_0, a_1, a_2, \ldots, a_{t-1}$ are known finite constants in F_p and $s = a_0 = f(0)$ is the secret information value which is to be kept private by the dealer.
2. For the given public information $x_j, x_k \in F_p$ of the participant P_j, dealer uses polynomial (3.1) to calculate the share $f(x_j)$ of P_j where $x_j \neq x_k$ for $j \neq k$.
3. Finally, dealer use the conventional communication methods discussed in [13, 24, 25] to distribute share $f(x_j)$ to each participants P_j.

Phase-II: Secret Reconstruction Phase: This phase (see in Fig. 1) consists of t-loops of which first two loops are explained in details. Rests of the loops are similar to the Loop 2.

Loop 1: This loop involving dealer and first participant P_1 consists of the following steps:

1. For the sharing of secret information $|A_q\rangle$ among the participants, dealer selects the following three mutually unbiased bases U, V and W as

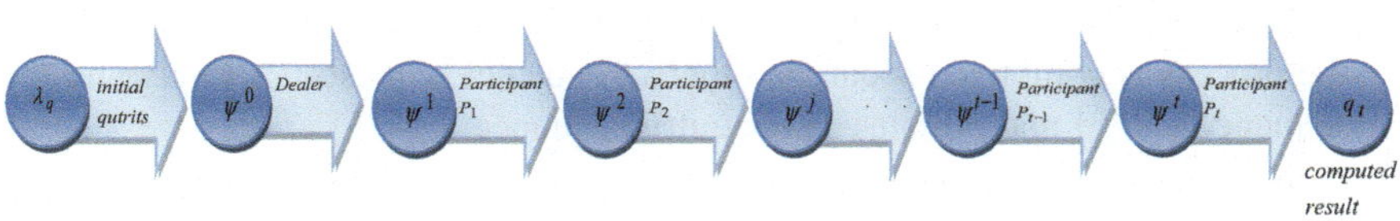

Fig. 1 Sequential operations on each qutrit in Conventional Information Sharing (λ_q is first phase value of qutrit $|A_q\rangle$)

$$U = \left\{ |u_1\rangle = \frac{1}{\sqrt{3}}(|0\rangle + |1\rangle + |2\rangle), |u_2\rangle = \frac{1}{\sqrt{3}}\left(|0\rangle + \omega|1\rangle + \omega^2|2\rangle\right), \right.$$
$$\left. |u_3\rangle = \frac{1}{\sqrt{3}}\left(|0\rangle + \omega^2|1\rangle + \omega|2\rangle\right) \right\} \tag{3.2}$$

$$V = \left\{ |v_1\rangle = \frac{1}{\sqrt{3}}(|0\rangle + \omega|1\rangle + \omega|2\rangle), |v_2\rangle = \frac{1}{\sqrt{3}}(|0\rangle + \omega|1\rangle + |2\rangle), \right.$$
$$\left. |v_3\rangle = \frac{1}{\sqrt{3}}(|0\rangle + |1\rangle + \omega\,|2\rangle) \right\} \tag{3.3}$$

$$W = \left\{ |w_1\rangle = \frac{1}{\sqrt{3}}\left(|0\rangle + |1\rangle + \omega^2|2\rangle\right), |w_2\rangle = \frac{1}{\sqrt{3}}\left(|0\rangle + \omega^2|1\rangle + |2\rangle\right), \right.$$
$$\left. |w_3\rangle = \frac{1}{\sqrt{3}}\left(|0\rangle + \omega^2|1\rangle + \omega^2|2\rangle\right) \right\} \tag{3.4}$$

where $\omega = e^{\frac{2\pi i}{3}}$ and $\omega^2 = e^{\frac{4\pi i}{3}}$.

2. Dealer constructs a random sequence of qutrits $\left\{|A_q\rangle : q = 1, 2, \ldots, m\right\}$ where each qutrit $|A_q\rangle$ has one state in $\{U,\ V,\ W\}$. Using this terminology we can write each qutrit $|A_q\rangle$ as

$$|A_q\rangle = \frac{1}{\sqrt{3}}\left(|0\rangle + e^{i\mu_q}|1\rangle + e^{iv_q}|1\rangle\right) \tag{3.5}$$

where $(\mu_q, v_q) \in \left\{(0,0), \left(0, \frac{2\pi}{3}\right), \left(\frac{2\pi}{3}, 0\right), \left(0, \frac{4\pi}{3}\right), \left(\frac{4\pi}{3}, 0\right), \left(\frac{2\pi}{3}, \frac{4\pi}{3}\right), \left(\frac{4\pi}{3}, \frac{2\pi}{3}\right), \left(\frac{2\pi}{3}, \frac{2\pi}{3}\right), \left(\frac{4\pi}{3}, \frac{4\pi}{3}\right)\right\}$.

3. Dealer operates unitary phase operation $U(\psi^0)$ on every element of qutrit sequence $\{\,|A_q\rangle\}$ to get a new transform qutrit sequence $\left\{|A_q^0\rangle = U(\psi^0)|A_q\rangle\right\}$ where $\psi^0 = -\frac{2\pi s}{p}$ and s are the private secret information value.
4. Next dealer adds some decoy qutrits randomly picked from the bases $\{U,\ V,\ W\}$ into the sequence $\{|A_q^0\rangle\}$ to get a new sequence $\{|B_q^0\rangle\}$ involving decoy qutrits.
5. After noting the place of every decoy qutrits, dealer transmits the expanded sequence $\{|B_q^0\rangle\}$ to participant P_1.
6. Dealer measures the error probability by juxtaposing the estimated results of P_1 with the qutrit sequence $\{|A_q^0\rangle\}$.
7. If the error ratio is smaller than the threshold value, then dealer divulged that the process is completed securely, otherwise, dealer requests P_1 to give up the sequence $\{|B_q^0\rangle\}$ and initiates a new qutrit sequence.
8. After securely received sequence $\{|B_q^0\rangle\}$, P_1 isolates the decoy qutrits from it to get the initial transformed sequence $\{|A_q^0\rangle\}$.

Loop 2: This loop involving participants P_1 and P_2 consist of the following steps:

1. After calculating $\delta_1 = f(x_1)\prod_{r=2}^{t} \frac{x_r}{x_r - x_1} \bmod p$, participants P_1 applies unitary transformation $U(\psi^1)$ on every element of qutrit sequence $\{\,|A_q^0\rangle\}$ to get a new

transformed qutrit sequence $\{|A_q^1\rangle = U(\psi^1)|A_q^0\rangle\}$ where $\psi^1 = \lambda_1 + \frac{2\pi\,\delta_1}{p}$ is chosen by P_1.

2. Next P_1 adds the some decoy qutrits (randomly picked from U , V and W) into the sequence $\{|A_q^1\rangle\}$ to get a new sequence $\{|B_q^1\rangle\}$ involving decoy qutrits.
3. After noting the place of every decoy qutrit, dealer sends the sequence $\{|B_q^1\rangle\}$ to P_2.
4. After the $P_2's$ cooperation of accepting the qutrit sequence $\{|B_q^1\rangle\}$, P_1 divulged the places and related measuring bases U , V and W of the decoy qutrits.
5. After estimating decoy qutrits of $\{|B_q^1\rangle\}$ in related bases U , V and W, P_2 publicizes his/her estimated results.
6. Participant P_1 measures the error probability by juxtaposing the estimated results of P_2 with the qutrit sequence $\{|A_q^1\rangle\}$.
7. If the error ratio is smaller than the threshold value, then P_1 divulged that the process is completed securely, otherwise, P_1 requests P_2 to give up the sequence $\{|B_q^1\rangle\}$ and initiates a new qutrit sequence.
8. After securely received sequence $\{|B_q^1\rangle\}$, P_2 removes the decoy qutrits from it to get the initial transformed sequence $\{|A_q^1\rangle\}$.

In a similar way, other loops can be explained.

After applying unitary phase operation by last participant P_t on the qutrit sequence received by participant P_{t-1}, the state of the qutrit sequence is given by

$$|A_q^t\rangle = U(\psi^0)U(\psi^1)\ldots U(\psi^t)|A_q\rangle$$

$$|A_q^t\rangle = \frac{1}{\sqrt{3}}\left(|0\rangle + e^{i\left[\mu_q+\sum_{j=1}^{t}\lambda_j+\frac{2\pi}{p}\left(\sum_{j=1}^{t}\delta_j-s\right)\right]}|1\rangle + e^{i\left[\upsilon_q+\sum_{j=1}^{t}\lambda_j+\frac{2\pi}{p}\left(\sum_{j=1}^{t}\delta_j-s\right)\right]}|2\rangle\right) \tag{3.6}$$

$$= \frac{1}{\sqrt{3}}\left(|0\rangle + e^{i\left[\mu_q+\sum_{j=1}^{t}\lambda_j\right]}|1\rangle + e^{i\left[\upsilon_q+\sum_{j=1}^{t}\lambda_j\right]}|2\rangle\right) \tag{3.7}$$

Every participant segregates his operation into three classes U , V and W according to the choices of $(\mu_q, \upsilon_q) \in \{(0, 0), (\frac{2\pi}{3}, \frac{4\pi}{3}), (\frac{4\pi}{3}, \frac{2\pi}{3})\}$, $(\mu_q, \upsilon_q) \in \{(\frac{2\pi}{3}, \frac{2\pi}{3}), (\frac{2\pi}{3}, 0), (0, \frac{2\pi}{3})\}$ and $(\mu_q, \upsilon_q) \in \{(\frac{4\pi}{3}, \frac{4\pi}{3}), (0, \frac{4\pi}{3}), (\frac{4\pi}{3}, 0)\}$, respectively. Each participant intimates the dealer about his/her classification of operation for each qutrit through classical communication. According to classes of each participant's phase operation on each qutrit dealer calculate the base in which last participant P_t measures the qutrit state. In particular, the measurement base will be U for $(\mu_q, \upsilon_q) \in \{(0, 0), (\frac{2\pi}{3}, \frac{4\pi}{3}), (\frac{4\pi}{3}, \frac{2\pi}{3})\}$, it will be V for $(\mu_q, \upsilon_q) \in \{(\frac{2\pi}{3}, \frac{2\pi}{3}), (\frac{2\pi}{3}, 0), (0, \frac{2\pi}{3})\}$ and for the choice of $(\mu_q, \upsilon_q) \in$

$\{(\frac{4\pi}{3}, \frac{4\pi}{3}), (0, \frac{4\pi}{3}), (\frac{4\pi}{3}, 0)\}$ measurement base will be W. After the measurement, participant P_t divulges each result q_t, $q = 1, 2, \ldots, m$ through classical communication methods [13, 24, 25].

3.2 State Sharing Scheme for QKD

This scheme is proposed for state sharing in QKD and it is identical to conventional secret sharing scheme described in the previous subsection. Similar to conventional secret sharing scheme this scheme also consists of two phases, of which, share distribution phase is exactly same as phase-I of conventional secret sharing scheme. The secret reconstruction phase (see in Fig. 2) for the proposed scheme consists of the following steps:

1. Dealer constructs a random sequence of qutrits $\{|A_q\rangle : q = 1, 2, \ldots, m\}$ and applies unitary phase operation $U(\psi^0)$ on every qutrit of the sequence to get a new sequence $\left\{ |A_q^0\rangle : |A_q^0\rangle = \alpha_q|0\rangle + \beta_q e^{\psi^0}|1\rangle + \gamma_q e^{\psi^0}|2\rangle \right\}$ of qutrit as where $\psi^0 = -\frac{2\pi s}{p}$.
2. Dealer inserts some decoy particles (randomly chosen from the bases U , V and W) into the sequence $\{|A_q^0\rangle\}$ to get a new sequence $\{|B_q^0\rangle\}$ consisting decoy qutrits.
3. After noticing the position of every decoy qutrits, dealer sends the sequence $\{|B_q^0\rangle\}$ to participant P_1. After the $P_1's$ corroboration of accepting the qutrit sequence $\{|B_q^0\rangle\}$, dealer divulged the positions and related measuring bases U , V and W of the decoy particles.
4. After estimating decoy qutrits of $\{|B_q^0\rangle\}$ under the bases U , V and W, P_1 publicizes his/her estimated results.
5. Dealer measures the error probability by juxtaposing the estimated results of P_1 with the qutrit sequence $\{|A_q^0\rangle\}$.
6. If the error ratio is smaller than the threshold value, then dealer divulged that the process is completed securely, otherwise, dealer requests P_1 to give up the sequence $\{|B_q^0\rangle\}$ and initiates a new qutrit sequence.
7. After securely received sequence $\{|B_q^0\rangle\}$, P_1 separates the decoy qutrits from it to get the initial transformed sequence $\{|A_q^0\rangle\}$.

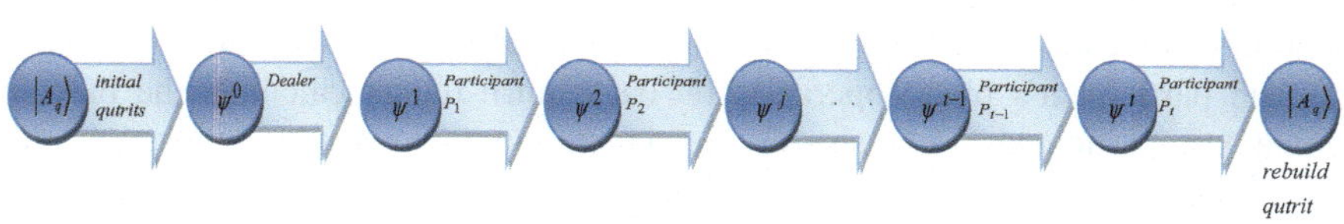

Fig. 2 Sequential operations on each qutrit in Quantum state sharing

A similar process consisting step 1 to 7 occurs between participants P_1 and P_2, P_2 and P_3, …, P_{t-1} and P_t under the unitary operation $U(\psi^1)$, $U(\psi^2)$, …, $U(\psi^t)$, respectively, where $\psi^j = \frac{2\pi\delta_j}{p}$ and δ_j are defined by Eq. (2.5). Thus, the original quantum state, reconstructed by the corroboration of all the t-participants and the dealer, is given by

$$|A_q^t\rangle = \frac{1}{\sqrt{3}}\left(\alpha_q|0\rangle + \beta_q e^{\frac{2\pi i}{p}\left(\sum_{j=1}^{t}\delta_j - s\right)}|1\rangle + \gamma_q e^{\frac{2\pi i}{p}\left(\sum_{j=1}^{t}\delta_j - s\right)}|2\rangle\right) \tag{3.8}$$

Finally, using the Corollary 2.3 in Eq. (3.8), we get the required initial state $|A_q\rangle$.

4 Correctness of the Proposed Scheme

4.1 For Conventional Secret Sharing Scheme

The correctness of this scheme is immediately implied by Lemma 3.2.1 given in Sect. 3.2.

4.2 For quantum State Sharing Scheme

In view of Lemma 2.1, after applying unitary phase operations $U(\psi^0)$ and $U(\psi^1)$, $U(\psi^2)$,..., $U(\psi^t)$ by the dealer and all t-participants P_1,P_2,…,P_t, respectively, the final state of the original qutrit $|A_q\rangle$ (consisting of secret information), is given by

$$\begin{aligned} |A_q^t\rangle &= U(\psi^t)\,U(\psi^{t-1})\,\ldots\,U(\psi^2)\,U(\psi^1)\,U(\psi^0)\,|A_q\rangle \\ |A_q^t\rangle &= U(\psi^t + \psi^{t-1} + \cdots + \psi^2 + \psi^1 + \psi^0)\,|A_q\rangle \end{aligned} \tag{4.1}$$

Substituting the values of ψ^k, with $k = 0, 1, 2, \ldots, t$, in (4.1) and then summing all the terms inside U, we get

$$|A_q^t\rangle = U\left(\frac{2\pi}{p}\left(\left(\sum_{j=1}^{t}\delta_j\right) - s\right)\right)|A_q\rangle \tag{4.2}$$

Now using Corollary 2.3 and definition of unitary transformation in Eq. (4.2), we obtain the required secret information as

$$\left| A_q^t \right\rangle = \left| A_q \right\rangle \tag{4.3}$$

This vindicates the correctness of the proposed scheme for quantum state sharing.

5 Concrete Illustration of the Proposed Scheme

We will justify our proposed schemes with the help of the following example consisting of a (3, 5) threshold QSS scheme over the finite field F_{17}. In this example, we have $t = 3$, $n = 5$ and $p = 17$.

Phase-I: Key Distribution Phase (common to both schemes): This phase completes in the following steps:

1. Dealer chooses a random polynomial $f(x) = 13 + 5x + 12x^2$ of degree two over F_{17}, with the secret information $s = 13 = f(0)$.
2. Using the polynomial $f(x) = 13 + 5x + 12x^2$ with the relation $y_j = f(x_j)$, dealer calculates y_j (with $j = 1, 2, 3, 4, 5$) as a share for each participant P_j, as follows:
   $y_1 = f(x_1 = 2) = 71 \pmod{17} = 3$, $y_2 = f(x_2 = 3) = 136 \pmod{17} = 0$
   $y_3 = f(x_3 = 4) = 225 \pmod{17} = 4$ $y_4 = f(x_4 = 5) = 338 \pmod{17} = 15$
   and $y_5 = f(x_5 = 6) = 475 \pmod{17} = 16$.
3. Finally, dealer use the conventional communication methods of [13, 24, 25] to distribute the shares $y_1 = 3$, $y_2 = 0$, $y_3 = 4$, $y_4 = 15$ and $y_5 = 16$ to the participants P_1, P_2, P_3, P_4 and P_5, respectively.

Phase-II (a): For conventional information sharing: If participants P_1,P_3 and P_4 wish to reconstruct the secret information s, then this phase is accomplished by the following steps:

1. Dealer constructs a random sequence of qutrits $\left\{ \left| A_q \right\rangle : q = 1, 2, \ldots, m \right\}$ where each qutrit $\left| A_q \right\rangle$ has one state in $\{U, V, W\}$.
2. Now dealer puts $\left(\mu_q, \upsilon_q \right) = \left(\frac{4\pi}{3}, \frac{2\pi}{3} \right)$ in Eq. (3.5) to obtain

$$\left| A_q \right\rangle = \frac{1}{\sqrt{3}} \left(|0\rangle + \omega^2 |1\rangle + \omega |2\rangle \right) \tag{5.1}$$

   and chooses $\psi^0 = -\frac{2\pi s}{p} = -\frac{26\pi}{17}$ to perform unitary operation on the above qutrit state (5.1).
3. Using the relation $\delta_j = y_j \prod_{r=1, r \neq j}^{3} \frac{x_r}{x_r - x_j} \bmod 17$, used in Lagrange's interpolation method, participants P_1,P_3 and P_4, respectively, calculate δ_1, δ_3 and δ_4 as follow:

$$\delta_1 = f(x_1) \prod_{r=2}^{t} \frac{x_r}{x_r - x_1} \bmod 17 = f(2).\frac{4}{4-2}.\frac{5}{5-2} \bmod 17 = 10 \quad (5.2)$$

$$\delta_3 = f(x_3) \prod_{r=1,\, r \neq 3}^{t} \frac{x_r}{x_r - x_3} \bmod 17 = f(4).\frac{2}{2-4}.\frac{5}{5-4} \bmod 17 = 14 \quad (5.3)$$

$$\delta_4 = f(x_4) \prod_{r=1,\, r \neq 4}^{t} \frac{x_r}{x_r - x_4} \bmod 17 = f(5).\frac{2}{2-5}.\frac{4}{4-5} \bmod 17 = 6 \quad (5.4)$$

4. After choosing $\lambda_1 = \frac{2\pi}{3}$ and using his/her δ_1, participant P_1 calculates $\psi^1 = \lambda_1 + \frac{2\pi\delta_1}{p} = \frac{2\pi}{3} + \frac{20\pi}{17}$ to perform unitary operation on qutrit state (5.1). Similarly, $\lambda_3 = 0$ and $\lambda_4 = \frac{4\pi}{3}$ are selected by the respective participants P_3 and P_4 to calculate ψ^3 and ψ^4 as follow: $\psi^3 = \lambda_3 + \frac{2\pi\delta_3}{p} = \frac{28\pi}{17}$ and $\psi^4 = \lambda_4 + \frac{2\pi\delta_4}{p} = \frac{4\pi}{3} + \frac{12\pi}{17}$.
5. After performing unitary operations on the qutrit state (5.1) by the dealer, and all the participants P_1, P_3 and P_4, the initial state can be reconstructed as follows:

$$U(\psi^0)\, U(\psi^1)\, U(\psi^3)\, U(\psi^4) \left|A_q\right\rangle = U\left(\psi^0 + \psi^1 + \psi^3 + \psi^4\right) \left|A_q\right\rangle \quad (5.5)$$

Finally, substituting the values of ψ^0, ψ^1, ψ^3, ψ^4 and using definition of unitary transformation in Eq. (5.5), we get

$$U(\psi^0)\, U(\psi^1)\, U(\psi^3)\, U(\psi^4) \left|A_q\right\rangle = \left|A_q\right\rangle \quad (5.6)$$

which is the required initial state (5.1).

Phase-II (b): For Quantum Information Sharing: It completes in the following steps:

1. Dealer chooses the same value of ψ^0 and a same random sequence of qutrits as in step-2 of phase-II (a).
2. By the cooperation of dealer with $\psi^0 = -\frac{2\pi s}{p} = -\frac{26\pi}{17}$, participants P_1, P_3 and P_4, respectively, choose $\psi^1 = \frac{20\pi}{17}$, $\psi^3 = \frac{28\pi}{17}$ and $\psi^4 = \frac{12\pi}{17}$ to rebuild the original qutrit state (5.1) as follow:

.

$$U(\psi^0)\, U(\psi^1)\, U(\psi^3)\, U(\psi^4) \left|A_q\right\rangle = U\left(\psi^0 + \psi^1 + \psi^3 + \psi^4\right) \left|A_q\right\rangle \quad (5.7)$$

Substituting the values of ψ^0, ψ^1, ψ^3 and ψ^4 in Eq. (5.7), we get

$$U(\psi^0)\,U(\psi^1)\,U(\psi^3)\,U(\psi^4)\,|A_q\rangle = U\left[\frac{-26\pi}{17} + \frac{20\pi}{17} + \frac{28\pi}{17} + \frac{12\pi}{17}\right]|A_q\rangle$$
$$= U(2\pi)\,|A_q\rangle \quad (5.8)$$

Finally, using definition of unitary transformation in Eq. (5.8), we get the required initial state (5.1).

6 Security Analysis

In this section, we will analyze how secure are our proposed schemes. The possible attacks on our proposed schemes are described as.

6.1 Intercept-And-Resend Attack

Since both the proposed schemes use decoy particles to protect from the forgery by an eavesdropper, therefore, an attacker can try to intercept-and-resend attack. For this, if he/she gets successful in obtaining the transmitted information, then he/she can only intercept the quantum sequence with knowing the states of the sequence. Subsequently, he/she fails to resend an exact copy of the sequence due to uncertainty and no-cloning principle of quantum mechanics. Besides this, an eavesdropper fails to know the states and places of decoy particles which results the attack an increase in an error and can be exposed with the probability $1-\left(\frac{1}{9}\right)^d$, where d is the number of decoy particles and for a very large number of decoy particles error detection expression $\left(1-\left(\frac{1}{9}\right)^d\right)$ converges to 1.

6.2 PNS Attack [22]

We know that PNS attack can be applied on realistic photon source which emit weak coherent pulses. Actually, these laser pulses produce very small number of photons which distributed in Poisson fashion. In fact, most laser pulses split no photons while very few laser pulses split more than one photon. In later case, an eavesdropper can gather the extra photons and send the remnant single photon to the receiver, which results the PNS attack. Generally, the attacker collects these extra photons in quantum memory until the receiver discovers the rest photon and the sender publishes the measuring bases U , V and W. In the mean time, an eavesdropper can measure his/her qutrits in the accurate bases and obtained secret information without precluding detectable errors.

6.3 Participants Attack

In conventional information sharing scheme, the first participant P_1 may be want to detect the information without the corroboration of other participants. Since, dealer has applied unknown unitary phase operation $U(\psi^0)$ on every element of the qutrit sequence, therefore, if participant P_1 gets successful to pick the correct base in the estimation of each qutrit $|A_q^0\rangle = \frac{1}{\sqrt{3}}\left(|0\rangle + e^{i(\mu_q+\psi^0)}|1\rangle + e^{i(\upsilon_q+\psi^0)}|2\rangle\right)$, then he/she will obtain $(\mu_q + \psi^0,\ \upsilon_q + \psi^0)$ in place of $(\mu_q,\ \upsilon_q)$ itself. Truly speaking, due to $\psi^0 = -\frac{2\pi s}{p}$ and $s \in F_p$, participant P_1 find out $(\mu_q,\ \upsilon_q)$ with the probability $\frac{1}{p}$ without knowing the exact value of $(\mu_q,\ \upsilon_q)$.

Since $(\mu_q, \upsilon_q) \in \{(0,0), (\frac{2\pi}{3}, 0), (0, \frac{2\pi}{3}), (0, \frac{4\pi}{3}), (\frac{4\pi}{3}, 0), (\frac{2\pi}{3}, \frac{4\pi}{3}), right\}$ $(\frac{4\pi}{3}, \frac{2\pi}{3}), (\frac{2\pi}{3} \frac{2\pi}{3}), (\frac{4\pi}{3}, \frac{4\pi}{3})\}$and p greater than 9, therefore, participant P_1 has the probability of $\frac{1}{9}$ to get the exact value of each $(\mu_q,\ \upsilon_q)$. Subsequently, the participant P_1 obtains no extra information about each $(\mu_q,\ \upsilon_q)$.

This shows that our schemes resist intercept-and-resend attack, PNS attack and participants attack.

7 Conclusion

Two quantum secret sharing schemes, using Lagrange's interpolation method, are proposed for conventional and quantum key distribution system. Security analysis of the proposed schemes exhibit that our schemes are immune to the possible existing attacks, namely intercept-and-resend, PNS and participants attack. Furthermore, a concrete example is given in Sect. 5 which indicates the correctness of the proposed schemes. The proposed schemes are more secure and efficient for sufficiently large value of p because the three mutually unbiased bases U , V and W endow less opportunity to make forgery on transmitted particles during conventional and quantum state sharing phase. Departing from qubit-based QSS scheme, our scheme has an important characteristic that it can be used to encode the large amount of classical data or information because the quantum state in the proposed scheme is studied in three-dimensional Hilbert space. More generally speaking, a three-dimensional-based quantum state is known as qutrit instead of qubit (which is based on two-dimensional Hilbert space). The proposed schemes are studied theoretically, there design and implementation in practical quantum channel indeed may require further research work.

Acknowledgements The authors would like to thanks the anonymous reviewers for their fruitful and valuable suggestions to enhance the clarity of the present manuscript.

References

1. Shamir A (1979) How to share a secret. Commun ACM 22(11):612–613
2. Blakley GR (1979) Safeguardimng cryptographic keys. In: Proceedings of AFIPS National Computer conference, New York, vol 48, pp 313–317
3. Gao T, Yan FL, Li YC (2009) Quantum secret sharing between m-party and n-party with six states. Sci China G, Phys Mech Astron 52(8):1191–1202
4. Kumar M (2015) An efficient secret sharing scheme for quantum key distribution. Int J Adv Res Sci Eng 4(3):318–324
5. Kumar M (2017) A verifiable threshold quantum secret sharing scheme using interpolation method. Int J Adv Res Comp Sci Softw Eng 7(7):42–47
6. Shamsoshoara A (2019) Overview of Blakley secret sharing scheme. https://arxiv.org/pdf/1901.02802.pdf
7. Hillery M, Buzek V, Berthiaume A (1999) Quantum secret sharing. Phys Rev A 59(3)
8. Deng F, Li X, Li C, Zhou P, Zhou HY (2005) Multiparty quantum-state sharing of an arbitrary two- particle state with Einstein-Podolsky-Rosen pair. Phys Rev A 72(4)
9. Qin H, Dai Y (2016) d-dimensional quantum state sharing with adversary structure. Quantum Inf Process 15(4):1689–1701
10. Wang X-J, An L-X, Yu X-T, Zhang Z-C (2017) Multilayer quantum secret sharing based on GHZ state and generalized Bell basis measurement in multiparty agents. Phys Lett A 381(38), 3282–3288
11. Liao C, Yang C, Hwang T (2014) Dynamic quantum secret sharing scheme based on GHZ state. Quantum Inf Process 13(8):1907–1916
12. Qin H, Tang WKS, Tso R (2018) Rational quantum secret sharing. Sci Rep 8(11115)
13. Qin H, Tso R, Dai Y (2018) Multi-dimensional quantum state sharing based on quantum Fourier transform. Quantum Inf Process 17(48)
14. Matsumoto R (2019) Strongly secure quantum ramp secret sharing constructed from algebraic curves over finite fields. Quantum Phys 13(1), 1–10
15. Habibidavijani M, Sanders BC (2019) Continuous-variable ramp quantum secret sharing with Gaussian states and operations. New J Phys 21(113023)
16. Qin H, Zhu X, Dai Y (2015) (t, n)-Threshold quantum secret sharing using the phase shift operation. Quantum Inf Process 14(8):2997–3004
17. Hao C, Wenping M (2017) (t, n)-threshold quantum state sharing scheme based on linear equations and unitary operation. IEEE Photon J 1(9)
18. Yang YG, Teng YW, Chai HP, Wen QY (2011) Verifiable quantum (k, n)-threshold secret key sharing. Int J Theor Phys 50(3):792–798
19. Lu C, Miao F, Meng K (2018) Threshold quantum secret sharing based on single qubit. Quantum Inf Process 17(64):1–13
20. Chen X-B, Sun Y-R, Xu G, Yan Y-X (2019) Quantum homomorphic encryption scheme with flexible number of evaluator based (k, n)-threshold quantum state sharing. Inf Sci 501, 172–181
21. Bechmann-Pasquinucci H, Peres A (2000) Quantum cryptography with state systems. Phys Lett 85:3313
22. Kumar M, Dubey SS (2019) (t, n)-threshold quantum secret sharing scheme of an arbitrary one-qutrit based on linear equation. Int J Sci Tech Res 8(10):334–340
23. Dukkipati RV (2015) Applied numerical methods using MATLAB. New Age International Publishers, New Delhi
24. Cai QY, Li WB (2004) Deterministic secure communication without using entanglement. Chin Phys Lett 21:601–603
25. Deng FG, Long GL (2004) Secure direct communication with a quantum one-time pad. Phys Rev A 69:052319
26. Singh PK, Bhargava BK, Paprzycki M, Kaushal NC, Hong WC (2020) Handbook of wireless sensor networks: issues and Challenges in Current Scenario's. Advances in Intelligent Systems and Computing, vol 1132. Springer, Cham, Switzerland, pp 155–437

27. Singh PK, Kar AK, Singh Y, Kolekar MH, Tanwar S (2020) Proceedings of ICRIC 2019. Recent Innovations in computing, Lecture Notes in Electrical Engineering, vol 597. Springer, Cham, Switzerland, pp 3–920
28. Poongodi M, Hamdi M, Sharma A, Ma M, Singh PK (2019) DDoS detection mechanism using trust-based evaluation system in VANET. IEEE Access 7:183532–183544. https://doi.org/10.1109/ACCESS.2019.2960367

The Modified Algorithm of Quantum Key Distribution System Synchronization

Anton Pljonkin and Sandeep Joshi

Abstract This paper describes the trends in the development of quantum communication systems. As part of the study, we have analyzed the operation of fiber-optic autocompensation system of quantum key distribution with phase coding of photon states. We have shown that the algorithm of autocompensation system synchronization may be vulnerable to a certain type of attacks. We have also provided you with a description of the developed modified synchronization algorithm. The developed algorithm has enhanced security against unauthorized data retrieval from the quantum communication channel. This paper also outlines the quantitative characteristics of the developed algorithm.

1 Introduction

The main problem with the transfer of sensitive information is the distribution of the private key between two remote users. Users have the same strings of random bits that they use as a cryptographic key. Classical cryptography cannot provide absolute security during data encryption. To ensure absolute secrecy in the cryptographic system, you must fulfill certain conditions: the key must be completely random, its length must be greater than or equal to the length of the encoded message, and you can use the key sequence only once.

In classical cryptography [1, 2], to ensure secrecy during message transmission, we use methods, protocols, and encryption algorithms, the security of which is limited to the computing resources of the attacker. The physical solution of the problem of ensuring secrecy during the distribution of the key relies on the principles of quantum cryptography and implies single particle (photon) quantum state coding. Here, the

A. Pljonkin (✉)
Southern Federal University, Taganrog, Rostov Region, Russian Federation
e-mail: pljonkin@mail.ru

S. Joshi
Manipal University Jaipur, Jaipur, India
e-mail: sandeep.joshi@jaipur.manipal.edu

P. K. Singh et al. (eds.), *Evolving Technologies for Computing, Communication and Smart World*, Lecture Notes in Electrical Engineering 694,
https://doi.org/10.1007/978-981-15-7804-5_29

secrecy of transmission and the impossibility of unauthorized access to messages relies on the laws of quantum physics, as opposed to classical cryptography methods that rely on mathematical laws and are potentially easy to decipher.

The quantum key distribution (QKD) relies on the basic statements: you cannot clone the unknown quantum state; you cannot extract information about non-orthogonal quantum states without perturbation; any measurement an attacker makes leads to the change of data carrier quantum state.

Quantum key distribution is implemented in quantum communication systems. Autocompensation two-pass fiber-optic systems of quantum key distribution (QKDS) with phase coding of photon states have already proved their efficiency and stable performance. Such QKD systems operate under the BB84 protocol and implement the "plug&play" technology. Note that this type of system needs to be implemented in fiber-optic communication lines, i.e., through the optical fiber. Today, we are witnessing a development of AAC systems for open space. When we transmit data through the atmosphere by means of optical radiation, we apply other coding methods. Moreover, the quantum key distribution protocols work on a different principle (for example, they use polarization coding of the optical impulse). Implementation of such protocols is via a single-pass scheme (one-way protocols of quantum cryptography).

Nevertheless, the study of the QKD autocompensation systems with phase coding of photon states and their algorithms is a relevant scientific and an applied problem.

2 Synchronization in the Quantum Key Distribution System

Just like any other communication system, QKD system cannot operate without synchronization. Here, synchronization in the autocompensation system of quantum key distribution refers to determination of optical impulse detection moment. With an accuracy of 10 ns, the transceiver system must determine the time of registration of the reflected optical impulse. Registration (detection) of impulses is through two avalanche photodiodes operating in a linear mode. Keep in mind a well-known scheme of the two-pass QKD autocompensation system [3].

A fiber-optic communication line (quantum channel) connects two stations and a shared communication channel (e.g., Ethernet) [4–6]. The laser diode of the QKD transceiver forms low-energy impulses with a wavelength of 1550 nm, a frequency of 800 Hz, and an amplitude of 140 mV. The impulses follow the beam splitter and then pass to different arms of the interferometer. After that, impulses appear on the polarization splitter and follow the quantum channel [7]. The time diagram of the impulses in the quantum channel shows that the impulses pass with a time delay of 50 ns. A small optical delay line in the interferometer arm causes the time delay. When entering the coding station, a beam splitter separates the impulses at a ratio of 10/90%. Most of the impulses go to the detector, and a smaller part follows through the fiber optic elements to the Faraday mirror. After reflection, the impulses return to the transceiver where avalanche photodiodes detect them.

After thoroughly analyzing the synchronization process, we have found that the transmission of optical impulses takes place in multiphoton mode. The use of a powerful optical impulse provides a high probability that the synchronization signal will be correctly detected. Experimental studies have shown that a multiphoton mode in the synchronization process is a vulnerability. An attacker can divert a part of the optical radiation in the synchronization process and obtain data on the impulse remirror time. Attackers may use this information to destabilize the quantum cryptography protocol. Note that the results of full-scale tests have detected no presence of an attacker in the quantum communication channel.

3 The Algorithm of Optical Impulse Detection

Synchronization in the QKD system involves three stages. The first stage takes place at the initial start of the system when the quantum channel length is unknown. According to the maximum length of the quantum channel, we calculate the limit impulse repetition period T_{pulse} (to exclude the pile-up of impulses). The impulse repetition period is divided into time windows N_w with a duration τ_w with a fair ratio equal to $T_{\text{pulse}} = N_w \cdot \tau_w$. Each time window is divided into a ratio of $\tau_w/3$.

At the first stage of synchronization, the transceiver generates strobing impulses for avalanche photodiodes. The strobing impulse activates the detector for a $\tau_w/3$ time. In this case, the delay time t_d relative to the moment of the sync impulse radiation increases sequentially. To calculate the delay time, use $t_{d(n-1)} = (\tau_w/3)_n$ dependency. The formula shows that for the first time window, the detection delay is equal to zero. Note that in real operating conditions, we do not apply the time delay $t_d = 0$ since the QKD system has a "blind zone" of the return signal. The latter is because the signal needs to pass a double optical path at least inside the QKDS stations.

Physically, the search for the synchronization signal is through successive switching of the counter by intervals (time windows). In each analyzed interval, we register the number of counts (photodetector operations). Since photodetectors operate in linear mode, they do not require time to recover. Thus, it takes one period to analyze one time window. If you know the optical radiation propagation speed in the fiber, the impulse repetition rate, the number of time windows, and the duration of the time window, you can calculate the total analysis time of the first synchronization stage.

We allocate one time window N_{wsn} with the maximum number of operations (i.e., the time window, which conditionally contains a greater number of photoelectrons—PE) from the entire period of the sequence. As a result of the first stage of synchronization, the QD system determines the coarse time delay (in ns), which corresponds to the time N_{wsn-1} [8].

At the second stage of synchronization, we calibrate the sync impulse detection more accurately [9, 10]. The procedure of analysis is similar, i.e., sequential polling of time windows with fixing of operations in each of them. The difference is that at the second stage, we only analyze the time interval $T \in [N_{wsn-1}|N_{wsn+1}]$, whereby $N_{wsn} = 1\,\text{ns} \rightarrow T = 3\,\text{ns}$. Each N_{wsn} interval is divided into equal time intervals n_{sn}, while $n_{sn} = N_{wsn}/17$ and the total number of analysis intervals at the second stage of synchronization is 51. Note that the number of checks (s) of each time interval at the first two synchronization stages is 800. As a result of the second stage of synchronization, we determine the time interval n_{sn} with the maximum number of registered PEs. Then, the system proceeds to the verification stage. In the third stage, we analyze a previously detected interval n_{sn}. The significant feature of the test phase is that the found interval is then divided into 5–6 subintervals and each of them needs to be analyzed tens of thousands of times. The latter procedure eliminates false detection of the signal due to multiple analysis. Note that the duration of the desired synchronization interval is 10 ns.

For the QD system to operate properly, the synchronization is necessary every once in a while. This is due to the physical properties of the optical fiber. For example, in real operating conditions, the influence of external factors on the quantum channel can cause changes in its parameters. Often there are cases when the temperature effect entails a micro-increase in the length of the optical fiber. This change is not visible to the eye, however, the desired interval (duration 10 ns) is shifted and the system ceases to function. During the operation of the quantum cryptography protocol, the verification phase of synchronization starts after each iteration of key generation.

As mentioned above, as a result of experimental studies, we have found that the synchronization process of autocompensation QKDS is vulnerable [11, 12]. The analysis of QDS energy model made it possible to calculate the power loss of optical radiation in all areas of the fiber-optical signal propagation path. The result has shown that during synchronization, the signal optical impulse contains more than 10^4 photons at the reverse propagation from the coding station to the transceiver. We have found that avalanche photodiodes in the synchronization process operate in a linear mode, and there are no error correction and power control procedures.

4 The Improved Algorithm

The characteristics of single-photon avalanche photodiodes (SPAP) used in QD systems differ from those of an ideal single-photon photodetector. Firstly, SPAP registers only one (the first) photon per $\tau_w/3$. Secondly, in case of photon reception fact registration or dark current impulse (DCI) for SPAP, one requires a certain time d_{time} to restore the operating state.

Considering the mentioned features, we have developed an improved synchronization algorithm that offers greater protection against unauthorized access. The algorithm is applicable to fiber-optic autocompensation systems of quantum key distribution with phase coding of photon states. The first feature of the algorithm is the application of photon-attenuated synchronization signals. The latter ensures the protection of time interval detection process from the Trojan horse attack and unauthorized removal of optical radiation part from the quantum communication channel. The second feature of the algorithm is to reduce the time required to detect the signal time interval. Note that the probability of detection does not decrease with decreasing time.

At the first stage of synchronization, SPAP function in Geiger mode. Detection time delay is $t_d = 0$. The absence of an initial delay means that the analysis of the first window in the first time frame begins immediately after the optical impulse is sent by the radiation source. The refractive index of optical radiation in the optical fiber core is $n = 1.49$. Then the speed of propagation of optical signals in the FOCL is $v = \frac{c}{n} = \frac{3\times10^5}{1.49} = 2.01 \times 10^5$ km/s, where s—the speed of light in a vacuum. The permissible length of the quantum channel between the transceiver and coding station of the QKDS is 100 km. Taking into account the back passage of the optical signal, the distance is doubled. To exclude the overlap of two counter impulses during the passage of the optical path, it is necessary to take into account the permissible value of the repetition period $T_s = 200/\left(2.01 \times 10^5\right)$. The optical impulse duration is $\tau_s = 1$ ns. We have found that in the process of detecting the signal time window at the first stage of synchronization, the optimal ratio is $\tau_w = 2 \cdot \tau_s$.

While searching the perfect will be the case when you consistently poll all time windows. SPAP applied in QD systems that rely on InGaAs has their own features. In Geiger mode, the main parameters are the quantum efficiency of the photocathode and the recovery time of the operating mode. We have described the influence of quantum efficiency on the registration of optical radiation by the probability dependence. The recovery time d_{time} is the time interval during which SPAP is inactive after registration of PE or DCI. For example, you can control recovery time for SPAP id230 applied in QDS in a programmatic manner within the range of $1\,\mu\text{s} < d_{\text{time}} < 100\,\mu\text{s}$.

If you activate the count mode of SPAP single photons at time intervals greater than d_{time}, you can analyze several time windows for one time frame. For example, when $d_{\text{time}} = 100\,\mu\text{s}$, we get the maximum quantum efficiency of the photodiode and the minimum frequency of DCI occurrence. At the same time, we analyze several time windows for the follow period. Thus, if we set the strobing interval to 200 μs, it helps to analyze five time windows for the follow period of the optical impulse. *In this case, it does not matter whether you have registered PE or DCI in the time window or not since it does not affect time frame analysis time.*

To calculate the time delay, use the following formula:

$$t_d = (T_s/5) \cdot (An - 1) + \tau_w \cdot (Bn - 1), \tag{1}$$

where A is the activation number of SPAP for one follow period; B is the frame number. In paper [8] described a graphical description of the time delay calculation.

Thus, at the first stage, for one frame, we poll five time windows. For example, for the first frame $t_{d1} = 0$, $t_{d2} = 200$ ns, …, $t_{d4} = 800$ ns. For the second time frame, $t_{d1} = 2$ ns, $t_{d2} = 202$ ns, …, $t_{d4} = 802$ ns. After polling all time windows, we have an array of values. Then, we choose two time windows with the maximum number of operations. After that, we double-check the sequence of determined windows. You successfully complete the first check if two windows are in adjacent time intervals. Otherwise, we should compare the quantitative values of the registered PEs in these windows. You successfully complete the second check if the difference between the values is more than 75%. These actions help to exclude false operations of SPAP. If the checks fail, then you have incorrectly detected the time window. The result of the first stage of synchronization is the detection of the signal time interval $\tau_w|2\tau_w$. Everyone knows that the influence of dispersion and re-reflection of the signal leads to distortion of the impulse shape. The latter assumes two cases of distribution of the signal impulse (an array of values of the registered PEs) on the time domain: $\tau_s \in \tau_w u \tau_s \in 2\tau_w$. I.e., is located strictly in one time window or belongs to two adjacent time intervals. The result of the first stage is the detection of a time window with the maximum number of registered PEs, where $N_{ws} \in \{\tau_{ws} - \tau_w/2|\tau_{ws} + \tau_w/2\}$. The latter means that the total duration of the desired interval is $2\tau_w$. The algorithm of the second and third stages is similar to the classical search algorithm. I.e., the determined interval $2\tau_w$ is divided into subintervals to be analyzed multiple times. To proceed to the verification stage of synchronization, you should allocate the time interval with the maximum number of operations.

In [12], we have found an expression to calculate time window correct detection probability taking into account the use of the ideal PE counter

$$P_D = \sum_{n_{w.N}=1}^{\infty} \frac{(\overline{n_{w.N}})^{n_{w.N}}}{n_{w.N}!} \cdot \exp[-\overline{n_{w.N}}] \cdot P_{d.N}\{n_{w.N}\} \tag{2}$$

here

$$P_{d.N}\{n_{w.N}\} = \left(\sum_{n_{d.N}=0}^{n_{w.N}-1} \frac{\overline{n_{d.N}}^{n_{d.N}}}{n_{d.N}!} \cdot \exp(-\overline{n_{d.N}}) \right)^{N_w - 1} \tag{3}$$

represents the probability of registering no more than $(n_{w.N} - 1)$ DCI in all $(N_w - 1)$ noise time windows during the analysis, provided that we have registered $n_{w.N}$ PE and DCI in the signal time window. The simulation results have shown that at $N_w > 1024$ the average number of DCI $(\overline{n_{d.N}})$ during the time window analysis is 0.00008. This allows summing up in the formula only at 2 $n_{d,N}$ values equal to 0 and 1. Then,

$$\begin{aligned} P_D &= \exp(-N_w \cdot \overline{n_{d.N}} + \overline{n_{d.N}})\overline{n_{w.N}} \cdot \exp(-\overline{n_{w.N}}) \\ &\quad + \left[1 - \exp(-\overline{n_{w.N}}) - \overline{n_{w.N}} \cdot \exp(-\overline{n_{w.N}})\right] \cdot (1 + \overline{n_{d.N}})^{N_w - 1}. \end{aligned} \tag{4}$$

The discrepancy in the calculation results according to Formulas (1)–(3) does not exceed 0.02%. Note that the registration condition is no higher than one PE and/or DCI and is a property of SPAP. The latter proves the possibility of using the expression (3) to calculate the probability of correct detection of the signal time window in the low-energy mode, provided that $\overline{n_{w.N}} \ll 1$.

Note that the proposed algorithm does not take into account the influence of dispersion and quantum efficiency on the detection probability. Moreover, when implementing the algorithm, one should take into account the level of standard losses in the optical fiber. The latter is necessary to calculate the attenuation coefficient in the controlled attenuator of the coding station. You can implement the technical solution to the problem of determining the loss in FOCL using reflectometry [13–17]. In experimental studies, we have used this method for the preliminary determination of the FOCL length and losses in the quantum communication channel.

5 Conclusion

The proposed synchronization algorithm has greater security against unauthorized information retrieval. Security is implemented with low-energy mode in the synchronization process. We have analyzed the operation of fiber-optic autocompensation system of quantum key distribution with phase coding of photon states. Moreover, we have demonstrated that the algorithm of autocompensation system synchronization is vulnerable to a certain type of attacks. Aside from that, we have provided you with a description of the improved synchronization algorithm. And then, we have shown that the developed algorithm has greater security against unauthorized information retrieval from the quantum communication channel. In summary, we have outlined the quantitative characteristics of the developed algorithm.

Acknowledgements Work is performed within the grant of President of Russian Federation for state support of young Russian scientists MK-2338.2018.9 "Creation of an automated algorithm for integrating quantum keys into the data network while providing enhanced security against unauthorized access to the quantum communication channel" and supported by Pljonkins international corp.

References

1. Gagliardi RM, Karp S (1995) Optical communications. Wiley, New York, 1976; trans: Russian, 1978; trans: Japanese, 1979, 2nd edn.
2. Gisin N, Ribordy G, Tittel W, Zbinden H (2002) Quantum cryptography. Rev Modern Phys 74(1):145–195
3. Rumyantsev KE, Pljonkin AP (2016) Preliminary stage synchronization algorithm of autocompensation quantum key distribution system with an unauthorized access security. In: International conference on electronics, information, and communications (ICEIC), Vietnam, Danang, pp 1–4. https://doi.org/10.1109/ELINFOCOM.2016.7562955
4. Kurochkin V, Zverev A, Kurochkin J, Riabtzev I, Neizvestnyi I (2012) Quantum cryptography experimental investigations. Photonics 5:54–66

5. Lydersen L, Wiechers C, Wittmann C, Elser D, Skaar J, Makarov V (2010) Hacking commercial quantum cryptography systems by tailored bright illumination. Nat Photonics 4:686
6. Rumyantsev KY, Pljonkin AP (2015) Synchronization of quantum key distribution system using single-photon pulses registration mode to improve the security. Radiotekhnika 2:125–134
7. Rumyantsev KE, Pljonkin AP (2014) Eksperimentalnye ispytaniya telekommunikatsionnoy seti s integrirovannoy sistemoy kvantovogo raspredeleniya klyuchey. Telekommunikatsii 10:11–16
8. Pljonkin AP, Rumyantsev KY (2016) Single-photon synchronization mode of quantum key distribution system, India, New Delhi, pp 531–534. https://doi.org/10.1109/ICCTICT.2016.7514637
9. Vidick T, Watrous J (2016) "Quantum Proofs" Foundations and Trends® in Theor Comput Sci 11(1–2):1–215. https://doi.org/10.1561/0400000068
10. Rumyantsev KY (2011) Quantum key distribution systems: monograph. SFedU, Taganrog, p 264
11. Stucki D, Gisin N, Guinnard O, Ribordy G, Zbinden H (2002) Quantum key distribution over 67 km with a plug & play system. New J Phys 4:41.1–41.8
12. Pljonkin A, Singh PK (2018) The review of the commercial quantum key distribution system. In: 2018 fifth international conference on parallel, distributed and grid computing. IEEE. https://doi.org/10.1109/PDGC.2018.8745822
13. Pljonkin AP (2019) Vulnerability of the synchronization process in the quantum key distribution system. Int J Cloud Appl Comput 9(1). https://doi.org/10.4018/IJCAC.2019010104
14. Plenkin A, Rumyantsev K, Rudinsky E (2017) Comparative analysis of single-photon synchronization algorithms in the quantum key distribution system. In: Proceedings of 2017 IEEE east-west design and test symposium. https://doi.org/10.1109/EWDTS.2017.8110047
15. Singh PK, Bhargava BK, Paprzycki M, Kaushal NC, Hong WC (2020) Handbook of wireless sensor networks: issues and challenges in current scenario's. In: Advances in intelligent systems and computing, vol 1132. Springer, Cham, pp 155–437
16. Singh P, Paprzycki M, Bhargava B, Chhabra J, Kaushal N, Kumar Y (2018) Futuristic trends in network and communication technologies. FTNCT 2018. In: Communications in computer and information science, vol 958, pp 3–509
17. Singh PK, Kar AK, Singh Y, Kolekar MH, Tanwar S (2020) Proceedings of ICRIC 2019, Recent innovations in computing, vol 597, Lecture Notes in Electrical engineering. Springer, Cham, pp 3–920

Person Re-Identification by Analyzing Dynamic Variations in Gait Sequences

Sandesh V. Bharadwaj and Kunal Chanda

Abstract Gait recognition is a biometric technology that identifies individuals in a video sequence by analysing their style of walking or limb movement. However, this identification is generally sensitive to appearance changes and conventional feature descriptors such as Gait Energy Image (GEI) lose some of the dynamic information in the gait sequence. Active Energy Image (AEI) focuses more on dynamic motion changes than GEI and is more suited to deal with appearance changes. We proposed a new approach, which allows recognizing people by analysing the dynamic motion variations and identifying people without using a database of predicted changes. In the proposed method, the active energy image is calculated by averaging the difference frames of the silhouette sequence and divided into multiple segments. Affine moment invariants are computed as gait features for each section. Next, matching weights are calculated based on the similarity between extracted features and those in the database. Finally, the subject is identified by the weighted combination of similarities in all segments. The CASIA-B Gait Database is used as the principal dataset for the experimental analysis.

Keywords Active energy image · Affine moment invariant · Gait analysis · Image segmentation · Person re-identification

---

Center for Development of Advanced Computing, Kolkata.

S. V. Bharadwaj (✉)
Indian Institute of Information Technology, Design and Manufacturing, Kancheepuram, Chennai 600127, India
e-mail: esd15i005@iiitdm.ac.in

S. V. Bharadwaj · K. Chanda
Center for Development of Advanced Computing, Kolkata 700091, India

P. K. Singh et al. (eds.), *Evolving Technologies for Computing, Communication and Smart World*, Lecture Notes in Electrical Engineering 694,
https://doi.org/10.1007/978-981-15-7804-5_30

1 Introduction

In the modern world, reliable recognition of people has become a fundamental requirement in various real-time applications such as forensics, international travel and surveillance. Biometrics have been applied to criminal identification, patient tracking in hospitals, and personalization of social services. Gait recognition is a biometric recognition technique that recognizes individuals based on their walk cycle, and has been a topic of continued interest for person identification due to the following reasons:

1. First, gait recognition can be performed with low-resolution videos with relatively simple instrumentation.
2. Second, gait recognition can work well remotely and perform unobtrusive identification, especially under conditions of low visibility.
3. Third, gait biometric overcomes most of the limitations that other biometric identifiers suffer from such as face, fingerprint and iris recognition which have certain hardware requirements that add to the cost of the system.
4. Finally, gait features are typically difficult to impersonate or change, making them somewhat robust to appearance changes.

Gait-based person recognition experience varying degrees of success due to internal and external factors, such as:

1. Low frame rates leading to incomplete gait extraction.
2. Incorrect or failure in identification of individuals due to

(a) Partial visibility due to occlusion.
(b) View variations. (lateral movement or coming at an angle)
(c) Appearance Changes. (wearing a bag/coat/jacket).

The rest of the paper is organized as follows. Section 2 covers related work and publications in person recognition, detailing the various methodologies implemented for gait feature extraction. Section 3 summarizes the modified approach we have implemented in our research. Section 4 explains the experimental results obtained and section 5 concludes our paper. Our main contribution is:

1. Focusing on dynamic variations embedded in gait sequences using active energy images.
2. Image segmentation of active energy images and extracting affine moment invariants as feature descriptors.
3. Assigning dynamic weights for each segment to attribute useful features to higher priority and vice versa.

2 Related Work

In the last two decades, significant efforts have been made to develop robust algorithms that can enable gait-based person recognition on real-time data. Modern gait recognition methods can be classified into two major groups, namely model-based and motion-based methods.

In model-based methods, the human body action is described using a mathematical model and the image features are extracted by measuring the structural components of models or by the motion trajectories of the body parts. These models are streamlined based on assumptions such as pathologically normal gait.

Model-based systems consist of a gait sequence,a model or models, feature extractor, and a classifier. The model can be 2-dimensional or 3-dimensional, which is useful for tracking a moving person. While this method is robust to problems of occlusion, noise, scale and rotation, system effects such as viewpoint invariance and effects of physiological, psychological and environmental changes are major limitations in implementing a model-based recognition system.

Motion-based methods consider the human gait cycle as a sequence of images and extract binary silhouettes from these images. Motion-based approaches are not influenced by the quality of images and have the added benefit of reduced computational cost compared to model-based approaches. There are usually some post-processing steps performed on the binary images to extract static appearance features and dynamic gait features.

The baseline algorithm proposed by Sarkar et al. [1] uses silhouettes as features themselves, scaling and aligning them before use. Carley et al.[2] proposed a new biometric feature based on autocorrelation using an end-to-end trained network to capture human gait from different viewpoints. Bobick and Davis [3] proposed the motion-energy image (MEI) and motion-history image (MHI) to convert the temporal silhouette sequence to a signal format.

Kolekar and Francis [4] For a given sequence I_t, MHI at pixel coordinates (x,y) and time t is defined as

$$\mathrm{MHI}(x, y, t)_\tau = \begin{cases} \tau, & I(x, y)_t = 1 \\ \max(\mathrm{MHI}(x, y, t-1)_\tau - 1, 0), & I(x, y)_t = 0 \end{cases} \tag{1}$$

where τ is a threshold value designed to capture maximum action in the image. Kolekar and Francis [4] MEI is obtained by binarizing the MHI. Given a binary image sequence $J(x, y, t)$ which contains regions of motion, MEI is defined as follows

$$\mathrm{MEI}(x, y, t) = \bigcup_{i=0}^{\tau-1} J(x, y, t-i) \tag{2}$$

Han and Bhanu [5] used the idea of MEI to propose the Gait Energy Image (GEI) for individual recognition. GEI converts the spatio-temporal information of one walking cycle into a single 2D gait template, avoiding matching features in temporal sequences.

Given a pre-processed binary gait silhouette sequence $I_t(x, y)$ at time t, the GEI is computed as:

$$G(x, y) = \frac{1}{N} \sum_{t=0}^{N-1} I_t(p, q) \tag{3}$$

GEI is comparatively robust to noise, but loses dynamic variations between successive frames. In order to retain dynamic changes in gait information, Zhang, Zhao and Xiong [6] proposed an active energy image (AEI) method for gait recognition, which focuses more on dynamic regions than GEI, and alleviates the effect caused by low quality silhouettes.

Current research on gait representation include Gait Entropy Image (GEnI) [7], frequency-domain gait entropy (EnDFT) [7], gait energy volume (GEV) etc., which focus more on dynamic areas and reducing view-dependence of traditional appearance-based techniques.

3 Proposed Methodology

In this project, we proposed a motion-based approach to person identification, using *affine moment invariants* (AMIs) as feature descriptors. *Active Energy Image* (AEI) are generated from the subject silhouette sequence, which retains the dynamic variations in the gait cycle.

The approach can be summarized in the following steps:

1. Extract AEI from the gait sequence and divide the image into multiple segments.
2. Extract AMI from each segment to use as gait features. The database consists of affine moment invariants of multiple people who wear standard clothing with no accessories.
3. AEI of test subject is also segmented like that of the dataset and AMIs are computed.
4. Matching weights are estimated at each area based on similarity between features of the subject segments and the database.
5. The subject is predicted by the weighted integration of the similarities of all segments. k-Nearest Neighbor classifier is used to classify the test subject. $k = 1$ for our experimental analysis.

3.1 Active Energy Image

In gait recognition, two types of information—static and dynamic—are extracted from the gait sequence. According to [5], GEI is efficient in extracting static and

Fig. 1 Gait silhouette sequence of individual

Fig. 2 Difference images between consecutive significant frames of gait sequence

Fig. 3 Active energy image

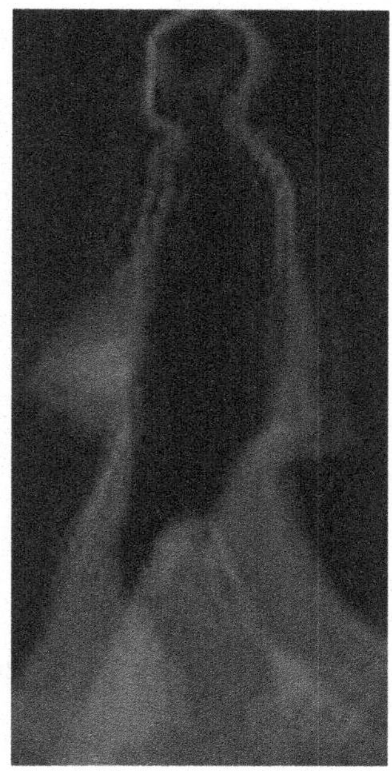

dynamic information, both implicitly and explicitly. However, since GEI represents gait as a single image, there is a loss of dynamic information such as the fore-and-aft frame relations. Active Energy Image (AEI) feature representation can be used to solve the above problems.

Given a pre-processed binary gait silhouette[6] $i = i_0, i_1, \ldots, i_N - 1$, where f_j represents the jth silhouette (Fig. 1), N is the total number of frames in the sequence, the difference image between frames is calculated as follows (Fig. 2):

$$I_j = \begin{cases} i_j(p, q), & j = 0 \\ ||i_j(p, q) - i_{j-1}(p, q)||, & j > 0 \end{cases} \tag{4}$$

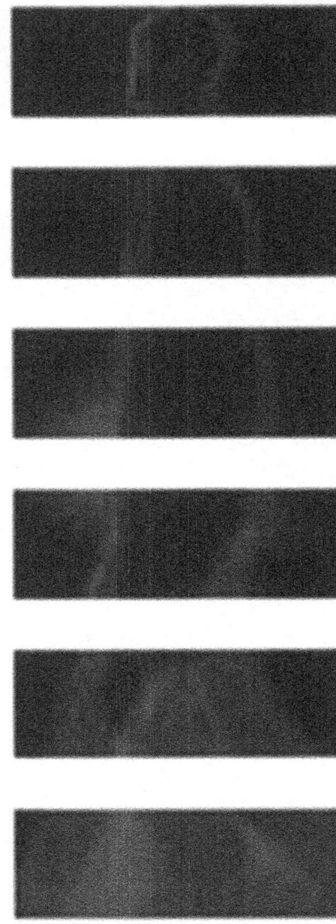

Fig. 4 Segmented AEI ($K = 6$)

The AEI is defined as (Fig. 3):

$$A(p, q) = \frac{1}{N} \sum_{j=0}^{N-1} I_j(p, q) \quad (5)$$

where N is the total number of difference images used to compute the AEI.

The AEI representation is divided into 'K' segments as shown in Fig. 4.

3.2 *Affine Moment Invariants*

Image moments and moment invariants

Image moments are weighted averages of the image pixels' intensities or a function of such moments, usually chosen due to some attractive property or interpretation. Some properties of images derived through image moments include area, centroid, and information about the image orientation.

Image moments are used to derive **moment invariants**, which are invariant to transformations such as translation and scaling. The well-known ***Hu moment invariants*** were shown to be invariant to translation, scale and rotation. However, Flusser [8] showed that the traditional set of Hu moment invariants is neither independent nor complete.

Affine moment invariants (AMIs), proposed by Flusser and Suk [9] are moment-based descriptors, which are invariant under a general affine transformation. The invariants are generally derived by means of the classical theory of algebraic invariants [10], graph theory [16], or the method of normalization. The most common method is the use of graph theory.

The moments describe shape properties of an object as it appears. For an image, the centralized moment of order *(a + b)* of an object *O* is given by

$$\mu_{ab} = \sum \sum_{(x,y)\in O} (x - x_{cg})^a (y - y_{cg})^b A(x, y) \tag{6}$$

Here, x_{cg} and y_{cg} define the center of the object, calculated from the geometric moments m_{ab}, given by $x_{cg} = \frac{m_{10}}{m_{00}}$ and $y_{cg} = \frac{m_{01}}{m_{00}}$.

The affine transformation is expressed as

$$u = a_0 + a_1 x + a_2 y \tag{7}$$

$$v = b_0 + b_1 x + b_2 y \tag{8}$$

Suk [11] we have used 10 AMIs ($\mathbf{A} = (A_1, A_2, \ldots, A_{10})^T$), 5 of which are shown below:

$$\begin{aligned}
A_1 &= \frac{1}{\mu_{00}^4}(\mu_{20}\mu_{02} - \mu_{11}^2) \\
A_2 &= \frac{1}{\mu_{00}^{10}}(\mu_{30}^2\mu_{03}^2 - 6\mu_{30}\mu_{21}\mu_{12}\mu_{03} + 4\mu_{30}\mu_{12}^3 + 4\mu_{03}\mu_{21}^3 \\
&\quad -3\mu_{21}^2\mu_{12}^2) \\
A_3 &= \frac{1}{\mu_{00}^7}(\mu_{20}(\mu_{21}\mu_{03} - \mu_{12}^2) - \mu_{11}(\mu_{30}\mu_{03} - \mu_{21}\mu_{12}) \\
&\quad +\mu_{02}(\mu_{30}\mu_{12} - \mu_{21}^2)) \\
A_4 &= \frac{1}{\mu_{00}^{11}}(\mu_{20}^3\mu_{03}^2 - 6\mu_{20}^2\mu_{11}\mu_{12}\mu_{03} - 6\mu_{20}^2\mu_{02}\mu_{21}\mu_{03} \\
&\quad +9\mu_{20}^2\mu_{02}\mu_{12}^2 + 12\mu_{20}\mu_{11}^2\mu_{21}\mu_{03} \\
&\quad +6\mu_{20}\mu_{11}\mu_{02}\mu_{30}\mu_{03} - 18\mu_{20}\mu_{11}\mu_{02}\mu_{21}\mu_{12} \\
&\quad -8\mu_{11}^3\mu_{30}\mu_{03} - 6\mu_{20}\mu_{02}^2\mu_{30}\mu_{12} + 9\mu_{20}\mu_{02}^2\mu_{21}^2 \\
&\quad +12\mu_{11}^2\mu_{02}\mu_{30}\mu_{12} - 6\mu_{11}\mu_{02}^2\mu_{30}\mu_{21} + \mu_{02}^3\mu_{30}^2) \\
A_5 &= \frac{1}{\mu_{00}^6}(\mu_{40}\mu_{04} - 4\mu_{31}\mu_{13} + 3\mu_{22}^2)
\end{aligned}$$

3.3 Estimation of Matching Weights

Iwashita et al. [12] Matching weights are calculated by first whitening the AMI in the database and the subject for each segment. $d^k_{n,s}$, which is the L2 norm between the unknown subject features and those of all known persons in the database, is computed as follows:

$$d^k_{n,s} = ||^w A^k_{\text{SUB}} -^w A^k_{DB_{n,s}}|| \tag{9}$$

where $^w A^k_{\text{SUB}}$ and $^w A^k_{DB_{n,s}}$ are the whitened AMI of the test subject and of a known person in the database respectively.

(n, s and k) are $1 \leq n \leq N$ (N is number of persons in database), $1 \leq s \leq S$ (S is number of sequences of each person), and $1 \leq k \leq K$ (K is number of divided areas). $d^k_{n,s}$ is calculated in the Euclidean norm.

The AMIs are whitened by:

1. Applying *principal component analysis (PCA)* on the AMIs and projecting them to a new feature space.
2. Normalizing the projected features based on the corresponding eigenvalues.

To estimate matching weights, we use the similarity between subject features and database features; high matching weights are set to areas with less appearance changes, and low matching weights set to those with more appearance changes.

Steps to estimate matching weights:

1. At each segment k, select sequences where $d^k_{n,s} < \bar{d}_{\min}$. These selected sequences are considered to have high similarity with the subject. $\bar{d}_{\min}$ is defined as:

$$\bar{d}_{\min} = \min_n \bar{d}^k_n \tag{10}$$

$$\bar{d}^k_n = \frac{1}{S}\sum_{s=1}^{S} d^k_{n,s} \tag{11}$$

   where S is the total number of sequences of the person being used. We consider that in each segment, if a Minimum of one sequence of a person in the database is selected, then the matching scores of all sequences of that person are also high.
2. The non-selected sequences are all considered to be low similarities, so the distances of these sequences are redefined as $d_{\max}$ ($d_{\max} = \max_{n,s,k} d^k_{n,s}$)
3. These steps are applied for all segments and the total distance for all segments is calculated by $D_{n,s} = \sum_{k=1}^{K} d^k_{n,s}$. Subject is identified by nearest-neighbor method.

3.4 Principal Component Analysis

The curse of dimensionality shows that as the number of features increase, it gets harder to visualize the training set and then work on it. Sometimes, many of these features are correlated, and hence redundant. This leads to a need for reducing the complexity of a model to avoid overfitting.

Principal Component Analysis (PCA) [13] is a popular algorithm used for dimension reduction. Proposed by Karl Pearson, PCA is an unsupervised linear transformation technique used in identifying patterns in data based on the correlation in features. It projects the data onto a new subspace along the direction of increasing variance with equal or lesser dimensions.

The PCA Algorithm consists of the following steps:

1. First the mean vector is computed.
2. Assemble all the data samples in a mean adjusted matrix, followed by creation of the covariance matrix.
3. Compute the Eigenvectors and Eigenvalues following the basis vectors.
4. Each sample is represented as a linear combination of basis vectors.

Figure 5 compares the distribution of image data before and after PCA dimensionality reduction.

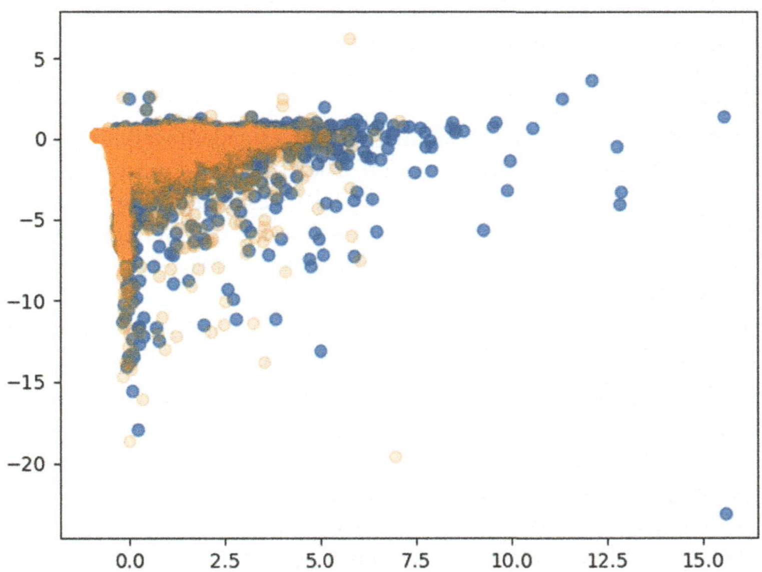

Fig. 5 PCA on gait data extracted from CASIA-B Dataset. Blue datapoints represent original image data and yellow datapoints represent the modified data after dimensionality reduction

3.5 K-Nearest Neighbor Classifier

The k-Nearest Neighbor (kNN) algorithm is one of the simplest Machine Learning algorithms based on Supervised Learning technique. It calculates the similarity between the new case/data and available cases and classifies the new case into the category closest to the existing classes.

The kNN algorithm stores the dataset during the training phase; during the testing phase, when it gets new data, it classifies that data into a category that is much similar to the new data. The key steps are as follows:

1. The number of neighbors (k) are selected.
2. The Euclidean distance of k number of neighbors is calculated.
3. The k nearest neighbors as per the calculated Euclidean distance are selected.
4. Among these k neighbors, the number of the data points in each category are counted.
5. The new data points are assigned to that category for which the number of neighbors is maximum.

4 Experimental Results and Observations

Our proposed method was applied to the CASIA-B Gait Dataset. A large multi-view gait database created in January 2005, it contains 124 subjects captured from 11 views. Three variations, namely view angle, clothing and carrying condition changes, are separately considered.

In the experiment, we used the lateral view (90°) standard walking sequences to analyze the prediction rate of the algorithm. The CASIA-B Dataset consists of six standard walking sequences for each person. CCRs are calculated by dividing the six sequences of each subject into two sets; $124 \times S$ sequences were used for training ($S = 3, 4, 5$) and the rest were used for testing. All input images from the sequences are resized to 128×64 images for uniformity.

The total number of AMIs M were varied between 1 and 10 using PCA dimensionality reduction and the parameter K (number of divided areas) from 5 to 30. In case of $K = 23$ and $M = 5$, the proposed method shows the highest accuracy of **91.13%** (Table 1). It is observed that increasing the number of image segments over a certain limit reduces the accuracy of the classification algorithm. This is likely due to the loss of features as the number of segments increases.

We also compared our proposed algorithm with existing methodologies that have been tested on the CASIA-B Gait Dataset, shown in Table 2. Our approach is competitive, which shows the effectiveness of evaluating dynamic variations for recognition.

Table 1 Experimental results of our algorithm on the CASIA-B Gait Dataset

Train/test split	No. of segments	Classification accuracy (%)
0.5	10	71
	20	78.76
	23	**82.00**
	30	79.57
0.66	10	76.61
	20	85.08
	23	**85.89**
	30	74.68
0.83	10	82.26
	20	90.32
	23	**91.13**
	30	87.90

Bold signifies best result of our approach in given setting

Table 2 Comparison of classification accuracy on CASIA-B Gait Dataset

Method	Rank-1 classification (%)
Method 1 [12]	97.70
Method 2 [14]	95.52
AEI+2DLPP [6]	98.39
STIPs+BoW [15]	94.5
Proposed method	91.13

5 Conclusion

This paper describes a modified approach to gait-based person re-identification using AEIs. AEIs were generated to locate and identify the dynamic motion regions in a walking sequence, which were then resized and segmented laterally. Then we extracted AMIs as descriptors from each segment, applying PCA analysis to reduce the dimensionality and maximize variance in the feature space. The reduced features were assigned matching weights based on the similarity between the test subjects and the database, which were then classified using the combination of similarities in the segments. The experimental results demonstrate that our proposed method provides results which rival that of other gait recognition methods. Our future work will focus on improving the performance of our algorithm and evaluating its performance on walking sequences from arbitrary viewpoints.

References

1. Sarkar S, Phillips PJ, Liu Z, Robledo Vega I, Grother P, Bowyer K (2005) The humanid gait challenge problem: data sets, performance, and analysis. IEEE Trans pattern Anal Mach Intell 27:162–77
2. Carley C, Ristani E, Tomasi C (2019) Person re-identification from gait using an autocorrelation network. In: CVPR workshops
3. Bobick A, Johnson A (2001) Gait recognition using static, activity-specific parameters 1:1–423
4. Kolekar M, Francis T (2018) Intelligent video surveillance systems: an algorithmic approach. Chapman and Hall/CRC
5. Han J, Bhanu B (2006) Individual recognition using gait energy image. IEEE Trans Pattern Anal Mach Intell 28:316–322
6. Zhang E, Zhao Y, Xiong W (2010) Active energy image plus 2dlpp for gait recognition. Sig Proc 90:2295–2302
7. Ahmed I, Rokanujjaman M (2016) Gait-based person identification considering clothing variation 5:28–42
8. Flusser J (2000) On the independence of rotation moment invariants. Pattern Recogn 33:1405–1410
9. Flusser J, Suk T (1993) Pattern recognition by affine moment invariants. Pattern Recogn 26:167–174
10. Marxsen S, Hilbert D, Laubenbacher R, Sturmfels B, David H, Laubenbacher R (1993) Theor Algebraic Invariants. Cambridge University Press, Cambridge Mathematical Library
11. Suk T (2005) Tables of affine moment invariants generated by the graph method
12. Iwashita Y, Uchino K, Kurazume R (2013) Gait-based person identification robust to changes in appearance. Sensors (Basel, Switzerland) 13:7884–7901
13. Jolliffe IT (2002) Principal Compon Anal. Springer Series in Statistics. Springer-Verlag, New York
14. Liu Z (2016) Gait recognition using active energy image and gabor wavelet 10:1354–1358
15. Kusakunniran W (2014) Attribute-based learning for gait recognition using spatio-temporal interest points. Image Vis Comput 32
16. Suk T, Flusser J (2004) Graph method for generating affine moment invariants 2:192 – 195

Big Data Analytics and Machine Learning Technologies for HPC Applications

Sukeshini, Priyanka Sharma, Mohit Ved, Janaki Chintalapti, and Supriya N. Pal

Abstract High-performance computing (HPC) has long been pivotal for carrying out data and compute-intensive large-scale scientific simulations, analytic workloads for advanced scientific progress and product innovations, in turn making HPC infrastructure more essential and valuable than ever. The convergence of HPC and big data technologies/machine learning (ML)/ deep learning (DL) is prescribed owing to the multifold growth of data across all HPC domains. HPC application experts are either already working or looking forward towards ML solutions to their applications, as it is proven successful in many scientific and commercial domains. Modern CPU/GPU-based HPC architectures are equipped with support for ML/DL promising the adaptability of AI capabilities in the HPC territory. Artificial intelligence (AI) enabled neuromorphic chips to add another ladder towards the convolution of AI on HPC. In this paper, we bring forth the merits of AI on HPC infrastructure for scientific applications in the shortlisted domains, viz weather and climate, astrophysics, agriculture, and bioinformatics. The paper discusses the current scenarios of wide adoption and merits of AI in the said domains. The survey lists the applications that are well received by the user communities for their performance, handling big unstructured data, improved results, etc., while solving the domain-specific problem.

Keywords Artificial intelligence · Machine learning · Deep learning · High-performance computing · Weather and climate · Astrophysics · Agriculture · Bioinformatics

Sukeshini · P. Sharma (✉) · M. Ved · J. Chintalapti · S. N. Pal
Centre for Development of Advanced Computing, Bangalore, India
e-mail: priyankas@cdac.in

M. Ved
e-mail: mohitv@cdac.in

J. Chintalapti
e-mail: janaki@cdac.in

S. N. Pal
e-mail: supriya@cdac.in

P. K. Singh et al. (eds.), *Evolving Technologies for Computing, Communication and Smart World*, Lecture Notes in Electrical Engineering 694,
https://doi.org/10.1007/978-981-15-7804-5_31

1 Introduction

HPC [1] is amalgamation of computing power delivering higher speed and performance for solving grand challenges and large-scale simulations in various scientific domains and for faster businesses solutions. The well-established de-facto technology, HPC has been embraced by government agencies, academics, enterprises, and scientific researchers' communities for economic viability to innovate breakthrough product and services. Be it the vertically scaled up supercomputers or the horizontally scaled-out, clusters of servers running in parallel to create supercomputing-class throughput, HPC has spread across enterprises. HPC researchers and developers can focus on being productive by quicker iterations of their empirical work, when they break free from the cycle of concept → design → code → execute → result. The usage and impact of HPC touches almost every facet of daily life.

Big data has moved way past the initial V's identified—volume, variety, velocity, and veracity. With the addition of every new V's—variability, validity, vulnerability, volatility, visualization, value brings in new dimensions to our understanding of the complexity of big data. Data-driven decision-making has picked up significantly over the last decade. The organizations and businesses are dealing with ever-increasing data constantly. Organizations with HPC base are diligently including big data requirements to be relevant in today's time. Eventually providing fundamental insights into the understanding and analyzing the current societal and economical ecosystem while predicting its future behavior. In this digital era, the data deluge cannot be overlooked; data has been predicted to reach or exceed 163 zettabytes by 2025. Given the ever-expanding infrastructure capabilities, HPC becomes the obvious and indispensable choice for analyzing the enormous data for greater insights and leading us to make informed decisions.

Despite been around for decades due to the lack of widespread familiarity, AI has finally been demystified and coming out from the laboratory into the real world for the implementation in a big way. AI embarks upon solving hard and meaningful real-life challenges, surpassing human performance across broad as well as specific, complex domains. The process of deriving accurate and fast decisions without human involvement or intervention requires the use of invaluable potential of ML, a subset of AI. New methodologies for data synthesis are required for combating data challenges. In this AI era, for the rampant breakthroughs across various realms are directly or indirectly related to ML, the credit goes to HPC for providing compute at affordable costs. The convergence of HPC, big data, and ML is transforming the world into a better place.

2 Survey of Key Domains

The advancement in AI beckons new HPC infrastructure solutions, as the computing landscape is being transformed by big data analytics. Our objective is to understand the dominos effect of HPC, big data, and AI in various domains, draw attention to the diversity of applications and industries migrating from HPC to AI and/or using AI with HPC to unleash enormous opportunities in scientific research and businesses. As AI is being integrated with and deployed into a variety of sectors, we have shortlisted and surveyed the following domains, viz. weather and climate, astrophysics, agriculture, and bioinformatics. In the following sections, we shall look into the popular HPC applications and their bindings in the selected domains along with their merits for migrating from HPC to AI.

2.1 Weather and Climate

AI can assist in making precise weather prediction a truth. Weather and climate prediction is a perfect application for AI, because the current and historical weather data together form a huge date mine, which when fed to ML algorithms can be very effective at combining past occurrences with future predictions.

To use AI in weather prediction, historical weather data is fed into an algorithm that learns using DL techniques and makes predictions based on its learning over the past data. DL techniques have already been proven to be successful in areas like video analytics, natural language processing (NLP), image and speech recognition, and it can be effectively applied to the weather and climate as well. Given the fact that weather features consist of a complex number of data markers throughout the globe, the solution to it makes weather prediction a highly compute- and data-intensive application. Additionally, data quality and labeling are very crucial components of DL algorithms since the efficiency of DL algorithms highly depends upon the inputs they are trained on. Nevertheless, the combined strength of AI and DL systems can be used to absorb and analyze these multi-dimensional datasets much more efficiently and quickly than what has been achieved so far. With many experiments being conducted at various organizations, results are beginning to emerge. An article [2] published by the American Meteorological Society (AMS) shows how AI and ML techniques can improve the capability to extract precise insights, and provide timely information and support to weather forecasters by analyzing the profuse amounts of weather data.

Weather scientists in India have successfully conducted experiments using several ML techniques in predicting various weather and climate phenomena. Dr. K. C. Tripathi and Dr. I. M. L. Das, University of Allahabad and Dr. A. K. Sahai, Indian Institute of Tropical Meteorology, Pune, have published their work [3], with the objective to predict the anomalies in Sea Surface Temperature of the Indian Ocean, exploring with artificial neural network. To carry out the prediction, twelve networks (one for each month) were trained on area average SST values. A fully connected multilayer feed-forward neural network model having one hidden layer was established. Input SST was taken from Reynolds SST [4] having data from 1 Jan 1950 to 1 Dec 2001. Out of the 50 years data, 38 values were used as estimation set for training, 7 values for cross-validation, and last 5 years values were used for testing. The performance of the networks was also compared with linear multivariate Regression model, and it was observed that ANN showed better results than Regression when the present anomalies are dependent on the past anomalies in a nonlinear manner.

Another work is carried out by Sahai et al. [5]. In this work, they have implemented artificial neural network to carry out the prediction of All India Summer Monsoon Rainfall. Their results illustrate that the monthly rainfall during the monsoon season can be predicted with adequate lead time and good skill, in turn increasing the likelihood of developing customized neural network configurations for predicting monsoon rainfall on suitably defined regional scale.

2.2 Astrophysics

Astrophysical and cosmological phenomena involve diverse physical processes at a massive range of spatial and temporal scales, due to which it becomes virtually impossible to reconstruct them in a laboratory. Along with observational and theoretical astronomy, numerical simulations in astronomy are one of the effective methods to compare observations, implement, and study new theoretical models, mandatory for any astrophysics experiments. For these astronomical phenomena to be scrutinized computationally, the applications must support modeling them at a gargantuan scale, efficiently on HPC platforms of ever-increasing parallelism and complexity. HPC infrastructure can use simulations to construct specific astronomical objects or even the Universe itself for comparing it with the observed Universe. AI is expected to help stimulate and guide the development of the next-gen techniques and methods used in astronomy. AI in astrophysics is in higher demand [6]. Few of the serious challenges posing the analyzing abilities of an astrophysicist, but not limited to, are:

- Inherent large-scale computations of multifaceted complex systems of equations, intractable with analytic techniques,
- Exponential increase in the amount of data gathered by scientific experiments,
- Huge amount of time to understand and process collected data, to list a few.

Overview of software and programming languages used in astrophysics. Table 1 lists the top and widely used software tools and packages [7]. Top applications of astrophysics on HPC are as follows: GAlaxies with Dark matter and Gas intEracT (GADGET), ChaNGa (Charm N-body GrAvity) Solver, General Astrophysics Simulation System (GenASiS), Lambda Cold Dark Matter (Lambda-CDM/ΛCDM)

Table 1 Popular software tools/packages in astrophysics

Image reduction and analysis facility (IRAF) SAO ds9	Astronomy S/W collection Scisoft 7.5 • SolarSoft: System utilities (common programming and data analysis environment), integrated software libraries and databases for solar physics • HEAsoft: software suite consisting of X-ray astronomical data analysis packages like FTOOLS/FV, XIMAGE, XRONOS, XSPEC and XSTAR • XMM-SAS: collection of tasks, scripts, and libraries, specifically designed to reduce and analyze data collected by the XMM-Newton observatory • CIAO X-ray data analysis software for Chandra Observatory • HIPE Herschel data access, analysis, and visualization software • Common astronomy software applications (CASA) package for data post-processing needs of radio telescopes and interferometers • Miriad: Radio interferometer data reduction package of particular interest to users of the Australia Telescope Compact Array (ATCA) • KARMA: Toolkit for IPC, authentication, encryption, graphics display and user interface; IRAF, MIDAS, STSDAS, VO-tools, etc.
Skycat tool	
Graphical astronomy and image analysis tool (GAIA)	
Astronomical image processing system (AIPS)	
Common astronomy software applications (CASA)	
Source extractor (SExtractor)	
Python Crater Detection Algorithm (PyCDA)	
Packages from Chandra X-ray observatory (CXO)	
Chandra Interactive Analysis of Observations (CIAO) • Chandra imaging and plotting system (ChIPS) • Sherpa • Chandra ray tracer (ChaRT) • MARX • Chandra calibration database (CALDB) • Chandra source catalog (CSC) • Chandra grating-data archive and catalog (TGCat)	
Interactive software	
Worldwide telescope	
Tool for OPerations on Catalogues and Tables (TopCat)	

Model, Illustris Project, **The Millennium Simulation** Project, **Eris Simulation**, Octo-Tiger, **Enzo-P and** The yt Project.

Merits of AI in Astrophysics. AI has been applied to astrophysical problems in the past with mixed outcomes, the remarkable aspect being self-learning and understanding what features to look for, formulate insights or predict outcomes, implying the availability of the training dataset that can teach your ML/DL algorithms, as a result how AI can assist you.

1. Astronomical data are used in ML for clustering them into planets, stars, type of galaxies, etc. Clustering algorithm in unsupervised learning finds good use in astronomy [8].
2. AI is used in robots and rovers for study of planets [9]. They are given intelligence specially to activate when there is no line-of-sight (LOS) communication, no input from human, with Earth due to eclipse or rover going other side of planet and the front side guard earth, then robot has to anticipate and act upon in the unexpected situation (or) sight.

3. Facilities like the Large Synoptic Survey Telescope (LSST) and the Wide Field InfraRed Telescope comes online, data volumes will increase in leaps and bounds. DL framework allows astronomers to identify and categorize astronomical objects in enormous datasets with more fidelity than ever.
4. WASP-12b [10] exoplanet was identified using random forest modality for ML and can be adapted to recognize similar patterns in spectral data from other exoplanets without alteration.
5. Enhanced efficiency: Example, in the Square Kilometer Array (SKA) [11], humans eyes cannot possibly locate, let alone assess the huge data generated. Even though a lot of chunk of SKA data generated may be 'noise' or temporary files, it is still worth considering what could be in the discarded data. AI provides greater clarity and certainty to the data you are keeping and discarding. AI could be used to create information, filling in blind spots in our observations of the Universe.

2.3 Agriculture

Agricultural production is absolutely dependent on various unpredictable environmental changes, variables, timeframes, differing scales in space, levels of permanence and automation. Owing to its subjective and vague nature, agriculture's sustainability is an intrinsically 'fuzzy' problem with countless gray regions. AI has been embraced and adopted for farming, agricultural products, and services. Cognitive computing is one of the most disruptive technologies in agriculture, as it can comprehend, learn, train, and counter different conditions; eventually leading to increase in production, efficiency, and quality. New terminologies and concepts embraced are Precision Agriculture [12], Smart Agriculture [13], Digital Agriculture [14], Knowledge-based Agriculture [15], etc. The journey from ML-driven farms to full-fledged AI systems handles the integration of automated data recording, data analysis, ML/DL algorithms decision-making into an interconnected system, farming practices, mathematical simulations in various aspects of plant growth, right from quality of soil, climatic conditions, fertilizers, pesticides, pest control, use of specific machineries at different stages, harvesting, storage, etc., to achieve superior varieties and better yields, without doing costly field trials that can harm the environment. The current situation demands the adoption of suitable technologies at each stage of agriculture.

Most popular applications of AI in agriculture can broadly be categorized into:

- Agricultural Robots: Building and programming autonomous robots to tackle critical agricultural work such as crop harvesting at a higher volume and faster pace than human)
- Crop and Soil Monitoring: Using computer vision, DL algorithms and models to process data collected by IoT devices/drones and/or software-based technology to scrutinize crop and soil health
- Predictive Analytics: ML algorithms and models track and envisage various environmental impacts on crop yield such as weather changes.

C-DAC's Contribution in Agriculture

1. As part of HPC Solutions and Services, C-DAC provides consultancy services for successfully realizing Bio-clustering and Portal for National Agricultural Bio-Grid (NABG) [16]: Mission funded by World Bank through Indian Agricultural Statistics Research Institute (IASRI), New Delhi, supports HPC clusters dispersed across 5 locations. C-DAC has developed web portal for integrating the HPC clusters to NABG.
2. NEtwork relationship Using causal ReasONing (NEURON) [17]: C-DAC's Bioinformatics team, Pune developed NEURON in Jan 2017, used in studying 70,000 varieties of rice crops in collaboration with the Indian Council of Agricultural Research (ICAR).
3. A Workflow Environment for High Throughput Comparative Genomics, Anvaya [18], developed by C-DAC for Advanced Supercomputing Hub for Omics Knowledge in Agriculture (ASHOKA) [19] supercomputer.

Overview of Software and Programming Languages used in Agriculture. Satellite/aerial images immensely help in identifying environmental changes, weather prediction, disaster management, crop assessment, biomass density, and other remote-sensing applications. Remote sensing software processes images (image classification, atmospheric correction, and even radar decomposition) and provides solutions to local or global issues. GRASS GIS and SAGA GIS are most popular and widely used GIS/RS software in agriculture domain (Table 2).

Merits of AI in Agriculture. The huge potential of HPC, big data, AI, and related technologies can be exploited in agriculture for deriving the following merits:

1. Facilitate IoT to attain its highest potential: IBM offers Watson IoT [20] platform for handling humongous data from sensors/drones, using AI technologies for preprocessing and processing, correlating data from multiple sources and formats for extracting prospective insights for better actions and/or recommendations for better production.
2. Insights from image recognition: Customized agriculture for scanning fields, monitoring crops, and analyzing near real-time health of the plants for identifying problematic area along with possible solutions by strategically and optimally apply IoT and Computer vision.
3. Labor force and skill set: Tackles decrease in workforce by handling few operations remotely, automating few processes, preventing few risks by prior identification, etc., as a result helping the farmers to take informed, correct and rapid decisions. We should focus on imparting right mix of technological and agricultural skill sets for the future workforce.
4. Maximize Return on Investments (RoI) in Agriculture: AI could predict the best crop and seed choices for specific weather, climate, and soil conditions best suited for farmer's budget/need. Analyzing and correlating the seeds reaction to different soil types, irrigation levels, local infestations, probabilities of crop diseases, previous best productions, demand–supply requirements, trends in market, prices, etc., to ensure best RoI on the crops.

Table 2 Popular open source GIS/RS software used in agriculture

Popular open source GIS Software	Popular open source RS Software
GRASS GIS	The Sentinel Toolbox (Sentinel-1, Sentinel-2, Sentinel-3)
System for Automated Geo-scientific Analyses (SAGA) GIS	ORFEO Toolbox (OTB): Optical and Radar Federated Earth Observation
QGIS semiautomatic classification plugin (SCP)—Formerly Quantum GIS; Remote sensing plug-ins available	InterImage (for Object-Based Image Analysis (OBIA))
Integrated Land and Water Information System (ILWIS)	Opticks
gVSIG (with remote sensing capabilities)	OSSIM: Open Source Software Image Map
Whitebox GAT	PolSARPro
uDIG (u—user-friendly interface; D—Desktop; I—Internet oriented consuming standard (WMS, WFS or WPS); G—GIS-ready for complex analytical capabilities)	E-foto
MapWindow	
GeoDa	
Diva GIS	
FalconView	
OrbisGIS	
OpenJump GIS—(Formerly called JAVA Unified Mapping Platform (JUMP) GIS)	

5. Agricultural Chatbots: Virtual conversational assistants for personalized interactions with farmers and assist them by answering their queries, providing recommendations or advice related to their specific problems pertaining to agriculture. Chatbots employs ML techniques and understand natural language for further outreach, especially benefiting the farmers in rural areas.

2.4 Bioinformatics

Bioinformatics involves study of distributed molecular data in the form of amino acids, DNA, RNA, peptides, and proteins. The sheer quantity and depth of data demand the creation of efficient knowledge/information extraction methods that can cope with the size and complexity of the data collected [21]. Some of the universal tools like R, S-Plus, Python, Perl, Biolinux, BioPig, BioSpark, FastDoop, GATK, SeqPig, SparkSeq, HadoopBAM are used extensively by the biologists. The following software list is categorized according to sub-domains of Bioinformatics.

We are listing the few widely used tools that are HPC enabled and/or using big data technologies, AI or ML.

Popular (AI/Big Data) Bioinformatics Tools for HPC

1. Genomics is focused on studying all aspects of genome. Dynamic programming, heuristic methods, or hidden Markov model-based algorithms are used to recognize patterns in DNA sequences. Various machine learning algorithms, both supervised and unsupervised, are used for identification of novel classes of functional elements, etc. Table 3 lists some widely used software in genomics that are implemented using parallel programming paradigms such as OpenMP, MPI to work in an HPC environment. Some of the tools such as HBlast, CloudBLAST are implemented to work on Hadoop environment using MapReduce framework.
2. Proteomics: Proteomics is the study of the proteomes. ML is applied to proteomics data for classification of samples and identification of biomarkers and to aid in diagnosis, prognosis, and treatment of specific diseases. Table 4 lists Proteomics software that uses big data technologies/AI.
3. Molecular Simulation studies: The complex and time-consuming calculations in molecular simulations are particularly suitable for a machine learning; Table 5 lists some of widely used software for studying molecular simulation that are HPC enabled/uses ML.

Table 3 Genomics software

Category	Software
Popular genomics tools for various work	Kmulus, ClustalW, MEME, Phylip, AlignACE, cross match, EMBOSS, RAxML, Elph, Weeder, Primer3, Seqclean, SIGNALP
Gene finding software	TMHMM, Genscan, Glimmer, Prodigal
Sequence alignment tool	FastA, MUMmer, Clustal, MGA, LAGAN, mauve, Murasaki, CloudAligner, BLAST, HBlast, CloudBLAST
Repeat finding tools	mRep, Phobos, repeatMasker, TRF, tandemscan
SNP identification	BlueSNP, CrossBow, Falco
Genome assembly	CloudBrush, Contrail
NGS data analysis	GATK
Additional applications	S-Chemo, GRIMD, BioDoop

Table 4 Proteomics software

Category	Software
Proteomics encyclopedia	EPD [22]
Gene finding software	LC–MS/MS [23], PRIDE, Firmiana

Table 5 Molecular simulation software

Category	Software
Molecular dynamics tools for HPC	AMBER, CHARMM, GROMACS, Sybyl, Accelrys NAMD, LAMBDA – H-BAT
Ab-initio tools	MOPAC, GAMESS, Gaussian
Docking software	DOCK6

4. Systems Biology: Network biology includes the reconstruction and study of large-scale biological endogenous networks, and the design and construction of small-scale synthetic gene networks [24]. Systems biology is benefitting from ML largely in the identification of network architectures [25]. The variety of network inference approaches is vast, and the relevant feedback can be seen here [26, 27]. The reverse-engineering approaches discussed in this report has demonstrated a remarkable ability to learn input data patterns to generate biologically important gene regulatory networks with interesting applications for the identification of drug response drivers or disease phenotypes [28, 29].
5. Microarrays: Techniques for analyzing microarrays for interpreting data produced from experiments on DNA (Gene Chip Analysis), RNA, and protein microarrays allow researchers to investigate the large number of gene expression and, in a few cases, the entire genome of an organism in a single experiment. HPC-based software for MDA: Cluster.
6. Text mining: Biological data mining and knowledge discovery are possible through the detection and extraction of text present in rich biomedical literature. Information retrieval, text classification, named-entity recognition, relation extraction, text clustering, and summarization are a few kinds of text mining techniques. Python packages such as NLTK and Beautiful Soup are good with text extraction and preprocessing. ML algorithms like Naive Bayes and support vector machine (SVM) work good for classification. Many DL algorithms are used for sequence-centric mining methods such as recurrent neural networks (RNN), long short-term memory (LSTM), Markov models.

Merits of AI in Bioinformatics. ML applications are becoming omnipresent in biology and include not just genome annotation [30], but also predictions of protein binding [31], the identification of key transcriptional drivers of cancer [32], predictions of metabolic functions in complex microbial communities [33], and the characterization of transcriptional regulatory networks [34], etc. In short, any task under the auspices of ML where a pattern can be learned and then applied to a new datasets. A major advantage is that ML approaches can sift through trends that would otherwise be overlooked across volumes of data. ML plays a critical role in identifying predictive trends in complex biological systems in the era of big data in biological and biomedical research.

C-DAC's contribution in Bioinformatics. The development and usage of any software that depends upon the nature of research of the institute are engaged in. Prestigious institutes like the Indian Institute of Technology (IIT), Indian Institute of Science (IISc), Indian Institutes of Information Technology (IIIT), National Institute of Technology (NIT), Government Labs, and many private labs contribute extensively to the ever-growing HPC and Bioinformatics software by making use of Big Data Analytics, ML technologies for the software development and deployment. The activities of Bioinformatics Resources and Applications Facility (BRAF) Group at C-DAC [35] aims to acquire in-depth knowledge and understanding of the different strata of bio-complexity and thus to include a whole spectrum of data analyzes and essential consumables for research. The assembly of genome sequences, analysis of microarray data, structure-based drug discovery, protein folding, molecular dynamics simulation using advanced molecular modeling techniques are the few research activities listed, along with developing a whole range of research tools, databases, and related resources to address current and future challenges in bioinformatics.

3 Merits of AI in HPC

HPC, big data analytics, and AI are all complementary. The initial AI adopters have implemented the best HPC approaches and techniques. AI is a new tool for the HPC community that can help research, science, and business community gain more mileage from the data they collect. AI can be good catalyst in innovating faster and accurate solutions. Applications involving domain experts, HPC experts, and AI experts together can benefit from the multi-disciplinary nature of each other by finding a far more efficient solution to a difficult problem. The general merits of using AI in HPC for most of the domains are listed as follows:

- Spectacularly increase efficiencies: ML/DL algorithms and techniques are applied for analyzing huge data in shorter time frames. It is computationally expensive to run simulations with a high degree of precision.
- Comprehend deep insights for faster actions: Choosing where to apply ML/DL algorithms and techniques depends on the nature and complexity of the problem to be solved in human-friendly formats.
- AI-based models can sometimes replace computationally intensive simulation tasks: AI can produce results comparable to numerically intense code, using only a fraction of the CPU cycles, thus enabling larger, better, and faster models.
- Next industrial revolution is marriage of HPC and AI: AI is going to transform HPC in every industry. Enterprises with analytical techniques skills are quick to adopt ML. DL depends on large sparse matrices and has more in common with conventional HPC users than numerical algorithms.
- New products, solutions, and services can be brought to market quickly: Enterprises have many opportunities to transform leading-edge research into new and faster products and services that address everyday needs.

4 Migration of HPC to AI

AI has been around since Alan Turing's publication of 'Computing Machinery and Intelligence' in the 1950s, when the computing power and the massive datasets needed to meaningfully run AI applications weren't easily available. With the current developments in computing technology and the deluge of data, researchers can broaden their horizons to explore AI capabilities. Big data, AI, and HPC all require strong compute and performance capabilities, existing HPC organizations are embracing AI to gain efficiency and cost factors by converging their applications on one common infrastructure. To accelerate innovation and discoveries along with return on systems investments, the following few approaches can be applied for converging:

- When both HPC infrastructure and AI applications exist independently: Focus on advanced architecture for both HPC and AI. Select the latest infrastructure-optimized AI frameworks, such as TensorFlow, Caffe, and MXNet, best suited for running your AI applications on HPC infrastructure.
- Embed AI to running HPC modeling and simulation applications: Identify the HPC modules for applying AI techniques for faster results and deriving better insights.
- Embed HPC simulations into AI: Recognize when AI uses simulations to expand training data or give supervised labels to unlabeled data.

5 Conclusion

The exponentially increase in domain-specific data, owing to big data and IoT devices, necessitates a new paradigmatic approach. AI has taken the world by storm. Growing adoption of AI has spurred interest in HPC beyond research institutions. The survey paper summarizes the findings of the team for identifying requirements of big data and AI in HPC application domains, viz. weather and climate, astrophysics, agriculture, and bioinformatics. This serves as a useful guide for the design and deployment of AI on supercomputing platforms. Adoption of AI in these domains is in its infancy, and we cannot predict where it will end. The convergence of AI and HPC will transform the technology landscape and touch almost every industry over the next decade.

References

1. MeitY GoI website: https://meity.gov.in/content/high-performance-computinghpc
2. McGovern A (2017) Using artificial intelligence to improve real-time decision making for high impact Weather, https://journals.ametsoc.org/doi/pdf/10.1175/BAMS-D-16-0123.1

3. Tripathi KC, Das ML, Sahai AK (2006) Predictability of sea surface temperature anomalies in the Indian Ocean using artificial neural networks. Indian J Mar Sci 35(3):210–220
4. Reynolds RW, Smith TM (1995) High-resolution global sea surface temperature climatology. J Clim 8:1571–1583
5. Sahai AK, Soman MK, Satyan V (2000) All India summer monsoon rainfall prediction using an artificial neural network. Clim Dynam 16(4):291–302
6. Yan LXY (2017) Machine learning for astronomical big data processing. IEEE Visual Communications and Image Processing (VCIP)
7. Momcheva I, Tollerud E (2015) Software use in astronomy: an informal survey, [astro-ph.EP]
8. Cisewski J (2014) Clustering and classification in astronomy, https://astrostatistics.psu.edu/RLectures/clustering_classification_Jessi.pdf
9. NASA Science, MARS website: A.I. Will Prepare Robots for the Unknown: https://mars.nasa.gov/news/2884/ai-will-prepare-robots-for-the-unknown/
10. Marquez-Neila P, Fisher C, Sznitman R, Heng K (2016) Supervised machine learning for analysing spectra of exoplanetary atmospheres, [astro-ph.EP]
11. Mosiane O, Oozeer N, Bassett BA (2016) Radio frequency interference detection using machine learning. IEEE Radio and Antenna Days of the Indian Ocean (RADIO)
12. Shadrin D, Menshchikov A, Somov A, Bornemann G, Hauslage J, Federovo M (2019) Enabling precision agriculture through embedded sensing with artificial intelligence. IEEE Trans Instrum Meas
13. Mat I, Kassim MRM, Harun AN, Yusoff IM (2018) Smart agriculture using internet of things. In: IEEE conference on open systems (ICOS)
14. Tang S, Zhu Q, Zhou X, Liu S, Wu M (2005) A conception of digital agriculture. In: IEEE international geoscience and remote sensing symposium
15. Ahmad T, Ahmad S, Jamshed M (2015) A knowledge-based Indian agriculture: with Cloud ERP arrangement. In: International conference on green computing and internet of things (ICGCIoT)
16. National Agricultural Bio-computing Portal Website: https://webapp.cabgrid.res.in/biocomp/
17. Neuron, C-DAC Website: https://www.cdac.in/index.aspx?id=pk_itn_spot1056
18. Anvaya, C-DAC Website: https://www.cdac.in/index.aspx?id=bio_products
19. Rai A, Angadi UB, Lal SB, Initiative of ASHOKA in Indian Agricultural Research: https://www.plant-phenotyping.org/lw_resource/datapool/_items/item_179/i4_angadi_ashoka.pdf
20. IBM Watson IoT Platform Website: https://www.ibm.com/in-en/internet-of-things/solutions/iot-platform/watson-iot-platform
21. Zhaoli, Machine learning in bioinformatics. In: Proceedings of 2011 international conference on computer science and network technology, Harbin, 2011, pp 582–584
22. Brenes A, Afzal V, Kent R, Lamond AI (2018) The encyclopedia of proteome dynamics: a big data ecosystem for (prote)omics. Nucleic Acids Res 46(D1):D1202–D1209. https://doi.org/10.1093/nar/gkx807
23. Kantz ED, Tiwari S, Watrous JD, Cheng S, Jain M (2019) Deep neural networks for classification of LC-MS spectral peaks. Anal Chem. https://doi.org/10.1021/acs.analchem.9b02983
24. Camacho DM, Collins KM, Powers RK, Collins JJ, Next-generation machine learning for biological networks, Camacho, Diogo M. et al. Cell 173(7):1581–1592
25. Butte AJ, Kohane IS (2000) Mutual information relevance networks: functional genomic clustering using pairwise entropy measurements. Pac Symp Biocomput, pp 418–429
26. De Smet R, Marchal K (2010) Advantages and limitations of current network inference methods. Nat Rev Microbiol 8:717–729
27. Marbach D, Costello JC, Küffner R, Vega NM, Prill RJ, Camacho DM, Allison KR, Kellis M, Collins JJ, Stolovitzky G (2012) DREAM5 consortium; wisdom of crowds for robust gene network inference. Nat Methods 9:796–804
28. Akavia UD, Litvin O, Kim J, Sanchezarcia F, Kotliar D, Causton HC, Pochanard P, Mozes E, Garraway LA, Pe'er D (2010) An integrated approach to uncover drivers of cancer. Cell 143:1005–1017

29. Costello JC, Heiser LM, Georgii E, Gönen M, Menden MP, Wang NJ, Bansal M, Ammad-ud-din M, Hintsanen P, Khan SA et al (2014) NCI DREAM community. A community effort to assess and improve drug sensitivity prediction algorithms. Nat Biotechnol 32:1202–1212
30. Leung MKK, Delong A, Alipanahi B, Frey BJ (2016) Machine learning in genomic medicine: a review of computational problems and data sets. Proc IEEE 104:176–197
31. Alipanahi B, Delong A, Weirauch MT, Frey BJ (2015) Predicting the sequence specificities of DNA- and RNA-binding proteins by deep learning. Nat Biotechnol 33:831–838
32. Califano A, Alvarez MJ (2017) The recurrent architecture of tumour initiation, progression and drug sensitivity. Nat Rev Cancer 17:116–130
33. Langille MGI, Zaneveld J, Caporaso JG, McDonald D, Knights D, Reyes JA, Clemente JC, Burkepile DE, Vega Thurber RL, Knight R et al (2013) Predictive functional profiling of microbial communities using 16S rRNA marker gene sequences. Nat Biotechnol 31:814–821
34. Djebali S, Davis CA, Merkel A, Dobin A, Lassmann T, Mortazavi A, Tanzer A, Lagarde J, Lin W, Schlesinger F et al (2012) Landscape of transcription in human cells. Nature 489:101–108
35. C-DAC's Bioinformatics Research Applications: https://www.cdac.in/index.aspx?id=hpc_sa_braf_access
36. Peres G, HPC in astronomy: overview and perspectives: Mem. S.A.It. vol 73, p 23

Breast Cancer Classification Using Transfer Learning

Animesh Seemendra, Rahul Singh, and Sukhendra Singh

Abstract Cancer is one of the most lethal forms of the disease. And in females, breast cancer is the most common cancer which could even lead to death if not properly diagnosed. Over the years, a lot of advancement can be seen in the field of medical technology but when it comes to detecting breast cancer biopsy is the only way. Pathologists detect cancer by using histological images under the microscope. Inspecting cancer visually is a critical task; it requires a lot of attention, skill and is time-consuming. Therefore, there is a need for a faster and efficient system for detecting breast cancer. Advancements in the field of machine learning and image processing lead to multiple types of research for creating an efficient partially or fully computer monitored diagnosis system. In this paper, we have used histological images to detect and classify invasive ductal carcinoma. Our approach involves convolutional neural networks which are a very advanced and efficient technique when dealing with images in machine learning. We compared various famous deep learning models, and we used these pre-trained CNN architectures with fine-tuning to provide an efficient solution. We also used image augmentation to further improve the efficiency of the solution. In this study, we used VGG, ResNet, DenseNet, MobileNet, EfficientNet. The best result we got was using fine-tuned VGG19 and with proper image augmentation. We achieved a sensitivity of 93.05% and a precision of 94.46 with the mentioned architecture. We improved the F-Score of the latest researches by 10.2%. We have achieved an accuracy of 86.97% using a pre-trained DenseNet model which is greater than the latest researches that achieved 85.41% [30] accuracy.

Keywords Breast cancer · Deep learning · Invasive ductal carcinoma (IDC)

A. Seemendra (✉) · R. Singh · S. Singh
JSS Academy of Technical Education, Noida, India
e-mail: animeshseemendra1997@gmail.com

R. Singh
e-mail: rahulsingh2240@gmail.com

S. Singh
e-mail: sukhendrasingh@gmail.com

P. K. Singh et al. (eds.), *Evolving Technologies for Computing, Communication and Smart World*, Lecture Notes in Electrical Engineering 694,
https://doi.org/10.1007/978-981-15-7804-5_32

1 Introduction

Breast cancer in women is the most common form of cancer [1]. The American Cancer Society [2] reports that breast cancer leads to deaths of 40,610 women and 460 men in 2017 in the USA. Another report by the American Cancer Society reveals in every 8 women 1 which is diagnosed with invasive breast cancer and 1 in 39 women die from breast cancer [3]. In the developed countries like USA and UK, breast cancer survival rate is relatively high but in poor and developing countries like India, survival rates are much less major reasons being lack of awareness, delay in diagnosis, and costly screenings. Though there has been an increase in survival rates due to the advancement of technology, a number of cases have also increased over the years. Invasive ductal carcinoma (IDC) constitutes about 80% of all breast cancer cases [4]. Doctors use various techniques to detect IDC such as physical examination, mammography, ultrasound, breast MRI, and biopsy. The biopsy is often done to check the suspicious mammogram. It involves examining the abnormal-looking tissue under a microscope by a pathologist. Such an examination requires a lot of time, skill, and precision. Therefore over the years, many computer-aided systems are being developed to make the work of a pathologist easier and efficient.

Since the advancements in machine learning and image processing a lot of research is oriented towards creating a better and efficient model to detect breast cancer. Machine learning is a field of computer science that uses data or past experiences to generate or predict outcomes for unseen data [5–8, 37]. Multiple methods have been applied to create an efficient machine learning model for detecting breast cancer. Classification using standard machine learning techniques often requires feature selections. Some of them used image segmentation such as thresholding along with the SVM classifier to create a classification model [9–12]. Preprocessing is an important task in machine learning which depends upon the type of data. Some studies used mammography images; therefore, a common preprocessing step was cropping and resizing [9–12]. Different studies used different image preprocessing techniques such as adaptive histogram equalization followed by segmentation, image masking, thresholding, feature extraction, and normalization before training the classifier [9], morphological operations for image preprocessing followed by an SVM classifier [13] and high-pass filtering followed by a clustering algorithm [14]. Some other preprocessing and classification steps involved were median filtering and fuzzy-C clustering with thresholding [15], region-based image segmentation followed by SVM classifier [16] and ostu-based global thresholding method followed by radial basis neural network for classification [17].

Later advancements in deep learning and availability of high computational GPU lead to multiple studies in this sub-domain of machine learning. Deep neural networks do not require any feature selection or extensive image preprocessing techniques. These networks mimic human brain neurons and automatically does feature selections [18]. The use of neural network architectures in breast cancer classification leads to the state-of-the-art results. Convolutional neural networks proved to be one of the most efficient models when working with images and videos. CNN maintains the structural architecture of the data while learning; this makes it achieve a higher accuracy over traditional machine learning methods [19]. Therefore, multiple studies for breast cancer detection revolve around convolutional neural networks [20–24]. Some of the studies combined deep neural architectures for feature selection and standard machine learning classifiers like SVM for the classification task [22]. Other approaches involve using transfer learning as a key part. Transfer learning is a technique to using knowledge of another network trained on a different dataset and use it in with your data. Various studies have used ResNet50, VGG16, VGG19, etc., models for breast cancer detection and have shown impressive results [25–28]. Studies mentioned in [29] have also used CNN and deep learning architectures for transfer learning to classify invasive ductal carcinoma. With the proposed methods, they have achieved an accuracy of 85.41%. We worked to improve these results.

Many of these approaches have used a multi-classification dataset that contains more than one class of breast cancer. Also, most of the approaches were applied to a mammographic dataset. The mammographic dataset can detect any abnormalities in the tissues which have to further go with biopsy. Our approach is on a binary classification dataset which addresses the problem of Invasive ductal carcinoma. The effective classification on the dataset will help in easing the task of the pathologist. In this paper, we have approached the problem by using various deep learning architectures through transfer learning. We applied image undersampling and image augmentation to handle imbalance dataset and to increase the accuracy of the model by providing necessary image transformation parameters. The objective of this paper is (a) to compare the performances of various famous convolutional deep learning architectures through transfer learning and achieve high efficiency in less computational time on the given dataset (b) to increase the efficiency in the IDC classification task over previous works.

2 Approach

In our approach, we used transfer learning and image augmentations on the given dataset. The following sections include steps applied to the dataset to detect invasive ductal carcinoma.

2.1 Undersampling and Data Preparation

In medical imaging, dataset class imbalance is a common problem; there are generally more images of negative results than positive results of disease. Similarly, in our dataset, there is a huge imbalance of the data between IDC(−) and IDC(+) classes. Such imbalance data could mislead the model in learning one type of class more than the other. Therefore, we used resampling techniques to create balanced data.

2.2 Image Augmentation

Image augmentation is a technique that is used to generate more data by applying certain processing to the images. It helps to create unseen yet valuable data which could help in further increase the accuracy. The common image augmentation techniques are zoom, flips, rotation, etc. These techniques help the model to learn variation in the dataset and not to get constraint in one particular type of format. It is also a powerful technique to solve the lack of data problem. We applied image augmentation techniques to add variation to the dataset which could further increase the efficiency of the model.

2.3 Transfer Learning

Transfer learning is a technique of using models that were trained on different datasets on your data with some tuning. We used various state-of-the-art CNN architectures that were trained on the ImageNet dataset. The idea was that the starting layers of CNN are used to capture the high-level features such as texture and shape which are not dependent on the data; therefore, utilizing those trained weights and fine-tuning it according to our model could help us achieve good results. The various deep learning architectures we used are present in Sect. 3.

3 Deep Learning Architectures

The various deep learning architectures used are mentioned below. While training, each of these architectures was fine-tuned so that they give the most efficient solution. While training validation set was used to prevent overfitting of the data. The weights were stored when the minimum loss was found, and the same weights were used to test the model efficiency on the test set. All the following architectures were trained on the ImageNet dataset [30] and achieved remarkable results.

3.1 VGG16 and VGG19

The VGG16 and VGG19 models were sequential CNN models with 3×3 convolutional layers stacked upon one another. The architecture contained max-pooling layers to reduce the volume as the layer increases finally the fully connected network layer with 4096 nodes followed by a 1000 node layer with a softmax activation function [31]. VGG16 and VGG19 are slow to train and have large weights themselves.

3.2 *ResNet50*

ResNet50, unlike VGG models, is a non-sequential model. This is a collection of CNN stacked together with residual networks added to each layer. The output of each layer of CNN layer is added with the actual input of that layer [32]. ResNet50 contains 50 weight layers and is faster to train with VGG networks.

3.3 *DenseNet*

DenseNet involves some advancements over ResNet50 networks, instead of adding feature maps DenseNet network concatenates output feature maps with input feature maps. Each layer output is concatenated with the outputs of all the previous layers, thus creating a dense architecture [38].

3.4 *EfficientNet*

This model performed so well on the ImageNet dataset that it was able to achieve 84.4% top 1 ranking. It crossed the state-of-the-art accuracy with 10 times better efficiency. The model was smaller and faster. Width, depth, and image resolution were scaled, and the best result was observed [33]. EfficientNet has shortcuts that directly connect between the bottlenecks and a fewer number of channels than expansion layers.

3.5 *MobileNet*

MobileNet is a lightweight model. Instead of performing combined convolutions on the three channels of colors and flattening, it applies convolutions on each color channel [34]. They are ideal for mobile devices.

4 Experiment and Results

4.1 *Datasets*

The dataset from which our data was derived was a 162 whole mount slide images of the breast cancer specimens scanned at 40× [35, 36]. The derived dataset contained 25,633 positive samples and 64,634 negative samples. Each image is of size (50 ×

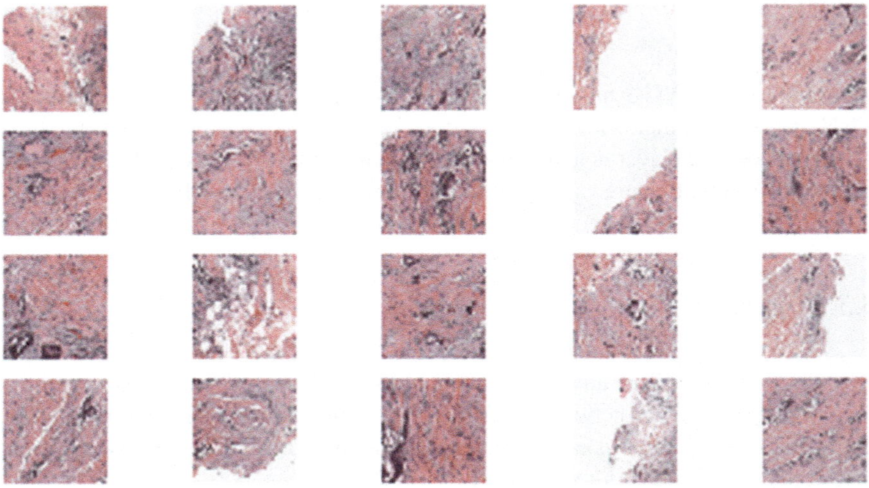

Fig. 1 Sample images from the dataset

50) extracted from the original dataset (Fig. 1). The image filename is in the form of z_xX_yY_classD.png. For example, 55634_idx8_x1551_y1000_class0.png where z is the patient ID (55634_idx8), X represents the x-coordinate and Y represents y-coordinate from which the patch is cropped and D indicates the class of the data. Class0 being IDC(−) and Class 1 being IDC(+).

4.2 Data Preparation

We performed undersampling on our dataset, i.e., removing samples from classes to make it more balanced (Fig. 2). The undersampling was done at random without replacement to create a subset of data for the target classes.

The final distribution contains 25,367 positive samples and 25,366 negative samples of data. Final data split into the train, test, and valid sets containing 31,393, 7849, 11,490 files, respectively. The class names were extracted from the names of the files and were one-hot encoded, i.e., binary class was represented by two features instead of one; for example, class0 was represented by an array of [1, 0] and class1 was represented by [0, 1] so that for deep learning model will give probabilities for the two indexes of the array and whichever is maximum that index will be the class of our array after which image augmentation techniques were applied such as zoom, rotation, width shift, height shift, height shift, shear, flip.

Now many deep learning model architectures were used to provide transfer learning. They were fine-tuned to give change their domain to our dataset.

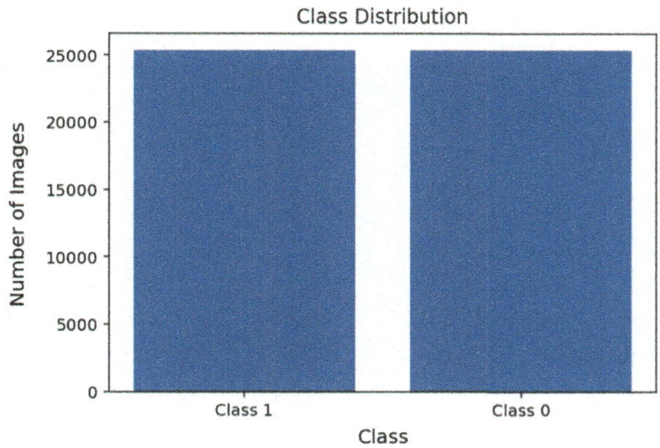

Fig. 2 Final distribution of class after undersampling

4.3 Results

The experiments were performed on Kaggle Notebook with its GPU with Keras API of Tensorflow. NVidia K80 GPU was used to perform experiments on the dataset. We used various famous deep learning architectures for transfer learning (VGG16, VGG19, ResNet50, DenseNet169, DenseNet121, DenseNet201, MobileNet, and EfficientNet). All these models were trained on the ImageNet dataset, and over the years, they have achieved high accuracy of the dataset. We have used only CNN architectures of these models and then added layers according to the requirement of the complexity of the dataset. The first set of experiments were conducted without any data augmentation, and then data argumentation was added to see the change. The evaluation metrics used to compare results were F1-score, recall, precision, accuracy, specificity, sensitivity. They are used to evaluate our model using true positives, true negatives, false positives, and false negatives.

DenseNet169 was trained for 60 epochs with a global pooling layer and two dense layers each followed by a dropout layer of 0.15 and 0.25, respectively, with ReLU activation added after the frozen CNN layers. The last layer was a dense layer with two classes and a softmax function. DenseNet 121, ResNet50, VGG19, and VGG16 were trained for 60, 60, 100, 100 epochs, respectively. The architecture of the last layers was made different than DenseNet169. We added batch normalization having 1e-05 epsilon value and 0.1 momentum followed by a dense layer with 512 nodes and a dropout of 0.45. After that, a dense layer with 2 nodes representing each class was added with softmax activation. The experiment results without data augmentation can be seen in Table 1, the highest value in each column is highlighted (in bold).

Data augmentation was further added to push the accuracy even further with more different types of images in the dataset. Augmentation was all general, i.e., zoom range of 0.3, rotation range of 20, width shift range, sheer range, and height shift range was all set to 20 and also horizontal flip was set as true. With the mentioned augmentation, all previous models were used with SGD optimizer, global pooling

Table 1 Results without data augmentation

Models	Sensitivity	Specificity	Accuracy	Precision	Recall	F-Score
DenseNet169	83.37	88.14	84.43	89.29	81.37	85.15
DenseNet121	83.42	86.61	84.94	87.22	83.42	85.28
ResNet50	81.21	**89.38**	84.82	**90.61**	81.21	85.65
MobileNet	85.83	85.26	85.54	85.13	85.83	85.48
VGG16	85.16	87.49	86.29	87.83	85.16	86.54
VGG19	85.40	87.71	**86.52**	88.10	85.40	**86.73**
EfficientNet b0	**87.60**	79.92	83.32	77.63	**87.60**	82.31
EfficientNet b4	85.36	84.55	84.95	84.38	85.25	84.87
EfficientNet b5	86.78	81.40	83.88	79.94	86.78	83.22

Bold signifies highest value in respective column

layer followed by 32-node and 64-node layer with 0.15 and 0.25 dropout layer, respectively, and ReLU was used as an activation function. Results can be seen in Figs. 3 and 4.

Result of the experiments can be seen in Table 2.

By using different deep architectures for transfer learning, we were able to achieve higher accuracy and recall with our models over previous claimed methods. Table 3 compares our best model's efficiency with respect to the efficiency claimed by previous works.

We presented F-Score as evaluation metrics for comparison with previous methods. We can see that our approach and method improved the F-Score by 10.2% compared to the latest researches.

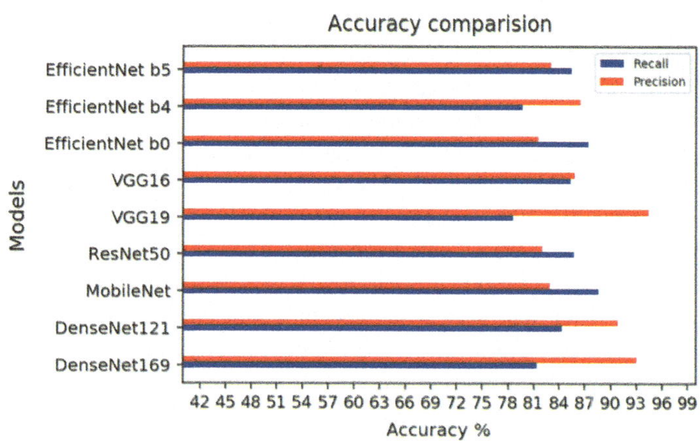

Fig. 3 Precision and recall versus fine-tuned models

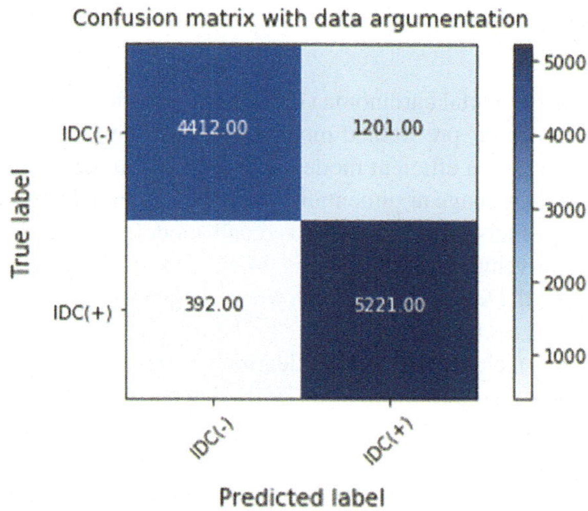

Fig. 4 Confusion matrix of DenseNet 169 with data augmentation

Table 2 Results with data augmentation

Models	Sensitivity	Specificity	Accuracy	Precision	Recall	F-Score
DenseNet169	81.29	91.84	85.80	93.01	81.29	86.76
DenseNet121	84.30	90.10	**86.97**	90.87	84.30	**87.46**
MobileNet	**88.57**	83.92	86.10	82.89	**88.57**	85.64
ResNet50	85.71	82.79	84.19	82.00	85.71	83.84
VGG19	78.51	**93.05**	84.30	**94.46**	78.51	80.82
VGG16	85.32	85.68	85.50	85.75	85.32	85.53
EfficientNet b0	87.36	82.71	84.88	81.56	87.36	84.36
EfficientNet b4	79.64	85.17	82.16	86.44	79.62	82.89
EfficientNet b5	85.35	83.44	84.37	82.98	85.35	84.14

Table 3 Our score versus existing deep learning approaches

Studies	Method	F1-Score
Paper [29] (2019)	Convolution Neural Networks trained from scratch with AdaDelta optimizer and image augmentation	0.8528
Our Result with DenseNet121	Transfer Learning with deep learning architectures (DenseNet121) with SGD optimizer and image augmentation	**0.8746**

5 Conclusion

We classified invasive ductal carcinoma (IDC) using deep learning. In our study, we took advantage of various pre-trained models and used their knowledge and added some fine-tuning to get an efficient model. We first tried undersampling techniques to balance classes then image argumentation followed by transfer learning. We were able to get a high precision and moderate recall model with VGG19 with image argumentation. The value of precision was 94.46, and recall was 78.51. We got an accuracy of 86.97 with DenseNet121 which when compared with other latest research that got 85.41% [29] accuracy.

Hence, we can conclude that transfer learning which is the simplest technique available in deep learning frameworks can be used to detect dangerous diseases such as cancer. We have also shown that deep learning can perform well while detecting IDC, and therefore, it is an improvement over manual segmentation. Deep learning has given us the freedom to detect the disease on smaller datasets with small size images from which only deep learning models can infer useful information. An increase in the dataset will improve the results. Hence, in the future, we are planning to work on the full dataset with full-size slide images and will work with more advanced techniques such as GAN networks to obtain better results.

References

1. American Institute of Cancer Research. https://www.wcrf.org/sites/default/files/Breast-Cancer-2010-Report.pdf
2. American Cancer Society. https://www.cancer.org/content/dam/cancer-org/research/cancer-facts-and-statistics/breast-cancer-facts-and-figures/breast-cancer-facts-and-figures-2017-2018.pdf
3. American Cancer Society. https://www.cancer.org/content/dam/cancer-org/research/cancer-facts-and-statistics/breast-cancer-facts-and-figures/breast-cancer-facts-and-figures-2019-2020.pdf
4. Breast Cancer Website. https://www.breastcancer.org/symptoms/types/idc
5. Dhall D, Kaur R, Juneja M (2020) Machine learning: a review of the algorithms and its applications. In: Singh P, Kar A, Singh Y, Kolekar M, Tanwar S (eds) Proceedings of ICRIC 2019. Lecture Notes in Electrical Engineering, vol 597. Springer, Cham. https://doi.org/10.1007/978-3-030-29407-6_5
6. Pillai R, Oza P, Sharma P (2020) Review of machine learning techniques in health care. In: Singh P, Kar A, Singh Y, Kolekar M, Tanwar S (eds) Proceedings of ICRIC 2019. Lecture Notes in Electrical Engineering, vol 597. Springer, Cham. https://doi.org/10.1007/978-3-030-29407-6_9
7. Jondhale SR, Shubair R, Labade RP, Lloret J, Gunjal PR (2020) Application of supervised learning approach for target localization in wireless sensor network. In: Singh P, Bhargava B, Paprzycki M, Kaushal N, Hong WC (eds) Handbook of wireless sensor networks, issues and challenges in current scenario's, advances in intelligent systems and computing, vol 1132. Springer, Cham. https://doi.org/10.1007/978-3-030-40305-8_24
8. Singh YV, Kumar B, Chand S, Sharma D (2019) A hybrid approach for requirements prioritization using logarithmic fuzzy trapezoidal approach (LFTA) and artificial neural network (ANN). In: Singh P, Paprzycki M, Bhargava B, Chhabra J, Kaushal N, Kumar Y (eds) Futuristic trends

in network and communication technologies. FTNCT 2018, communications in computer and information science, vol 958. Springer, Singapore. https://doi.org/10.1007/978-981-13-3804-5_26
9. Breast Cancer Classification using Image Processing and Support Vector Machine. https://pdfs.semanticscholar.org/d414/5b40d6a65b84e320a092220dc8e6cc54a7dc.pdf
10. Sudharshan PJ et al (2019) Multiple instance learning for histopathological breast cancer image classification. Expert Syst Appl 117:103–111. https://doi.org/10.1016/j.eswa.2018.09.049
11. Rejani YA, Selvi ST (2009) Early detection of breast cancer using SVM classifier technique. Int J Comput Sci Eng
12. Pathak R (2020) Support vector machines: introduction and the dual formulation. In: Advances in cybernetics, cognition, and machine learning for communication technologies. Lecture Notes in Electrical Engineering, vol 643. Springer, Singapore. https://doi.org/10.1007/978-981-15-3125-5_57
13. Naresh S, Kumari SV (2015) Breast cancer detection using local binary patterns. Int J Comput Appl 123(16):6–9
14. Guzman-Cabrera R, Guzaman-Supulveda JR, Torres-Cisneros M, May-Arrioja DA, Ruiz-Pinales J, Ibarra-Manzano OG, AvinaCervantes G, Parada GA (2013) Digital image processing technique for breast cancer detection. Int J Thermophys 34:1519–1531
15. Kashyap KL, Bajpai MK, Khanna P (2015) Breast cancer detection in digital mammograms. In: IEEE international conference in imaging systems and techniques, pp 1–6
16. Oliver A, Marti J, Marti R, Bosch A, Freixenet J (2006) A new approach to the classification of mammographic masses and normal breast tissue||. In: International conference on pattern recognition, pp 1–4
17. Kanojia MG, Abraham S (2016) Breast cancer detection using RBF neural network. In: IEEE conference on contemporary computing and informatics, pp 363–368
18. Goodfellow I, Bengio Y, Courville A, Deep learning book
19. Gu J, Wang Z, Kuen J, Ma L, Shahroudy A, Shuai B, Liu T, Wang X, Wang L, Wang G, Cai J, Chen T (2015) Recent advances in convolutional neural networks. arxiv: 1502.07108
20. Selvathi D, Poornila AA (2017) Deep learning techniques for breast cancer detection using medical image analysis. In: Biologically rationalized computing techniques for image processing applications, pp 159–186
21. Ragab DA, Sharkas M, Marshall S, Ren J (2019) Breast cancer detection using deep convolutional neural networks and support vector machines. PeerJ 7:e6201. https://doi.org/10.7717/peerj.6201
22. Shen L, Margolies RL, Rothstein JH, Fluder E, McBride R, Sieh W (2017) Learning to improve breast cancer detection on screening mammography. arxiv: 1708.09427
23. Zou L, Yu S, Meng T, Zhang Z, Liang X, Xie Y (2019) A technical review of convolutional neural network-based mammographic breast cancer diagnosis. In: Computational and mathematical methods in medicine. https://doi.org/10.1155/2019/6509357
24. Ciresan CD, Giusti A, Gambardella ML, Schmidhuber J (2013) Mitosis detection in breast cancer histology images with deep neural networks. In: International conference on medical image computing and computer-assisted intervention
25. Le H, Gupta R, Hou L, Abousamra S, Fassler D, Kurc T, Samaras D, Batiste R, Zhao T, Dyke AL, Sharma A, Bremer E, Almeida SJ, Saltz J (2019) Utilizing automated breast cancer detection to identify spatial distributions of tumor infiltrating lymphocytes in invasive breast cancer
26. Wu N et al (2019) Deep neural networks improve radiologists performance in breast cancer screening. In: Medical imaging with deep learning conference
27. Rakhlin A, Shvets A, Iglovikov V, Kalinin AA (2018) Deep convolutional neural networks for breast cancer histology image analysis. In: International conference on image analysis and recognition
28. Shen L, Margolies LR, Rothstein JH, Fluder E, McBride RB, Sieh W (2017) Deep learning to improve breast cancer early detection on screening mammography. arxiv: 1708.09427

29. Romano AM, Hernandez AA (2019) Enhanced deep learning approach for predicting invasive ductal carcinoma from histopathology images. In: International conference on artificial intelligence and big data
30. Deng J, Dong W, Socher R, Li L, Li K, Fei-Fei L (2009) ImageNet: a large-scale hierarchical image database. In: IEEE conference on computer vision and pattern recognition
31. Simonyan K, Zisserman A (2015) Very deep convolutional network for large-scale image recognition, arxiv: 1409.1556
32. He K, Zhang X, Ren S, Sun J (2015) Deep residual learning for image recognition, arxiv: 1512.03385
33. Tan M, Le QV (2019) EfficientNet: rethinking model scaling for convolutional neural networks, arxiv: 1905.11946
34. Howard AG, Zhu M, Chen B, Kalenichenko D, Wang W, Weyand T, Andreetto M, Adam H (2017) MobileNets: efficient convolutional neural networks for mobile vision applications, arxiv: 1704.04861
35. Janowczyk A, Madabhushi A (2016) Deep learning for digital pathology image analysis: a comprehensive tutorial with selected use cases
36. Cruz-Roaa A et al (2014) Automatic detection of invasive ductal carcinoma in whole slide images with convolutional neural networks. In: Proceedings of SPIE—the international society for optical engineering, vol 9041
37. Kral P, Lenc L (2016) LBP features for breast cancer detection. In: IEEE international conference on image processing, pp 2643–2647
38. Huang G, Liu Z, Maaten LD, Weinberger QK (2016) Densely connected convolutional networks, arxiv: 1608.06993

A Novel Fusion Framework for Salient Object Detection Based on Support Vector Regression

Vivek Kumar Singh and Nitin Kumar

Abstract Salient object detection is highly influenced with a variety of salient features extracted from an image. It is a challenging task in saliency detection to find the suitable contribution of each salient feature for salient object detection. In this paper, we propose a novel fusion framework for salient object detection based on support vector regression (SVR) that effectively combines salient features for computing saliency score. Firstly, we extract salient features using a state-of-the-art saliency method. Secondly, training samples are generated that include positive and negative samples. These samples are produced by applying two different thresholds on salient features and these thresholds are also obtained from salient features. Thirdly, the training samples are fed into support vector regression to learnt SVR model for obtaining a set of learnable parameters. SVR saliency score of the input image is computed by using learnt SVR model with salient features of the input image. Finally, SVR saliency score is refined by post processing operations to obtained enhanced saliency map. Experimental results are shown efficacy of the proposed framework against compared 10 state-of-the-art saliency methods on two publicly available salient object detection datasets viz. ASD and ECSSD. The proposed framework achieves 1.7% and 19.4% improvement of Recall and MAE score on ASD dataset and 11.1% and 3.2% improvement of Recall and F-measure score on ECSSD dataset from the compared highest performing method.

Keywords Unsupervised · Majority voting · Post processing · Saliency map

V. K. Singh (✉) · N. Kumar
National Institute of Technology, Uttarakhand, India
e-mail: vivek.kumarsingh@nituk.ac.in

N. Kumar
e-mail: nitin@nituk.ac.in

P. K. Singh et al. (eds.), *Evolving Technologies for Computing, Communication and Smart World*, Lecture Notes in Electrical Engineering 694,
https://doi.org/10.1007/978-981-15-7804-5_33

1 Introduction

Human visual system (HVS) is capable of filtering images according to their significant role in grabbing visual attention. This process assists human brain to quickly and effectively process and analyze a natural image. Saliency detection is a method to simulate HVS selective attention mechanisms using computational resources. In this automated process, saliency value is specifically calculated for each image element i.e. pixel or superpixel, which unfolds the degree of visual attention and identifies salient regions in the image. The set of these methods that extract saliency value from the input image is an important and attractive research area in the field of computer vision. Saliency detection can be applied as prepossessing step in various computer vision applications such as image segmentation [5], image retrieval [14], image collection browsing [16] etc.

Existing Saliency detection methods can be typically categories into two groups: supervised methods and unsupervised methods. Supervised saliency detection methods [8, 11, 20, 23, 29] contain training process and require numerous training images. These methods are more effective and learn salient features from training images. However, such methods are less economical and the performance of these methods depend on training images. In contrast, unsupervised saliency methods [19, 25, 27] are simple and computationally effective due to no requirement of training images. Typically, these methods require some priors knowledge about salient object(s) such as contrast prior, background prior etc. Effectiveness of such methods depends on the reliability of chosen priors in order to characterize the representation of salient object. In recent years, various saliency fusion methods [3, 4, 18] have been proposed that integrate different saliency feature maps to generate more accurate saliency map.

In this paper, we proposed a novel fusion framework for salient object detection based on unsupervised learning of support vector regression. Here, a support vector regression model is learnt for each individual image and learning process does not required any human annotation. For learning process, foreground (i.e. positive) and background(i.e. negative) labels are obtained based on various salient features extracted from the input image. Further, post processing orations are applied on the saliency score obtained from support vector regression in order to generate more accurate saliency map. The main contributions of our work are summarized as follows:

1. An effective and efficient salient features fusion framework is proposed to exploit saliency features by defining the fusion framework in which salient features are combined in an appropriate manner and use post processing operations for generation of saliency map.
2. Extensive experiments are conducted to demonstrate superiority of the proposed framework against 10 recent state-of-art-methods over two publicly available datasets viz. ASD [1] and ECSSD [24].

Rest of the paper is organized as follows. Brief discussion of state-of-the-art saliency detection methods in Sect. 2. Section 3 introduce the details of the proposed framework. Section 4 is devoted to the results and comparative analysis. Conclusion and future work are drawn in Sect. 5.

2 Related Work

In last few decades, many saliency detection methods have been proposed by various researchers and the performance is increasing gradually. These methods can be grouped into three categories: supervised saliency detection methods, unsupervised saliency detection methods, and saliency fusion methods [3, 13]. In supervised saliency detection method, Liu et al. [11] proposed method which learnt weights of various features using Condition Random Field (CRF) and combined to calculate final saliency value of each element of image. Recently, deep learning techniques are also exploited in saliency detection. Hou et al. [8] suggested a short connections based Holisitcally-Nested Edge Detector (HED)architecture that extract various robust multi scale feature maps in order to produce effective saliency map. Wang et al. [23] proposed a global Recurrent Localization Network (RLN) that extracts contextual features and refines convolutional blocks for saliency detection. Tang et al. [20] construct Convolutional Neural Networks (CNNs) that learnt structural information via adversarial learning to compute saliency. Unsupervised learning is economically better than supervised learning, Zhang et al. [27] proposed a saliency learning framework without using human annotated labeled map in which saliency is learnt through multiple noisy labeling obtained by weak and noisy unsupervised handcrafted saliency methods. Zeng et al. [25] proposed game-theoretic based salient object detection methods in an unsupervised manner to generate effective saliency map. Singh and Kumar [19] proposed a feature fusion approach based on linear regression without utilizing the ground labels.

Recently, saliency fusion mechanisms of salient features which refine and enhance saliency detection process have been attracted a lot of interest. Singh et al. [18] employed Constrained Particle Swarm Optimization (C-PSO) algorithm to obtain better weights vector for combing various salient features for saliency detection. Chen et al. [3] proposed saliency fusion based on DS Evidence theory at pixel level across variety of saliency detection methods. Chen et al. [4] proposed a saliency integration framework that integrates salient features using deep learning approach.

3 Proposed Model

In this paper, we propose a novel and efficient salient features fusion framework which combines various features rationally to produce better saliency map. The framework exploits support vector regression (SVR) for learning a set of learnable

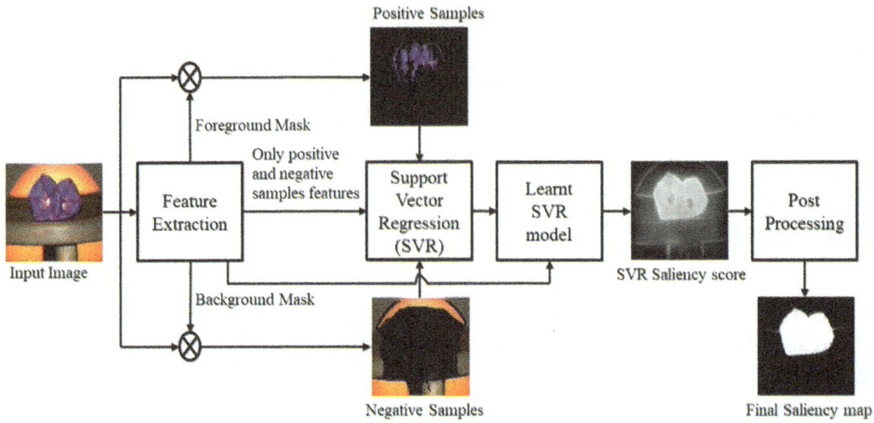

Fig. 1 The flowchart of the proposed framework

parameters which are later used for integrating variety of salient features to generate final saliency map. During learning process, SVR algorithm does not need human annotated labels instead, the framework generates rough labels from salient features which are used in the learning process. After SVR training, the framework generates a saliency score corresponding to the input image by using learnt SVR model and extracted salient features. The saliency score obtained from learnt SVR model is further refined using post processing operations. A flowchart of proposed framework is illustrated in Fig. 1.

As shown in Fig. 1, the proposed framework extracts variety of salient features namely multi-scale contrast, center-surround histogram, and color spatial distribution from the input image using the method proposed by Liu et al. [11]. Then, two binary masks namely foreground mask and background mask are obtained from the salient features using majority voting and two different thresholds. These masks are utilized for generating rough training samples which include positive and negative samples of image elements. Afterwords, training samples are fed into a support vector regression (SVR) to learn the SVR model. Subsequently, salient features of all image elements are fed into learnt SVR model to get the SVR saliency map. Further, SVR saliency map is refined to enhance reliability post processing operation is applied for generation of final saliency map.

3.1 *Training Samples Generation*

For training samples, salient features are utilized to produce rough labels of positive and negative samples for learning process. The positive labels are assigned to the image elements by applying majority voting on binarized maps of salient features. A binary map is computed from salient feature using modified adaptive thresholding as suggested by Achanta et al. [1]. Let i-th salient feature is denoted as $\mathbf{f}_i$ and its corresponding binary map is denoted as $\mathbf{b}_i$. The threshold t_i is calculated as follows [1]:

$$th_i = \frac{2}{Img_w \times Img_h} \sum_{x=1}^{Img_w} \sum_{y=1}^{Img_h} \mathbf{f}_i(x, y) \qquad i = 1, 2.., n \tag{1}$$

$$t_i = th_i \times (1 + \alpha) \tag{2}$$

where Img_w and Img_h are width and height of the given image and α is the foreground label control parameter. Using the threshold t_i binary map $\mathbf{b}_i$ is obtained from salient feature $\mathbf{f}_i$ as follows:

$$\mathbf{b}_i\,(x, y) = \begin{cases} 1 & \text{if } \mathbf{f}_i\,(x, y) \geqslant t_i \\ 0 & \text{otherwise} \end{cases} \tag{3}$$

Similarly, we obtained binary maps $\{\mathbf{b}_1, \mathbf{b}_2, ..., \mathbf{b}_n\}$ from extracted salient features $\{\mathbf{f}_1, \mathbf{f}_2, ..., \mathbf{f}_n\}$ using thresholds $\{t_1, t_2, ..., t_n\}$ respectively. Afterwords, the binary maps contains 0 and 1 values at each image element location where 0 depicts background regions and 1 depicts foreground regions. Majority voting is employed on binary maps to obtain a reliable binary labels (i.e. reliable binary map $\mathbf{b}_r$) of each image elements as follows:

$$\mathbf{b_r}(x, y) = \begin{cases} 1 & \text{if } \sum_{i=1}^{n} \mathbf{b}_i(x, y) > n/2 \\ 0 & \text{otherwise} \end{cases} \tag{4}$$

The reliable binary map $\mathbf{b}_r$ is used to select only positive samples from the image elements. All the image elements whose location in reliable binary map $\mathbf{b}_r$ contains 1 are selected as positive samples. It is mathematically denoted as follows:

$$ps = \{e | \mathbf{b}_r(e) = 1, \quad and \quad \forall e \in \mathbf{I}\} \tag{5}$$

where ps is positive samples, e is image element, and $\mathbf{I}$ is input images. For negative samples ns, the threshold value t_{ns} is determine as follows:

$$t_{ns} = \frac{\gamma}{n} \sum_{i=1}^{n} th_i \tag{6}$$

$$\mathbf{b_{ns}} = \sum_{i=i}^{n} f_i < t_{ns} \tag{7}$$

where b_{ns} is background oriented binary map that contain 1 and 0 values where 1 denotes background and 0 denotes foreground and γ is the background label control parameter. Negative samples collected are image elements corresponding background regions of background oriented binary map b_{ns}. This process is mathematically formulated as follows:

$$ns = \{e|\mathbf{b}_{ns}(e) = 1, \quad and \quad \forall e \in \mathbf{I}\} \tag{8}$$

The training set $\{\mathbf{x}_i, y_i\}_{i=1}^{n_s}$ include all the positive samples ps and negative samples ns with their salient features vector and labels where $\mathbf{x}_i$ and y_i are salient features vector and label of i-th training sample and $n_s = |ps| + |ns|$ is number of training samples. Here, positive and negative labels are set as 1 and -1 respectively.

3.2 Support Vector Regression for Saliency Computation

The saliency score of each image element from salient features is determined using support vector regression (SVR) [21]. SVR is used to optimize the generalized error bound so as to achieve generalized performance. The proposed framework exploits SVR with linear kernel to learn a SVR model using training set $\{\mathbf{x}_i, y_i\}_{i=1}^{n_s}$ of given image $\mathbf{I}$. Then, SVR saliency score $\mathbf{S}_{ss}(\mathbf{x}_i)$ is calculated for i-th image element by using learnt SVR model and it is mathematically represented as follows [21]:

$$\mathbf{S}_{ss}(\mathbf{x}_i) = \mathbf{w}^T\varphi(\mathbf{x}_i) + b \tag{9}$$

where $\varphi(.)$ describes a nonlinear space transformation, $\mathbf{w}$ is the weight vector and b is the bias. Equation (9) can be represented using Lagrange multipliers (i.e., δ_j and δ_j^*), and the linear kernel function $k(\mathbf{x}_i, \mathbf{x}_j) = \varphi(\mathbf{x}_i)^T\varphi(\mathbf{x}_j)$ as follows [21]:

$$\mathbf{S}_{ss}(\mathbf{x}_i) = \sum_{j=1}^{n_s}(\delta_j - \delta_j^*)k(\mathbf{x}_i, \mathbf{x}_j) + b \tag{10}$$

3.3 Post Processing

In order to enhance saliency detection performance, the framework refines the saliency score $\mathbf{S}_{ss}$ with the help of set of operation suggested in [26]. First, a morphological smoothing with a kernel of width ω is employed on S_{ss} that performs a reconstruction-by-dilation operation followed by a reconstruction-by-erosion [22] to generate smoothed saliency map $\mathbf{S}_s$. This operation smooths S_{ss} simultaneously preserving valuable boundaries information. The ω is defined as follows [26]:

$$\omega = \theta \sqrt{m_{\mathbf{S}_{ss}}} \tag{11}$$

where θ is a control parameter and $m_{\mathbf{S}_{ss}}$ is mean of $\mathbf{S}_{ss}$ saliency score map. Subsequently, normalized $\mathbf{S}_s$ is exploited for better contrast enhancement between foreground and background by modifying equation (11) of [26] as follows:

$$\mathbf{A} = (\mathbf{S}_s - (\sigma.m_{\mathbf{S}_s})).c_l \tag{12}$$

$$\mathbf{S} = \frac{e^{\mathbf{A}}}{e^{\mathbf{A}} + 1} \tag{13}$$

where $m_{\mathbf{S}_s}$ is mean of $\mathbf{S}_s$, c_l and σ are control parameters respectively.

4 Experimental Setup and Results

In this section, we provide the experimental evidence to evaluate the performance of the proposed framework on two different characteristics publicly available salient object datasets i.e. ASD [1] (1000 images) and ECSSD [24] (1000 images). ASD [1] is the most widely used dataset includes natural images while ECSSD [24] is contained semantically meaningful but structurally complex images. The proposed framework performance compares against 10 state-of-the-art saliency methods such as *viz.* SUN [28], SeR [17], CA [7], SEG [15], SIM [12], Liu [11], SP [10], SSD [9], LDS [6] and U-FIN [19]. The quantitative examination is performed with five performance measures i.e. Recall, Precision, Receiver Operating Characteristics (ROC), F-Measure and Mean Absolute Error (MAE) for provided strang validation of the proposed framework. Precision and Recall are obtained from the saliency map (**Sm**) and corresponding ground truth (**Gt**) by inferring their overlapped regions. F-measure is formulated as a weighted combination of Precision and Recall for comprehensive evaluation. The parameter β is keep fixed with 0.3 in entire experiments as given in [1] to largely consider Precision instead of Recall. ROC is illustrated with the help of false positive rate (FPR) and true positive rate (TPR) where false positive rate (FPR) and true positive rate (TPR) on the x-axis and y-axis in the plot respectively. The TPR and FPR find using a sequence of thresholds which are varied between the range of [0, 1] with equal steps. The detailed mathematical representation of these metrics are given in [2]. The experimental parameter such as $\alpha = 0.10$, $\gamma = 1$, $\theta = 50$, $c_l = 10$, and $\sigma = 2.8$ are set empirically.

4.1 Performance Comparison with State-of-the-art Methods

Figure 2 illustrates the qualitative outcomes of the proposed framework and also depicts 10 compared state-of-the-art saliency methods. The columns are (from left

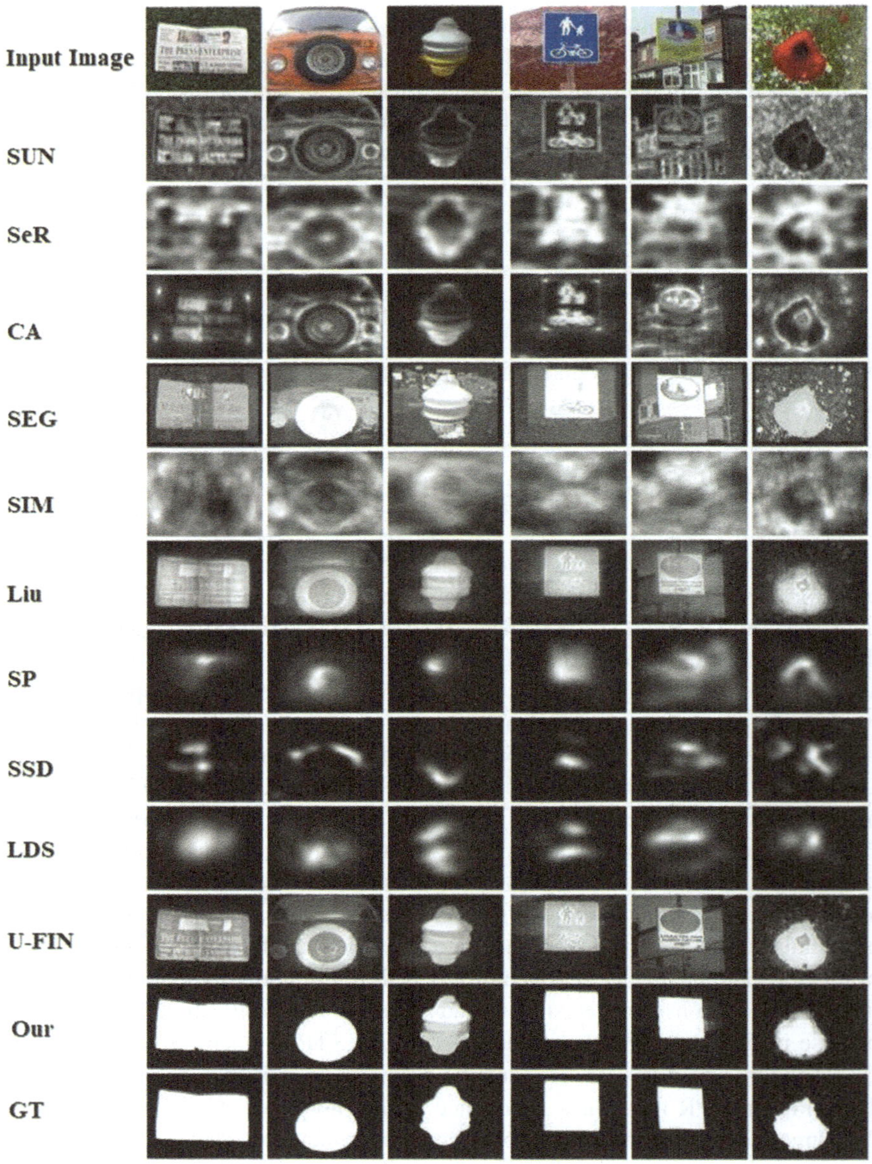

Fig. 2 Visual results of the proposed framework with compared 10 state-of-the-art methods on ASD [1] and ECSSD [24] datasets

to right) shows first-third and fourth-sixth input images from ASD [1] and ECSSD [24] datasets respectively. The visual samples shown in Fig. 2 demonstrate a variety of characteristics such as objects near to image boundary, complex background, and heterogeneous foreground. It can be apparent that few compared state-of-the-art saliency methods such as SP [10], SSD [9], LDS [6], SIM [12] and SeR [17] fail to produced good saliency score whereas SUN [28] simply highlights boundaries of salient object but fails to suppress background and highlights regions inside an object e.g. second, fifth, and sixth columns. Although, SEG [15]fails to suppress background on all images and Liu [11] generates better saliency score on simple images e.g. third column while fails to smoothly highlights foreground on heterogeneous foreground image e.g. first and fifth columns. U-FIN [19] successfully and partially suppresses background on simple and complex background images e.g. first and second columns respectively. In contrast, the proposed framework obtains a better saliency score on all visual samples uniformly.

In order to support the performance of proposed framework, we evaluate proposed framework and compared state-of-the-art saliency methods in terms of *Precision*, *Recall*, *F-measure*, *ROC curve*, and *MAE*. The qualitative results are shown in Figs. 3-7 respectively. Precision scores are shown in Fig. 3. The proposed framework is comparable with top performer U-FIN [19] and outperforms other compared methods on ASD [1]. Similarly, it is comparable with U-FIN [19] and Liu [11] on ECSSD [24]. In Fig. 4, recall scores are shown where proposed method outperforms all the compared methods on both datasets. The F-measure scores are shown in Fig. 5. It can be observed that the proposed framework obtains better score than all compared methods ECSSD [24] and it is comparable with top scorer U-FIN [19] and better than Liu [11] on ASD [1] dataset. ROC plots on two datasets are depicted in Fig. 6. The proposed framework is comparable with U-FIN [19] and Liu [11] and outperforms with other methods. MAE scores are shown in Fig. 7. The proposed framework outperforms on ASD [1] dataset while it is nearly equal to top scorer method SP [10] and LDS [6].

4.2 Computational Time

The computational time of the proposed framework and compared state-of-the-art saliency methods on the ASD [1] dataset are given in Fig. 8. This salient object detection dataset contains images with size is 400×300. The execution timings have been calculated on a desktop PC that configured with following specification: Intel(R)Core(TM)i7-4770 CPU@3.40GHz. As shown in Fig. 8, the proposed framework is faster than U-FIN [19], CA [7], and SEG [15] while SSD [9], LDS [6], SP [10], SeR [17], and SIM [12] are better than proposed framework and it is comparable with Liu [11].

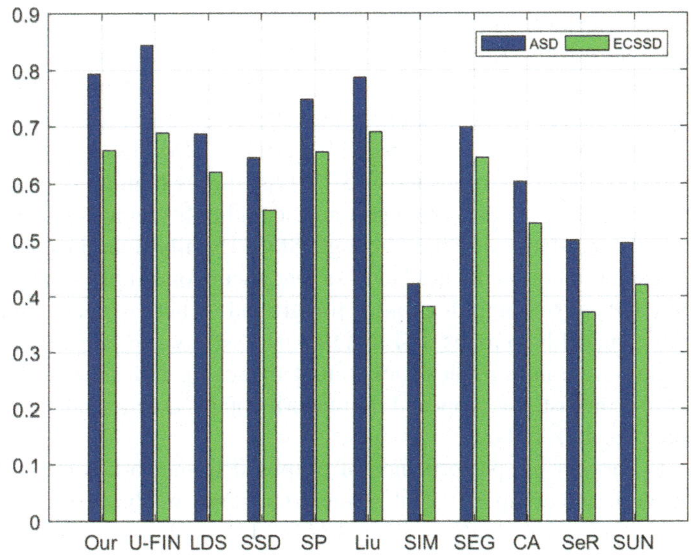

Fig. 3 Precision scores of the proposed framework with compared 10 state-of-the-art methods on ASD [1] and ECSSD [24] datasets

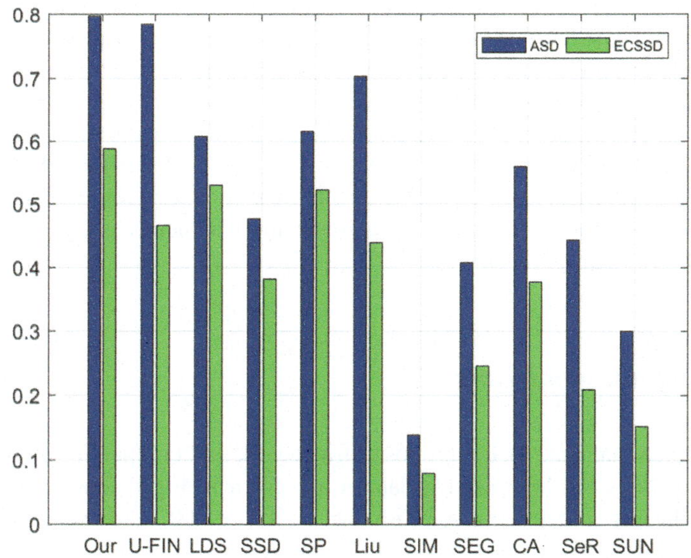

Fig. 4 Recall scores of the proposed framework with compared 10 state-of-the-art methods on ASD [1] and ECSSD [24] datasets

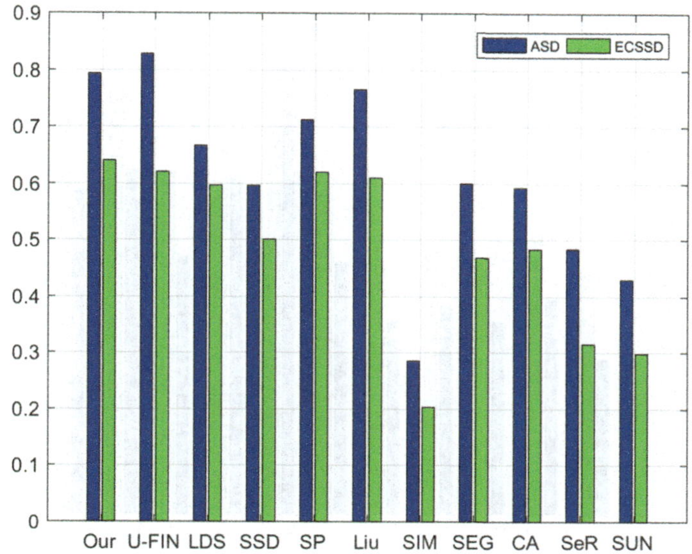

Fig. 5 F-measure scores the proposed framework with compared 10 state-of-the-art methods on ASD [1] and ECSSD [24] datasets

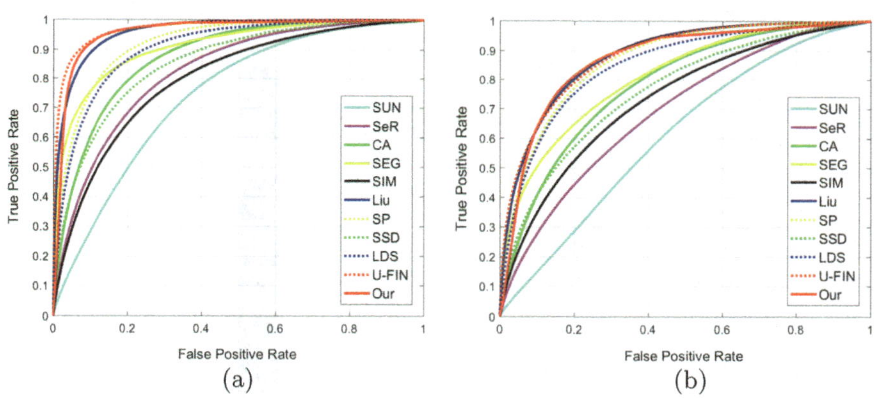

Fig. 6 ROC on the two widely used datasets: **a** ASD [1] **b** ECSSD [24]

5 Conclusion and Future Work

In this paper, we propose a fusion framework that integrates salient features based on support vector regression. Initially, some salient features are extracted from an input image using an existing saliency method. Afterwords, training samples are collected from the image elements with two categories namely positive and negative samples. These samples are selected by applying two thresholds on the salient features and these thresholds are also computed from the salient features. Then, the training sam-

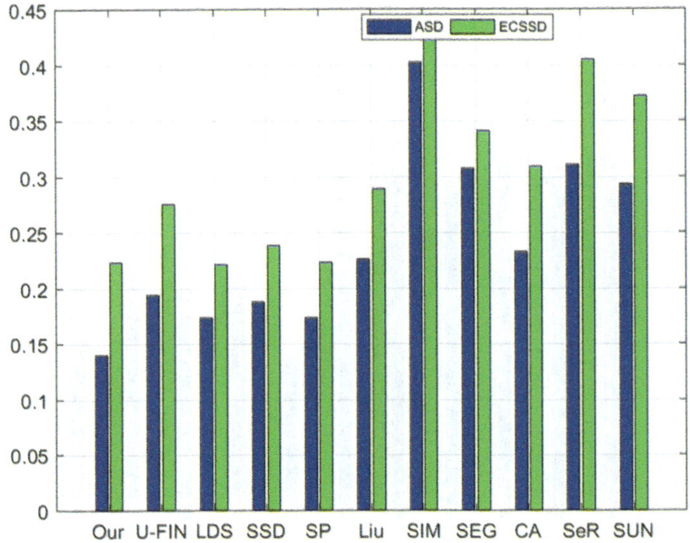

Fig. 7 MAE scores of the proposed framework with compared 10 state-of-the-art methods on ASD [1] and ECSSD [24] datasets

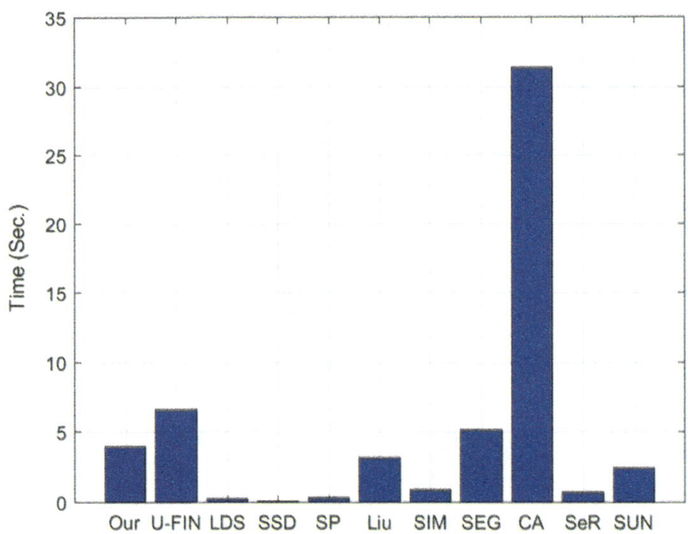

Fig. 8 Computational time (in seconds) analysis of the proposed framework with compared 10 state-of-the-art methods on ASD [1] dataset

ples are fed into support vector regression to obtain a learnt SVR model. This model is used with salient features to determine SVR saliency score of each input image element. Finally, SVR saliency score is further enhanced by applying a sequence of post processing operations to obtained improved saliency map. Experimental results of the proposed framework demonstrate efficacy against compared 10 state-of-the-art saliency detection methods on two publicly available datasets. In future work, we shall extend this work with other machine learning approach which is more effective.

References

1. Achanta R, Hemami S, Estrada F, Süsstrunk S (2009) Frequency-tuned salient region detection. In: IEEE international conference on Computer Vision and Pattern Recognition (CVPR 2009), pp 1597–1604. No. CONF
2. Borji A, Cheng MM, Jiang H, Li J (2015) Salient object detection: a benchmark. IEEE Trans Image Process 12(24):5706–5722
3. Chen BC, Tao X, Yang MR, Yu C, Pan WM, Leung VC (2018) A saliency map fusion method based on weighted ds evidence theory. IEEE Access 6:27346–27355
4. Chen Q, Liu T, Shang Y, Shao Z, Ding H (2019) Salient object detection: integrate salient features in the deep learning framework. IEEE Access 7:152483–152492
5. Donoser M, Urschler M, Hirzer M, Bischof H (2009) Saliency driven total variation segmentation. In: 2009 IEEE 12th international conference on computer vision, IEEE, pp 817–824
6. Fang S, Li J, Tian Y, Huang T, Chen X (2017) Learning discriminative subspaces on random contrasts for image saliency analysis. IEEE Trans Neural Netw Learn Syst 28(5):1095
7. Goeferman S (2010) Context-aware saliency detection. In: Proceedings of the IEEE conference computer vision and pattern recognition, pp 2376–2383
8. Hou Q, Cheng MM, Hu X, Borji A, Tu Z, Torr PH (2017) Deeply supervised salient object detection with short connections. In: Proceedings of the IEEE conference on computer vision and pattern recognition, pp 3203–3212
9. Li J, Duan LY, Chen X, Huang T, Tian Y (2015) Finding the secret of image saliency in the frequency domain. IEEE Trans Pattern Anal Mach Intell 37(12):2428–2440
10. Li J, Tian Y, Huang T (2014) Visual saliency with statistical priors. Int J Comput Vis 107(3):239–253
11. Liu T, Yuan Z, Sun J, Wang J, Zheng N, Tang X, Shum H (2011) Learning to detect a salient object. IEEE Trans Pattern Anal Mach Intell 33(2):353
12. Murray N, Vanrell M, Otazu X, Parraga CA (2011) Saliency estimation using a non-parametric low-level vision model. In: 2011 IEEE conference on Computer Vision and Pattern Recognition (CVPR), IEEE, pp 433–440
13. Pang Y, Yu X, Wang Y, Wu C (2019) Salient object detection based on novel graph model. J Vis Commun Image Representation 65:102676
14. Papushoy A, Bors AG (2015) Image retrieval based on query by saliency content. Digital Signal Process 36:156–173
15. Rahtu E, Kannala J, Salo M, Heikkilä J (2010) Segmenting salient objects from images and videos. In: Computer Vision—ECCV 2010. Springer, Heidelberg, pp 366–379
16. Rother C, Bordeaux L, Hamadi Y, Blake A (2006) Autocollage. In: ACM transactions on graphics (TOG), vol 25. ACM, pp 847–852
17. Seo HJ, Milanfar P (2009) Static and space-time visual saliency detection by self-resemblance. J Vision 9(12):15–15
18. Singh N, Arya R, Agrawal R (2014) A novel approach to combine features for salient object detection using constrained particle swarm optimization. Pattern Recogn 47(4):1731–1739

19. Singh VK, Kumar N (2019) U-fin: unsupervised feature integration approach for salient object detection. In: First international conference on advanced communication & computational technology, Accepted. Springer, Heidelberg (2019)
20. Tang Y, Wu X (2019) Salient object detection using cascaded convolutional neural networks and adversarial learning. IEEE Trans Multimedia
21. Vapnik V, Golowich SE, Smola AJ (1997) Support vector method for function approximation, regression estimation and signal processing. In: Advances in neural information processing systems, pp 281–287
22. Vincent L (1993) Morphological grayscale reconstruction in image analysis: applications and e cient algorithms. IEEE Trans Image Process 2(2):176–201
23. Wang T, Zhang L, Wang S, Lu H, Yang G, Ruan X, Borji A (2018) Detect globally, refine locally: a novel approach to saliency detection. In: Proceedings of the IEEE conference on computer vision and pattern recognition, pp 3127–3135
24. Yan Q, Xu L, Shi J, Jia J (2013) Hierarchical saliency detection. In: Proceedings of the IEEE conference on computer vision and pattern recognition, pp 1155–1162
25. Zeng Y, Feng M, Lu H, Yang G, Borji A (2018) An unsupervised game-theoretic approach to saliency detection. IEEE Trans Image Process 27(9):4545–4554
26. Zhang J, Sclaroff S, Lin Z, Shen X., Price B, Mech R (2015) Minimum barrier salient object detection at 80 fps. In: Proceedings of the IEEE international conference on computer vision, pp 1404–1412
27. Zhang J, Zhang T, Dai Y, Harandi M, Hartley R (2018) Deep unsupervised saliency detection: a multiple noisy labeling perspective. In: Proceedings of the IEEE conference on computer vision and pattern recognition, pp 9029–9038
28. Zhang L, Tong MH, Marks TK, Shan H, Cottrell GW (2008) Sun: a Bayesian framework for saliency using natural statistics. J Vision 8(7):32–32
29. Zhao R, Ouyang W, Li H, Wang X (2015) Saliency detection by multi-context deep learning. In: Proceedings of the IEEE conference on computer vision and pattern recognition, pp. 1265–1274

Implementing Delta Checks for Laboratory Investigations Module of Hospital Management Systems

Ashutosh Kumar, Sumit Soman, and Priyesh Ranjan

Abstract The laboratory investigations module is a core component of the Hospital Management Information System (HMIS) which deals with the management of patient's laboratory tests and results. This module is also crucial for patient health care as the results of investigations are an important source of data for evaluating the patient's treatment plan. Errors in investigation results can led to misdiagnosis or erroneous treatment being administered to the patient; hence, this module requires mechanisms to detect such errors and generate alerts. Due to high patient loads and high volume of investigations being performed at large hospitals, such mechanisms need to be implemented in electronic health record (EHR) systems. This paper presents the implementation and evaluation of the delta check for quality control in laboratory investigations that has been deployed for a tertiary care hospital in India. The paper presents the proposed mechanism and the process of its integration in an operational HMIS, as well as an analysis of the implementation of this mechanism.

Keywords Laboratory Investigations · Hospital Management Information Systems · Delta Checks · Quality Control.

1 Introduction

Hospital Management Information Systems (HMIS) have been widely used to digitize various clinical and non-clinical workflows in hospitals. The objectives of developing such systems include increasing operational efficiency and improving the qual-

A. Kumar (✉) · S. Soman · P. Ranjan
Centre for Development of Advanced Computing, Noida, Uttar Pradesh, India
e-mail: ashutoshk@cdac.in
URL: https://www.cdac.in

S. Soman
e-mail: sumitsoman@cdac.in

P. Ranjan
e-mail: priyeshr@cdac.in

P. K. Singh et al. (eds.), *Evolving Technologies for Computing, Communication and Smart World*, Lecture Notes in Electrical Engineering 694,
https://doi.org/10.1007/978-981-15-7804-5_34

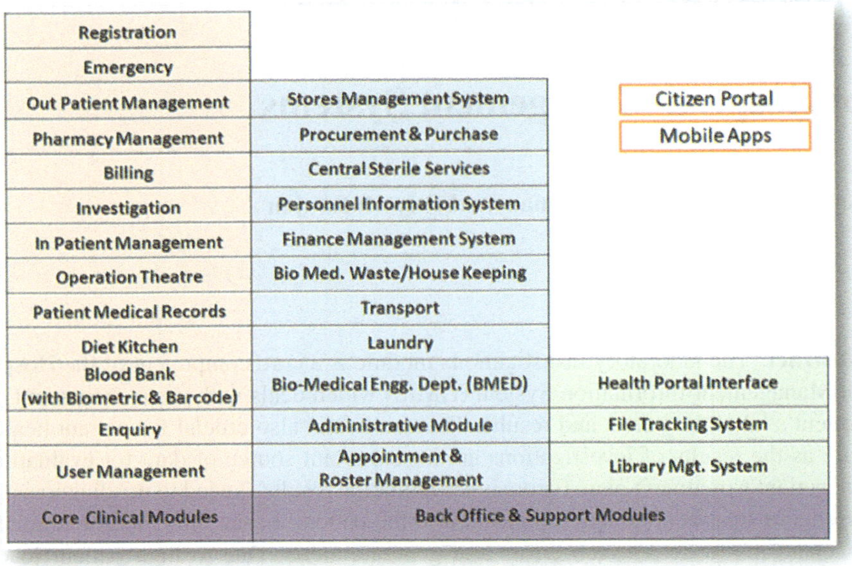

Fig. 1 Summary of clinical and non-clinical modules in the HMIS

ity of healthcare services being delivered, while also allowing for efficient data and resource management. CDAC has been involved with the design, development and implementation of HMIS at various hospitals across India. This HMIS is compliant with various electronic health record (EHR) standards including SNOMED Clinical Terminology (SNOMED CT) [1], LOINC for laboratory tests, medical equipment interfacing [2], generation of Continuity of Care Document (CCD) [3] and a standard-compliant blood bank management system [4]. The HMIS also uses a phonetic-based health identifier system for patient identification [5] and digital payment solutions [6]. A summary of the modules of the HMIS is shown in Fig. 1.

Among the clinical modules of HMIS, the laboratory investigations module is an important component which deals with various processes involved in raising, conducting and reporting of laboratory investigations. These include pathological as well as imaging based investigations, and the module is configured according to the facilities available at the specific hospital. In hospitals with high patient loads, the investigation module handles a large number of investigation requests and reports, especially since multiple investigation requests are often raised for even for a single patient. Though the investigation reports conventionally indicate anomalous test results explicitly, it is often possible that errors during conducting the investigation or during reporting of results may lead to anomalous entries in reports. In this context, delta checks have been often used as a method to flag such anomalous test result values, based on evaluating the patient's previous test results for the test parameters and flagging based on current results exceeding specified thresholds.

This paper discusses the delta check function and its variants as well as its implementation in the laboratory investigations module of the HMIS. The module is part of the HMIS that is operational at various hospitals and serves various clinical laboratories at these medical facilities for end-to-end management of laboratory investigation workflows. Since these medical facilities handle a large volume of investigations due to high patient loads, the approach presented in this paper is useful for application developers and service providers to easily incorporate an automated alert generation system for the investigation module.

The rest of the paper is organized as follows. Section 2 reviews the literature available in this domain and presents the motivation for the implementation discussed in this paper. This is followed by a discussion of delta function and its variants, including computation of parameters, in Sect. 3. The details of the investigation module workflow in the Hospital Management Information System (HMIS) are presented in Sect. 4, and the implementation of Delta function for this module is detailed in Sect. 5. Finally, conclusions and future work are mentioned in Sect. 6.

2 Literature Review and Motivation

There have been several works in literature that have used the delta function-based approaches for quality control in investigation results. A comprehensive review of the uses of delta function, including the pros and cons, has been presented in the recent work by Randell et al. [7]. Miller [8] has compared the use of composite CBC delta (CCD), mean red blood cell volume (MCV) and logical delta check (LDC) for the complete blood cell (CBC) count investigations for identifying erroneously labeled laboratory specimens and shows that the LDC is useful for the proposed objective. A similar analysis for CBC tests based on area under the curve for the receiver operating characteristic (AUC-ROC) was done in [9]. A variant of the delta check function called as the weighted cumulative delta check (wCDC) was proposed and evaluated for detecting specimen mix-up by [10]. They evaluated the method on a laboratory tests dataset of 32 items and found that it was efficient for identifying errors in specimen labeling. An analysis of the effect of ranges chosen in computing the delta check limits has been presented by Ko et al. [11] for a dataset of 1,650,518 paired results for 23 general chemistry test results collected across two years. An analysis on the effect of time interval in delta check has been presented by Tan et al. [12].

Classical delta checks were implemented on univariate (or unianalyte) data; however, there have also been multiple studies that have advocated the use of multianalyte delta function checks, where the delta check is implemented on a combination of tests (or analytes). This has been a more complex and challenging problem since the dimensionality of the analysis space grows and methods to detect errors are based on identifying correlations. A recent study of creatinine delta checks has been done by Gruenberg et al. [13] on a dataset of 23,410 creatinine results and concluded that a large number of results flagged by the delta check were due to pathology or dialysis-related causes. Work by Rosenbaum and Baron [14] has shown that machine learning approaches trained on data annotated using multianalyte delta checks have resulted in high accuracy detection of *"Wrong Blood In Tube" (WBIT)* errors.

Most of the available works in literature have focused on retrospective analysis of implementation of delta checks on laboratory investigation datasets. Further, in several datasets, erroneous test values have been synthetically added in order to generate an annotated dataset. This paper focuses on the live implementation of delta checks on the laboratory investigation module of an operational HMIS in a tertiary care hospital in India. The basic implementation of the delta check function has been done, and its working in flagging cases from a representative example has been discussed.

3 Delta Function and Its Variants

The delta check implementation in its basic form can be computed as either the absolute change or percentage change in a test parameter value of a patient between a specified time interval. In addition, there are other variants in literature that have been used as well [7]. We briefly review the various representations in the following subsections.

3.1 Absolute Delta Value

This is computed as

$$\Delta_{abs} = x_{current} - x_{previous} \tag{1}$$

where Δ_{abs} is the absolute delta value computed as the difference between the current test parameter value $x_{current}$ and the previous test parameter value $x_{previous}$.

3.2 Percentage Delta Value

The delta value may also be computed in terms of percentage that is expressed relative to the previous test parameter value as follows

$$\Delta_{abs}^{\%} = \frac{x_{current} - x_{previous}}{x_{previous}} \times 100 \tag{2}$$

3.3 Absolute Rate Difference Delta

The absolute rate difference (ARD) Delta Δ_{ARD} is computed as

$$\Delta_{ARD} = \frac{x_{current} - x_{previous}}{t_{current} - t_{previous}} \quad (3)$$

where $t_{current}$ and $t_{previous}$ denote the current and previous time instants respectively, and their difference represents the time interval between samples in suitable units (depending on the test parameter).

3.4 Percentage Absolute Rate Difference Delta

Analogous to percentage delta function, the ARD can also be computed as a percentage as follows

$$\Delta^{\%}_{ARD} = \left(\frac{x_{current} - x_{previous}}{x_{previous}}\right) / (t_{current} - t_{previous}) \times 100 \quad (4)$$

3.5 Estimating Delta Check Thresholds

There have also been multiple approaches to estimating the threshold values for estimation of the delta function. Common among them are approaches that involve manual setting of the thresholds by domain experts (pathologists or clinicians), or estimation by percentile-based methods that involve computation of frequency distribution of differences from among a set of identified samples of analytes.

Another method is estimation by computation of rate/reference change value (RCV), which is computed as

$$RCV = \sqrt{2} \times z \times \sqrt{(CV_I^2 + CV_A^2)} \quad (5)$$

Here, CV_I denotes the intra-individual variability obtained from available sources and CV_A is the analytical imprecision for an analyte. Depending on the nature of analysis, e.g., for two-tailed analysis, value of z is chosen as significant (1.96, probability 95%) or highly significant (2.58, probability 99%).

4 Investigation Module in HMIS

The following processes are available in the investigation module in HMIS. These are also summarized in Fig. 2.

1. **Online Requisition Raising**: This process is used to raise test(s) for patients. Tests can be raised either individually or in groups. Once the tests are raised, it is given a unique requisition number.
2. **Sample Collection**: Once the tests are raised, samples are collected at different collection centers in the hospital. Samples can be of different types, for example, serum, urine, etc. At the end of this process, every sample is given a sample number.
3. **Packing List Generation**: Based on the sample type and lab, samples are packaged and sent for sample acceptance.
4. **Sample Acceptance**: Here the veracity of a sample is checked. After that, the sample is either accepted or rejected. There can be different reasons for rejecting a sample, for example, sample is clotted, etc.
5. **Patient Acceptance**: There are some tests for which there is no need to collect samples. Those are performed directly on patients, for example, X-ray or other imaging investigations. After its processing, the result is entered through Result Entry process.

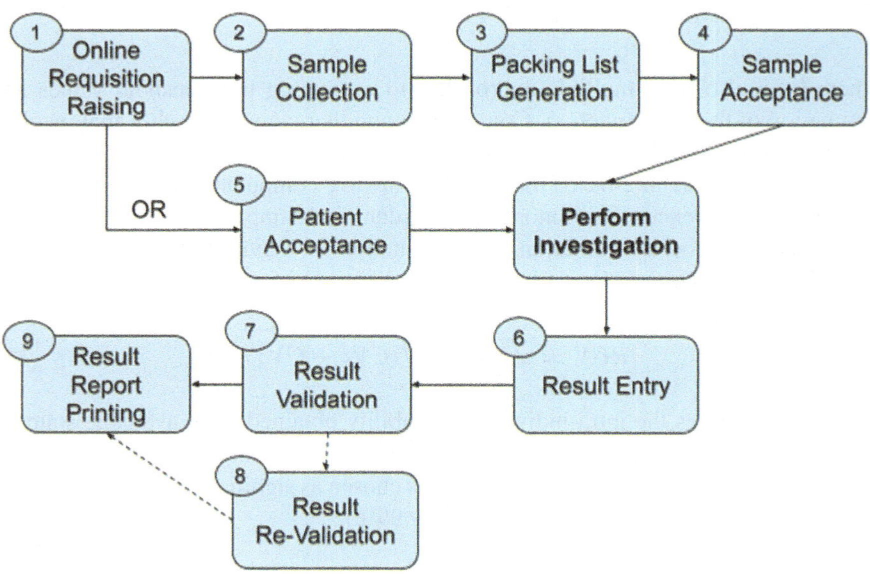

Fig. 2 Processes involved in the investigation module

6. **Result Entry**: Once the samples are processed, result is entered in the HMIS application either manually or through machines. Result Entry process is used to enter test results manually. A test usually consists of different parameters. Test result is entered parameter-wise.
7. **Result Validation**: As the name suggests, result entered through Result Entry is validated at Result Validation Process.
8. **Result Re-validation**: A double check might be needed for some tests. Result Re-validation is used for those tests to re-validate some test parameters.
9. **Result Report Printing**: Once the process of validation is complete, report of test is generated in portable document format (PDF). The report can be viewed as well as downloaded through Result Report Printing process.
10. **Operational/Statistical Reports**: Functionality is provided to generated different Operational/Statistical Reports, for example, lab-wise status of test, requisition listing report, etc.

5 Implementation of Delta Function

Investigation module of HMIS application deals with test results of patients. Various tests can be raised for a given patient through 'Online Requisition Raising' process. After the completion of sample collection, result is processed and entered in the HMIS application through 'Result Entry' process. After entry, result is validated at 'Result Validation' process. The functionality of delta check has been implemented at 'Result Validation' process.

For each test parameter, these attributes are defined beforehand:

- Delta check threshold (Δ_{thresh}): The delta check value (the difference between current result and previous result should not be greater than this value)
- Time period (t_Δ) : Time period within which delta check is to be applied. (This check will not applied if current result and previous result are beyond this time period).

At the 'Result Validation' process, this check is applied for each parameter of a test (for which the above attributes have been provided).

For any parameter of any test, if the difference between current result of that test parameter and its previous result is greater than the delta check value and the time period between which this difference has been calculated is within the provided range, then an alert is shown to the operator. Based on that the operator can decide whether he wants to go forward and save the test results or do a re-check in order to establish the cause of the alert. This is summarized in Algorithm 1.

Data: Test instances $T_i|i = 1 : M$ with parameters $p_j|j = 1 : N$, thresholds $\Delta^j_{thresh}|j = 1 : N$, time intervals $t^j_\Delta||j = 1 : N$
Result: Parameters $\hat{p}$ for which delta check generates alerts
Initialize $\hat{p} = \{\phi\}$;
`// For all parameters`
for *j=1:N* **do**
    `// For all test instances check time period`
    for $\forall(T^i_{j^1}, T^i_{j^2}) \leq t^i_\Delta|i = 1 : M$ **do**
        `// Delta check for parameters`
        if $\Delta(p_i^{j^1}, p_i^{j^2}) \geq \Delta^i_{thresh}$ **then**
            `// Add to alert generation list`
            Add $\hat{p} \leftarrow \{\hat{p}|p_i\}$;
        else
            continue;
        end
    end
end

Algorithm 1: Algorithm for implementation of delta check. The algorithm takes the test instances with parameters along with time intervals and thresholds respectively, and returns the list of parameters for which the delta function is evaluated and an alert needs to be generated.

5.1 User Interfaces & Alerts

In this subsection, we show an illustrative example to demonstrate the working of the delta function evaluation in the Investigation module. Table 1 shows parameter-wise test results for 'Liver Function Test' when it is raised for the first time for a given patient.

Table 2 represents parameter-wise test results for 'Liver Function Test' when it is raised again for the second time for the same patient within a time frame.

In Table 2, we can see that an extra column named 'Diff Value / Delta value'. 'Diff Value' represents parameter-wise difference between current value and previous value. 'Delta Value' represents the threshold for that parameter. The row is emphasized if 'Diff Value' is greater than 'Delta Value.' Alerts are generated in the user interface and the user is provided the option to re-check the test results and ascertain the reason behind the parameter value being flagged, and making subsequent correction (if necessary) prior to result report printing. The user interface generating the delta check alert is shown in Fig. 3 as an example.

Table 1 Representative values of test parameters

Parameter name	Parameter value	Reference range
Serum AST	30	≤40U/L
Serum ALT	35	≤40U/L
Serum Alkaline Phosphatase	90	≤30 U/L
Serum Bilirubin (Total)	1	≤1.1 mg/dL
Serum Bilirubin (Conjugated)	0.2	≤0.2 mg/dL
Total Protein	5	6–8 g/dL
Albumin	3	3.5–5.2 g/dL

Table 2 Representative values of delta function determined for test parameters

Parameter name	Parameter value	Reference range	Diff value/Delta value
Serum AST	35	≤40U/L	5.0 / 10.0
Serum ALT	40	≤40U/L	5.0 / 8.0
Serum Alkaline Phosphatase	130	≤130 U/L	**40.0 / 20.0**
Serum Bilirubin (Total)	1	≤1.1 mg/dL	0.0 / 1.0
Serum Bilirubin (Conjugated)	1.4	≤0.2 mg/dL	**1.2 / 1.0**
Total Protein	30	6–8 g/dL	**25.0 / 20.0**
Albumin	4	3.5–5.2 g/dL	1.0 / 2.0

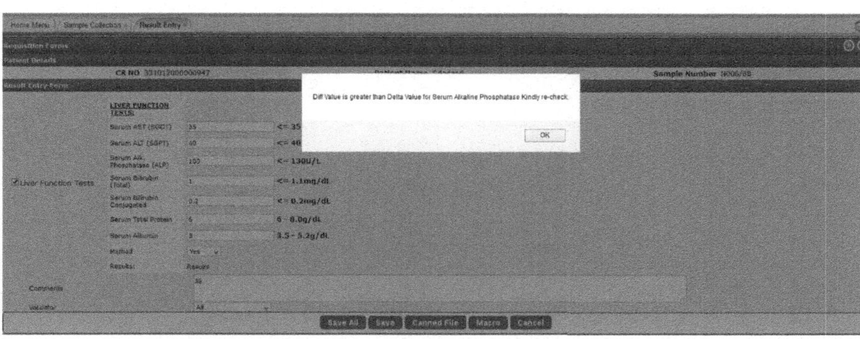

Fig. 3 Working of Delta Alert in HMIS

5.2 Post-Alert Reconciliation

Depending on the test parameters for which the delta check algorithm generates alerts, the laboratory technician/pathologist or clinician needs to take corrective action to reconcile the erroneous result after suitable verification and establishing the reason for the alert. The sample workflow for alert reconciliation with delta check is summarized in Fig. 4.

As the first step, the analyte specimen and result reported are reviewed, and any anomalies found are rectified. If no reason is determined, the patient medical records are examined for consistency in the detected trend with patient's past investigation results or diagnosis. Checks with any other samples of the patient drawn around the same time of conducting that investigation are also examined, if available. These samples (current and previous) are analyzed for their integrity, and any findings are reported. In case there is similarity with previous trends and the result for which the alert has been generated has been established as being invalid, the alert is rejected and reason for the same is recorded in the system. Alternatively, the clinician or laboratory that conducted the investigation is contacted for further analysis, which may involve clinical or machine-related issues. In all cases, the workflow concludes

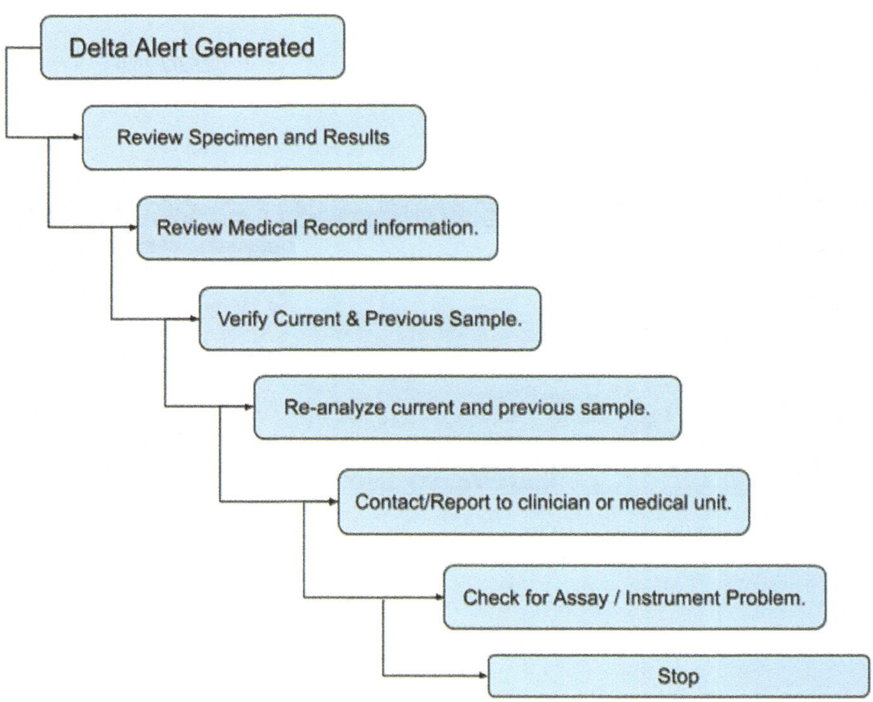

Fig. 4 Post-alert reconciliation workflow (adapted from [7])

with recording the reason for the alert generation, as well as corrective action which may include correction in reported results, sample recollection or suitable corrective action on a case-specific basis.

6 Conclusions and Future Work

This paper presented the implementation of delta checks for generation of alerts in test parameter values for laboratory test results within the investigation module of a HMIS. This mechanism allows the system to flag results that are anomalous in nature, thereby enabling the users to identify the cause of error and take corrective action. The delta check-based alert is useful in this module as it is a simple and efficient mechanism that can be used for a large number of investigation results without the need for manual intervention. It would also potentially reduce errors in reporting and subsequently ensure correct diagnosis and treatment. The basic delta check mechanism discussed in this paper has been implemented at various deployments of HMIS and is undergoing operational trials. Future work involves evaluating other models for delta function estimation, as well as the use of automated methods to determine thresholds (which may also be chosen adaptively) that could be used to make the alert generation system more efficient.

References

1. Srivastava PK, Soman S, Sharma P (2017) Perspectives on SNOMED CT implementation in Indian HMIS. In: Proceedings of the SNOMED CT Expo 2017. SNOMED International
2. Srivastava S, Gupta R, Rai A, Cheema AS (2014) Electronic health records and cloud based generic medical equipment interface. arXiv preprint arXiv:1411.1387
3. Srivastava S, Soman S, Rai A, Cheema A, Srivastava PK (2017) Continuity of care document for hospital management systems: an implementation perspective. In Proceedings of the 10th International Conference on Theory and Practice of Electronic Governance. ACM, pp 339–345
4. Cheema AS, Srivastava S, Srivastava PK, Murthy BK (2015) A standard compliant blood bank management system with enforcing mechanism. In: 2015 International Conference on Computing, Communication and Security (ICCCS). IEEE, pp 1–7
5. Soman S, Srivastava P, Murthy BK (2015) Unique health identifier for India: an algorithm and feasibility analysis on patient data. In: 2015 17th international conference on E-health networking, application and services (HealthCom). IEEE, pp 250–255
6. Ranjan P, Soman S, Ateria AK, Srivastava PK (2018) Streamlining payment workflows using a patient wallet for hospital information systems. In: 2018 IEEE 31st international symposium on Computer-Based Medical Systems (CBMS). IEEE, pp 339–344
7. Randell EW, Yenice S (2019) Delta checks in the clinical laboratory. Crit Rev Clin Lab Sci 56(2):75–97
8. Miller I (2015) Development and evaluation of a logical delta check for identifying erroneous blood count results in a tertiary care hospital. Arch Pathol Lab Med 139(8):1042–1047
9. Balamurugan S, Rohith V (2019) Receiver operator characteristics (roc) analyses of complete blood count (cbc) delta. J Clin Diagnostic Res 13(7)

10. Yamashita T, Ichihara K, Miyamoto A (2013) A novel weighted cumulative delta-check method for highly sensitive detection of specimen mix-up in the clinical laboratory. Clin Chem Lab Med 51(4):781–789
11. Ko D-H, Park H-I, Hyun J, Kim HS, Park M-J, Shin DH (2017) Utility of reference change values for delta check limits. Am J Clin Pathol 148(4):323–329
12. Tan RZ, Markus C, Loh TP (2019) Impact of delta check time intervals on error detection capability. Clin Chem Lab Med (CCLM)
13. Gruenberg JM, Stein TA, Karger AB (2018) Determining the utility of creatinine delta checks: a large retrospective analysis. Clin Biochem 53:139–142
14. Rosenbaum MW, Baron JM (2018) Using machine learning-based multianalyte delta checks to detect wrong blood in tube errors. Am J Clin Pathol 150(6):555–566

Convolutional Neural Network-Based Automatic Brain Tumor Detection

Vishal K. Waghmare and Maheshkumar H. Kolekar

Abstract In today's world, many imaging techniques are available such as computed tomography (CT), magnetic resonance imaging (MRI) and ultrasound images to assess the tumor in a brain, breast, lung, prostate, liver. We have specifically used the brain MRI images in this paper. Brain tumors are the most prevalent and insistent disease because of which our life span is getting short. Hence, reliable and automatic classification machination is necessary to avert the death rate of individuals. In this paper, we have implemented the model to classify brain tumor automatically by using CNN as they are compatible to perform image recognition tasks, including image classification, detection, segmentation. We have implemented various convolutional neural network (CNN) architectures for classification and detection of tumors using multiple models like basic CNN, fine-tuned VGG-16. This work will increase the accuracy of detecting brain tumors and using different models, and we are getting significant advancement in accuracy. The future scope of this work is that we can apply for ovarian, breast, lung, skin tumors. Also, we will extend it to detect the actual size and shape of the tumor.

Keywords Brain tumor · CNN · MRI · Data augmentation · VGG16

1 Introduction

The computer-assisted exploration for better understanding of images has been a longstanding question in the medical imaging field. Study says that brain tumor is marked as one of the most fatal and dangerous cancers among adults and children. Therefore, early detection and classification of brain tumors into their specified grade becomes essential to treat the tumor efficaciously [1]. Magnetic resonance imaging

V. K. Waghmare (✉) · M. H. Kolekar
Electrical Engineering Department, Indian Institute of Technology Patna, Patna, India
e-mail: vishalkwaghmare@gmail.com; 1811ee13@iitp.ac.in

M. H. Kolekar
e-mail: mahesh@iitp.ac.in

P. K. Singh et al. (eds.), *Evolving Technologies for Computing, Communication and Smart World*, Lecture Notes in Electrical Engineering 694,
https://doi.org/10.1007/978-981-15-7804-5_35

(MRI) technique is used widely which facilitates the prognosis and diagnosis of brain tumors in many anatomical, functional, also for organic disorders of nerves and the nervous system. The sequences which we are getting from standard MRI are used to find the difference between various types of brain tumors using contrast texture and visual quality analysis of the soft tissue. It creates unique pictures based on a pixel by pixel. There is a lot of change between MRI and CT images. As we know, CT uses the ionizing radiation of X-rays and MRI uses an attractive growing field to arrange the atomic charge of hydrogen.

In simple way, brain tumor can be defined as a grouped tissue that is arranged by the gentle addition of many asymmetrical cells. The samples of normal brain, benign tumor and malignant tumor are shown in Fig. 1a, b, respectively. Generally, tumor occurs when a cell gets abnormal formation within the brain. The case study says that this has become a substantial cause for the death of many people. Therefore, we have to be very serious about the diagnosis of a brain tumor. Early detecting and diagnosing (It is becoming a vital issue for providing improved treatment.) the brain tumor can save a life by giving proper treatment. Detecting these cells becomes a hard problem due to the creation of tumor cells in the brain. This becomes necessary for comparison of the brain tumor with the help of MRI treatment [2].

The motivation behind the study is in the field of medical image analysis; research on brain tumors is one of the most prominent ones. Today, brain tumor occurs approximately to 250,000 people in a year globally which makes up less than 2% of cancers [3]. Reducing the pressure on human judgment builds a user interface which can identify the cancerous cells reducing the death rate by early detection. Brain tumors do not differentiate, and it may affect all ages, genders and race. This year 80,000 people were being spotted with a primary brain tumor. In medical imaging, we can find more than 120 different kinds of the central brain and central nervous system tumors [4].

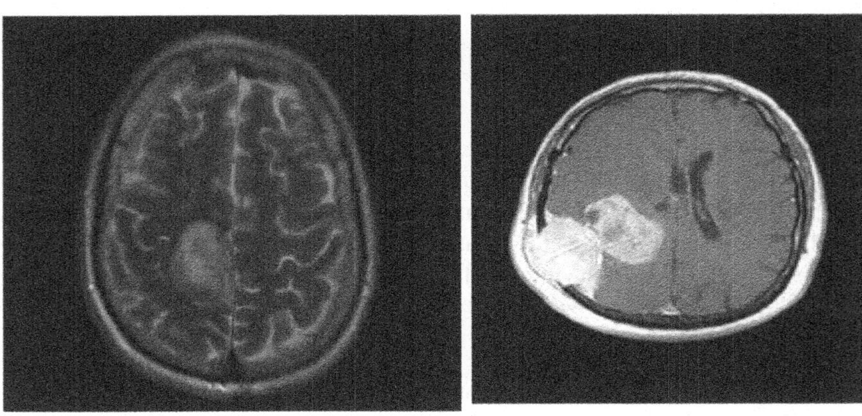

(a) (b)

Fig. 1 **a** Benign tumor, **b** malignant tumor

In our paper, we have proposed a unique deep learning algorithm which classifies brain tumor types, categorized by the standard of the World Health Organization (WHO) [3]. For the operative analysis of MRI images, it is needed for segmentation of the tumor from MRI images and then by doing augmentation that segmented tumor is augmented by different techniques. Furthermore, at last, we can classify the tumor by using fine-tuned CNN model. Machine learning algorithms such as Bayesian belief network [5], hidden Markov model [6], support vector machine [7] are very popular in object detection and classification task. But performance of these approaches depends on use of efficient features. The deep learning algorithms have capacity to execute feature engineering on its own. They need lot of domain expertise and human intervention. This research paper is planned as follows: In Section 2, the literature survey for brain tumor classification is explicated. Section 3 discovers the proposed model which is focusing on preprocessing, CNN architecture, data augmentation and fine-tuned VGG-16 architecture. Section 4 gives all details of the dataset, also the experimental results by comparing with other methods. Furthermore, the last section concludes and tells about future work.

2 Background

Nowadays, in the biomedical image analysis area, many researchers are contributing to different sub-fields of medical imaging, and they have done a lot of work. Under this headline, our focus will be on past studies of brain tumor classification. When we have a looked on past research on this medical imaging field, we are coming to know that most of these pre-existing work is for the automatic segmentation for brain tumor region of MRI images. Lately, several researchers are presenting diverse methods to discover and classify the brain tumor in MRI images.

In [8], the fuzzy c-means segmentation is incorporated to distinct the tumor and non-tumor region of the brain. Also, the wavelet feature is extracted by using a multilevel discrete wavelet transform (DWT). Eventually, deep neural network (DNN) is applied for brain tumor classification with a high degree of correctness. The forementioned practice is compared with linear discriminant analysis (LDA), KNN and sequential minimal optimization (SMO) classification. An accuracy rate of 96.97% was obtained through DNN-based brain tumor classification. However, the complexity was pretty high and performance was very poor.

In [9], a different bio-physio-mechanical tumor growth modeling is introduced to scrutinize the step-by-step development of tumor patients. It was being applied for gliomas and solid tumors with separate margins to capture the remarkable tumor mass effect. The discrete and continuous methods are used simultaneously to model a tumor growth. The proposed scheme provided the likelihood to segment tumor-bearing brain images based on atlas-based registration tacitly. The technique was only for brain tissue segmentation and the problem with this method is its computational time is very high.

In [10], new brain tumor segmentation is introduced, which we can call as a multi-modal brain tumor segmentation structure. They have combined diverse segmentation algorithms in order to accomplish good performance than the existing method; the only drawback is its complexity is high. In [11], the survey of brain tumor segmentation is presented discussing different segmentation methods such as region-based segmentation, threshold-based segmentation, Margo Random Field (MRF), atlas-based segmentation, deformable model, geometric deformable model. In it, the accuracy, precision and resolution are analyzed for all the methods.

In [12], they have proposed an innovative CNN centered multigrade brain tumor classification system with segmenting tumor regions from an MRI image using deep learning models. They have performed widespread data augmentation to meritoriously train the proposed model. Data augmentation is a tactic that allows practitioners to expressively upturn the variety of data accessible for training models, deprived of indeed gathering new data. This is because of the unavailability of datasets dealing with MRI images for the multigrade brain tumor classification, and at last, a pre-trained CNN model (VGG-16) is fine-tuned using augmented data for the classification brain tumor.

Although there are various algorithms for brain tumor classification, there some limitations that need to be taken into consideration while working with a brain tumor. For classification and segmentation, ambiguities arise for binary classification of brain tumor. Also, the deficiency of data is a crucial challenge to attain accurate results. Therefore, we have performed a data augmentation. For a better understanding and clarity, the classification which we are doing needs to be multi-class, which will classify the brain tumor into its appropriate types, i.e., grades. Deep learning algorithms try to learn high-level features from data in an incremental manner. They eliminate the need of domain expertise and hard core feature extraction. Hence, in this paper, we proposed tumor classification algorithm using fine-tuned CNN model by using data augmentation technique to attain good results.

3 Methodology

The tumor detection stages start with a preprocessing method, then a feature extraction and reduction and finally at the deep learning algorithm for classification.

3.1 Preprocessing

Preprocessing of image before segmentation is grave for accurate detection of tumor. In this stage, we perform noise and artifacts reduction and perfecting of edges. There are minute chances of noise being present in the modern MRI images. Therefore, the main task of the preprocessing is to sharpen edges in the image. There are many imaging modalities like x-ray, ultrasonography, single-photon emission computed tomography (SPECT) and MRI. We have specifically used MRI images [13].

MRI is a technique which uses the radio frequency signals to get the image of the brain. This imaging technique is our focusing technique. It is a widely used technique which smooths the prognosis and diagnosis of brain tumors for numerous neurological diseases and conditions. We may use the standard MRI sequences are for the differentiation of different types of brain tumors. Furthermore, we do this because of their visual qualities and contrast texture analysis within soft tissues [12]. This is creating a unique picture based on their pixel by the pixel value. Especially, MRI is helpful in oncological (growth), neurological (mind), musculoskeletal imaging because it is offering much more notable complexity between the various delicate body tissues. We are using the brain MRI images mainly to identify the tumor and its tumor progress is modeling procedure. With this information, we are detecting the tumor and its treatment processes mainly [14–16].

The MRI images can be classified into these types:

- T1W (T1-Weighted): Allow an easy annotation of the healthy. Make the brain tumor border bright.
- T2W (T2-Weignted): Edema region can appear brighter than other.
- FLAIR: This type we used because it separates the edema region from the CSF.

There are two types of noise we encountered when we were dealing with MRI images that are mainly salt-and-pepper noise (It appears like randomly sparse white, black or both pixels over the image.) and Gaussian noise (This is caused by the random fluctuations in the signal.).

3.2 Convolution Neural Network

Human brains can be carved out with the design and implementing the neural networks which are predominantly used for classification, data clustering, vector quantization and pattern resemblances techniques. It can be labeled as feedback, feed-forward, and closed-loop-based feedback neural networks also known as recurrent network based on their interconnections. These feed-forward networks can be further classified as a single layer and multilayer networks. The single-layer network consists of input and output layer only, without having hidden layers in between. However, the multilayer comprises of all the three layers [2].

In a regular neural network, images could not be scaled but, in the convolutional neural network, images are scalable (we can change their size), it can take three-dimensional input volume (length, width, height) to three-dimensional output volume. Recently, deep learning has developed out as a powerful tool for the scope of its varied application in domains such as image processing or classifications, speech recognition, natural language processing, computer vision, authentication system with a high level of accuracy. The CNN is an of multilevel perceptron with some regularization to it. It employs convolution operations which is a linear mathematical operation, hence the name. It is a multilayered network consists of an input layer, hidden layers and output layer. The hidden layers start with convolution layer followed by pooling layers with suitable batch normalizations and activation functions. Pooling layer is used to diminish the dimensionality of the image (mostly max-pooling is preferable). Though it is said loosely, that convolution operation is performed in convolution layers but cross-correlation is performed between convolution kernel and image pixels. The image is segregated into small regions and important features are extracted. The end layers consist of a fully connected layer or dense layers with appropriate nonlinear activation function to generate out the label score between 0–1 based on probabilities [8–10].

The contemporary neural network makes use of nonlinear functions as activation functions which allows producing multifaceted mappings between the inputs and outputs for a network. These allow stacking of multiple layers in a sequence to build deep neural network. Series of layers in hidden layers of neurons are necessary for learning and modeling datasets with high complexity and dimensionality. Using nonlinear activation functions, every overall operation could be executed in a neural network.

There are various alternatives for the nonlinearity, such as the rectified linear unit (ReLU), sigmoid, softmax, hyperbolic, tangent, etc. In our work, ReLU and softmax activation function are used. ReLU can be mathematically defined as

$$y = \max(0, x)$$

ReLU has a derivative function which allows for backpropagation and converges fast. Softmax activation function results in a vector that shows the probability distributions of a series of possible outcomes (Fig. 2).

Fig. 2 Rectified linear unit (ReLU)

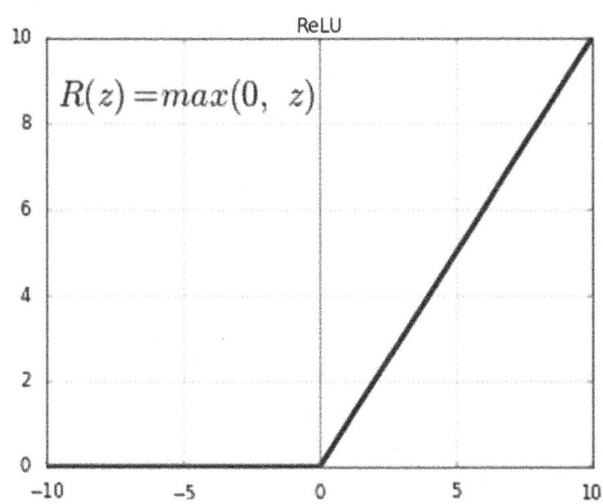

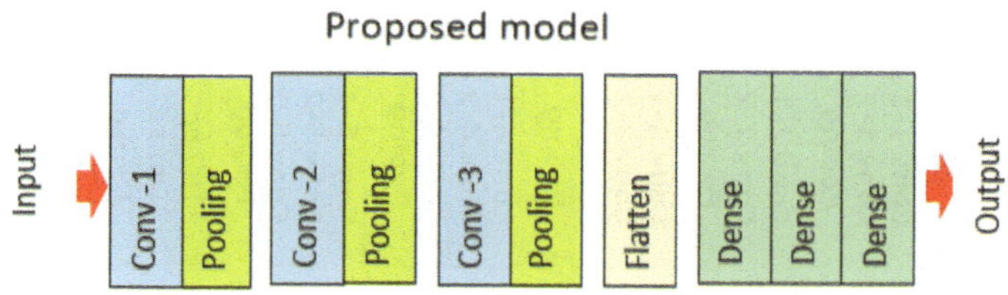

Fig. 3 CNN model architecture

The basic block diagram of the basic architecture CNN is depicted in Fig. 3. The classification of this CNN architecture is divided into two stages; they are training and testing stages. Brain tumor images are classified as normal and abnormal brain image and also on type of brain tumor present. The steps performed in building a prediction model include preprocessing of the raw data, feature extraction and classification with the use of appropriate loss function. In the preprocessing, MRI images were resized to change the dimension to make the dataset uniform. Finally, CNN is used for automatic brain tumor classification.

3.3 *Data Augmentation*

In simple words, the data augmentation is defined as the process of increasing the amount and diversity of data. High dimensionality and quality datasets are crucial for bringing resources into effective action of various deep learning models. The challenge in our work was lack of adequate data in the available dataset to feed to the deep learning model and get good accuracy. Hence, to achieve this better accuracy, we stretched the available data with the help different augmentation techniques such as flipping, skewness, rotation. There are many techniques like emboss, edge detection, sharpening and Gaussian blur can be used for the noise invariance [1].

3.4 *VGG-16 CNN Architecture*

There is something called "transfer learning" comes into the picture. It enables to use the pre-trained models by making some changes to other models which are used by other people. Therefore, a pre-trained model can be defined as; it is a model which is created by someone else to solve a similar kind of problem. In another way, instead of building a model from scratch to solve some problem, you use the model trained on other problem as a starting point [17].

We can better understand this by the following instance: Let us consider that you want to shape a self-learning car, and for this, you can spend many years on building a decent image recognition algorithm from scratch. Also you can take the

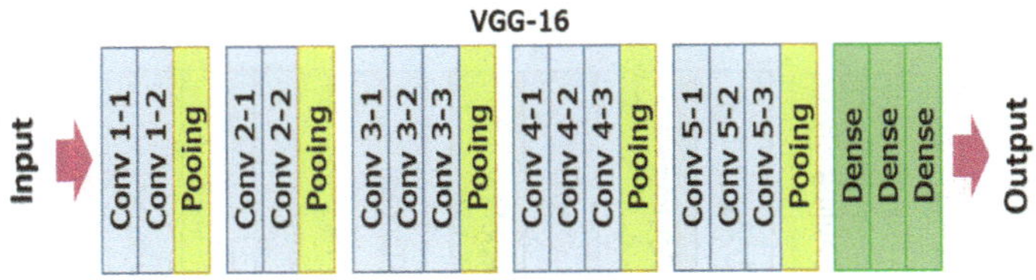

Fig. 4 Fine-tuned VGG-16 CNN architecture

inception model which is called as a pre-trained model from google which was built on ImageNet data for identifying images in those pictures. We know a pre-trained model may not be 100% precise for our application, but it saves enormous hard work required to reinvent the wheel. In simple words, this pre-trained model is just like applying knowledge of recognizing a car to recognize a truck [18, 19].

The use of pre-trained models increases the accuracy of basic CNN model-based classification. This directed us to switch to the use of pre-trained models where we are not training the entire architecture but we are training only a few layers. Therefore, we have used VGG16 model which is pre-trained on the ImageNet dataset and provided in the keras library for use [1].

In proposed methodology, fine-tuned VGG-16 CNN architecture is used for classifying the brain tumor. VGG-16 architecture constitutes 16 weighted layers. In this, 13 layers are convolutional layers and rest three layers are fully connected (i.e., dense) layers. In the VGG-16 architecture, there are two convolutional layers which are trailed by max-pooling layer. Like this, same combinations are present. This can be clearly shown in Fig. 4. For next three layers also, the combination of convolutional layers trailed by max-pooling we can observe [20].

As already discussed, we are using a pre-trained model which will study the similarity and shapes in the data. Hence, we alter the weights of only fully connected layers not of other layers.

4 Results and Discussion

In this section, we have presented the experimental evaluation for our proposed model to classify brain tumor. Therefore, we have conducted experimentations on two datasets, i.e., basic dataset and brain tumor dataset. We divided these datasets as 60%, 20%, and 20% for training, cross-validation, and testing sets. We know given datasets are too small; therefore, we are using data augmentation technique for increasing data set by using different parameters. The statistics of both datasets before and after augmentation is tabulated in Tables 1 and 2. The experiments we have assessed using Nvidia Quadro K2000 2 GB GDDR5 Graphics Card and deep learning.

Table 1 Statistics details of basic dataset (images)

Tumor (present/absent)	Before augmentation	After augmentation
Tumor present	135	5400
Tumor absent	98	3920

Table 2 Statistics details of brain tumor dataset (images)

Type of tumor	Before augmentation	After augmentation
Meningioma	884	35,360
Glioma	1677	67,080
Pituitary tumor	1143	45,720

4.1 Basic Dataset

This dataset consists of 235 MR images which are further classified into two different categories such as tumor present and tumor absent. Since the number of images in this dataset is too small, we have applied the data augmentation technique for increasing this number. In deep learning for better understanding of the image, we need large database. Therefore, we have created a large number of images. The details of these newly created images are presented in Table 1.

4.2 Brain Tumor Dataset

This dataset has 3704 T1-weighted contrast-enhanced MR images. The resolution of these images was not uniform, but by using the resizing algorithm, we have resized them to 512 $\times$ 512-pixels. Also, we have used data augmentation technique to get more images. The details of these newly created images are available in Table 2. We have performed a data augmentation because of which the number of images has increased from 3704 to 148,160.

4.3 Evaluation with Other Models

This proposed algorithm is tested after creating large number of images using data augmentation technique for both datasets. The obtained results are good enough to be considered in a real context. There is a significant advancement in overall accuracy in brain tumor classification which is increased from 90.03 to 95.71%. Therefore, these results have proved that how huge data improves the accuracy; also, it has proved how abundant data can be useful for deep learning.

Theoretically, sensitivity is test ability to correctly identify ill patient who has the condition. Mathematically, it can be expressed as:

$$\text{Sensitivity} = \text{Number of true positives}/(\text{number of true positives} + \text{number of true negatives})$$

Specificity gives the test's ability to correctly reject healthy patients without a condition. Mathematically, this can be written as (Fig. 5):

$$\text{Specificity} = \text{Number of true negatives}/(\text{number of true positives} + \text{number of true negatives})$$

We have compared our results of brain tumor classification by using sensitivity, specificity, and accuracy, which are given in Table 3. The performance of the proposed approach is better than Cheng et al. [4] and Sajjad et al. [1]. Our proposed algorithm flabbergasted the present techniques and attained the value of 95.71%, 83.28% and 93.37% for accuracy, sensitivity and specificity, respectively.

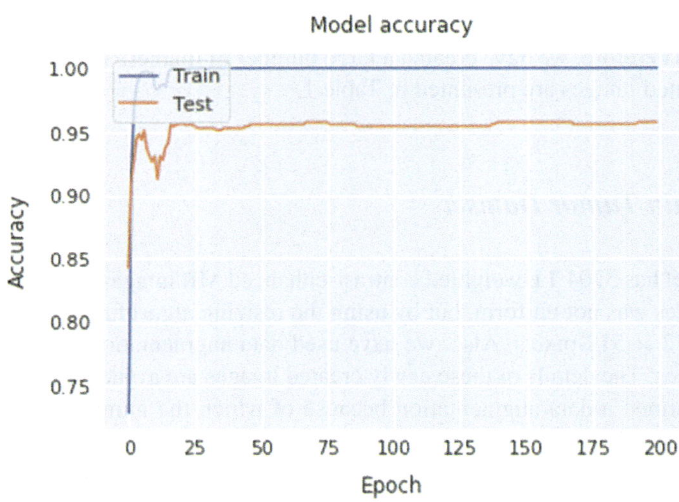

Fig. 5 Results of proposed fine-tuned VGG-16 CNN architecture for brain tumor dataset

Table 3 Evaluation with other models for brain tumor dataset

Model	Accuracy (%)	Sensitivity (%)	Specificity (%)
Cheng et al. [4]	91.28	81	92
Sajjad et al. [1]	94.58	88.41	96.12
Proposed	95.71	83.28	93.37

5 Conclusion

We have proposed a unique deep learning algorithm for brain tumor classification algorithm. The steps of proposed brain tumor classification approach are: (1) preprocessing, (2) augmenting the data and (3) classification using a fine-tuned VGG-16 CNN architecture. We improved the accuracy of classifying the tumor by applying data augmentation technique and CNN. This research work can be made as a part of the computer-aided diagnosis tool to be used by non-radiologist clinician in the identification of tumor and its types.

The proposed system can also be stretched to in image modalities such as CT, positron emission tomography (PET) and ultrasound. Also, it can be extended to diagnose diseases related to other tumors such as liver cancer, prostate cancer and uterine cancers. In future, we will extend our work for detecting the actual size and shape of the tumor using various deep learning models by keeping the efficiency and accuracy in mind.

References

1. Sajjad M, Khan S, Muhammad K, Wanqing Wu, Ullah A, Baik SW (2019) Multi-grade brain tumor classification using deep CNN with extensive data augmentation. J Comput Sci 30:174–182
2. Ari A, Hanbay D (2018) Deep learning based brain tumor classification and detection system. Turkish J Electr Eng Comput Sci 2275–2286
3. McGuire S (2016) Health organization, international agency for research on cancer, world cancer report 2014. Adv Nutr 7:418–419
4. Cheng J, Huang W, Cao S, Yang R, Yang W, Yun Z (2015) Enhanced performance of brain tumor classification via tumor region augmentation and partition. PloS ONE 10
5. Kolekar MH (2011) Bayesian belief network based broadcast sports video indexing. Multimedia Tools Appl 54(1):27–54
6. Dash DP, Kolekar MH, Jha K (2019) Multi-channel EEG based automatic epileptic seizure detection using iterative filtering decomposition and Hidden Markov Model. Comput Biol Med 116:103571
7. Kolekar MH, Dash DP (2015) A nonlinear feature based epileptic seizure detection using least square support vector machine classifier. TENCON 2015–2015, IEEE Region 10 Conference, pp 1–6
8. Mohsen H, El-Dahshan EA, El-Horbaty EM, Salem AM (2017) Classification using deep learning neural networks for brain tumors. Future Comput Inform J 1–4
9. Bauer S, May C, Dionysiou D, Stamatakos G, Buchler P, Reyes M (2012) Multi-scale modeling for image analysis of brain tumor studies. Trans Biomed Eng 59(1):25–29
10. Menze B, Reyes M, Van Leemput K (2015) The multimodal brain tumor image segmentation benchmark (BRATS). IEEE Trans Med Imaging 1993–2024
11. Liu J, Wang J, Wu F, Liu T, Pan Y (2014) A survey of MRI-based brain tumor segmentation methods. Tsinghua Sci Technol 19(6):578–595
12. Kolekar MH, Kumar V (2017)Biomedical signal and image processing in patient care. IGI Global Publisher
13. Pereira S, Pinto A, Alves V, Silva C (2016) Brain tumor segmentation using convolutional neural networks in MRI images. IEEE Trans Med Image 35(5):1240–1251

14. Havaei M, Davy A, Warde-Farley D, Biard A, Courville A, Bengio Y, Pal C, Jodoin PM, Larochelle H (2017) Brain tumor segmentation with deep neural networks. Med Image Anal 35:18–31
15. Ghosal D, Kolekar MH (2018) Music genre recognition using deep neural networks and transfer learning. In: Proceedings of the Interspeech," Hyderabad, India, 2018, 2–6 Sept 2018, pp 2087–2091
16. Bhatnagar S, Ghosal D, Kolekar MH (2017) Classification of fashion article images using convolutional neural networks. In: 2017 fourth international conference on image information processing (ICIIP). IEEE, 2017, pp 1–6
17. Ching T, Himmelstein DS, Beaulieu-Jones BK et al (2017) Opportunities and obstacles for deep learning in biology and medicine. bioRxiv 2017, 142760
18. Greenspan H, van Ginneken B, Summers RM (2016) Guest editorial deep learning in medical imaging: overview and future promise of an exciting new technique. IEEE Trans Med Imaging 35(5):1153–1159
19. Litjens G et al (2017) A survey on deep learning in medical image analysis. Available https://arxiv.org/abs/1702.05747
20. Singh PK, Kar AK, Singh Y, Kolekar MH, Tanwar S (2020) Proceedings of ICRIC 2019, recent innovations in computing. Lecture Notes in Electrical Engineering, vol 597. Springer, Cham, pp 3–920

An Efficient Communication Protocol for Energy-Constraint IoT Devices for Environment Monitoring Applications

Nabajyoti Mazumdar, Debasish Chouhan, Dipankar ch Barman, Bhargav Bordoloi, Debakanta Gogoi, Bishnu Prasad Saikia, Suman Sau, and Saugata Roy

Abstract The Internet of Things impacts our everyday lives in many ways, from small smart devices to large industrial structures. WSN is a part of the IoT topology. It is the foundation of many IoT applications like surveillance, monitoring, defense technology, etc. Since cloud-assisted IoT services may be very energy demanding for energy-constrained sensor nodes, an edge-computing architecture is preferred. A typical WSN consists of tiny devices known as nodes, and they have limited computational power. To improve the energy utilization, the clustering is recognized as an significant method in safeguarding the energy of WSNs. Clustering strategies concentrate on conflict resolution arising from inadequate data transmission. Energy-efficient clustering methods are suggested in this paper to improve WSN's lifespan. The proposed clustering methods are (i) hierarchical clustering and (ii) distributed in nature. The proposed protocols to some extent give better output than the existing one but still, it will require some refinement. For the simulation purpose, we have used Python programming environment in Windows 10 machine, and in the hardware implementation, we have used Ubimote as our main device.

Keywords Clustering · Energy · IoT · Routing · WSN

N. Mazumdar (✉)
Department of Computer Science and Engineering, Central Institute of Technology Kokrajhar, Kokrajhar, India
e-mail: nabajyoti@cse.ism.ac.in

D. Chouhan · D. Barman · B. Bordoloi · D. Gogoi · B. P. Saikia
Department of IT, Central Institute of Technology Kokrajhar, Kokrajhar, India

S. Sau
S'O'A Deemed to be University, Odisha, India

S. Roy
Department of CSE, Indian Institute of Technology (ISM), Dhanbad, India

P. K. Singh et al. (eds.), *Evolving Technologies for Computing, Communication and Smart World*, Lecture Notes in Electrical Engineering 694,
https://doi.org/10.1007/978-981-15-7804-5_36

1 Introduction

Internet is a very advanced technology which is adapting and evolving daily according to the users need. In today's world, everyone owns a supercomputer in the form of a smart gadget. These devices are embedded with sensors and provide wide range of application and services such as online gaming, virtual reality, UHD video streaming, etc. Already 7 billion things were connected to the Internet, and experts from Gartner declared that by 2020 nearby 26 billion devices will be connected [1–3]. Cloud computing is an efficient computing platform for data processing; however, it is not being able to match the explosive growth of data that is being generated by these devices. The real-time and flexible requirement of data processing is bringing an unsolvable problem to traditional cloud storage. The drawbacks of cloud computing are being covered by edge computing. Here, an edge device is placed between the cloud and the user, which is at a minimum distance from the user. This process helps in offloading some of the computing operations from the cloud and helps in the real-time application where time plays a prominent role. Figure 1 shows an IOT model with cloud and edge devices.

In recent years, WSN has evolved as a primary environment monitoring infrastructure for IoT applications. A WSN contains a large number of sensor devices, connected via wireless communication. However, the sensor nodes in WSN are battery-operated; thus, they have limited energy which incurs problems while transmitting huge amounts of data. As it is known that transmitting data consumes more energy than processing it. Therefore, this will result in high energy consumption and eventually reducing the service life. These nodes are scattered in different paths for data transmissions, and a huge network is required to maintain this very challenging device. The energy of the entire network must remain for a longer period, otherwise the whole system will collapse. Thus, incorporating edge computing helps the sensor nodes to communicate their data at a near distance edge device rather than cloud. Further, studies have shown that cluster-based communication technology can balance the energy consumption of sensor nodes and extend the lifetime of WSN. WSN has been used in various dimension like atmosphere monitoring, intrusion detection, etc. Figure 2 shows the clustered WSN model. Many protocols have been experimented to increase the lifetime of WSN, but the cluster-based model is more effective. In a wireless sensor network, the sensor nodes are divided into set of groups called clusters. Each clusters has a leader or a controller called the CH which collects the cluster information and sends it to the BS directly or via other cluster heads. The cluster-based approach has many advantages:

1. It helps in reducing the energy consumption as only the CH is involved in transmitting the data to the BS.
2. Routing becomes easily manageable because only the CH need to be monitored.

However, this technique too has some disadvantages. The CH not only aggregates the datas from its cluster members but also other CH within its transmission area. This creates some extra load on the CH resulting in early depletion of energy in comparison with non-CH nodes. Another problem that occurs in this approach is the hotspot problem where nodes closer to BS depletes energy early. The hotspot problem

Fig. 1 An IOT model

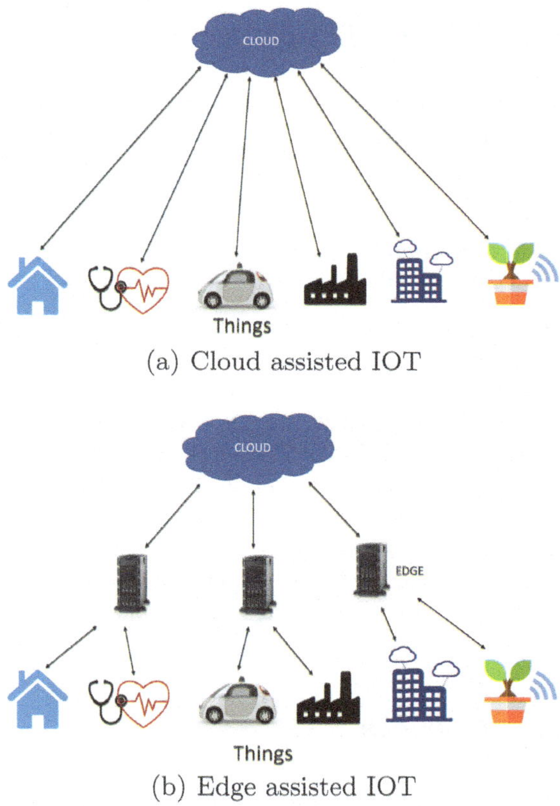

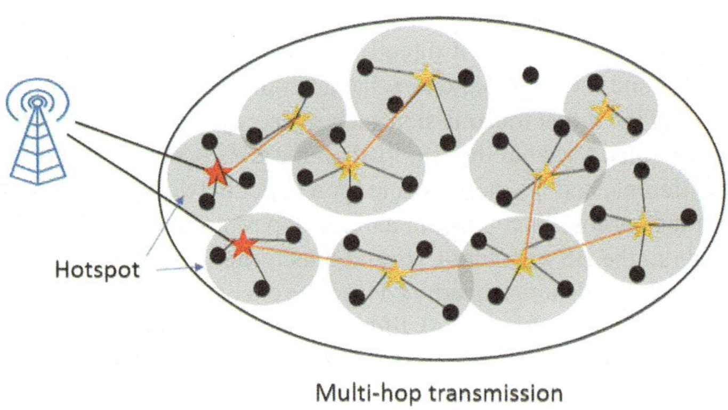

Fig. 2 A clustered WSN model

produces a sensing hole in the network due to the CH's early power depletion. As a result, the network data cannot reach the base station.

2 Literature Survey

Numerous protocols have been proposed for wireless sensor networks related to clustering and routing [4–6]. Some of the existing articles are presented below:

2.1 Clustering in WSN

LEACH [7] is the most popular clustering method for WSN, where CHs were elected based on a probability function. LEACH has several shortcomings such as (1) a less energy node may become the CH due to randomization and (2) the CHs may not be uniformly placed over the network. To overcome these drawbacks, different algorithms such as PEGASIS and HEED[8] have been proposed. In PEGASIS, each node only communicates with the nearest neighbor and takes turns to transmit data to the base station, thus reducing the amount of energy spent per round. HEED is one of the most powerful protocols for cluster-based routing in WSN. It is an energy-efficient distributed clustering approach which makes use of the sensor nodes residual energy as a primary parameter for CH selection and clustering.

2.2 Routing in WSN

There are mainly two types of routing algorithms in cluster-based WSNs viz single-hop routing and multi-hop routing. In a single-hop routing protocol, each CH after collecting its cluster data sends it directly to the BS, hence minimizing the transmission latency and message/packet exchange complexity. Though there are some disadvantages of the direct transmission such as energy consumption of the CHs are higher if it is located far from the BS, load on the CHs is not balanced and reduces the network lifetime, and it is not scalable for a larger network. HEED protocol uses single-hop transmission. In contrast in a multi-hop routing protocol, a CH forwards its cluster data to the BS via other CHs and thus forming a routing path till the BS. Since a CH sends the data to a nearby CH, so the energy consumption of the CHs is less and hence increasing the network lifetime and it is feasible for a larger network. In DFCR [9], the routing path is constructed based on the energy of the CHs and their distance from the BS. Though the multi-hop routing approach suffers from a bottleneck problem called hotspot problem. In this regard, several multi-hop, cluster-based routing protocols are introduced to form the routing path from the CHs to the BS based on the distance from the BS, average residual energy of the network, minimized hop count. To overcome the hotspot problem, unequal cluster size strategy is discussed. Such that smaller-sized clusters and the number of CHs are present nearer to the BS to share the load.

3 Preliminaries

Here, along with the terminology, the network model that is being considered in the proposed algorithm is presented.

3.1 Network Model

The network is homogeneous WSN where $S = S_1, S_2, S_3, \ldots S_n$ n numbers of the sensor node are randomly deployed over a goal area (Fig. 3). The distance between nodes S_i,S_j is denoted by $Dist(i, j)$. The energy model used in this paper is similar to [7]. The initial and residual energy of a node is represented by $Energy_{init}(S_i)$ and $Energy_{res}(S_i)$, respectively.
Following are different terminologies used:

- $Cand_{CH}(S_i)$ represents the candidate CHs of sensor node S_i.
- $Relay(S_i)$ is the relay node of S_i used to communicate to a CH.
- $Deg(S_i)$ is the node degree of sensor node S_i which is equal to its number of neighbors.

4 Proposed Work

In this paper, we proposed a distributed cluster-based routing protocol for energy efficiency and to enhance the network lifetime. The following subsections discusses the different phases of the proposed algorithm.

4.1 Start Phase

In this phase, each node determine the neighbors and its distance from the neighbors. Initially, the BS broadcast a wake-up message to its network to start all the sensor nodes. After receiving the wake-up message, all sensor nodes broadcast a control message with their IDs, their co-ordinates within a range equal to the cluster radius. A sensor node S_j which receives the control message of S_i updates its neigh-

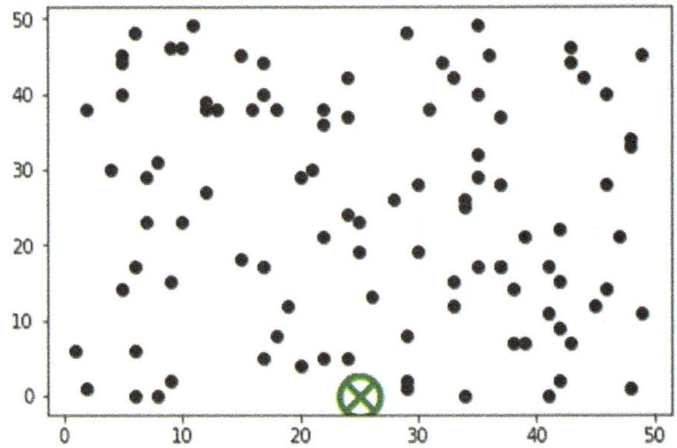

Fig. 3 Sensor nodes deployment

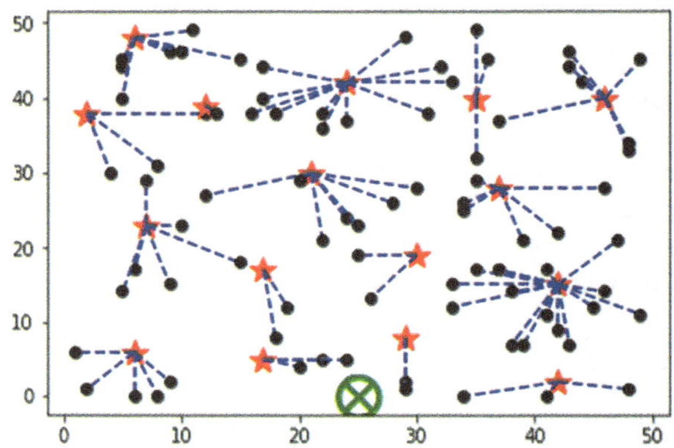

Fig. 4 Clustering

bor list by adding S_i and calculates distance between them. Each node calculates its node degree equal to the number of its neighbors.

4.2 Clustering Phase

Here, CH election and cluster formation is carried out. Here, each node initiates a timer T_d using Eq. 1 before electing itself as a CH.

$$T_d(S_i) = \alpha \times (1 - \frac{Energy_{res}(S_i)}{Energy_{init}(S_i)}) + \beta \times (1 - \frac{Deg(S_i)}{n}) + \gamma \times (1 - \frac{Dist_{BS}(S_i)}{Dist_{max}}) \quad (1)$$

where α, β, γ are random weighted values assigned to all the nodes such that $\alpha + \beta + \gamma = 1$.

When the time delay for a node expires, it broadcasts a CH_ADV_MSG message within its cluster radius containing its node degree, distance from the BS, and energy information . While still running its timer if another node receives the CH_ADV_MSG message, then it stops its timer and adds the ID of the sender to its $CAND_{Ch}$ list and itself becomes a non-CH node. A non-CH node chooses its CH as the node having higher residual energy from its $CAND_{Ch}$ list (Fig. 4).

Following algorithm is used for CH selection and cluster formation:

Algorithm 1: Cluster formation

Data: A randomly deployed sensor nodes in WSN
Result: A clustered WSN
for *each node* S_i **do**
  | Compute delay time (T_i) using equation 3
end
for *each node* S_i **do**
  if $(T_i == 0)$, *i.e., Delay time expires* **then**
    | S_i becomes CH and broadcasts *CH_ADVERTISE* message within its cluster radius
  end
  if S_j *receives CH_ADVERTISE message from* S_i **then**
    | S_j stops its timer and update $CH(S_j) = S_i$
  end
end

4.3 CH-to-CH Routing Path Formation

Here, a routing path is formed from every CH to the BS through other CHs. For this process, every CH is assigned a HOP count based on its distance from the BS. The HOP count calculation process is carried out from the CHs which are nearest to the BS till the CHs residing at the furthest from the BS. Initially, all CHs set their HOP count to infinity. The BS initiates the HOP count calculation process by broadcasting a HOP_selection message within the maximum transmission range of a node and all the CHs within that range get assigned a HOP count of 1. Then again, the 1-HOP CHs broadcast a HOP_selection message along with their residual energy and their distance from the BS upto the maximum transmission range assigning a HOP count 2 to the CHs in that range. This process continues until every CHs get assigned a HOP count. During this process, if a CH receives multiple HOP_selection message, then the CH updates its HOP count only for the message which is coming from the other CH having least HOP count. Thus, each CH is get assigned a HOP count. A CH Ch_i maintains a Cand_Next_HOP_CH list containing the IDs of all the CHs for which it receives the HOP_selection message and which are at the same HOP as Ch_i or at the lower HOP than Ch_i (Fig. 5).

After assigning the HOP count to every CH, the actual data routing path from CH-to-CH is formed.

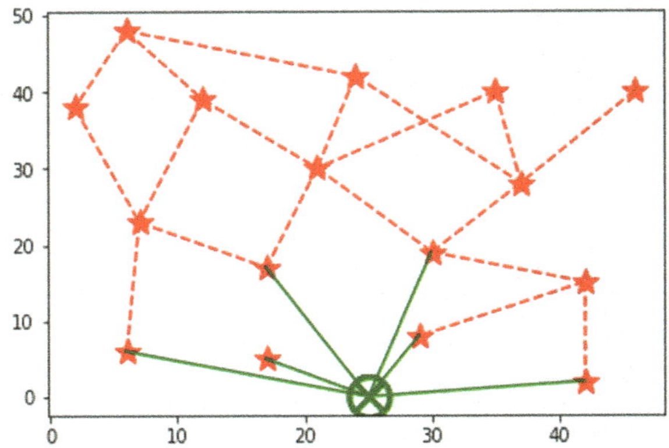

Fig. 5 Hop selection and possible next_Hop

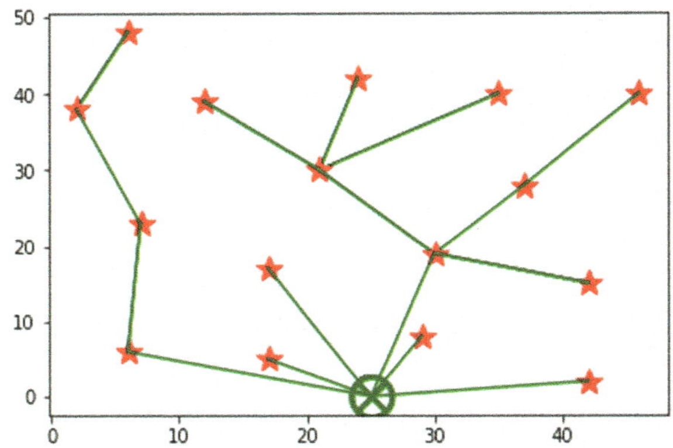

Fig. 6 Data routing path

4.3.1 Data Routing Path Formation:

Data routing path formation process starts from the CHs having the highest HOP count. For that, every CH starts a timer according to equation:

$$T(Ch_i) = e^{\left(\frac{1}{HOP(Chi)}\right)} \times T_2 \quad (2)$$

where $Dist_{BS}$ is the distance between CH_i and BS. $Load_i$ is the number of data.

Among the highest HOP count, CHs whose timer expires first broadcasts a Find_Msg message to all its candidate Next_HOP CHs. Then, those candidate Next_HOP CHs after receiving the Find_Msg message send an Ack with their location, current load $Load(Ch_i)$, residual energy, and distance from the BS to the sender CH. Upon receiving the Ack message from the candidate Next_HOP CHs, a CH Ch_i chooses its Next_HOP based on the residual energy, manageable load, and a average distance from it. For this, each CH evaluates a cost to make a trade-off between the residual energy and load on the candidate Next_HOP CHs and its distance from those candidate Next_HOP CHs. The cost evaluation function is given by

$$Cost(i, j) = \frac{Dist(i, j)}{E_{res}(i) \times Load(Ch_j)} \tag{3}$$

Then, the CH Ch_i sends a CH_Selected message to its Next_HOP CH which corresponds to minimal $cost(i, j)$ value and accordingly the selected Next_HOP CH Ch_j updates its load by adding the load of Ch_i and sends an ACK to the Ch_i. Thus, the routing path for the higher HOP CHs is formed first and this process continues till the HOP one CHs (Fig. 6).

Following algorithm is used for data routing:

Algorithm 2: Routing Algorithm

---

for *each CH* Ch_i **do**
  | Calculate HOP count
end
for *each CH* Ch_i **do**
  | Compute a delay time $T(Ch_i)$ using equation 4
end
for *each CH* Ch_i **do**
  if ($T(Ch_i) == 0$) *i.e., Delay time expires* **then**
    Broadcast a FIND message within range R_{max},
    if *(a CH* Ch_j *receives FIND message AND* $HOP(Ch_j) \leq HOP(Ch_i)$*)* **then**
      | Ch_j sends Ack message to Ch_i and Ch_i calculates $Cost(Ch_i, Ch_j)$
    end
  end
  if $Cost(Ch_i, Ch_j) \leq Cost(Ch_i, Ch_k)$ **then**
    | Update $NEXT\_HOP(Ch_i) = Ch_j$ and sends ACK to Ch_j
  end
  Ch_j updates its load
end

4.4 Data Forwarding Phase

In this phase, transmission of each node's data to their respective CH and then from the CHs to the BS through their Next_Hop CH is considered. The CMs sent their data to the cluster head which is aggregated by CH to the BS through other CHs. Inter-cluster data forwarding starts from the CH having maximum HOP count to the one having least HOP count, and finally, entire network data is gathered at the BS. A CH sends its data to the BS via its Next_Hop CH. Every CH updates its Next_Hop CH from its Next_Hop candidate CH based on the distance and the residual energy after every round.

The overall network data transmission to the BS is shown in Fig. 7.

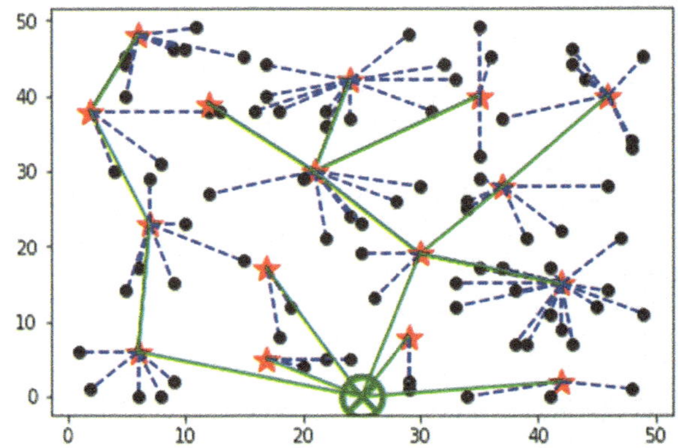

Fig. 7 Data transmission

5 Implementation and Performance Evaluation

5.1 Simulation Setup

The simulation is performed by Python language on the Intel i5 8th gen processor having 8GB RAM running on Windows 10 operating system in the Anaconda 3.5.2.0 environment using the Spyder IDLE. 100 nodes are randomly deployed in an area of $50 \times 50\, m^2$.

We have evaluated the performance of the proposed algorithm considering 1000 rounds. Initially, each node has 1 joule of energy.

5.2 Hardware Setup

5.2.1 Ubimote:

See Fig. 8.

Ubimote is an IEEE802.15.4 compliant wireless sensor mote that is designed and developed by CDAC. It has a chip, Chip with ARM Cortex M3 microcontroller with 32 kB RAM, 512 kB of programmable flash And clock speeds up to 32 MHz. It has an ISM band RF antenna with RF frequency range 2.4 GHz. This board supports a 20-pin connector. This mote can be powered through USB or batteries. We hare used Ubimote as the base station and end device as cluster head and cluster member[10] (Fig. 9).

Figure 10 refers to the average residual energy of the three approaches for each round.

Figure 11 refers to the total numbers of alive nodes in each round for the three algorithms.

Figure 12 shows the network lifetime for different number of nodes.

Fig. 8 Ubimote

Fig. 9 Hardware implementation

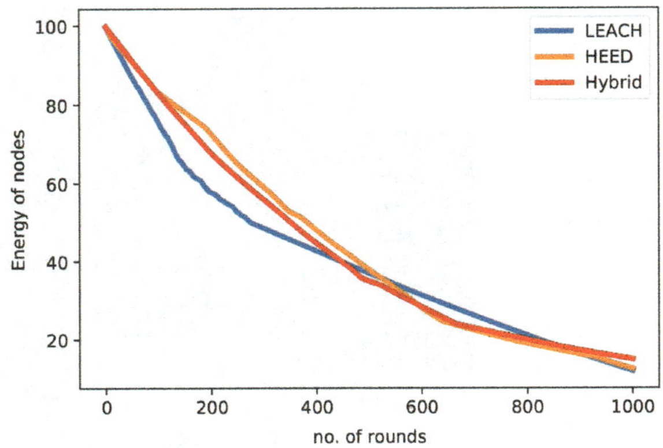

Fig. 10 Energy versus no. of rounds

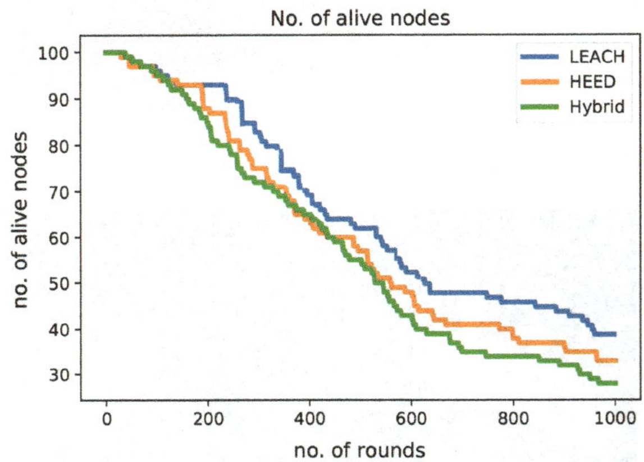

Fig. 11 No. of alive node versus no. of rounds

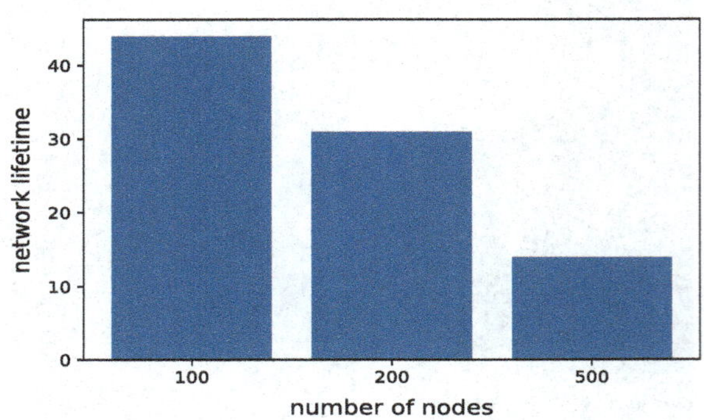

Fig. 12 Network lifetime for various number of nodes

6 Conclusion

IoT is the network of physical objects, and it is used for connecting and exchanging data over the Internet. WSN is the foundation of IoT application. In our work, we have discussed an energy-efficient routing protocol for energy-constraint IoT devices. We have studied several energy-efficient communication protocols such as LEACH and HEED. We have considered LEACH and HEED as the primary protocols to study and to set against our proposed approach. The proposed approach uses the residual energy of each node, node degree, and residual energy for selecting the CH for each round. The non-CH nodes join a CH based on its residual energy. Finally, the routing path is formed according to the hop counts of CHs and a trade-off between their average energy and their load. We have also tried to implement our work on hardware using Ubimote. The simulation results show efficiency of the proposed one over the existing protocols. In future, many aspects can be added to this work like mobile sink-based energy-efficient routing.

References

1. Singh A, Mahapatra S (2020) Network-based applications of multimedia big data computing in iot environment. In: Multimedia big data computing for IoT applications. Springer, Heidelberg, pp 435–452
2. Borujeni EM, Rahbari D, Nickray M (2018) Fog-based energy-efficient routing protocol for wireless sensor networks. J Supercomputing 74(12):6831–6858
3. Naranjo PGV, Shojafar M, Mostafaei H, Pooranian Z, Baccarelli E (2017) P-sep: a prolong stable election routing algorithm for energy-limited heterogeneous fog-supported wireless sensor networks. J Supercomputing 73(2):733–755
4. Mazumdar N, Om H (2017) Ducr: distributed unequal cluster-based routing algorithm for heterogeneous wireless sensor networks. Int J Commun Syst 30(18):e3374
5. Mazumdar N, Roy S, Nayak S (2018) A survey on clustering approaches for wireless sensor networks. In: 2018 2nd International Conference on Data Science and Business Analytics (ICDSBA). IEEE, pp 236–240
6. Kumar SA, Ilango P, Dinesh GH (2016) A modified leach protocol for increasing lifetime of the wireless sensor network. Cybern Inf Technol 16(3):154–164
7. Heinzelman WR, Chandrakasan A, Balakrishnan H (2000) Energy-efficient communication protocol for wireless microsensor networks. In: Proceedings of the 33rd annual Hawaii international conference on System sciences. IEEE, p 10
8. Younis O, Fahmy S (2004) Heed: a hybrid, energy-efficient, distributed clustering approach for ad hoc sensor networks. IEEE Trans Mobile Comput 3(4):366–379
9. Azharuddin M, Kuila P, Jana PK (2015) Energy efficient fault tolerant clustering and routing algorithms for wireless sensor networks. Comput Electric Eng 41:177–190
10. Nanda K, Nayak K, Chippalkatti S, Rao R, Selvakumar D, Pasupuleti H (2012) Web based monitoring and control of wsn using wingz (wireless ip network gateway for zigbee). In: 2012 sixth International Conference on Sensing Technology (ICST). IEEE, pp 666–671

Efficient Pest Bird-Controlling Algorithm in Unmanned Agriculture System

Saugata Roy, Nabajyoti Mazumdar, Rajendra Pamula, and Divya Tarkas

Abstract The revolution of Internet of Things has made remarkable advancement toward precision agriculture. In this context, crop prevention from destructive bird attacks has become one of the crucial challenges under agriculture automation. Most of the conventional birds repelling strategies are not approved by farmers due to their inefficacy and lethal nature. The utilization of mobile scarecrows has been seen as a solution to the bird-repelling problem where the mobile scarecrow aims to serve the event requests with minimal delay. However, these existing state-of-the-art algorithms fail to perform efficiently for a delay-constrained application when multiple event requests are generated at a time in vast farmland. Moreover, the terrestrial movement of the mobile scarecrow is not feasible for the surface irrigation system. Considering the aforementioned issues, a pest bird-controlling protocol is proposed in this article which is based on the exploitation of unmanned aerial vehicle (UAV). The proposed protocol reduces the request serving delay significantly due to the aerial communication mode as well as it is widely accepted in any kind of irrigation system. A comparative experimental analysis manifests the efficacy of the proposed protocol with respect to metrics like overhead serving delay and success ratio.

Keywords Smart farming · Unmanned aerial vehicle · Effective radius · Serving point · Success ratio

S. Roy (✉) · R. Pamula · D. Tarkas
Department of CSE, Indian Institute of Technology (ISM), Dhanbad, India
e-mail: saugataroy15@gmail.com

N. Mazumdar
Department of IT, Central Institute of Technology Kokrajhar, Kokrajhar, India

P. K. Singh et al. (eds.), *Evolving Technologies for Computing, Communication and Smart World*, Lecture Notes in Electrical Engineering 694,
https://doi.org/10.1007/978-981-15-7804-5_37

1 Introduction

1.1 Background

India is the second-largest country that depends on agriculture as the primary source of income. It has been observed from the Indian economic survey that about 50% of the total manpower is engaged in agricultural activities [1]. When it comes to the growth of the crop yield, pest birds are one of the important concerns as they cause a major threat to fruits and edible crops like rice, millet, sorghum, peas, beans, etc. Generally, the pest birds attack in a flock throughout the year, but their major invasion time is late spring and early summer when tender leafy crops are easily found. Moreover, the fields located near roosting sites are prone to be damaged by these birds. Table 1 lists the major edible crops/fruits in India mostly damaged by the respective pest birds along with the damage extent [2–10]. The presented damage extent clearly exhibits that bird invasion inflicts a severe crop depredation which may result in poor agricultural economy; accordingly, the necessity of crop preventing strategies has emerged.

1.2 Crop-Preventing Strategies

The extent of crop depredation by pest birds can be mitigated by the conventional crop-preventing strategies that are categorized as follows:

1.2.1 Lethal Strategies

They involve bird (i) killing, (ii) trapping [11], (iii) shooting, (iv) egg & nest destruction and, exploiting (v) fumigation, (vi) poison baiting, etc. Lethal techniques are the absolute way to eliminate the bird invasion from the farmland, and hence, they were pervasively used by the farmers in earlier days. But nowadays, they are not approved all over the world as the destruction of avian species may violate the ecological balance of nature [12]. Consequently, the farmers continued the pursuit of eco-friendly bird-repelling solutions that will reduce crop depredation effectively.

1.2.2 Non-lethal Strategies

These strategies do not harm the avian species directly and thereby ensuring the conservation of ecological balance. Depending on the usage, they are classified as:

- Habitat manipulation: It indicates the removal of roosting, nesting, and feeding sites of the pest birds.

Table 1 Pest birds and their major damage on edible crops/fruits

Crop/Fruit	Bird species	Damage extent (%)	Reference
Crop Groundnut	Sparrows, baya weavers	26	Kale et al. [3]
Maize	Crows, doves, babblers	12–21	Kale et al. [2]
Mustard	Parakeets	63	Simwat and Sidhu [4]
Pearl millet	Sparrows, parakeets, baya weavers	10–100	Jain and Prakash [5] Dhindsa et al. [6]
Peas	Pigeons	42	Kale et al. [3]
Pulses	Doves, pigeons, parakeets, sparrows	66	Saini et al. [7]
Rice	Sparrows, baya weavers	41	Kale et al. [2]
Sorghum	Doves, pigeons, parakeets, sparrows	12–85	Kale et al. [2]
Sunflower	Crows, parakeets	22	Toor and Ramzan [8]
Wheat	Crows	17–20	Kale et al. [2]
Fruit			
Almonds	Parakeets	7	Kale et al. [2]
Ber	Parakeets	Not recorded	Lakra et al. [9]
Datepalm	Parakeets	Not recorded	Gupta [10]
Grapes	Mynas	12.5	Toor and Ramzan [8]
Guavas	Parakeets	20	Kale et al. [2]
Peach	Parakeets, crows	32	Kale et al. [2]

– Use of repellents: They involve visual, auditory, and chemical repellents to drive away from the birds which are concisely discussed below:

- Visual repellents: As an upgrade of traditional scarecrow (or scary ballons), farmers started using aluminum foils over the scarecrow (or the ballon) to make the surface shiny and reflective. Another well effective repellent is reflective scare tape [13] which drives the birds away from the farmland by reflecting the sunlight and a humming noise in the blowing wind.
- Auditory repellents: The major pitfall of visual repellents is that most of the time the pest birds get accustomed to them and carried on their destructive nature. Hence, as a better alternative, bird gard pro bird control [14] device is heavily used in recent day farmlands which exploits prerecorded bird distress sounds to scare away the birds. Despite having high efficacy, this strategy is not feasible in the fields surrounded by human lives.

- Chemical repellents: Since many years, chemical compounds (non-toxic and non-lethal) have been using on the crops to alleviate the crop damage from the best birds. The popular chemical compounds employed on the crops are trimethacarb, methiocarb, and curb [15].

However, most of the farmers are not happy with the foregoing conventional strategies due to their less effectiveness and laborious nature which eventually result in a poor annual crop yield. On the other hand, it is quite impracticable for the farmers to surveil the farmland continuously in order to safeguard the crops from the injurious birds. Thus, the notion of efficient unmanned farming is demanded by the recent day farmers to improve the quality and quantity of the crop yield. With the advent of Internet of Things (IoT), the objects, devices, machines, and even animals or people can stay connected via network and perform data exchanges in an unmanned way. One of crucial application of IoT is smart farming [16–20] which provides the sensing and communication technologies, robotics, automation, and data analytics to the farmers to facilitate their job of crop and soil health monitoring, irrigating, fertilizing, and so on. Smart farming enables the farmers to remotely monitor the whole farmland with the help of a set of wireless sensor nodes. These nodes embedded with a passive infrared (**PIR**) sensor are capable to detect any moving object like birds or wild animals trespassed in the field. The exploitation of mobile robot/scarecrow has been adopted by existing approaches [21, 22] to effectively control the pest bird attacks without any human interaction. As soon as any bird attack occurs on some crop, an event request (EV_REQ) message will be generated by the proximate sensor(s) to report the base station (**BS**). Upon receiving such message(s), the BS commands the mobile robot/scarecrow to serve the EV_REQs by following some BS-generated trajectory.

1.3 Motivation

In recent day smart farming applications, bird-repelling protocols reveal the following challenges:

- Serving Delay: Delay-constrained applications of IoT demand time-bound service as the pest birds can inflict rapid crop damage within a short period of time. Unfortunately, the existing mobile scarecrow-based protocols may suffer from high request serving latency when a large number of scattered EV_REQs are triggered in vast farmland. Thus, the protocols designed for bird repulsion should effectively handle multiple event requests with minimal serving latency.
- Feasibility: Surface irrigation is considered as the pervasive irrigation approach which floods the soil in the field by the gravity flow of the water. Unfortunately, such type of irrigation restricts the employment of the mobile scarecrow due to its terrestrial movement. Hence, the protocols designed for preventing the crops should aim at aerial communication mode, so that it would be feasible in any kind of irrigation system.

1.4 Contribution

Considering the aforementioned challenges, an eco-friendly unmanned aerial vehicle (**UAV**)-based pest bird-repelling protocol is proposed in this article. In a nutshell, the contribution of this article is given below:

- Performing an immediate detection of bird invasion in any position of the field.
- Exploiting a UAV-based bird-scaring strategy to alleviate the request serving latency significantly.
- Minimizing the UAV tour length by nominating an optimal set of serving points.
- Modifying the ant colony optimization (**ACO**) algorithm marginally in order to provide a priority-based service depending on both distance and damage extent value.

2 System Design and Preliminaries

2.1 Network Model

In this article, the whole farmland (region of interest (**RoI**)) is virtually partitioned into $k \times k$ number of equal-sized grid cells where for each cell a sensor node is manually deployed at the center point. Each sensor node is an *Arduino uno R3* microcontroller equipped with PIR sensor to competently detect the presence of any moving creature. For the sake of simplicity, coverage of only $(\frac{1}{4})^{th}$ part of the farmland is presented. The proposed smart farm also involves *Raspberry pi 4 model B* microprocessor as base station (BS) and *Altair aerial blackhawk* model UAV equipped with an ultrasonic bird repeller. The UAV take-off, landing, and its flight route is regulated by the BS. Figure 1 depicts the system model of the proposed approach where Fig. 2 represents the initial setup of the deployment scenario. It is to be noted that the grid size is taken as $= \sqrt{2} \times R_{sen}$ where R_{sen} is the sensing range of each sensor. This ensures full coverage of the farmland (Fig. 3).

2.2 Rudimentary Assumptions

- All deployed sensor nodes remain stationary throughout their lifetime.
- Each sensor is concerned about its geographical position by means of a global positioning system or any well-known localization algorithm [23].
- All sensors are homogeneous in nature, i.e., all are having equal battery power, sensing radius, and communication radius.
- All the sensors are well synchronized with respect to their timer values [24].

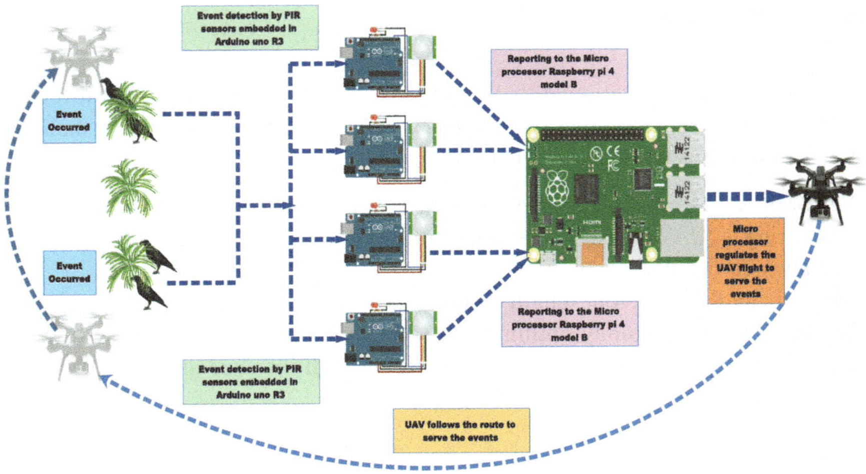

Fig. 1 System Model

- Intelligent UAV charging pad is utilized to charge the UAV without any human interaction.
- Any pest bird is scared and fly away as soon as it will be in the effective radius (subsection 3.2) of the UAV.

3 Proposed Work

This section presents an eco-friendly bird-repelling strategy by exploiting the aerial behavior of UAV. Once the network setup is finalized, the smart farm will be ready to notify the BS about any sort of bird invasion in the field, so that the BS can accomplish a subsequent bird-scaring action. The whole procedure can be divided into the following phases:

3.1 Phase I: Event Reporting

As soon as any bird invasion occurs, the proximate sensor(s) trigger(s) an event message (EV_REQ) to report the BS for requesting a necessary action (as shown in Fig. 3). The EV_REQ message contains the event (bird attack) location information (latitude and longitude) of the respective event. If duplicate EV_REQs have arrived at the BS by distinct sensors, BS eliminates the redundant one(s).

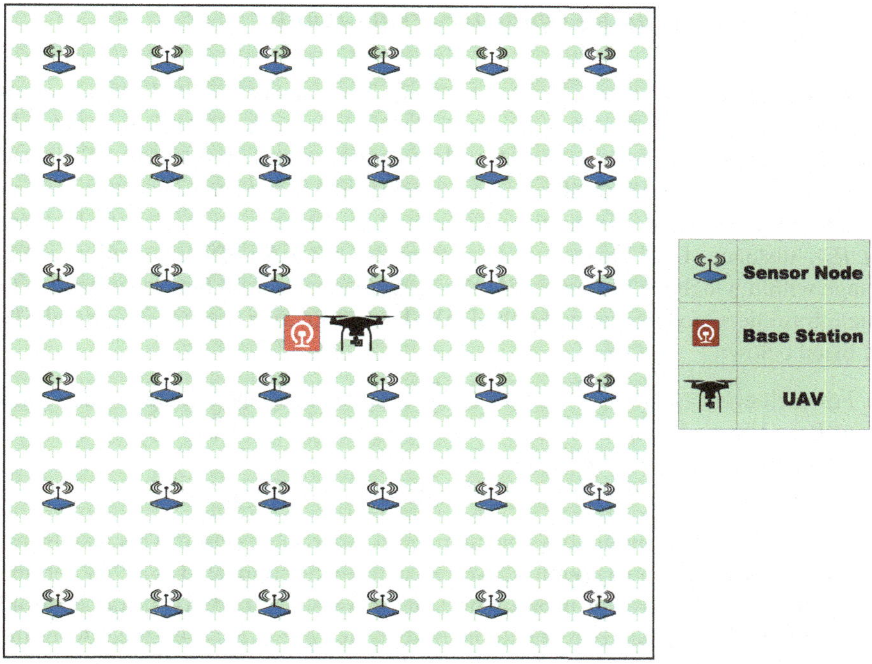

Fig. 2 Initial deployment scenario

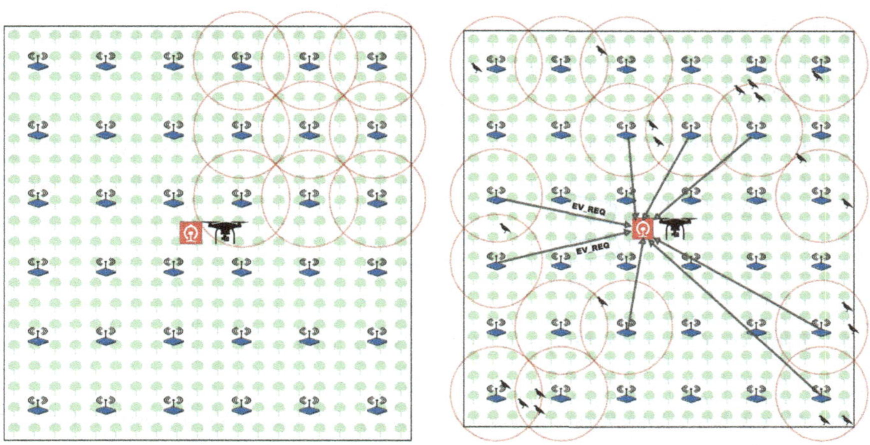

(a) Field coverage by the sensors (b) Birds attack and generated EV_REQs

Fig. 3 Phase I: Event reporting

3.2 Phase II: Finding Optimal Serving Points

Based on the event location information, BS nominates a set of optimal serving points (SP) exploiting which a UAV flight path is built where each SP will be visited only once by the UAV for bird repelling. In order to mitigate the serving delay, BS aims to minimize the number of SPs. Therefore, BS selects the SPs in such a way that visiting one SP may serve the maximum possible number of events (bird attack) which are at R_{eff} distance (UAV effective radius) from that SP. **UAV effective radius** is the range within which the presence of UAV guarantees successful bird repelling due to the ultrasonic sound generated by the UAV. The nomination process of optimal SPs is listed below:

- For each event ε_i, BS sets a virtual circle centered at ε_i and having radius equals to R_{eff}. Figure 4 presents a simulation scenario showing 20 number of events (η_ε) surrounded by a circle having radius R_{eff} (here 15 ft).
- BS verifies whether there is any circle intersection or not. If any,
  - BS seeks for maximum possible number of circle intersection. This results in a common overlapping region whose centroid will be nominated as a UAV serving point (SP). It is worth mentioning that the presence of UAV in the centroid point will competently repel all the birds which are at R_{eff} distance. After the nomination of an SP, BS removes the circles participated in the intersection.
  - For the leftover circles, BS does the same and generates another SP of the UAV. This continues till any circle intersection is left.
  - If no circle intersection is left, the remaining circles which have become disjoint or the circles which were disjoint from the beginning will nominate themselves as SPs.
- Finally, a number of optimal serving points (η_{sp}) is generated where $\eta_{sp} < \eta_\varepsilon$. Figure 4 exhibits 13 number of SPs covering 20 number of events (bird attack location) in R_{eff} distance.

3.3 Phase III: UAV Trajectory Construction

In this phase, BS establishes an optimal trajectory for the UAV to visit all the SPs exactly once and returns back to the UAV charging pad by following some sequence. It is to be noted that such a sequence should produce the shortest length UAV route among all possible route sequences in order to minimize the EV_REQ serving delay. Now, for η_{sp} number of SPs in the field, there are $\frac{(\eta_{sp}-1)!}{2}$ number of route possibilities as per the brute force approach. However, when the pest birds attack in a flock, with the increasing value of η_{sp}, the factorial value increases faster than all polynomial and exponential functions. This infers that the UAV path construction incurs a non-polynomial time solution. Hence, it is highly desirable to exploit an optimization

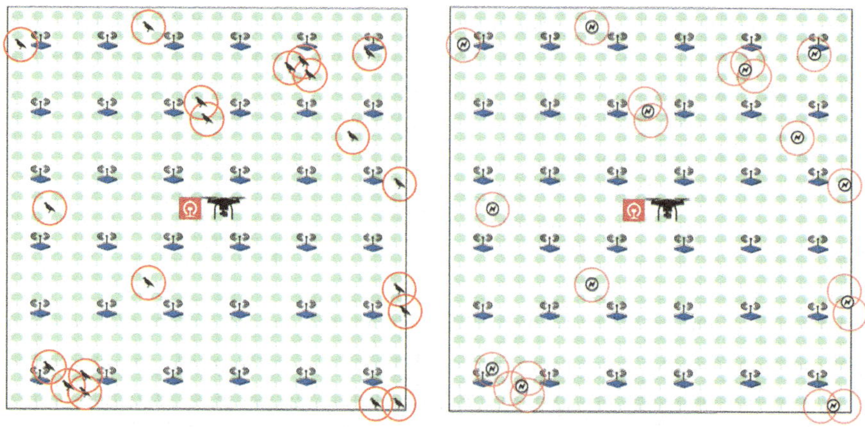

(a) UAV effective radius (b) Serving points

Fig. 4 Phase II: Finding serving points

strategy like ant colony optimization (**ACO**) algorithm to build a near-optimal UAV route in polynomial time. Utilizing ACO, the proposed protocol results a minimal length hamiltonian cycle (UAV flight path) from a fully connected graph $G = (V, E)$ where V is the set of all serving points (SP) obtained in 3.2 and E is the set of edges between all these SPs. However, unlike traditional ACO, this article marginally modifies the transition probability in order to produce a priority-based UAV service in terms of both distance (of the SP) and damage extent value (no of distinct event locations). The essential parameters used in modified ACO algorithm are described below:

- **Modified Transition Probability(Mod_$\mathbf{P^k_{ij}}$(t))**: It is the probability of how ant k will choose the SP j while sitting at SP i at time t

$$Mod\_P^k_{ij}(t) = \begin{cases} \frac{[\tau_{ij}(t)]^{\alpha}[D_{ij}]^{\beta}}{\sum_{l \in allowed_k}[\tau_{ip}(t)]^{\alpha}[D_{ip}]^{\beta}} & if\ j \in allowed_k \\ 0 & otherwise \end{cases} \tag{1}$$

  where $allowed_k$ is the set of SP(s) which is(are) not yet visited by ant k, $\tau_{ij}(t)$ is intensity of pheromone trail between SPs i and j at time t, D_{ij} is the donation value (Eq.3) of SP j from SP i, and α and β are the parameters to regulate the relative influence of trail versus donation value.
- **Donation Value ($\mathbf{D_{ij}}$)**: Each SP plays a significant role in constructing the UAV route. Based on the distance and damage extent rate, a donation value D_{ij} is assigned to each SP j from SP i. The lower the donation value D_{ij}, the higher the probability of the SP j being chosen by ant k sitting at SP i. It can be mathematically expressed as:

$$D_{ij} = W_1 \times \delta_{ij} + W_2 \times (1 - \eta_{\varepsilon_j}) \tag{2}$$

where δ_{ij} is the distance between SP i and j, η_{ε_j} is the number of events (bird attacks) occurred at SP j, W_1 and W_2 are weight factors to make a compromise between δ_{ij} and η_{ε_j}. But δ_{ij} and η_{ε_j} may produce different impact on D_{ij} value since they are measured in different scale. This issue can be resolved by min–max normalization technique as:

$$\delta_{ij}^{norm} = \frac{\delta_{ij} - \delta_{min}^{j}}{\delta_{max}^{j} - \delta_{min}^{j}} \quad and \quad \eta_{\varepsilon_j}^{norm} = \frac{\eta_{\varepsilon_j} - \eta_{\varepsilon_{min}}}{\eta_{\varepsilon_{max}} - \eta_{\varepsilon_{min}}}$$

where δ_{ij}^{norm} and $\eta_{\varepsilon_j}^{norm}$ are the normalized values of δ_{ij} and η_{ε_j}, respectively, scaled in the range [0, 1]. δ_{min}^{j} is minimum of all the distances from SP j to all allowed SPs (not yet visited by ant k). Similarly, δ_{max}^{j} is the maximum of all distances from j to all allowed SPs. $\eta_{\varepsilon_{max}}$ and $\eta_{\varepsilon_{min}}$ are the maximum and minimum no. of events occurred at some arbitrary SPs u and v. Hence, Eq. 2 can be rewritten as:

$$D_{ij} = W_1 \times \delta_{ij}^{norm} + W_2 \times (1 - \eta_{\varepsilon_j}^{norm}) \tag{3}$$

- **Pheromone Updating** ($\boldsymbol{\tau_{ij}(t + \eta_{sp})}$): Since there are η_{sp} number of SPs, after η_{sp} iterations, each ant completes a tour. Subsequently, pheromone trail $\tau_{ij}(t + \eta_{sp})$ on edge (i,j) at time $t + n_{sp}$ will be updated as

$$\tau_{ij}(t + n_{sp}) = \rho.\tau_{ij}(t) + \Delta\tau_{ij} \tag{4}$$

where $\tau_{ij}(t)$ is pheromone trail on edge (i,j) at time t, ρ is evaporation factor which regulates pheromone reduction and $\Delta\tau_{ij}$ is total change in pheromone trail between time t and $t + \eta_{sp}$ as shown below

$$\Delta\tau_{ij} = \sum_{k=1}^{l} \Delta\tau_{ij}^{k} \tag{5}$$

where l is the no. of ants and $\Delta\tau_{ij}^{k}$ is quantity per unit length of trail on edge (i,j) by k^{th} ant between time t and $t + \eta_{sp}$

$$\Delta\tau_{ij}^{k} = \begin{cases} Q/L_k & if\ ant\ k\ travels\ on\ edge(i,j)\ between\ t\ and\ t+n \\ 0 & otherwise \end{cases}$$

where Q is constant and L_k is tour length by ant k.

Fig. 5 exhibits the constructed UAV path followed by the modified ACO algorithm.

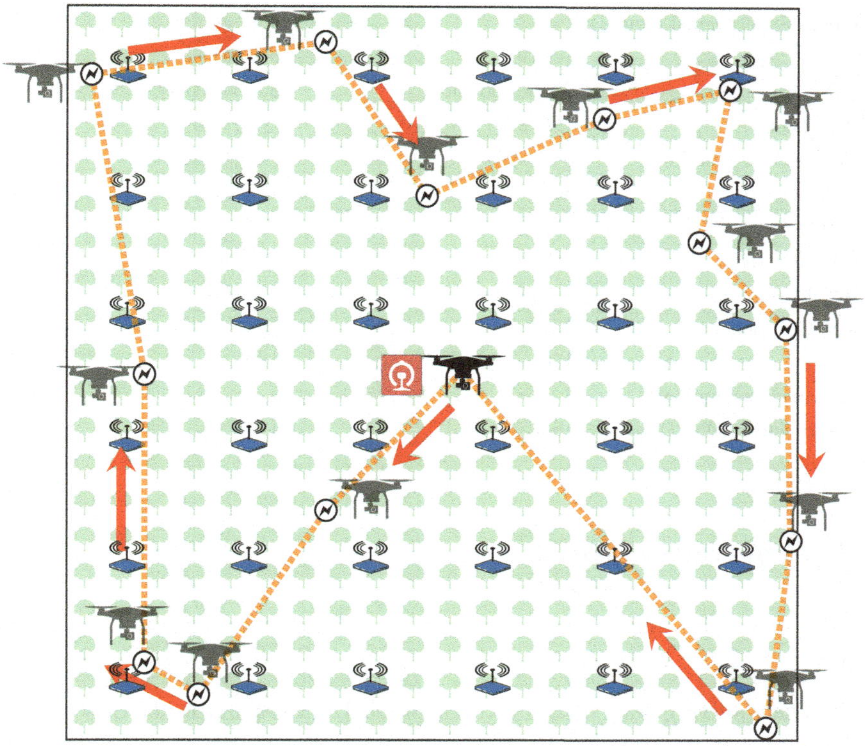

Fig. 5 Phase III: UAV trajectory construction

3.4 Phase IV: Request Serving

After building the optimal UAV trajectory, the BS regulates the UAV to traverse along the trajectory and halt at the optimal serving points (SP) in order to accomplish the bird repelling. At each SP, UAV pauses for a predefined time quantum (δt) and produces an ultrasonic sound to ensure the flee of any pest birds within its effective radius (R_{eff}).

4 Performance Evaluation

The efficacy of the proposed protocol is validated throughout an experimental analysis presented in this section. The simulation result shows that the proposed protocol outperforms the existing mobile robot/scarecrow-based approaches like Scareduino [22] and Mobile robot [21] in terms of overhead serving delay and serving success ratio. Table 2 lists the experimental setup parameters in two different scenarios:

Table 2 Experimental setup

	Scenario 1	Scenario 2
Farm size	$600 \times 600\ ft^2$	$600 \times 300\ ft^2$
BS location	(275,300)	(275,150)
UAV stationary location	(325,300)	(325,150)
No. of sensors	36	18
Sensing range(R_{sen})	71 ft	71 ft
UAV flight speed	25 mph	25 mph
UAV flight time	15–17 mins	15–17 mins
UAV flight range	800 ft	800 ft
Service time period (Δt)	180 s	180 s
Pause time quantum (δt)	3 s	3 s
UAV effective radius (R_{eff})	15 ft	15 ft

4.1 Performance Metrics

Based on the following metrics, the performance of the proposed protocol is evaluated:

– ***Overhead serving delay***: It denotes the additional time required by the UAV to serve the leftover events (EV_REQ) which were not served within the service time period (Δt). Due to the aerial service along with a priority-based route, the proposed UAV protocol can competently serve significantly larger number of events within Δt than the existing Scareduino [22] and Mobile robot [21] protocols whose motions are restricted due to terrestrial movement. Consequently, the UAV protocol attains much lower overhead serving delay. A comparison graph of the proposed UAV and other existing protocols is presented in Fig. 6 which manifests the efficacy of the proposed protocol.

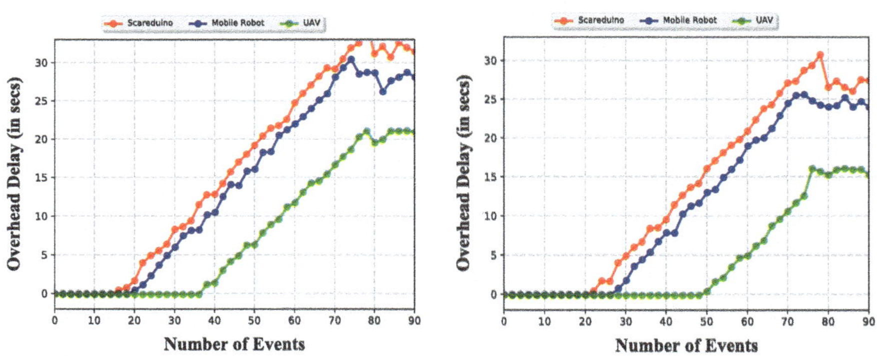

(a) Scenario 1 (b) Scenario 2

Fig. 6 Overhead serving delay for different scenarios

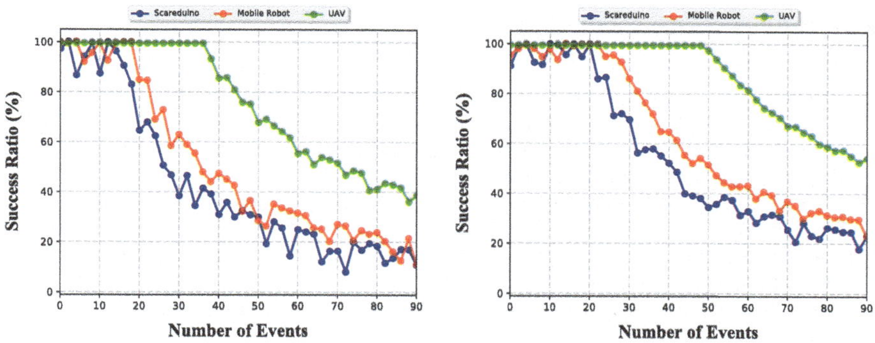

(a) Scenario 1 (b) Scenario 2

Fig. 7 Success ratio for different scenarios

– ***Success ratio***: This metric is measured as following:

$$\frac{no.\ of\ events\ successfully\ served\ within\ \Delta t}{total\ no\ \ of\ events\ (\eta_{\varepsilon})} \times 100\%$$

As our UAV protocol offers a priority-based serving route in aerial communication mode, it maintains a higher success ratio for comparatively larger number of EV_REQs than the related existing protocols. It is to be noted that varying the flight speed of UAV may affect the performance of the proposed protocol in a delay-sensitive application. This article has assumed a standard UAV speed as 25 mph ($\approx$ 36 feet/sec) which can fulfill approximately 36 and 49 events, respectively, within Δt for scenario 1 and scenario 2 (Fig. 7).

5 Conclusion

In this paper, a UAV-based pest bird-repelling protocol has been proposed which significantly mitigates the request serving latency. The proposed protocol aims to minimize the UAV tour length by nominating a set of optimal serving points where visiting each serving point guarantees the serving of maximum possible event requests. Furthermore, a modified version of the ACO algorithm been exploited to provide a priority-based service based on both distance and number of events occurred. The experimental results manifest the potential of the proposed protocol. In the future, we will study on finding the optimal number of UAVs in a delay-constrained smart-farming application. In addition, we will look into the serving load balancing among these multiple UAVs.

References

1. Madhusudhan L (2015) Agriculture role on Indian economy. Bus Econ J
2. Kale M, Balfors B, Mörtberg U, Bhattacharya P, Chakane S (2012) Damage to agricultural yield due to farmland birds, present repelling techniques and its impacts: an insight from the Indian perspective. J Agric Technol 8(1):49–62
3. Kale MA, Dudhe N, Kasambe R, Bhattacharya P (2014) Crop depredation by birds in Deccan Plateau, India. Int J Biodiversity
4. Simwat GS, Sidhu AS (1973) Feeding Habits of rose-ringed parakeet, psittacula-krameri (scopoli). Indian J Agric Sci 43(6):607–609
5. Jain MB, Prakash I (1974) Bird damage in relation to varietal differences in bajra crop. Ann Arid Zone
6. Atwal AS, Bains SS, Dhindsa MS (1983) Status of wildlife in Punjab. Indian Ecological Society, 1984. In: Seminar on Status of Wildlife in Punjab, Punjab Agricultural University
7. Saini HK, Saini MS, Dhindsa MS (1992) Pigeon damage to lentil: a preharvest assessment. Pavo 30:47–52
8. Toor HS, Ramzan M (1974) Extent of losses to sunflower due to rose-ringed parakeet, Psittacula krameri,(Scopoli) at Ludhiana (Punjab). J Res Punjab Agric Univ
9. Lakra RK, Kharub WS, SinghZ (1979) Relationships of shedding of 'ber'fruits and the incidence of 'ber'fruit fly and bird damage. Indian J Ecol
10. Gupta MR (1980) Resistance of some promising datepalm cultivars against rainfall and bird damage. Punjab Hortic J 20(1/2):74–77
11. Singh RP, Dungan GH (1955) How to catch crows. Allahabad Farmer 29:59–67
12. Dhindsa MS, Saini HK (1994) Agricultural ornithology: an Indian perspective. J Biosci 19(4):391–402
13. Bruggers RL, Brooks JE, Dolbeer RA, Woronecki PP, Pandit RK, Tarimo T (1986) All-India co-ordinated research project on economic ornithology. In: Hoque M (ed) Responses of pest birds to reflecting tape in agriculture. Wildlife Society Bulletin, pp 161–170
14. https://www.birdgard.com/bird-repellent-technology/
15. Issa MA, El-Bakhshawngi MIA, Lokma V (2019) Repellent effect of certain chemicals compounds on wild birds attacking wheat and cowpea under field conditions. Zagazig J Agric Res 46(4):1029–1038
16. Ryu M, Yun J, Miao T, Ahn I-Y, Choi S-C, Kim J (2015) Design and implementation of a connected farm for smart farming system. In: 2015 IEEE sensors. IEEE, pp 1–4
17. Kale AP, Sonavane SP (2019) IoT based smart farming: feature subset selection for optimized high-dimensional data using improved GA based approach for ELM. Comput Electronic Agric 161:225–232
18. Zamora-Izquierdo, Miguel A, Santa J, Martínez JA, Martínez V, Skarmeta AF (2019) Smart farming IoT platform based on edge and cloud computing. Biosyst Eng 177:4–17
19. Doshi J, Patel T, Bharti SK (2019) Smart farming using IoT, a solution for optimally monitoring farming conditions. Proc Comput Sci 160:746–751
20. Hasan M (2020) Real-time and low-cost IoT based farming using raspberry Pi. Indonesian J Electric Eng Comput Sci 17(1):197–204
21. Krishna, Lokesh K, Silver O, Malende WF, Anuradha K (2017) Internet of Things application for implementation of smart agriculture system. In: 2017 International Conference on I-SMAC (IoT in Social, Mobile, Analytics and Cloud)(I-SMAC). IEEE, pp 54–59
22. Lim LJ, Sambas H, MarcusGoh NC, Kawada T, JosephNg PS (2017) Scareduino: smart-farming with IoT. Int J Sci Eng Technol 6(6):207–210
23. Wang Z, Zhang B, Wang X, Jin X, Bai Y (2018) Improvements of multihop localization algorithm for wireless sensor networks. IEEE Syst J 13(1):365–376
24. Gherbi C, Aliouat Z, Benmohammed M (2016) An adaptive clustering approach to dynamic load balancing and energy efficiency in wireless sensor networks. Energy 114:64–662

Face Tracker-Assisted Multi-Person Face Recognition in Surveillance Videos

Tapas Saini, Priyanka Wawdhane, Anoop Kumar, and E. Magesh

Abstract Face recognition is a method to recognize people from image and video sources. It can be used for several applications by law enforcement agencies such as surveillance systems and person authentications. Multi-person face recognition is the task of identifying multiple people approaching camera view at same time. Performing real-time and accurate multiple person face recognition is highly challenging due to the uncontrolled face poses, illuminations and expressions. Due to these factors of unconstrained environment, the accuracy gets degraded across video frames. In this paper, we propose improved face recognition framework for multiple people in unconstrained movement/walking towards camera which is implemented on the NVDIA graphics processing unit (GPU)-enabled computer systems. We propose the use of a face tracker to save the GPU computational power and also to increase the overall accuracy of the system. The idea is to do the detection of the faces once through face detector model and subsequently use the face tracker for further frames until the face disappears from camera view. We do combined the use of this tracked faces information along with a face recognition model over same faces. This has achieved face recognition rate improvements on videos of people walking across camera.

Keywords Face recognition · Surveillance · Face detection · Precision · Recall · F-score · DLIB library · Face tracker

---

T. Saini (✉) · P. Wawdhane · A. Kumar · E. Magesh
Centre for Development of Advanced Computing, Hyderabad, India
e-mail: stapas@cdac.in

P. Wawdhane
e-mail: priyankaw@cdac.in

A. Kumar
e-mail: anoopkr@cdac.in

E. Magesh
e-mail: magesh@cdac.in

P. K. Singh et al. (eds.), *Evolving Technologies for Computing, Communication and Smart World*, Lecture Notes in Electrical Engineering 694,
https://doi.org/10.1007/978-981-15-7804-5_38

1 Introduction

Face recognition system is a technology capable of identification or verification of a person in a given digital image or a video frame from a video source. Modern face recognition algorithms are implemented in the form of machine learning (specifically deep learning) models. Deep learning is a promising machine learning methodology which has shown remarkable result in computer vision applications. There are multiple number of methods [1–3] for implementing facial recognition systems out of which current popular methods are based on deep convolutional neural networks (CNNs) models. These models work by extracting facial features also known as embeddings from given image and comparing with other faces (embedding) stored in a database. Every embedding forms a unique face signature of a particular human face (after face detection on a frame) evaluation of which requires high amount of computational power. Multiple face recognition when compared with the single face recognition is further challenging as it needs more computations power to process multiple faces per video frame. There are various requirements and expectations from face recognition software itself, as an example, there are face recognition systems that do not have very high requirements on the accuracy but have a requirement referred to the overall speed of the system. On the other side, there are systems where the primary requirement is accuracy. Both accuracy and real-time recognition are challenging to achieve on videos of subjects that do not co-operate with the recognition system. Query videos are acquired in an uncontrolled environment and as a result illumination and appearance of subjects can change. As said, every video is a sequence of images (called frames) with time gap of few milliseconds between two adjacent frames. Also, attempts to recognize faces in every independent video frame having people walking across cameras may give inconsistent matching results due to uncontrolled face poses. To reduce the computation and increase the accuracy of recognition for this scenario, a face tracker is used in purposed method. Face tracker helps reducing burden keeping track of multiple people in crowded environment and assist in better decision making for recognition. Face recognition is a field that was extensively studied within the past years; however, it's still an energetic space of analysis in area of research. The remaining paper is organized as follows: Sect. 2 reviews recent related work done on face detection algorithms, face recognition and face tracking. Our projected methodology is presented in Sect. 3. Experiments and results are presented in Sect. 4. Finally, Sect. 5 concludes this paper.

2 Related Works

In the following sections, we discuss typical algorithms used in face recognition and face tracking domain. The framework can be divided into three critical parts, viz., face detection, face recognition and tracking algorithms.

2.1 Face Detection Algorithms

Input frame can consist of one or more faces in different locations. All the faces have a general structure such as eyes, nose, forehead, chin and mouth. Extracting all the faces from the frame is also possible with the size and location of each face. Viola and Jones [4] have proposed the face detection algorithm. The algorithm scans a sub-window which is able to detect faces in a given input image. This algorithm contains three main ideas that can be executed in real time: the image integral, classifier learning with AdaBoost and the cascade structure. The approach is to rescale the image to a different size. Lastly, the algorithm runs the fixed size detector through these images. However, it can take much time to process due to the calculation of the different size images. Wu et al. [5] have proposes a multi-face alignment method. This algorithm focuses on identifying geometric structures of multiple faces in an image using clustering, same face features as a single cluster. Zhang et al. [6] have proposed a face detector which is independent of the human face pose variations. This framework is deep cascaded and does multiple tasks which does the face detection as well as face alignment which helps to boost up the framework performance using online hard sample mining strategy. This cascaded architecture made up of three stages of deep convolutional networks which predict the face location and landmark location.

King [7] has proposed a detector which detects faces in images with quiet robust to pose variations. The proposed method optimizes all sub-windows in an image instead of performing any sub-sampling. This method can be used in any object detection method which uses linear parameters to learn to improve the accuracy. Author has showed in experimental results that the detector shows considerable amount of performance gains on three datasets.

2.2 Face Recognition Algorithms

The most important part of face recognition system is to learn the distinct features of the human face to be able to differentiate between different human faces. It is very necessary to design a face recognition algorithm to withstand under dull or obscure lightening conditions. The most traditional methods of facial recognition like principle component analysis [8] may perform greatly in specific conditions. In recent years, face recognition based on convolutional neural networks have been proposed with different algorithms promising higher accuracy.

Liu et al. [1] have proposed a face recognition model which makes use of [6] multi-task cascaded detector to detect faces. This method uses total 512 face embeddings. This method expects the ideal face features to have smaller maximal intra-class distance and minimal inter-class distance under a chosen metric space. This method uses angular softmax loss which enables convolutional neural networks to learn angularly discriminative face features also the size of angular margin can be adjusted by a parameter m. Schroff et al. [2] have proposed a face recognition method which directly maps a face image to a compact Euclidean distance space where these distances directly correspond to the measure of face similarity. This method does face recognition using feature vectors. Romić et al. [3] have proposed an enhanced eigenfaces method for face recognition which uses the Viola–Jones method for face detection, an artificially expanded training set, as well as the CLAHE method for contrast enhancement. Eigenface is set of eigenvectors used in face recognition which is used to reduce the dimensionality of input images into smaller dimensions. Once we get smaller representation of our faces, we apply a classifier which takes the reduced-dimension input and produces a class label. King [9] has proposed a face recognition model which makes use of Euclidean distance to get similarity between pair of face images. According to this, the Euclidean distance between the same person image pair will be less and compared with the different people images. The framework has been tested and validated on labelled faces in wild (LFW) dataset.

However, in all these above-mentioned algorithms, it will be difficult for the algorithm to recognize people in moving motion as the embeddings will keep on changing while the person is walking. These mentioned algorithm results may also suffer from low lightening conditions and occlusions occurred during people movement.

2.3 Face Tracking Algorithms

Face tracking is a special case of object tracking method which detects and tracks the movement of human face in a digital video or frame. In the camera/video feed when the human face is first detected in the first frame, tracker is created and updated in every subsequent frame to follow the same face in every frame. Tracking the face helps in many advantages as such tracking the direction in which the face is looking, counting the number of human faces in a live video or video frame, and following a particular face as it moves in a video stream.

Danelljan et al. [10] have proposed an object tracker for robust scale estimation which uses a tracking-by-detection approach. The proposed method learns discriminative correlation filters for translation and scale estimation which are based on scale pyramid representation. This tracking method improves the performance compared to a fully comprehensive search which is generic to any tracking method. Savath Saypadith et al. have proposed the real-time multiple face recognition [11] algorithm which uses the framework that contains face detection based on convolutional neural network (CNN) with face tracking and state of the art deep CNN face recognition algorithm. This multiple face recognition framework is executed on the embedded graphics processing unit (GPU) system, i.e., NVIDIA Jetson TX2 board. Their experiment results stated that the projected system will acknowledge multiple faces at an equivalent time in real time. Lin [12] has proposed a recent method for face tracking and clustering an unknown number of human faces and maintaining their individual identities in unconstrained videos. The method track faces with slightly occluded and strong appearance changes in multiple shots resulting from significant variations of makeup, facial expression, and head pose and illumination. There are some methods [13–17] which proposes various face recognition technique based on multi-person face recognition and tracking using various methods like eigenfaces and Haar cascade. Instead of only using detector, we used the face tracker to keep track of the person in moving motion which is more useful than only using face detector for person recognition. To increase the accuracy of the face recognition and detection process, the tracking algorithm can be efficiently used. The tracking algorithm when used showed significant accuracy improvements in multi-person recognition.

3 Proposed Framework

We propose the method for multi-person face recognition using face tracker for increasing the algorithm accuracy. Many multiple face recognition algorithms in the past few years have been implemented for the faces which are steady and still or independent video frames. Vanilla face recognition algorithms are best suited for stationary images. These algorithms perform inconsistently on video frames having people walking around the camera view. Inconsistency can be clearly seen, when face orientations changes from front to non-front view. On one hand, front projection of person's face gets better recognition results, whereas, on the other hand, non-front face projection gives incorrect recognition results for same person in video. This paper mainly focuses on improving multiple face recognition on videos of people walking across the camera using a per face tracker. Our algorithm works on the assumption that people approach the camera from certain distance and finally they walk across the camera and go out of the camera view. In the scenario, when people are walking in a mob or group, the chances of occlusions are higher. As already said, the face features may vary depending on the face position, it is to be noted that frontal face position results in best face features and increase in deviation from front pose degrades the face features quality. We also do back tracking of the person from the last to first frame, and subsequently, we store the detected and correctly recognized person's face image in records. This helps to improve the recognition rate and reduce the recognition fluctuations caused by deviations from frontal face poses.

In the proposed method, initially, we are detecting faces with the help of face detection and alignment algorithm. As soon as we get that face we are creating a new tracker for each detector box. Detected and aligned faces will be the input for recognition algorithm. At last, we are applying the tracker along with the detected face. Figure 1. shows flow diagram of our proposed method for multi-person face recognition. Video input or video source—Video source is given to the algorithm in which we are continuously fetching the frames from input video source. For each frame, at first, we are running face detection and alignment algorithm. Once the face is detected we are creating a tracker for each particular detector box co-ordinate. The face tracker is maintained throughout tracking with the assigned unique face tracker ID for each face tracker.

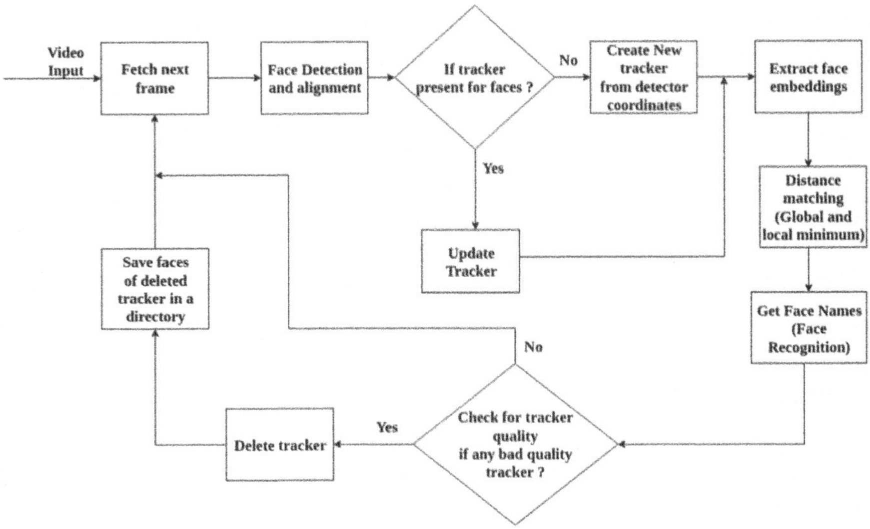

Fig. 1 Overview of the proposed framework

3.1 Extracting Features and Distance Calculation

After a face has been detected and for the detected face using the co-ordinate points, the face tracker is enabled. We are extracting the face features of this detected face using face landmark shape predictor. Once we extract the face features, we match this face features with our database face features. Comparison of these face features is done by Euclidean distance calculation. Here, for the tracked face, we are maintaining two separate distances global minimum distance and local minimum distance. Assuming the lowest distance will be the better match for the detected person, we will change the global minimum only when the local minimum distance will be less than global minimum. On the global minimum distance value, we are then recognizing the person. We are continuously checking the tracker quality or confidence if it is above a defined confidence threshold we are just fetching the next frame, otherwise, we are deleting the tracker. Before the person go out of camera scope, we are saving the person face from the first time in the frame he entered into the last frame. From the second time of face detection, we are first making sure if the tracker is already present for the detected face or not. If the tracker is present we are just updating the tracker to do further feature extraction. If the face tracker is not present we are initiating a new tracker for the detected face.

3.2 Face Detection and Alignment

Face detection is the prior and foremost step in this framework. The faces which are detected by detector model are then provided as an input to the face recognition deep neural network model. We have used DLIB mmod face detector [7] which is based on the convolutional neural network(CNN). The detector uses the method of max margin object detection technique which was specifically trained to detect faces of size at least 50 × 50. Instead of doing any sub-sampling, this method optimizes over all sub-windows of image. This face detector is having the ability of detecting human faces almost in all angles. The detector will give output in the form of bounding box/coordinates. We have used DLIB's 68 landmark/shape predictor to find the facial landmarks of each detected face. The facial landmarks are then fed to face recognition model to frontalize the face and extract unique face features.

3.3 Face Recognition Methodology

Unique face features are in the form of embedding vectors which can be compared with the other face feature embedding vectors in the form of Euclidean distance. We have used DLIB's CNN-based face recognition module, which is provided with the shape predictor which generates 68 face landmarks. DLIB's face recognition module with the help of 68 face landmarks can identify a picture of a human face by generating a 128-dimensional vector (from picture). Pair of pictures of the same person are close to each other and pair of pictures of different people are far apart in terms of mathematical Euclidean distance through their respective 128-dimensional vectors. The closeness of distance can be set by a threshold τ. Pair of 128-dimensional vectors, for whom the distance is below τ are inferred as belonging to same class (person). Pairs for whom distance is above τ are decided as belonging to different classes. Thus, performing face recognition by extracting 128-dimensional features vectors from face images and checking if their Euclidean distance is small enough is possible.

When setting a distance threshold τ as 0.6, the face recognition model [18] achieves an accuracy of 99.38% over the standard labelled faces in wild (LFW) face recognition benchmark. Accuracy of this model implies that, when given a pair of face pictures, the model will correctly state if the pair is of the same person or if it is from different people. Given an estimate of the distance margin/threshold value τ, face recognition can be done simply by calculating the distances between an input embedding vector and all embedding vectors in a database. The input is assigned the label (identity) of the database entry with the smallest distance if it is less than τ we label it or otherwise if the label is unknown. This procedure can even scale to giant databases as it may be easily parallelized. It additionally supports one-shot learning, as adding solely a single entry of a new identity may be sufficient enough to recognize new examples of that identity. For our experiments, the value τ is used as 0.55.

3.4 Distance Calculation

We are using Euclidean distance to get a match. For that, we have used Python's Numpy library. The face recognition model is trained with contrastive loss L that is minimized when the distance between pair of positive images is less and distance between pair of positive–negative images is more. At a given threshold, all possible embedding vector pairs are classified as either same identity or different identity and compared to the ground truth.

3.5 Improved Face Recognition Using Intelligent Face Tracking

The processing accuracy and the real-time recognition computations are the main problem we are targeting over. The continuous processing of face detection and face recognition without tracker is observed to take much time as compared with the use of tracker. The possible reason is, mmod face detection is GPU intensive and tracker is central processing unit (CPU) intensive. Algorithmically, tracking algorithm does updation in neighbourhood region of it previous coordinates, thus, leaves major region of frame unprocessed. Face detector, on the other hand, needs to process complete video frame/image in order to get all face coordinates. Once the face is detected, we are applying DLIB implemented correlation tracker [10] and recognize the face. We are updating and quality checking the tracker in each frame every time to validate if a face is there inside the box. If not we are deleting the tracker. For each newly detected face, we are creating a new tracker. We calculate the face embeddings of every tracked face and compare with embeddings in database to get minimum distance (with the best-matched embedding).

3.6 Local Minimum and Global Minimum

We fetch $\mathbf{D_M}$ minimum distance for every face and initialize $\mathbf{D_{ML}}^{(q)}$ with it. $\mathbf{D_{ML}}^{(q)}$(distance minimum local for query face embedding $\mathbf{q}$) and $\mathbf{D_{MG}}^{(q)}$(distance minimum global) across tracked faces of same person. Distance minimum local provides the minimum distance between face in current frame and stored faces in records. Distance minimum global provides the minimum distance among all the local minimum distances of tracked face for same person. The distance ($\mathbf{D_{MG}}^{(q)}$)

will only change when the local minimum is less than the global minimum.

$$\text{Update } \mathbf{D}_{\mathbf{MG}}^{(q)} \text{only when } \mathbf{D}_{\mathbf{ML}}^{(q)} < \mathbf{D}_{\mathbf{MG}}^{(q)} \tag{1}$$

Initially, after the feature extraction is completed from the detected faces, the face embedding are compared with the gallery/database stored face features. The distance at initial point will be the same for global minimum as well as local minimum. So far the current distance will be the local minimum. As a result, the label corresponding to $\mathbf{D_{MG}}^{(q)}$ will be chosen as **P**. For the second time, when the current local minimum distance changes, we compare the current local minimum with the global minimum distance. This will also enable to change the label to **R**. If current distance value is more, than the distance global minimum, the global minimum value will remain unchanged. As a result, this will preserve the last label which is **R** in our case. Once the person goes out the camera view, the label corresponding to last updated distance $\mathbf{D_{MG}}^{(q)}$ will be used for all the (tracked) faces of a person. Thus, the use of global minimum distance concept with tracking helped in more accurate results in our experiments (Fig. 2).

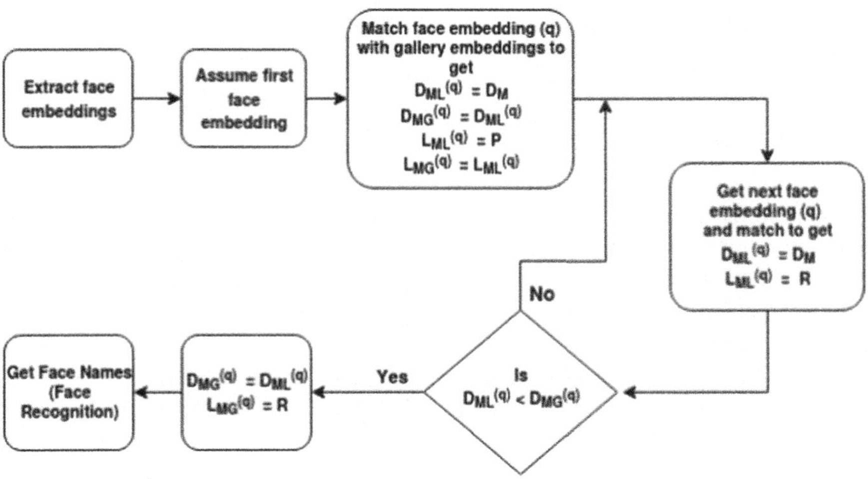

Fig. 2 Flow diagram explaining global and local minimum rule face recognition accuracy improvements. $\mathbf{D_{ML}}^{(q)}$ is distance minimum local for query embedding **q**, $\mathbf{D_M}$ is current distance minimum, $\mathbf{D_{MG}}$ is distance minimum global, $\mathbf{L_{ML}}$ is label minimum local corresponding to $\mathbf{D_{ML}}$, $\mathbf{L_{MG}}$ is label minimum global, **P** is identified person name, **R** is identified person name not same as **P**

4 Experimental and Results

4.1 Experimental Setup

Considering the computational time of the convolutional neural network architecture, we used high performance systems with NVIDIA GPUs.

4.2 Dataset

The dataset used in this paper consists of the people faces with different variations, lightening conditions, poses and occlusions. The ChokePoint dataset [19] is used in the proposed work, which consists of videos of 25 subjects (including male and female). The dataset video has a frame rate of 30 frames per second and the resolution of image is 800 × 600 pixels. In total, the dataset contains 48 video sequences with different indoor as well as daylight environment and 64,204 face images. Faces captured in such scenario will have variations in lightening conditions, sharpness, pose, as well as misalignment.

4.3 Experimental Results

The experimental results in this paper are given as performance of tracking and recognition rate for multiple faces. To show recognition improvements, we have tested the tracker-assisted face recognition algorithm with ChokePoint. The tracking algorithm performance as in processing time when compared to normal detection is faster. To measure the overall system performance, we have used precision, recall and F-score and recognition rate RR. The precision recall and F-score [20] calculations are: Precision is defined as the total number of true positives (T_P) over the summation of true positives and false positives (F_P). Recall is defined as the total number of true positives (T_P) over the summation of true positives and false negatives (F_N). F-score is evaluated as harmonic mean of precision and recall.

$$\text{Precision} = \frac{T_P}{T_P + F_P} \tag{2}$$

$$\text{Recall} = \frac{T_P}{T_P + F_N} \tag{3}$$

$$\text{F-score} = 2 * \frac{\text{Precision} * \text{Recall}}{\text{Precision} + \text{Recall}} \tag{4}$$

The experiment has been performed on 25 people walking across the camera. For every person, we have selected one exemplar face image and added in the database for matching with video faces. The results are given for 25 people separated by top ten results as shown in Tables 1 and 2 for recognition without tracking and with tracking, respectively.

The bounding box is located around the face with the recognized label (name) of the person and confidence of recognition with matched database (see Fig. 3).

Table 1 Result of precision, recall and F-score on ChokePoint dataset without using tracker

S. No.	Class ID	Precision	Recall	F-score
1	3	0.31	1	0.48
2	5	1	0.7	0.82
3	7	1	0.4	0.57
4	9	0.45	0.5	0.48
5	11	0.31	0.5	0.38
6	13	0	0	0
7	16	0.06	0.2	0.09
8	17	1	0.3	0.46
9	23	0.4	0.6	0.48
10	24	1	0.5	0.67
11	Others	0.30	0.29	0.23

Table 2 Result of precision, recall and F-score on ChokePoint dataset with use of tracker

S. No.	Class ID	Precision with tracking	Recall with tracking	F-score
1	3	1	1	1
2	5	1	1	1
3	7	1	1	1
4	9	1	0.8	0.89
5	11	0.67	1	0.8
6	13	0.89	0.8	0.84
7	16	0.67	1	0.8
8	17	1	1	1
9	23	1	1	1
10	24	1	1	1
11	Other	0.57	0.6	0.55

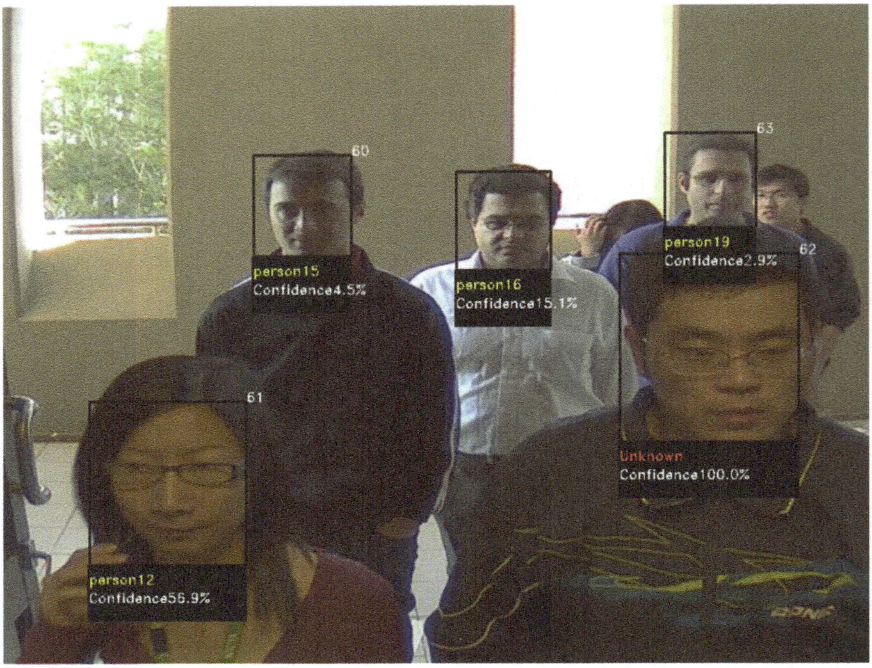

Fig. 3 Output of multi-person recognition

4.4 Recognition Performance

We have calculated the multiple people face recognition rate (RR) using below formula, considering total number of correctly recognized faces with respect to the total testing faces.

$$RR = \frac{\text{Total number of correctly recognized faces}}{\text{Total testing faces}} \times 100 \quad (5)$$

The result for multiple face recognition rates is given in Table 3.

Table 3 Result of algorithm for multiple face recognition with and without tracker

S. No.	Recognition method	Recognition rate (RR %)
1	Recognition without tracker	36
2	Recognition with tracker	74

5 Conclusion

In this paper, we proposed a method for multiple person face recognition when the people are moving in forward direction/walking towards the camera. Our approach used the combination of state of the art face recognition model with fast face tracker. The best of all embeddings of every tracked face is used to label the person name as identified. Best embedding is the embedding which has the best match (among all tracked embeddings) with gallery embeddings. The algorithm has been implemented on powerful NVIDIA GPUs-based system and clearly shows recognition improvements over untracked recognition. This tracker-assisted technique for consistent recognition increases the overall processing accuracy and recognizes multiple faces in real time.

Acknowledgements The authors thank Ministry of Electronics and Information Technology (MeitY), Government of India and Centre for Development of Advanced Computing (CDAC) for providing opportunity and resources to conduct this research.

References

1. Liu W, Wen Y, Yu Z, Li M, Raj B, Le S (2017) Sphereface: deep hypersphere embedding for face recognition. CVPR
2. Schroff F, Kalenichenko D, Philbin J (2015) Facenet: a unified embedding for face recognition and clustering. In Proc, CVPR
3. Romić K, Livada Č, Glavaš A (2016) “Single and multi-person face recognition using the enhanced eigen faces method” preliminary communication. Int J Electr Comput Eng Syst 7
4. Viola P, Jones MJ (2004) Robust real-time face detection. Int J Comput Vision 57(2):137–154
5. Wu Y, Cheng Z, Huang B, Chen Y, Zhu X, Wang W (2019) Foxnet: a multi-face alignment method. In: IEEE international conference on image processing (ICIP)
6. Zhang K, Zhang Z, Li Z, Qiao Y (2016) Joint face detection and alignment using multi-task cascaded convolutional networks. arXiv preprint 2016
7. King DE (2015) Max-margin object detection. CoRR
8. Yang J, Zhang D, Frangi AF, Yang J-Y (2004) Two dimensional PCA: a new approach to appearance-based face representation and recognition. IEEE Trans Pattern Anal Mach Intell 26(1):131–137
9. https://blog.dlib.net/2017/02/high-quality-face-recognition-with-deep.html
10. Hager DG, Khan F, Felsberg M (2014) Accurate scale estimation for robust visual tracking. In: BMVCM
11. Saypadith S, Aramvith S (2018) Real-time multiple face recognition using deep learning on embedded GPU system. In: Annual summit and conference 2018 Hawaii APSIPA proceedings, 12–15 Nov 2018
12. Lin CC, Hung Y (2018) A prior-less method for multi-face tracking in unconstrained videos. In: The IEEE conference on computer vision and pattern recognition (CVPR), pp 538–547
13. Narayan N, Sankaran N, Arpit D, Dantu K, Setlur S, Govindaraju V (2017) Person re-identification for improved multi-person multi-camera tracking by continuous entity association. In: IEEE conference on computer vision and pattern recognition workshops (CVPRW)
14. Juneja K, Gill NS (2015) A hybrid mathematical model for face localization over multi-person images and videos. In: 4th international conference on reliability, infocomtechnologies and optimization (ICRITO) (trends and future directions)

15. Mantoro T, Ayu MA, Suhendi (2018) Multi-faces recognition process using haar cascades and eigenface methods. In: 6th international conference on multimedia computing and systems (ICMCS)
16. Mao Y, Li H, Yin Z (2014) Who missed the class?—unifying multi-face detection, tracking and recognition in videos. In: IEEE international conference on multimedia and expo (ICME)
17. Audigier R, Dhome Y, Lerasle F (2015) Online multi-person tracking based on global sparse collaborative representations. In: Low fagot-bouquet, IEEE international conference on image processing (ICIP)
18. King DE (2009) Dlib-ml: a machine learning toolkit . Journal of Machine Learning Research 10:1755–1758
19. Wong Y, Chen S, Mau S, Sanderson C, Lovell BC (2011) Patch-based probabilistic image quality assessment for face selection and improved video-based face recognition IEEE biometrics. In: June workshop, computer vision and pattern recognition (CVPR) workshops, IEEE, pp 81–88
20. Ae B, Marco B, Julia E, Matthias K, Loos HS, Merkel M, Niem W (2008) A Review and comparison of measures for automatic video surveillance systems. EURASIP J Image Video Process 2008(1):1

Decision Support System on the Need for Veterinary Control of Passing Livestock and Farm Produce

Alexander Golubenkov, Dmitry Alexandrov, Sanjay Misra, Olusola Abayomi-Alli, Marcelo Leon, and Ravin Ahuja

Abstract The results of the development of a decision support system (DSS) that can be used by the customs. This system helps to inspectors at the border veterinary checkpoints (VCPR) to make decisions on the need for veterinary control of passing livestock and farm produce. The system makes the decisions by analyzing the known information about cargo with the use of logistic regression model. This module was developed using the Python programming language, mathematical libraries for data analysis, and framework Flask for creating Web applications. The paper also presents the results of a correlation analysis of factors with an identification of significant features, the results of testing the module, and the conclusions about its effectiveness.

Keywords Decision support system (DSS) · Logistic regression · Customs control

A. Golubenkov
Vladimir State University named after Alexander and Nikolay Stoletovs (VlSU), Vladimir, Russian Federation
e-mail: golubenkov1@yandex.ru

D. Alexandrov
National Research University Higher School of Economics (NRU HSE), Moscow, Russian Federation
e-mail: dvalexandrov@hse.ru

S. Misra (✉) · O. Abayomi-Alli
Covenant University, Ota, Ogun State, Nigeria
e-mail: sanjay.misra@covenantuniversity.edu.ng

O. Abayomi-Alli
e-mail: olusola.abayomi-alli@covenantuniversity.edu.ng

M. Leon
Universidad Nacional de Loja, Loja, Ecuador
e-mail: marceloleon11@hotmail.com

R. Ahuja
Shri Viskarma Skill University, Gurgaon, India
e-mail: ravinahujadce@gmail.com

P. K. Singh et al. (eds.), *Evolving Technologies for Computing, Communication and Smart World*, Lecture Notes in Electrical Engineering 694,
https://doi.org/10.1007/978-981-15-7804-5_39

1 Introduction

The movement of containers through the worldwide networks of container terminals and seaport is quite alarming [1] making the role of custom officers is very crucial in ensuring the proper decisions are been made. Customs inspectors who are working at border veterinary checkpoints of the country have to make decisions on carrying out the veterinary examination of goods coming from abroad and containing products of animal origin. Due to the large volume of cargo passing through the customs, the controlling bodies do not have the ability to inspect all shipments. In addition, the operation of a container terminal is complex and thereby involves a wide variety of real-time decision [2].

In this regard, there is a need to identify and inspect the most risky goods. Inspectors have to decide upon whether to carry out the examination of goods or not, immediately, in order to prevent congestion of cargo at the border.

At present, the decision support system developed at participation of the authors of this article based on the fuzzy logic apparatus is used to solve this problem [3, 4]. The choice of this apparatus is conditioned by the need for the inspector to verbally formulate the reason for the delay of cargo, and not all approaches provide a detailed explanation of the decision, such as neural networks. A risk factor is used for the determination of the veterinary danger levels of countries and companies. It is a numerical value that is assigned to each registered company, country, and type of product in the system. Their values can be changed by responsible employees of Rosselkhoznadzor in accordance with the current situation in the respective regions. Thus, decisions made can be explained, for example, by applying the rule: 'IF the risk factor of the company is 86%, we will send the goods for veterinary inspection'. However, this approach has the following disadvantages: a small number of deducible rules of fuzzy logic, the solution does not cover all factors available for accounting, there is no control over the accuracy of the proposed solutions, there is no accounting for statistical information with the results of inspections.

In this regard, it was necessary to improve the present DSS for imported goods, which would take into account these disadvantages. As a result the following system requirements are formulated:

1. Provide the possibility of entering statistical information about companies, countries, and the goods themselves. In addition, it is necessary to consider the number and reasons for inspections of such goods.
2. Provide the possibility of entering information of new goods into the system on which the decision has not yet been taken.
3. Provide the possibility of receiving the explanation of the made decision from the system. This problem is common to all decision support systems.
4. Provide the possibility of training the system based on examining the consequences of earlier made decisions (e.g., a cargo that has not been contaminated can be checked, or a cargo that later appeared to have infringed custom rules has been missed).

The rest of the paper is divided into sections where: Sect. 2 gave a detailed literature review and Sect. 3 discussed the proposed methodology. The system implementation and testing are discussed in Sect. 4 and the conclusion and future recommendation are summarized in Sect. 5.

2 Literature Review

The Several efforts have been proposed by researchers in improving decision support systems for ensuring ease inspections for custom officers in the area of veterinary controls or other related areas.

Author in [1] designed and integrated a security procedure for container inspection. The proposed approach was based on combining modeling and simulation of interactive tools for scenario testing with statistical tools. The simulation result shows that the effect of quadratic surface was better than the linear model. In addition, the integration of container inspection process gave a trade-off between technologies advances.

Murty et al. [2] proposed a decision support system for operations in a container terminal while Burgemeestre et al. [5] presented a comparative analysis of requirement engineering based on mental models of custom officials against the model of the developed tools. The paper concludes that mapping overlaps is essential in a regulatory context when developing models for decision support.

Permanasari et al. [6] introduced a decision support system for livestock disease diagnosis assistance. An Info 3-link decision support system was proposed which is divided into the information, the tester, and the decision maker. The study concluded that the proposed system was able to reduce time consumption.

Ramadhan et al. [7] proposed a geographical information system-based decision support system for e-livestock in Indonesia. The study utilized five modules and adopted hermeneutic circle principle for interpreting some specific sections to general knowledge section thereby helping government officials during decision making relating to livestock.

Antonov et al. [8] analyzed the performance of decision support system for livestock enterprise. The experimental result showed their outlined parameters are important in enhancing better decision for livestock.

Mnale et al. [9] presented the impacts of information system development of a mobile decision support system for monitoring real-time container terminal operations. The study adopted a design thinking approach for identifying possible requirements and in addition generated several prototypes. Authors claimed that the proposed method was able to improve the body of knowledge for decision support researchers and practitioners in information system development.

There are several other works in which decision support and expert systems are developed for solving problems in various application areas like health [10–13] and education [14].

3 Design Methodology

This section gives a detailed description of the proposed system design and its specification. To solve the problem of DSS, statistical data analysis method was proposed. A feature of such methods is their complexity due to the variety of forms of statistical regularity, as well as the complexity of the process of statistical analysis.

Since the data on the work of inspectors of veterinary checkpoints contain decisions made by them on incoming cargo, they can be used to create a model taught with the teacher. The decision task on a cargo is to determine which category each particular case belongs to base on its characteristics. A certain decision corresponds to each group. Since there are several such categories, there is a multiclass classification tasks. To solve such tasks, there are a number of methods [15]. When selecting the appropriate method, the following characteristics should be considered: accuracy, training time, linearity, number of influencing factors, and number of functions.

To solve multiclass classification tasks, the following methods should be used: data analysis based on the 'random decision forest,' logistic regression, neural networks, and the 'one versus all' method [16].

Considering performance accuracy, priority is placed on random decision forest. However, it should be noted that each decision must be justified by submitting to the inspector an appropriate report. Only logistic regression meets this requirement. This method of constructing a linear classifier allows estimate posteriori probabilities of belonging to classes. That is why this method has been chosen to create a decision support model.

In order to explain to the inspector, the reason for making a decision, we can rely on the weighting factors that a model arranges for each input of parameters at training. If we look at examples of creating systems of machine learning and processing large amounts of data, we will see that most of them are written in *R* or Python language. Python language is chosen for solving such tasks mainly because of the simplicity of the process of writing programs and the availability of fast mathematical libraries that allow you to create models and store large amounts of data in random access memory (RAM) during the development [17, 18].

A library for Python—NumPy was used for loading, processing, and storing data in RAM. It expands the language capabilities for working with arrays, adds the support for working with large multidimensional matrices, and also includes a number of fast high-level mathematical functions for operations with these arrays [19]. Sklearn library was used for teaching the model. Also, Jupyter Notebook was used while working on the mathematical model of the decision support module which is an interactive medium for creation of informative analytical reports.

3.1 Factor Model

During the development of the model, at first, a set of factors for analysis was created. The general form of the function can be described as follows:

$$\begin{cases} \text{risc}_{\text{country}} \\ \text{risc}_{\text{enterprise}} \\ \text{risc}_{\text{product}} \\ \text{allConsigment}_{\text{product}} \\ \text{checkedConsigment}_{\text{product}} \\ \text{inspectorDesicion} \\ \text{isVetChecked} \\ \text{isDenied} \\ \text{product} \\ \text{transportType} \\ \text{packageType} \\ \text{permissionType} \\ \text{enterpriseStatus} \end{cases}$$

where, risk factor for a country of origin,

$$0 \leq \text{risc}_{\text{country}} \leq 100$$

$\text{risc}_{\text{enterprise}}$—risk factor for manufacturing enterprise of products,

$$0 \leq \text{risc}_{\text{enterprise}} \leq 100$$

$\text{risc}_{\text{product}}$—risk factor for this type of product (e.g: a particular type of meat),

$$0 \leq \text{risc}_{\text{product}} \leq 100$$

$0 \leq \text{risc}_{\text{product}} \leq 100$—number of batches of products of a certain type that are currently received at the border post

$\text{checkedConsigment}_{\text{product}}$—number of batches of products of a certain type that are sent for veterinary inspection

inspectorDesicion—decision taken on the cargo by the inspector,

$$\begin{cases} \text{true, send for inspection} \\ \text{false, let it pass without the inspection} \\ \text{isVetChecked} - \text{decision of the system to send for veterinary inspection} \end{cases}$$

$$\begin{cases} \text{true, send for inspection} \\ \text{false, let it pass without the inspection} \end{cases}$$

isDenied—whether the cargo was detained as a result of veterinary inspection,

- true, cargo is detained
- false, cargo has no violation
- product—transported products

transportType—Type of vehicle by which cargo has been delivered.

- By sea
- By rail
  - cistern
  - container
- By air
- By car
  - van
  - Refrigerator

packageType – Type of product package,

- Barrel
- Pallet
- Bunch

permissionType – Permission type,

- Import
- Export
- Transit

enterpriseStatus – a category to which this enterprise is assigned to in the system.

4 System Implementation and Testing

This section described the implementation of the proposed system with discussion of the training and testing model.

4.1 Training of the Proposed System

It is supposed that after the training, the system will place weights for each factor. For next invokes, this system will consider the significant factors. Building the model of logistic regression and its training with the teacher requires a sufficiently large volume of the test set. All information of cargo arrivals at veterinary checkpoints is stored in the database (DB) of the system. The required data for 2016 was uploaded for working on a model using this database. The volume of set was 103,825 lines.

Before starting logical regression model training, it is necessary to prepare the data, in particular, to get rid of passes values in the columns, as this method is very sensitive to uncertainties. For this purpose, passes values in the columns of 'enterprises status' and 'package type' were replaced by zero. In order to increase the accuracy of the logistic regression model, it is required to put the data to the required form, indicating the type of columns. Also it is necessary to make sure that there is no correlation between factors.

The DataFrame class from the Pandas package for Python where all the data is stored represents the necessary tools. The diagram (see Fig. 1) shows a table, in which at the intersection of the lines and columns of factors is the corresponding correlation value.

The diagram demonstrates a pair of factors 'all Consignment products' and 'checked Consignment products,' 'inspector Decision' and 'isVetChecked' have a high correlation. This pattern is explained by the fact that the volume of tested batches directly depends on the total number of batches of this type of product. Also, the output value strongly affects whether the consignment was previously sent for veterinary inspection. In this case, it is necessary to rely solely on the results of veterinary checking. However, this characteristic is indicated for a small percentage of cargo.

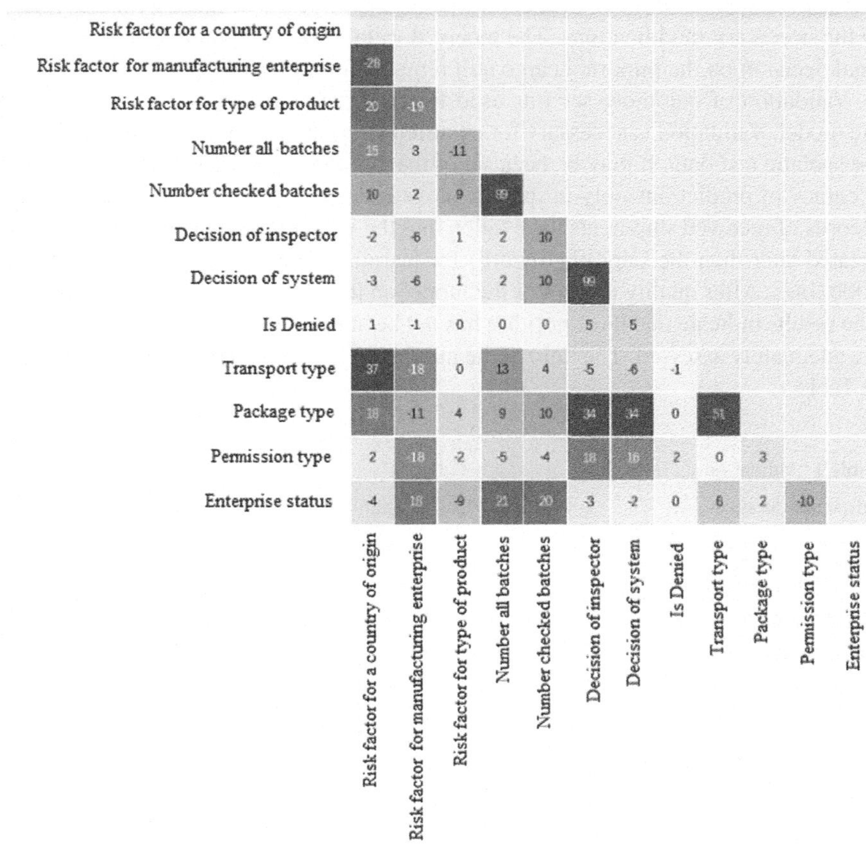

Fig. 1 Correlation model of factors

Despite the existence of a relationship between characteristics, they should not be deleted, since they have a significant effect on the decision being made. In addition, the subsequent selection of characteristics will get rid of some insignificant factors.

When using logistic regression as a learning algorithm, it is reasonable to use the following modifications: to add a new column for each categorical factor and put the ones in those records that belong to this category. The remaining lines will get zero value. As practice shows the accuracy of the model is significantly improved [20].

4.2 Testing the Proposed Model

The decision support model was tested in the process of its development. After applying various data modification algorithms, the percentage of discrepancy between the actual values of the predicted factors and the values counted by the model was calculated; as a result, it was determined that the model is able to predict the inspector decision with an accuracy of 95.1% on test data which was subjected to the necessary modifications. The received value is an acceptable result, since the final decision on the imported cargo still remains for the inspector.

Validation of one more set was used for final determination of the accuracy of the model. Validation is necessary for excluding the case when the model is adjusted for specific test data. It may be because of the fact that selected factors increase the accuracy of predictions only in specific cases occurring in the data for the test. The records of received shipments have fallen into the validation set. These records were created later than the sampling records for learning. The size of validation set was 1000 lines. After quality control of the model on this data, the accuracy was 95.1%. The results indicate that the algorithm has not been retrained and the selected factors are adequately assessed. The ratio of the number of real and predicted values is given in Table 1.

Table 1 Validation set test result

Inspector decision	Quantity in sampling	Predicted quantity
Import is approved	816	810
Export is approved	12	10
Transit is approved	14	13
Import is prohibited	4	3
Export is prohibited	0	1
Transit is prohibited	0	0
Send for veterinary inspection	152	161
Inspection is not possible for technical reasons	0	0
Cargo was detained	2	2

5 Conclusion and Future Work

The result of the developed decision support module fully meets the stated requirements. Further work for improving DSS will focus on developing methods for preliminary data processing and searching for opportunities to increase the accuracy of the model. It should be noted that the problem of correlation of factors was not solved during the process of developing the current algorithm. Eliminating the dependencies between input parameters will reduce their number, simplify the model, and increase its accuracy.

Acknowledgements We would like to thank all the colleagues at the Chair of Information Systems and Software Engineering of Vladimir State University named after Alexander and Nikolay Stoletovs for their feedback and useful recommendations that contributed to bringing this paper to its final form. We also acknowledge the support and sponsorship provided by Covenant University through the Center for Research, Innovation and Discovery (CUCRID).

References

1. Longo F (2010) Design and integration of the containers inspection activities in the container terminal operations. Int J Prod Econom 125(2):272–283
2. Murty KG, Liu J, Wan Y, Linn R (2005) A decision support system for operations in a container terminal. Decis Support Syst 39(3):309–332
3. Golubenkov A D (2016) Razrabotka sistemy podderzhki prinyatiya reshenij dlya sistemy tamozhennogo kontrolya [Development of a decision support system for the customs control system]. In Yanishevskaya A G (eds.). Informacionnye tekhnologii v nauke i proizvodstve: materialy III Vseros.molodezh.nauch.-tekhn. konf (Omsk, 8-9 fevr. 2016 g.) [Information Technologies in Science and Production: Materials of the Third All-Russian Youth Scientific and Technical Conference]. M-vo obrazovaniya i nauki Ros. Federacii, M-vo obrazovniya Omskoj obl., OmGTU (ed.). Omsk: Publ. OmGTU, pp 9–13
4. Golubenkov A D, Shevchenko D V (2017) Sravnenie podhodov k realizacii sistemy podderzhki prinyatiya reshenij dlya sistemy tamozhennogo kontrolya [Comparison of Approaches to Implementing the Decision Support System for the Customs Control System]. In Yanishevskaya A G (eds). Informacionnye tekhnologii v naukec i proizvodstve: materialy IV Vseros. molodezh. nauch.-tekhn. konf. (Omsk, 8-9 fevr. 2017 g.) [Information Technologies in Science and Production: Materials of the Forth All-Russian Youth Scientific and Technical Conference]. M-vo obrazovaniya i nauki Ros.Federacii, M-vo obrazovniya Omskoj obl., OmGTU (ed.). Omsk: Publ. OmGTU, pp 7–11
5. Burgemeestre B, Liu J, Hulstijn J, Tan Y-H (2009) Early requirements engineering for e-customs decision support: assessing overlap in mental models. In: Dagstuhl seminar proceedings. Schloss Dagstuhl-Leibniz-Zentrum für Informatik
6. Permanasari AE, Najib W (2012) Infolab 3-link as a proposed decision support system for assisting livestock disease diagnose. In: 2012 International conference on computer and information science (ICCIS), vol 1, IEEE, pp 176–179
7. Ramadhan A, Sensuse DI, Pratama MO, Ayumi V, Arymurthy AM (2014) GIS-based DSS in e-Livestock Indonesia. In: 2014 International conference on advanced computer science and information systems (ICACSIS). IEEE, pp 84–89

8. Antonov L, Privezentsev D, Orlov A (2015) Identification of the main parameters for animal performance assessment in decision support system of a livestock enterprise. In: 2015 International conference stability and control processes in memory of VI Zubov (SCP). IEEE, pp 508–509
9. Mnale F, Salama S, Park J, Eltawil AB (2017) MobDesctop: a mobile decision support application for monitoring real-time container terminals operations
10. Iheme P, Omoregbe N, Misra S, Ayeni F, Adeloye D (2017) A decision support system for pediatric diagnosis. In: Innovation and interdisciplinary solutions for underserved areas. Springer, Cham, pp 177–185
11. Thompson T, Sowunmi O, Misra S, Fernandez-Sanz L, Crawford B, Soto R (2017) An expert system for the diagnosis of sexually transmitted diseases–ESSTD. J Intell Fuzzy Syst 33(4):2007–2017
12. Azeez NA, Towolawi T, Van der Vyver C, Misra S, Adewumi A, Damaševičius R, Ahuja R (2018) A fuzzy expert system for diagnosing and analyzing human diseases. In: International conference on innovations in bio-inspired computing and applications, Springer, Cham, pp 474–484
13. Iheme P, Omoregbe NA, Misra S, Adeloye D, Adewumi A (2017) Mobile-bayesian diagnostic system for childhood infectious diseases. In: ICADIWT, pp 109–118
14. Sowunmi OY, Misra S, Omoregbe N, Damasevicius R, Maskeliūnas R (2017) A semantic web-based framework for information retrieval in e-learning systems. In: International conference on recent developments in science, engineering and technology, Springer, Singapore, pp 96–106
15. Knyaz' D V (2014) Analiz osnovnyh algoritmov klasterizacii mnogomernyh dannyh. Lambert Academic Publishing, p 64. ISBN 978-3659636493
16. Kritskij OL (2012) Mnogomernye statisticheskie metody. Lambert Academic Publishing, p 168. ISBN 978-3848486755
17. Idris I (2014) Python data analysis. Packt Publishing, p 348. ISBN 9781783553358
18. Python 3.5.10 documentation [Elektronnyi resurs]. https://docs.python.org/3.5/ (Accessed 07.08.2020)
19. NumPy: The fundamental package for scientific computing with Python [Elektronnyi resurs]. http://www.numpy.org/ (Accessed 07.08.2020)
20. Flach P (2012) Machine learning: the art and science of algorithms that make sense of data. Cambridge University Press, p 410

A Home Automation System Based on Bluetooth Technology Using an Android Smartphone

Elvis Ngerem, Sanjay Misra, Jonathan Oluranti, Hernán Castillo-Beltran, Ravin Ahuja, and Robertas Damasevicius

Abstract In this work, a simple home automation system is to be discussed and designed. It is a concept where devices can be controlled remotely from a central hub and also allow these devices share data. It is proposed to design and build a system for home automation using Bluetooth technology. The reason Bluetooth is being used is because, it is an easily accessible network, thereby making it cheap. The implementation process will involve a software design for the android smartphone and a circuit design for the operation of the devices. An Arduino microcontroller will be used to control the circuit, and an android smartphone will be used as the control hub. Both will be linked by a Bluetooth module. At the end of the exercise, an active system for the control of basic home appliances will be developed, having an easy to use UI "user interface." In the introduction, the background of the study will be discussed.

Keywords Home automation · Bluetooth module · Arduino microcontroller

---

E. Ngerem · S. Misra (✉) · J. Oluranti
Covenant University, Ota, Ogun State, Nigeria
e-mail: Sanjay.misra@covenantuniversity.edu.ng

E. Ngerem
e-mail: elvisngerem@ymail.com

J. Oluranti
e-mail: jonathan.oluranti@covenantuniversity.edu.ng

H. Castillo-Beltran
Universidad Internacional del Ecuador, Quito, Ecuador
e-mail: hernancastillobeltran@gmail.com

R. Ahuja
Shri Vishwakarma Skill University, Gurgaon, India
e-mail: ravinahujadce@gmail.com

R. Damasevicius
Kanus University of Technology, Kaunas, Lithuania
e-mail: robertas.damasevicius@ktu.lt

P. K. Singh et al. (eds.), *Evolving Technologies for Computing, Communication and Smart World*, Lecture Notes in Electrical Engineering 694,
https://doi.org/10.1007/978-981-15-7804-5_40

1 Introduction

Technology never stops improving. To be able to create a product by the usage of modern-day technology in order to make an impact in the lives of people is a huge step in the right direction [1]. Here, the layout/design and implementation of an efficient, cheap, and flexible home automation system will be discussed [2].

The world is embracing the wireless lifestyle, from wireless data transfer to Bluetooth headphones and speakers, to even wireless charging. We are moving to an era where everything can be operated remotely and without the use of cables or cords. In turn, this tends to relief people from all "entanglements" caused by wired systems and devices [1]. This technological advancement has influenced the automation and control of devices by simply connecting a smartphone or a tablet to that device wirelessly [3].

The whole idea of this home automation system is coined out of the concept of "Internet of Things" [4]. It is a network where devices can be interconnected to a control hub via the aid of software, sensors, and other connecting aids which help them interact in the system and exchange data. The beauty of this is that these devices can be your everyday devices like vehicles, smart doors, etc., basically any device that has a sensing unit embedded in its configuration. Simply put, it is connecting regular/everyday devices to the Internet.

Based on the fact that people tend to use their smartphones very often, it makes it even more convenient to create a system which uses just a smartphone to control home devices. Mobile devices can use GSM, Wi-Fi, ZIGBEE, and BLUETOOTH technologies [5], but in this project we will be using the Bluetooth because of its easy usability [5]. Another reason is because of the fact that it is an offline based network.

Android operating system (OS) is a versatile one which supports easy development of mobile apps using java script [6]. It also provides the liberty of developing an application to control the operations of the circuit. Most smartphones are already equipped with wireless connectivity features of which Bluetooth happens to be one of them [7]. This gives a seamless connection between the smartphone and the microcontroller to be used. The major reason the Bluetooth is used in this work is for the fact that it is a feature which is widely available in devices. Also, it has a range of 10 m to 20 m which means it can only be used within that range, and it also has a speed which could reach 3Mbps though that is solely dependent on the class of the Bluetooth device [8].

The literature review will give an insight on related works and the issues encountered. In the third section, we will discuss the design and implementation of the software. The last two sections will focus on the results obtained, recommendations, and conclusion drawn from the work.

2 Related Works

In home automation systems, there are two (2) ways of implementation. It could be done online or offline. Online, by making use of Wi-Fi or an Internet service and offline by the use of Bluetooth, infrared, etc. Both ways of implementation have its advantages and disadvantages. One of the major disadvantages of online implementation is Internet network failures. Others will be addressed further in the project depending on the setup utilized.

There are basically three classification of home automation or device switching: manual, semi-automatic, or fully automatic. In manual, physical contact has to be made by human using a switch. In semi-switching, human intervention is partial like using remote control, GSM message control, voice control, etc. Whereas, the fully automatic system device switching is done on pre-defined conditions. In this project, the semi-automatic switch is utilized having a programmable microcontroller and a human voice as the input for its control.

Several related works have been carried with the aim to achieve a fully functional system where home appliances could be controlled remotely. One of them is that done by Sikandar where a GSM-based approach was used [9]. A GSM modem is incorporated in the control system, this modem is actually what gives commands to the system for its operation.

The drawbacks faced in this are the absence of graphical user interface "GUI." This tends to make the control of such system cumbersome because the users have to always remember certain codes for specific actions [10]. Failures due to the operators of the network also causes issue which causes delays in messages, this makes it very unsuitable for distant usage and monitoring in real time [4].

In Yang et al. worked [11], the system used was ZIGBEE and 802.15.4 protocols in a chip (CC2430). Be that as it may, these authors centered just on the results obtained from the communication between the gateway and organizer, the end gadget and the sensors. The article does not give results for the control of the sensors and UI [12].

Piyare and Tazil proposed a home automation system by using Bluetooth and Arduino device. The Arduino and Bluetooth module was controlled using a Symbian OS cell phone application [1]. The issues faced by this method are the fact that Symbian OS phones are only compactible with python language scripts and doesn't work with java applications. These days, most smartphone applications are based on java script [5].

In the works of Suryadevara and Mukhopadhyay [13], a system utilizing a home monitoring framework can give improved well-being to the elderly. The methodology is based on the execution of sensing units for various variables, for example, humidity, temperature, and light. The monitoring process is all well characterized; although, the security features are not well defined or considered. Ransing and Rajput [14] built a brilliant home for the aged; it was based on a remote sensor. It used LabVIEW UI to monitor a few parameters, and it is also responsible for sending of a SMS message in the event of danger. In any case, just the temperature is displayed in the interface.

With the popularity of Internet, people now have routers, Wi-Fi, mobile hotspot, and other access points which aid them connect appliances to these networks and control them remotely [15]. This led to the development of Wi-Fi-based home automation which uses wireless local area network (WLAN) to manage the devices connected to the network [16, 17].

The issue with this method is the complexity involved in it remote access operation [18]. Similarly local web servers have been used to account for that but yet another drawback arises in high cost of the high-end computers, increased installation cost, large consumption energy, and space due to size [19]. Another major drawback is the interface applications which these servers run are very rigid, difficult to upgrade, and scale to support future demands [20].

What most of the various modes of implementation methods have in common is that they all require Internet connection, which is very good but as a result of that makes all of them prone to Internet network failures [1, 21]. A slight fluctuation in the Internet connection may cause a bridge in the data transfer between the devices. For that reason, this project is offline based using Bluetooth. This would not have a connection issue, and it is readily available in all android devices.

Bluetooth home automation is significant because it can be used for real-time purposes, and for the fact that android devices have display screens [3], the entire operation can be monitored visually [22]. Bluetooth also has its drawbacks of which the most obvious it the range which is just 10–20 m depending on the Bluetooth version in use and the devices too. So this makes it more suitable for close or relatively close usage [23].

Various configurations have been used for Bluetooth home automation [24] in the past using Arduino, but this project will also be using voice commands for it execution. The Arduino UNO will be programmed using Arduino IDE software [25].

Apart from above works, one can find several other android and mobile applications [26–30] for various purposes which are available in the literature.

3 Methodology and Design

Here, the hardware design and software layout will be discussed in detail. There are two major processes that will be done to bring the desired system to life [31]. These are the software and hardware implementation [32]. The hardware section comprises of three fundamental equipment Android smartphone, Arduino microcontroller, and Bluetooth module (hC-05). The microcontroller is programmed with the aid of Arduino (IDE). A mobile app which controls the operations of the Bluetooth device. The following figures and charts give a clear illustration of the operation of the circuit. The functional block diagram is given below showing the basic function devices.

For the hardware execution of the block diagram in Fig. 1, the following are required:

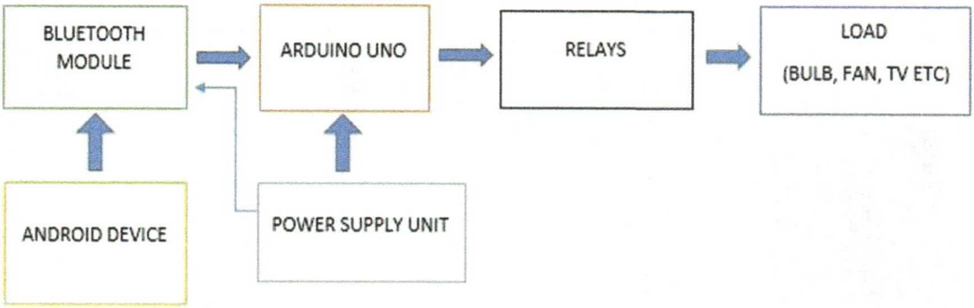

Fig. 1 Functional block diagram

- An Android device (Bluetooth enabled).
- Bluetooth module (HC-05).
- A microcontroller (Arduino UNO).
- 10 and 20 KΩ Resistors. (one each)
- 1 KΩ Resistor ×4.
- Diodes (1N4007 ×4).
- Bread board for testing.
- Jump cables (connecting cables).
- 12 V Power Supply.
- A switching device (12 v SSR Relay ×4).
- Load (LED bulb in this case).

The three most important components in this project are the microcontroller (Arduino UNO), Bluetooth module (HC-05), and the switching device.

Arduino is one of the most used microcontroller, and it could be used for a wide range of applications with the "UNO" being one of the most popular one. It is modeled from MicrochipATmega328P. It is has sets of pins which give room for both analog and digital input and output depending on the job required to be done. This also makes it possible for interfacing with various circuits.

The HC-05 is used as the Bluetooth device. This is provides the connectivity between the smartphone and the entire circuit. It connected to the microcontroller and powered a 5 V input on its VCC terminal and the grounded. Its connection can be seen in Fig. 2 "circuit diagram." Although it is limited to function within a range of 10 m [33]. For the sake of this project, the preferred switching device is a steady-state relay (SSR) [34]. It is used to control the load, switching it OFF and ON based on the signal it receives from the microcontroller [35].

3.1 Working Principle

A power supply unit of 12 v feeds the microcontroller which in turn, supplies a 5 v signal to power the Bluetooth module. The Bluetooth module acts as the link between

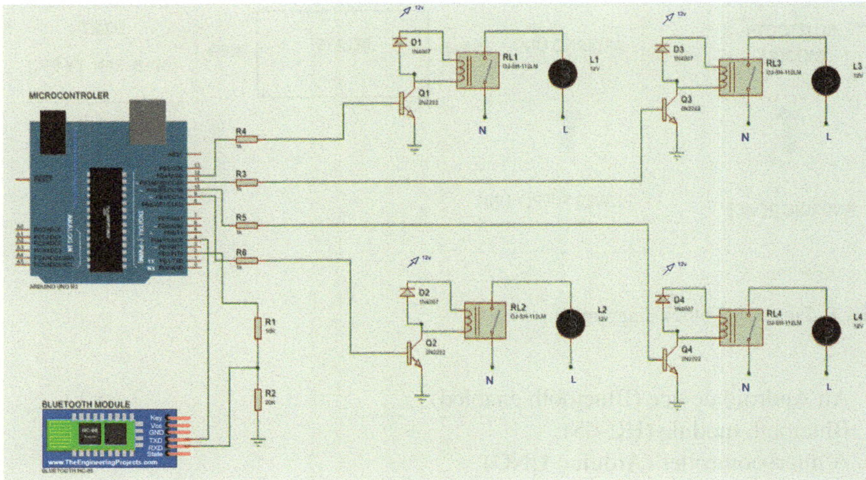

Fig. 2 Circuit diagram

the mobile device and Arduino UNO. The "AMR_voice" mobile app is installed on the smartphone which has a user interface for interaction with the microcontroller. The Arduino operates in line with the program which has been preloaded in it. The Arduino IDE software is used for the programming of the microcontroller. The program is called a sketch. Based on this program, the relay is controlled to make the appropriate switching to control the load.

3.2 *Software Requirements*

- Arduino IDE software: This software is used to program the microcontroller (Arduino device). The program is called a sketch which is uploaded into the device using a USB cord. This is what tells the microcontroller what to do when a given command is sent.
- Proteus eight professional software: This is the software where the circuit is designed and simulated to ensure it functions properly. The system is virtually tested here before being implemented in the real world. The advantage this gives is the fact that problems with the circuit can be detected and corrected before installation. This in turn saves cost that may arise if one of the components happens to get destroyed while testing.

The flowchart below shows the system operation, how the circuit works. The first step is to turn on the power supply, and then establish a connection between the Bluetooth module and the smartphone. After which, the desired commands are sent from the smartphone for control of the circuit. Based on the command sent, the microcontroller acts as programmed to monitor or control the home devices after which the status of the devices is displayed on the smartphone (Fig. 3).

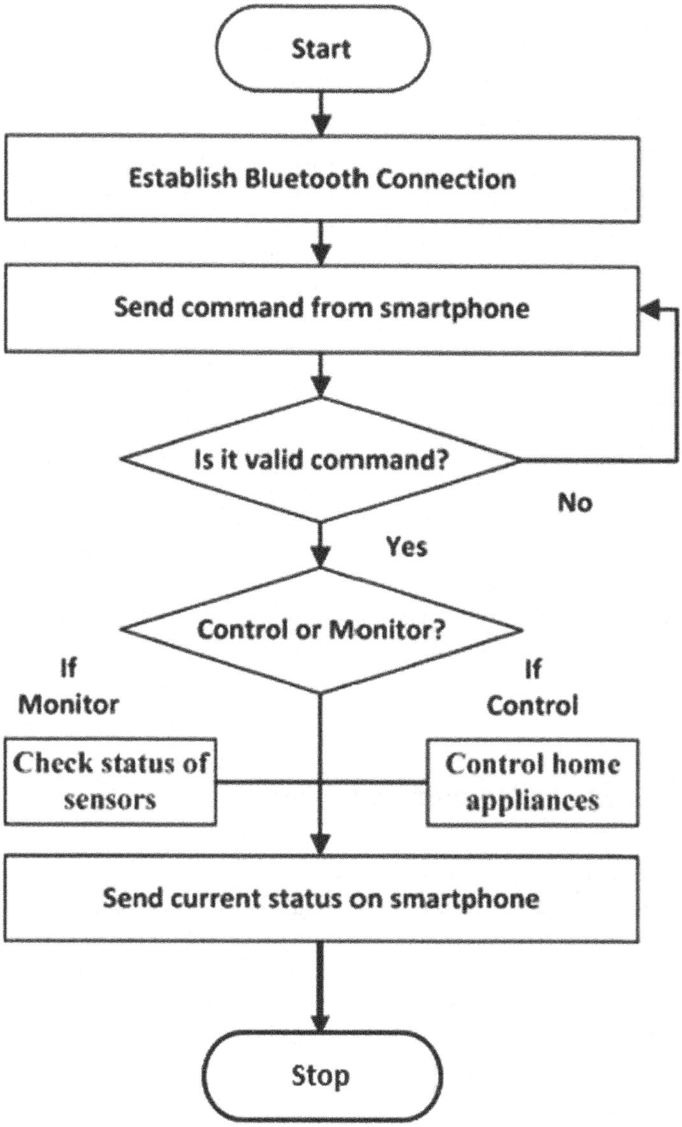

Fig. 3 System flowchart

Table 1 Voice commands, touch commands, and the responses gotten

S/N	Voice command	Touch command	Action performed	Indicators
1	Fan ON	1 ON	The fan comes ON	Red ON
2	Fan OFF	1 OFF	The fan goes OFF	Red OFF
3	Tube ON	2 ON	The lamp comes ON	Blue ON
4	Tube OFF	2 OFF	The lamp goes OFF	Blue OFF
5	TV ON	3 ON	The TV comes ON	Green ON
6	TV OFF	3 OFF	The TV goes OFF	Green OFF
7	Music ON	4 ON	The sound system comes ON	Yellow ON
8	Music OFF	4 OFF	The sound system goes OFF	Yellow OFF

4 Results

The software app of choice in this project is AMR_Voice. It provides both a voice control interface and a touch control interface for interaction with the Bluetooth device. Based on the design, we had four loads to be controlled (4 LED bulbs). Although this could also be four different kinds of loads depending on the design and purpose of usage. For the sake of this project, four LED lamps were used to represent four home appliances. Red, Blue, Green, and yellow were used as indicators to show that the circuit was fully functional.

Table 1 is drawn from the tests carried out from the different commands that were input from the mobile device, and it shows the operation of the circuit when it was tested.

5 Conclusion and Recommendations

The home automation system was fully functional, with the voice and touch commands working perfectly during tests. The circuit was also tested with Proteus and was found to be working fine. When the voice command is being used, it is advisable to be done in a quiet place because noise tends to make it difficult for the commands to be detected. This system works optimally within the range of 10 m, anything more may cause distortion in the signal this is due to the range of Bluetooth.

The aim if the project was achieved, because it was able to provide a convenient and low-cost system for control of devices remotely. It is engineered for convenience and can be used by anyone. In future modifications for the project, more devices could be added, and also, the voice commands could be in other languages other than English. Also, in future works, the system could be implemented on other platforms like iOS and windows device to broaden the scope of the project.

References

1. Piyare R, Tazil M (2011) Bluetooth based home automation system using cell phone. In: 2011 IEEE 15th International symposium on consumer electronics, Samabula, Suva, Fiji, pp 150–152
2. Sushant K, Solanki S (2016) Voice and touch control home automation: In: 3rd International conference on recent advances in information technology, pp 5–8
3. Prusty S et al (2017) Arduino based home automation using android. department of electronics and telecommunication engineering, DRIEMS, Cuttaac, pp 3–4
4. Abdulrahman TA, Isiwekpeni OH, Surajudeen-Bakinde NT, Otuoze AO (2016) Design, specification and implementation of a distributed home automation system. In: FNC/MobiSPC, pp 473–478
5. Asadullah M, Ullah K (2017) Smart home automation system using bluetooth technology. In: 2017 International conference on innovations in electrical engineering and computational technologies, pp 102–121
6. AL-Rousan M, Khiyal H (2004) Java-based home automation system. IEEE Trans Consum Electron 5(2):230–236
7. Pradeep G, Schlinhting H (2016) AdHoc low powered 802.15.1 protocol based systems. IEEE, pp 340–342
8. The official Bluetooth website from Bluetooth SIG: p. http://www.bluetooth.com. Accessed 12 Apr 2019
9. Sikandar M, Khiyal H, Khan A (2009) SMS based wireless home appliance control system (HACS) for automating appliances and security. Iss Inf Sci Inf Technol 6
10. Rozita T, KokWai C, Mok VH (2013) Smart GSM based home automation system. In: IEEE conference on system, process and control (ICSPC2013). Kuala Lumpur, Malaysia
11. Yang L, Ji M, Gao Z, Zhang W, Guo T (2009) Design of home automation system based on ZigBee wireless sensor network. In: First international conference on information science and engineering. Nanjing, pp 415–420
12. Armando G, Karim H (2018) Home automation architecture for comfort, security, and resource savings. IEEE, Ciudad Obrego´n, Mexico, pp 530–540
13. Suryadevara NK, Hasan B (2012) Wireless sensor network based home monitoring system for wellness determination of elderly, vol 2, IEEE, pp 450–463
14. Ransing RS, Rajput M (2015) Smart home for elderly care, based on wireless sensor network. In: 2015 International conference on nascent technologies in the engineering field (ICNTE). Navi, Mumbai, India, pp 201–212
15. Swamy N, Kuljaca O, Lewis F (2002) Internet-based educational control systems lab using net-meeting. IEEE Trans Edu 45(2):145–151
16. Karim H, Durgasree B (2012) Design and implementation of a Wi-Fi based home automation system 6:1856–1862
17. Aggarwal N, Singh P (2014) Design and implementation of a Wi-Fi based home automation system. Int J Electr Electron Eng 6(2):273–279
18. Jemiseye W, Duygu K (2014) Design and implementation of smart home.: B.Eng. Report, Department Electrical and Electronics Engineering, pp 43–57
19. Pratiksha DN, Jayashree GG, Pornima UK (2014) Design and implementation of cloud based home automation. In: Int J Eng Res Technol 3(2):2059–2062
20. Gogawale R et al (2016) Bluetooth remote home automation system. Int Eng Res J II(2):848–850
21. Chowdhry D, Paranjape R (2015) Smart home automation system for intrusion detection. In: 2015 IEEE 14th canadian workshop. Alberta, pp 234–245
22. Ramlee S, Ridza A et al (2013) Bluetooth remote home automation system. Int J Eng Sci (The TJES). New York
23. Ramlee RA, Othman MA, Leong HM, Ismail MM (2013) Smart home system using android application. In: 2013 International conference of information and communication technology (ICoICT). Bandung, pp 298–307

24. Sagar KN, Kusuam SM (2015) Home automation using internet of things. Int Res J Eng Technol (IRJET) 2(4):200–250
25. Dhiren T, Mohammed A, Al-Kuwari AH (2011) Energy conservation in smart home. In: 5th IEEE international conference on digital ecosystems and technologies. Daejeon
26. Jonathan O, Misra S, Ibanga E, Maskeliunas R, Damasevicius R, Ahuja R (2019) Design and implementation of a mobile webcast application with google analytics and cloud messaging functionality. In: Journal of physics: conference series, vol 1235, no 1. IOP Publishing, p 012023
27. Jonathan O, Ogbunude C, Misra S, Damaševičius R, Maskeliunas R, Ahuja R (2018) Design and implementation of a mobile-based personal digital assistant (MPDA). In: International conference on computational intelligence, communications, and business analytics. Springer, Singapore, pp 15–28
28. Iheme P, Omoregbe NA, Misra S, Adeloye D, Adewumi A (2017) Mobile-bayesian diagnostic system for childhood infectious diseases. In: ICADIWT, pp 109–118
29. Azeta AA, Omoregbe NA, Gberevbie DE, Ikpefan OA, Ayo CK, Misra S Williams JT (2016) Motivating effective ICT users' support through automated mobile edu-helpdesk system. Int J Pharmacy Technol
30. Sowumni OY, Misra S, Fernandez-Sanz L, Medina-Merodio JA (2016) Framework for academic advice through mobile applications
31. Mitali P, Ashwini B, Varsha P (2013) The design and implementation of voice controlled wireless intelligent home automation system based on Zigbee, 2(5):60–79
32. Koyuncu B et al (2014) PC remote control of appliances by using telephone lines. IEEE Trans Consum Electron Hong Kong, pp 90–98
33. Swamy N, Kuljaca O, Lewis F (2002) Internet-based educational control systems lab using net-meeting 45(3):155–161
34. Karand T, Sriskanthan N (2002) Bluetooth based home automation system. J Microprocess Microsystems 15(2):286–290
35. Benjamin E, Arul S (2014) Wireless home automation system using ZigBee 6(3):133–143

Author Index

P. K. Singh et al. (eds.), *Evolving Technologies for Computing, Communication and Smart World*, Lecture Notes in Electrical Engineering 694,
https://doi.org/10.1007/978-981-15-7804-5

The manufacturer's authorised representative in the EU is Springer Nature Customer Service Centre GmbH, Europaplatz 3, 69115 Heidelberg, Germany. If you have any concerns regarding our products, please contact ProductSafety@springernature.com

Printed and bound by CPI Group (UK) Ltd, Croydon, CR0 4YY
30/09/2026
02225288-0016